P9-ECP-990

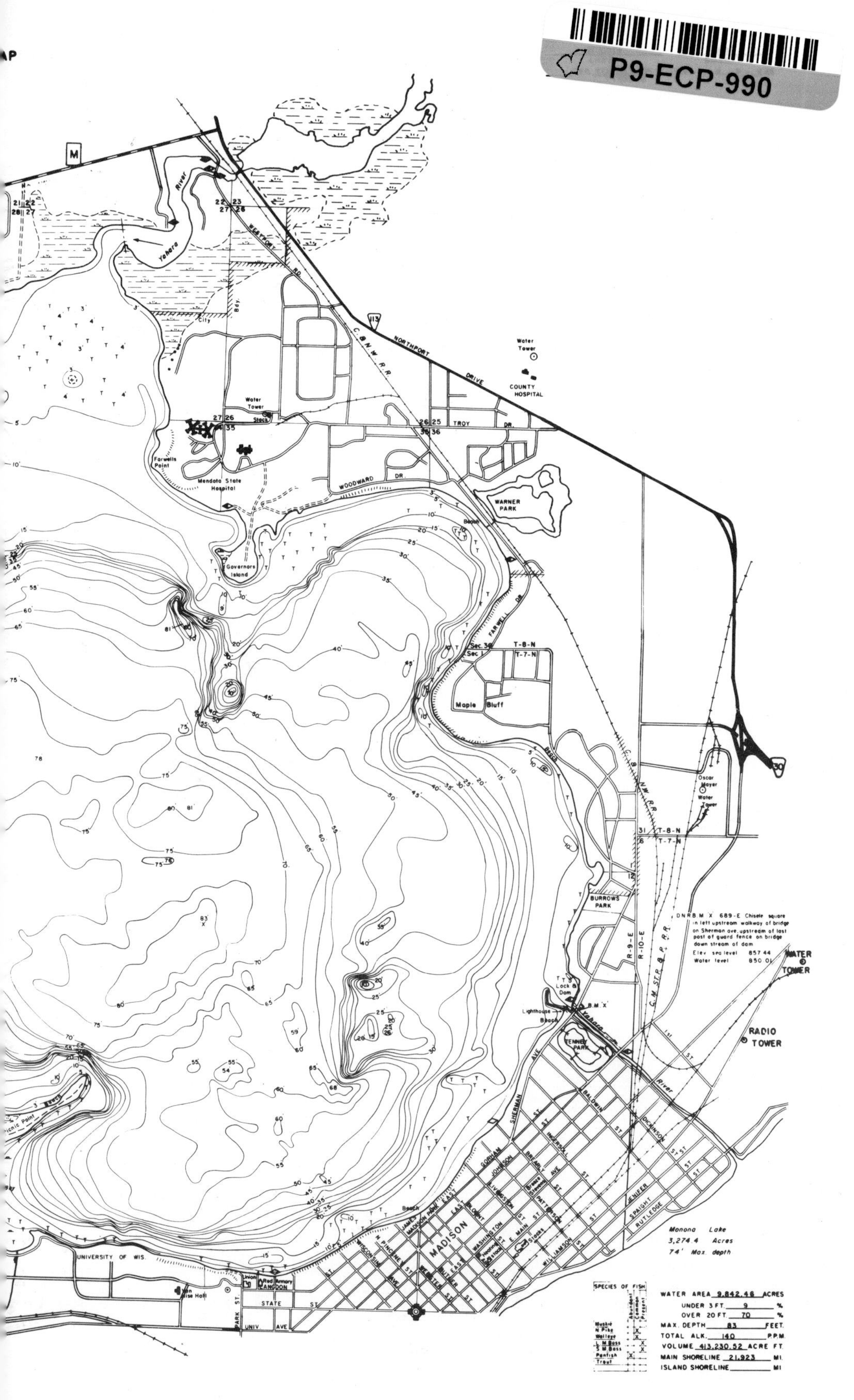

M
Yahara River
Westport Rd.
113
Northport Drive
C. & N.W. R.R.
Water Tower
County Hospital
City Bay
Water Tower
Stack
Troy Dr.
Farwells Point
Mendota State Hospital
Woodward Dr.
Warner Park
Beach
Governors Island
Farwell Dr.
Sec. 36
Sec. 1
T-8-N
T-7-N
Maple Bluff
Beach
Oscar Mayer Water Tower
30
R-9-E
R-10-E
Burrows Park
DNR B.M. X 689-E Chisele square in left upstream walkway of bridge on Sherman ave. upstream of last post of guard fence on bridge down stream of dam
Elev. sea level 857.44
Water level 850.01
Water Tower
Lock & Dam
B.M. X
Lighthouse
Beach
Tenney Park
Radio Tower
C.M. St.P. & P. R.R.
Yahara River
1st St
Sherman Ave
Baldwin St
Dickinson St
Ingersoll St
Gorham
Johnson
Brearly
Livingston
Paterson
Jenifer
Spaight
Rutledge
Williamson St
E. Main St
East Washington
Madison
Wisconsin Ave
N. Pinckney St
Webster St
Butler St
University of Wis.
Van Hise Hall
Union
Red Armory
Langdon
State St
Park St.
Univ. Ave
Picnic Point
Beach
Monona Lake
3,274.4 Acres
74' Max. depth
SPECIES OF FISH — Abundant / Common / Present
Muskie
N. Pike — Common X
Walleye — Common X
L. M. Bass — Present X
S. M. Bass — Present X
Panfish — Abundant X
Trout
WATER AREA 9,842.46 ACRES
UNDER 3 FT. 9 %
OVER 20 FT. 70 %
MAX. DEPTH 83 FEET.
TOTAL ALK. 140 P.P.M.
VOLUME 413,230.52 ACRE FT.
MAIN SHORELINE 21.923 MI.
ISLAND SHORELINE MI

Ecological Studies

Analysis and Synthesis

Edited by

W.D. Billings, Durham (USA) F. Golley, Athens (USA)

O.L. Lange, Würzburg (FRG) J.S. Olson, Oak Ridge (USA)

H. Remmert, Marburg (FRG)

Volume 55

Ecological Studies

Volume 46
Eutrophication and Land Use
By William M. Lewis, Jr., James F. Saunders III, David W. Crumpacker, and Charles Brendecke
1984. X, 202 p. 67 figures, cloth
ISBN 0-387-90961-3

Volume 47
Pollutants in Porous Media
The Unsaturated Zone Between Soil Surface and Groundwater
Edited by B. Yaron, G. Dagan, and J. Goldshmid
1984. XVI, 296 p., 119 figures. cloth
ISBN 0-387-13179-5

Volume 48
Ecological Effects of Fire in South African Ecosystems
Edited by P. de V. Booysen and N.M. Tainton
1984. XVI, 426p. 54 figures. cloth
ISBN 0-387-13501-4

Volume 49
Forest Ecosystems in Industrial Regions
Studies on the Cycling of Energy, Nutrients and Pollutants in the Niepolomice Forest, Southern Poland
Edited by W. Grodziński, J. Weiner, and P.F. Maycock
1984. XVIII, 277p. 116 figures. cloth
ISBN 0-387-13498-0

Volume 50
The Gulf of Agaba (Elat)
Ecological Micropaleontology
By Z. Reiss and L. Hottinger
1984. VIII, 360p., 207 figures. cloth
ISBN 0-387-13486-7

Volume 51
Soil Salinity under Irrigation
Processes and Management
Edited by I. Shainberg and J. Shalhevet
1984. X, 358p., 133 figures. cloth
ISBN 0-387-13565-0

Volume 52
Air Pollution by Photochemical Oxidants
Formation, Transport, Control and Effects on Plants
Edited by Robert Guderian
1985. XI, 346p., 54 figures. cloth
ISBN 0-387-13966-4

Volume 53
The Gavish Sabkha
A Model of a Hypersaline Ecosystem
Edited by G.M. Friedman and W. Krumbein
1985. approx. 512p., approx. 247 figures. cloth
ISBN 0-387-15245-8

Volume 54
Tidal Flat Ecology
An Experimental Approach to Species Interactions
By Karsten Reise
1985. approx. 160p., approx. 70 figures. cloth
ISBN 0-387-15447-7

Volume 55
A Eutrophic Lake
Lake Mendota, Wisconsin
By Thomas D. Brock
1985. XII, 308 p., 82 figures. cloth
ISBN 0-387-96184-4

Thomas D. Brock

A Eutrophic Lake

Lake Mendota, Wisconsin

With 82 Figures

Springer-Verlag
New York Berlin Heidelberg Tokyo

THOMAS D. BROCK
Department of Bacteriology
University of Wisconsin–Madison
Madison, Wisconsin 53706
U.S.A.

Library of Congress Cataloging in Publication Data
Brock, Thomas D.
A eutrophic lake.
(Ecological studies)
Includes index.
1. Limnology–Wisconsin–Mendota, Lake.
2. Lake ecology–Wisconsin–Mendota, Lake.
I. Title. II. Series.
QH105.W6B76 1985 574.5′26332′0977583 85-9915

57,930

© 1985 by Springer-Verlag New York Inc.

All rights reserved. No part of this book may be translated or reproduced in any form without written permission from Springer-Verlag, 175 Fifth Ave., New York, New York 10010, U.S.A. The use of general descriptive names, trade names, trademarks, etc., in this publication, even if the former are not especially identified, is not to be taken as a sign that such names, as understood by the Trade Marks and Merchandise Marks Act, may accordingly be used freely by anyone.

Media conversion by Impressions, Inc., Madison, Wisconsin.
Production coordinated by Science Tech, Inc., Madison, Wisconsin.
Printed and bound by Halliday Lithograph, West Hanover, Massachusetts.
Printed in the United States of America.

9 8 7 6 5 4 3 2 1

ISBN 0-387-96184-4 Springer-Verlag New York Berlin Heidelberg Tokyo
ISBN 3-540-96184-4 Springer-Verlag Berlin Heidelberg New York Tokyo

Preface

Lake Mendota has often been called "the most studied lake in the world." Beginning in the "classic" period of limnology in the late 19th century and continuing through the present time, this lake has been the subject of a wide variety of studies. Although many of these studies have been published in accessible journals, a significant number have appeared in local monographs and reports, ephemeral documents, or poorly distributed journals. To date, there has been no attempt at a synthetic treatment of the vast amount of work that has been published. One intent of the present book is to present a comprehensive compilation of the major early studies on Lake Mendota and to examine how they impinge on important present-day biological questions. In addition, this book presents a summary of field and laboratory work carried out in my own laboratory over a period of about 6 years and shows where correlations with earlier work exist.

The book should be of interest to limnologists desiring a ready reference to data and published papers on this important lake, to biogeochemists, oceanographers, and low-temperature geochemists interested in lakes as model systems for global processes, and to lake managers interested in understanding short-term and long-term changes in lake systems. Although the major thrust of the present book is ecological and environmental, sufficient background has been presented on other aspects of Lake Mendota's limnology so that the book should also be useful to nonbiologists.

One of the humbling realizations that came to me as I prepared this book

CAMROSE LUTHERAN COLLEGE
LIBRARY

was the great debt that current workers on Lake Mendota owe to earlier workers. Most of the studies of the recent period can be said to have amplified the results of earlier workers, but they have by no means superseded the earlier work. Most of the concepts that I have developed during my own studies on Lake Mendota can be viewed primarily as refinements of the work of the earlier period, rather than as conceptual advances. Only in such areas as nutrient loading and biogeochemical cycling can it be stated confidently that the work of contemporary scientists has led to new concepts. One thus stands in awe of the truly remarkable pioneering achievements that Edward A. Birge and Chancey Juday made in the field of limnology. They were far more than pathfinders; they created the early beginnings and continued on, developing limnology almost into a mature science.

Although my own work on Lake Mendota has been fairly extensive, it has depended greatly on the hard work of others. Of these, the most critical at an early stage were Dr. Allan Konopka and Dr. Robert Fallon, who suffered through the tentative stages of developing a practical methodology, and Patrick Remington and Katherine Olson, who worked out many of the field methods. Dr. Carlos Pedrós-Alió also made extensive contributions to methodology and approach, and also carried the whole research project forward significantly. In the later stages of this project, Dr. Vicki Watson played a crucial role in data analysis, making many contributions to the computerization of the data, as well as writing many of the FORTRAN programs needed for data analysis. Technical help, both in the laboratory and in the field, was provided at one time or another by the following individuals: Joan Sesing-Lenz, Ruth Kamrath, Charlene Knaack, John Gustafson, Gary Koritzinski, Mary Evenson, and Kristin Bergsland. Students or postdoctorates in the laboratory who contributed to the studies include: Kjeld Ingvorsen, Alexander J. B. Zehnder, Stephen Zinder, David M. Ward, Timothy Parkin, Tom Gries, David Winek, Bibiani Lay, and David Janes. Two scientists at other institutions who were associated with special phases of the work were A. H. Walsby and David R. Lee. Richard Lathrop of the Wisconsin Department of Natural Resources provided important data and insights on nutrient loading. Finally, my own understanding of limnology, especially the physical and chemical aspects, was greatly improved by numerous discussions over more than 10 years with Dr. Robert Stauffer. Individual chapters in this book have been read by Robert Stauffer, Vicki Watson, Carlos Pedrós-Alió, and Allan Konopka, for which I am very grateful. Any errors, of course, are my own responsibility.

The field and laboratory work on which this book is partly based has been supported by several research grants from the National Science Foundation, either to me directly, or to Dr. John Magnuson as Principal Investigator for the Northern Lakes Component of the Long-term Ecological Research Program. The manuscript has been typed in expert fashion by Patti Jungwirth, Nancy Gouker, and especially, Irene Slater. The graphs and figures were drawn or redrawn from published sources by Ed Phillips. Part of the cost of publishing the lengthy Appendix of raw data was covered by a research grant to me from the National Science Foundation (DEB82212459) and from The Graduate School of The University of Wisconsin-Madison.

The final production of this book involved the use of machine-readable text created on an NBI System 3000 incorporating a generic typesetting system of my own design. The typesetting was done by Impressions, Inc. using Penta-Page software and an Autologic digital typesetter. Assembly of the final copy was handled by Ed Phillips. Production was coordinated by Science Tech, Inc.

Contents

Lake Mendota, Wisconsin, February 1979. The photograph was taken looking north from a point on the south shore west of James Madison Park. About one-third of the lake is visible. The Yahara River enters from the upper right. The Wisconsin Student Association of the University of Wisconsin–Madison had constructed a replica of the Statue of Liberty as part of its winter festivities. Photograph courtesy of the University of Wisconsin News Service.

1. Introduction

The State of Wisconsin is well provided with lakes (Figure 1.1). There are about 9,000 lakes in the state, of which over 4,000 have an area exceeding 4 hectares (10 acres). Lake Mendota (43°4'37"N, 89°24'28"W) is the uppermost lake in a chain of lakes on the Yahara River, a river which is a branch of the Rock River, which itself is a branch of the Mississippi River (Figure 1.2). The other lakes in this chain, downstream from Lake Mendota, are Monona, Waubesa, and Kegonsa (Figure 1.3). As a group, these four lakes are often simply called "the Madison lakes". The relationship of the Lake Mendota drainage basin to the other drainage basins in the region is shown in the map of Dane County, Wisconsin which can be found in the End Papers.

Lake Mendota is frequently called "the most studied lake in the world." The lake has been a research magnet primarily because it is situated adjacent to a major research university, and one which has recognized the importance of basic studies in aquatic ecology. However, location has been only partly responsible for Lake Mendota's extensive research base. Of even more importance is the fact that Edward A. Birge, one of the founders of limnology, was an early Professor of Zoology at the University.

E.A. Birge (Figure 1.4) was a distinguished teacher, scholar, administrator, and scientist (Sellery, 1956). Born in Troy, New York in 1851, Birge received his Ph.D. degree from Harvard University and joined the faculty of the University of Wisconsin in 1875. He was Dean of the College of Letters and Science from 1891 until 1918, continuing throughout this whole time an active teaching

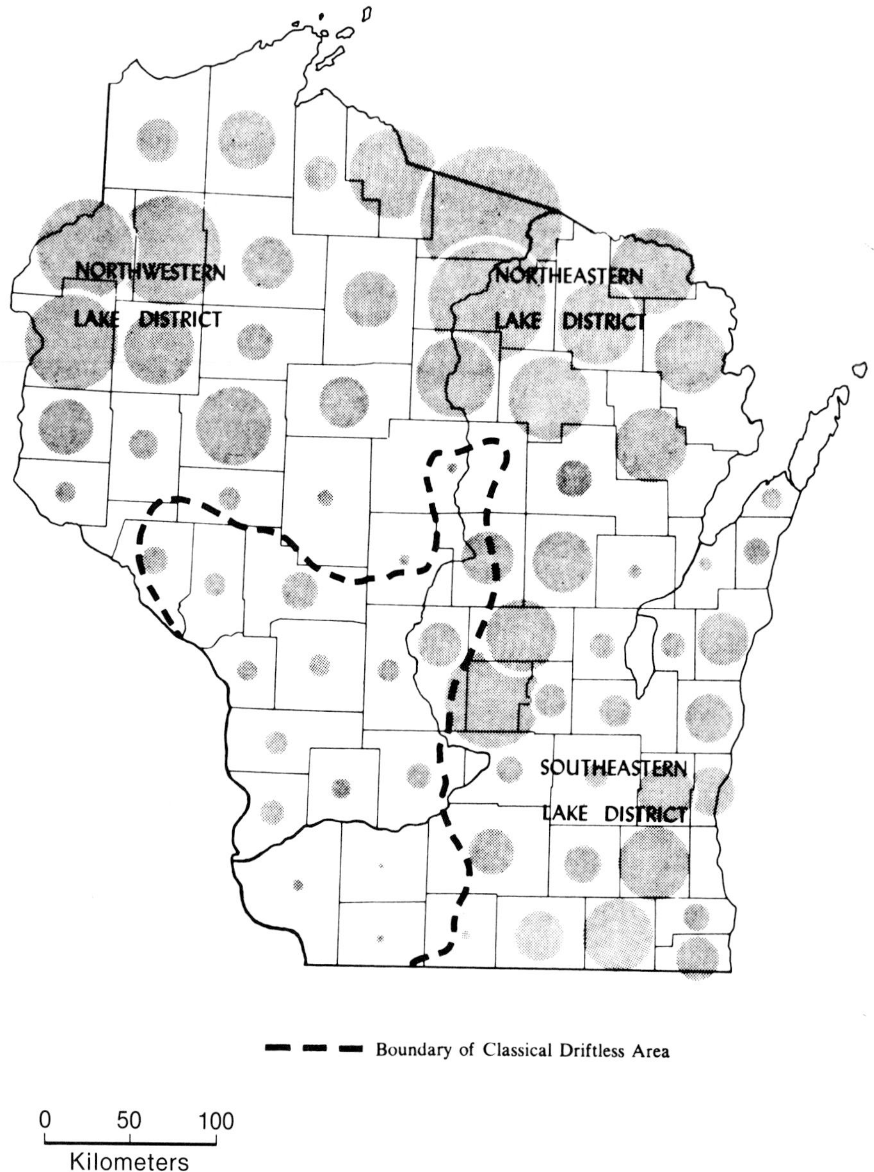

Figure 1.1. Density of lakes in Wisconsin by counties. The areas of the circles are proportional to the mean number of lakes per unit area. The dashed line represents the boundary of the unglaciated driftless area. From Frey (1963).

and research program. Birge became President of the University of Wisconsin in 1918, a position he held until he retired on September 1, 1925, at the age of 64. (He had also been Acting President of the University in the years 1900–

Figure 1.2. Major rivers of Wisconsin, and the location of Lake Mendota. The dashed line marks the approximate watershed divide between the Mississippi and Great Lakes drainages. From Frey (1963).

1903.) It was after Birge retired from the presidency in 1925 that he initiated the research studies in Northern Wisconsin which later became so well known (Mortimer, 1956). Birge died in 1950 at the age of 99; he was active scientifically at least until the age of 90 (Sellery, 1956).

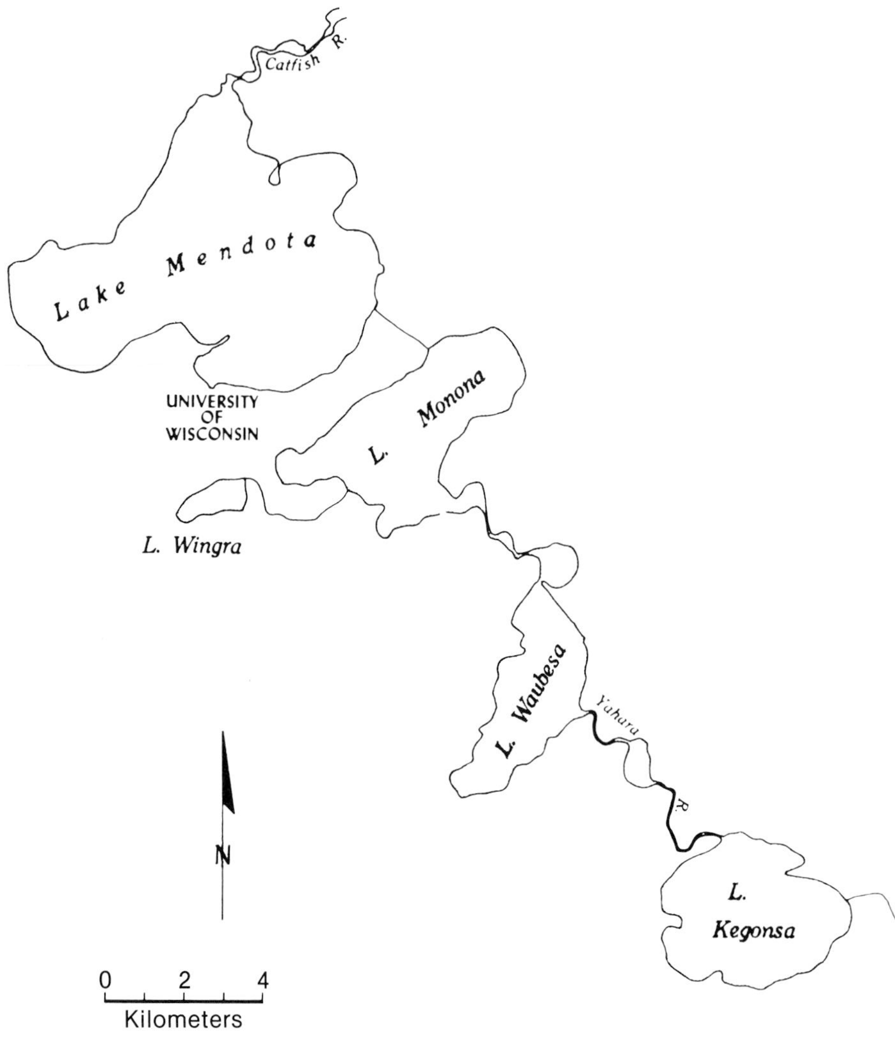

Figure 1.3. The lakes of the Yahara River chain at Madison, Wisconsin. From Frey (1963).

Much of Birge's research on lakes was carried out under the auspices of the Wisconsin Geological and Natural History Survey, which he directed from 1897 onward. The Survey provided a small fund for research, at a time when the College of Letters and Science had none, and enabled Birge to foster aquatic biological investigations. Lake Mendota became only a small part of the aquatic research of the Survey, since studies were done throughout the state, and elsewhere, but Lake Mendota was always available whenever a comparative study was under way, so that Lake Mendota data are scattered throughout many papers published on other lake systems.

Figure 1.4. Left, Edward A. Birge; right, Chancey Juday.

Although it was Birge who in the first instance initiated research on Lake Mendota, it is almost certain that most of the work on Lake Mendota would never have been carried to completion without the major participation of Birge's long-term associate, Chancey Juday (1871–1944). Juday (Figure 1.4) was hired by Birge as an assistant biologist of the Survey in 1900, and eventually became a distinguished Professor of Zoology. Until his death, Juday continued to work closely with Birge (Birge himself died after Juday). In Frey's extensive survey on Limnology in America (Frey, 1963), the Wisconsin work is divided into two parts, the Birge-Juday era and the modern era. Birge and Juday's influence was so great that virtually all work done under their auspices is referred to as work of Birge and Juday, even though a large number of other scientists (some of them quite distinguished), actually performed many of the studies.

Perhaps another reason that Lake Mendota has been so extensively studied is that it is considered a prime example of a eutrophic lake (Edmondson, 1969). Not only has this eutrophic condition stimulated scientists to find out why, it has elicited large and dynamic responses in the biota of the lake, making research studies even with simple equipment relatively easy to carry out. The first report of an algal bloom on Lake Mendota was that of Trelease (1889), based strictly on visual observations. Trelease's paper is entitled: "The 'Working' of the Madison Lakes". Trelease's observations are so precise and detailed that we have no difficulty in understanding the situation:

Every season a greenish-yellow scum occurs in greater or less quantity on Third and

> Fourth Lakes (Monona and Mendota), during the hot weather of summer, after the water has been calm for a number of days in succession. When but little of it is present, it appears as fine granules suspended in the light, often scarcely visible to the naked eye except as they reflect the light, when they call to mind the dancing motes in a beam of sunlight. Under the influence of a gentle, but continuous breeze, these particles are collected into fleecy masses, and driven ashore, so that they accumulate along the margin of the lake, forming a slimy scum which quickly putrefies, giving off a very disagreeable odor. During this change, its color changes to a decided blue-green, which stains the pebbles, sticks, etc., over which it is smeared. The appearance of this scum is sometimes spoken of as the working of the lakes, from a resemblance to a collection of a scum in cider, etc., when fermenting, or, as an American idiom expresses it, "working".

The word "working" was stated by Trelease to be analogous to the term "breaking" used in Great Britain or the words *Wasserblute, Fleur d'eau,* or *Flos aquae,* used in continental Europe.

Lake Mendota was not always a eutrophic lake. As will be discussed in detail later (see primarily Chapter 3), before settlement of the Madison area Lake Mendota was probably a rather oligotrophic marl lake. The term "cultural eutrophication" is frequently used to refer to the process of eutrophication which occurs as a result of human activities (Edmondson, 1969). The adjective "cultural" is used to distinguish human-induced eutrophication from the natural eutrophication which occurs when a lake undergoes a normal aging process. Lake Mendota is a classical example of a culturally-eutrophied lake. Much research has been done on Lake Mendota because of this fact, and the lake has attracted attention from the Madison community for many years because of its deteriorated condition.

The origin of the lake*

Lake Mendota is a typical glacial lake, a product of the last stage of glaciation in southern Wisconsin. The time of this last glaciation was about 15,000 years ago, at which time glacier ice was more than 300 meters thick over the Madison lakes. At this time, the Yahara River valley was probably over 1,000 meters deep, if measured from the highest hills in the area. The glacier ice scoured some material out of the lake basins, but it also deposited glacial debris in the old Yahara valley. The glacier ice retreated between 14,000 and 15,000 years ago, leaving behind a vast amount of water which had to drain away. Some of this water drained to the west, south, and southwest. As the edge of the glacier ice retreated approximately to the City of Middleton, a glacial lake was dammed to the west of it, and fine-grained lake sediments were deposited. The surface of this flat lake bed is visible today and is the source of water for Pheasant Branch, one of the tributaries of Lake Mendota.

* This section is based on material provided by Prof. David M. Mickelson of the Department of Geology, University of Wisconsin-Madison.

As glacier ice retreated toward the northwest, the ice margin entered what is now called Lake Mendota, and the water level in the lake dropped to an elevation about 4 meters higher than the present Lake Mendota. As ice continued to retreat from the Yahara basin, a large lake, probably continuous from the north side of Lake Mendota down to the present location of Stoughton (below Lake Kegonsa), was present at an elevation of 262 meters. The fairly flat surface underlain by lake sediments deposited in this lake is obvious throughout the Madison area.

It is not clear how long the lake remained at this elevation, but it was probably not more than 1,000 years (10,000 years before the present) before it dropped to an elevation of 260 meters, or about 2 meters higher than the present Lake Mendota. At this time, the present lakes became separated as they are now. Then, numerous marshes occurred in shallow-water areas and these areas remained marshes until drainage or filling by settlers. Since the fall of the lake level to 260 meters, progressive erosion of bedrock near the City of Stoughton has caused a progressive lowering of water levels in the lakes upstream.

The present lake level is not the natural level, however, since a dam was installed at the outlet to the Yahara River. The first dam was built in 1847–1850 and the present dam and boat locks were constructed in 1896 (Kanneberg, 1936). This latter dam raised the water level about 1.2 meters. The water level of Lake Mendota is controlled to attempt to alleviate flooding in the lower lakes area, and the target elevation is 259 meters although it may fluctuate as much as 1 meter.

The physical characteristics of the four lakes in the Yahara chain are discussed in Chapter 2 and the geological setting is discussed later in this chapter.

The naming of the Madison lakes

The four lakes of the Madison chain have been well known for many years. A famous American poet, Henry Wadsworth Longfellow, wrote the following poem on January 15, 1870:

Four limpid lakes, — four Naiades
Or sylvan deities are these,
  In flowing robes of azure dressed;
Four lovely handmaids, that uphold
Their shining mirrors, rimmed with gold,
  To the fair city in the West.

By day the coursers of the sun
Drink of these waters as they run
  Their swift diurnal round on high,
By night the constellations glow
Far down the hollow deeps below
  And glimmer in another sky.

Fair lakes, serene and full of light,

Fair town, arrayed in robes of white,
  How visionary ye appear!
All like a floating landscape seems
In cloud-land or the land of dreams,
  Bathed in a golden atmosphere!

One might wonder whether Longfellow saw the same lake as Trelease!

Lake Mendota seems to have inspired other poets. A famous poet of the 1920's and 30's, William Ellery Leonard, penned the following lines about 1913 (Leonard, 1940):

The shining City of my manhood's grief
Is girt by hills and lakes (the lakes are four),
Left by the ice-sheet which from Labrador
Under old suns once carved this land's relief,
Ere wild men came with building and belief
Across the midland swale. And slope and shore
Still guard the forest pathos of dead lore
With burial mound of many an Indian chief,
And sacred spring.

The aboriginal burial mounds around the edge of Lake Mendota, alluded to in this poem, are extensive and well known, and suggest that the American Indians found the shores of the lake favorable for living. Most of the springs, however, have dried up, victims of a lowering water table brought about by urbanization and extensive groundwater pumping by the City of Madison.

Although the word "Mendota" is an American Indian word (it means "the mouth of the river" in the Sioux Indian dialect), it was not the name that the Indians used. Cassidy (1945) has written that the first name which the Madison lakes bore was "Taychopera", a Winnebago Indian word which means "the four lakes". Another Indian name for Lake Mendota was "Wonk-shek-ho-mik-la" which in Winnebago means "where the man lies". However, the original settlers did not use the Indian word, but its English translation, "The Four Lakes." Simeon Mills, a settler who came to the Madison area in 1837, one year after the town was established, remarked that "These beautiful sheets of water were probably called The Four Lakes for the same reason that the principal divisions of the year are called the 'four seasons', because they are four in number". As Cassidy notes, nowhere else in the whole region was there a similar grouping of four sizable lakes strung along a river in an otherwise comparatively lakeless area. "These are *the* four lakes: the definitive article is significant."

In addition to being known collectively as The Four Lakes, each lake acquired an individual numerical name, First, Second, Third, and Fourth lakes, going up the chain from Kegonsa, through Waubesa, Monona, and Mendota. In fact, the name "Fourth Lake" for Lake Mendota was used by many local residents long after the name "Mendota" had come into wide use. The Yahara River was also known as Cat Fish River or "Myan-mek" in Indian language.

The present names were given to Mendota and Monona by Frank Hudson about 1849. Mr. Hudson was employed as a land surveyor, and was responsible for platting the land on which the University of Wisconsin was subsequently established. Mr. Hudson was fascinated with American Indian lore and legends and found the names Monona and Mendota in some written accounts. Thinking these names charming, he suggested that the lakes on each side of the City of Madison be given these names, and this suggestion was quickly agreed upon and adopted into law. As noted, "Mendota" is a Sioux Indian word meaning "the mouth of the river". A related word, "Manto-ka" was used by the Prairie Potawatomi Indians, meaning "snake maker", apparently from rattlesnakes found along the shores of prairie lakes. In 1855, the subject of giving the other two lakes Indian names was raised, and the names Kegonsa and Waubesa were found and adopted. All four names were made official with the adoption of a state law on February 14, 1855.

Ownership of Lake Mendota

Who owns Lake Mendota? The lake itself is owned by the State of Wisconsin, but the land bordering the lake is under the jurisdiction of various other governmental entities. Among the separate entities bordering the lake are the following major property owners:

City of Madison
University of Wisconsin-Madison
Village of Maple Bluff
Village of Shorewood Hills
County of Dane
City of Middleton
State of Wisconsin
Private owners

Although individual property owners have control of the shorelines abutting their property, certain aspects of the lake are controlled directly by governmental bodies. Recreational use on the lake was for many years controlled by the City of Madison even for those portions of the lake outside the city limits, but is now controlled by the County of Dane, which operates police boats. Life-saving activities are handled to some extent by the County police boats, but more extensively and importantly by the University of Wisconsin-Madison, which maintains a life-saving station on the southern shore and provides extensive monitoring during the warm season for boats in distress. The University of Wisconsin-Madison outing club, called "The Hoofers", operates signals which indicate safe or unsafe boating conditions on the lake, an especially important activity in a lake of such large size which receives such extensive use for sailing.

The water level on the lake is controlled by the County of Dane, although the original water level targets were established by the Wisconsin Railroad Commission.

Registration of boats and enforcement of state laws regarding boating is the responsibility of the Wisconsin Department of Natural Resources. This agency is also responsible for licensing weed control operations in the lake, and monitors the use of both chemical weed control agents and mechanical weed cutters. The weed cutters themselves are operated primarily by the County of Dane, although the University of Wisconsin-Madison has also operated weed cutters, primarily for research purposes.

The regional setting

A detailed bathymetric map of Lake Mendota is given in the End Papers and an abbreviated version is presented here (Figure 1.5).

Dane County, in which Lake Mendota is totally situated, is one of the richest agricultural counties in the United States, generally ranking among the top 10 counties in the U. S. in agricultural wealth. Dane County encompasses about 3,186 square kilometers (1,230 square miles). A detailed map of Dane County, showing the principal watersheds, is given in the End Papers. The eastern part of the county, which includes all of the Lake Mendota drainage basin (Figure 1.1, 1.2), has been extensively glaciated and consists of a slightly rolling plain of low hills with intervening wetlands drained by small streams or artificial ditches. The western part of the county, which is in the Wisconsin River drainage basin, was never glaciated (this is part of the Wisconsin "driftless" area), and consists of steep valleys and ridges drained by fast-flowing spring-fed streams.

The bedrock in the Lake Mendota area consists of Paleozoic sandstones and limestones. However, most of the deposits along the shores of Lake Mendota are Quaternary glacial deposits (Figure 1.6). A well drilled in the deep hole of the lake would have a log approximately as follows: water, 24 meters; sediment (black mud and marl), 10.7 meters; glacial drift, 39.6 meters; Eau Claire Sandstone, 67 meters; Mt. Simon Sandstone, 76.2 meters; Precambrian granite reached at about 220 meters below the lake surface (Bean, 1936). Outcroppings of bedrock in the Lake Mendota basin include Madison and Franconia Sandstone, both of Cambrian age, which are seen from the water in steep bluffs at cutting shores along the southern side of the lake between Picnic Point, Second Point and Eagle Heights (just west of Second Point), along the shore just east of the mouth of Pheasant Branch Creek, at Fox Bluff, Farwell's Point, Governor's Island, and Maple Bluff.

The climate of Dane County is typical of continental United States (Bartlett, 1905; National Oceanic and Atmospheric Administration, 1982). Warm, humid, relatively short summers are followed by long, cold, relatively dry winters. The annual temperature range is large, and short-term variations in temperature frequently occur. The mean annual temperature in Dane County is about 7.4°C (45°F). The warmest month, July, has an average of about 21.2°C (70°F), while the coldest month, January, has an average temperature of about −8°C (17°F). In a typical winter, 20 days may have temperatures of −17°C (0°F) or less.

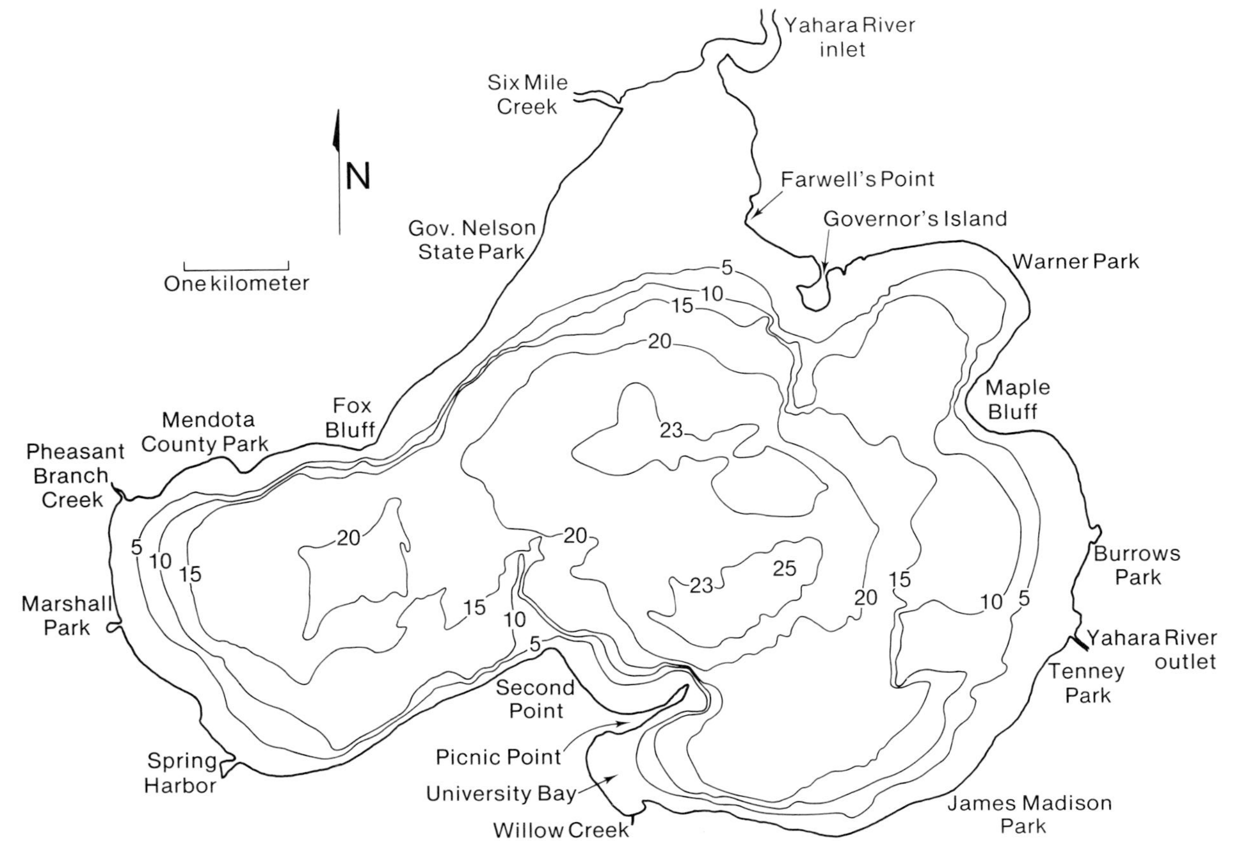

Figure 1.5. Lake Mendota, Wisconsin. Prepared from the detailed bathymetric map given in the End Papers.

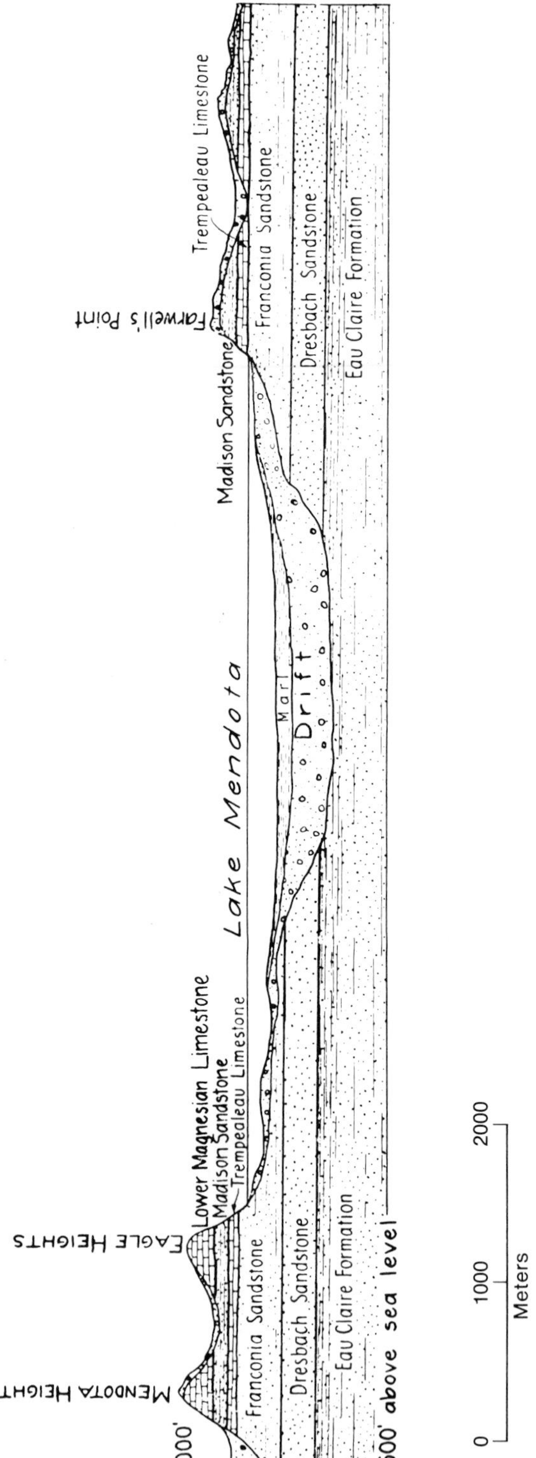

Figure 1.6. Cross-section of Lake Mendota, showing the wide pre-glacial valley in the Paleozoic sandstone. From Bean (1936).

Table 1.1. Land use in the Lake Mendota watershed

Category	Area, hectares	Percent of total
Agricultural	43,000	77
Urban	6,500	12
Woodlands	4,000	7
Wetlands	2,400	4

Source of data: Dane County Lake Quality Advisory Council Report–A framework for lake management. 1975. Available from Dane County Regional Planning Commission, Madison, Wisconsin.

Average annual precipitation is about 77.5 cm (31 inches), and about two-thirds of this precipitation falls during the five summer months from May through September, mostly in thunderstorms. Severe storms often occur from late fall through mid-spring, markedly affecting the circulation on Lake Mendota (Stauffer, 1980). Although the average annual precipitation is sufficient to make the area an excellent agricultural area, drought is not unknown (Mitchell, 1979), and wide yearly variation in precipitation occurs. Because of the extremely varied nature of the weather in the Lake Mendota area, runoff and nutrient-loading studies are difficult to carry out accurately, since many years of study are essential before an "average" figure can be obtained.

Land use in the Lake Mendota watershed is summarized in Table 1.1. As seen, the land in the drainage basin is predominantly rural, with over three-fourths of the land in farming. Almost all of this agricultural land was in forest at the time of settlement. At the present, woodlands occur primarily in the western unglaciated portion of the drainage basin, mostly on steep slopes on the tops of hills. A significant amount of land is in wetlands, the most extensive being Waunakee Marsh, which drains into Six Mile Creek. Although agricultural practices vary with market conditions, at present the two most extensive crops are maize (raised primarily as a cash crop) and hay (raised primarily for feeding of cattle).

In addition to its extensive agricultural base, Dane County is highly urbanized, being the second most populous county in Wisconsin. The home of both the State Capitol and the University of Wisconsin-Madison, Dane County has a population of over 300,000 people, of whom about half live in the City of Madison, one quarter live in smaller cities and villages, and another quarter live in unincorporated, primarily rural, areas. Nearly 80% of the county's work force hold jobs in government, trade, or service industries. There is relatively little manufacturing or heavy industry.

Although population growth in Dane County has been high over the past 75–100 years, this growth has been primarily urban. The establishment of the extensive agricultural base occurred much earlier, in the years from first settlement (1850's) through the 1880's. As far as Lake Mendota is concerned, the agricultural developments are probably the most significant, since they led to uncontrolled runoff of soil particles and input of massive amounts of nutrients into the lake (see Chapter 3). Despite extensive urbanization, Lake Mendota has never received significant input of domestic sewage, either treated or untreated, although Lake Monona and the lower lakes have been subjected to a heavy onslaught of this type, as discussed in Chapter 3.

References

Bartlett, J.L. 1905. The climate of Madison, Wisconsin. *Monthly Weather Review, December 1905.* 7 pp.

Bean, Ernest, F. 1936. Geological history of Lake Mendota. pp. 3–9. In: *Lake Mendota, origin and history.* The Technical Club of Madison, Madison, Wisconsin.

Cassidy, F.G. 1945. The naming of the "four lakes". *Wisconsin Magazine of History* 29: 7–24.

Edmondson, W.T. 1969. Eutrophication in North America. pp. 124–149 In *Eutrophication: Causes, Consequences, Correctives.* National Academy of Sciences, Washington, D.C.

Frey, D.G. 1963. *Limnology in North America.* University of Wisconsin Press, Madison. 734 pp.

Kanneberg, Adolph. 1936. The dam at the outlet of Lake Mendota. pp. 17–19. In: *Lake Mendota, origin and history.* The Technical Club of Madison, Madison, Wisconsin.

Leonard, W.E. 1940. *Two Lives. A Poem.* The Viking Press, New York. 109 pp.

Mitchell, V.L. 1979. Drought in Wisconsin. *Transactions of the Wisconsin Academy of Sciences, Arts and Letters* 67: 130–134.

Mortimer, C.H. 1956. E.A. Birge. An explorer of lakes. pp. 163–211 In: *E.A. Birge. A Memoir.* By G.C. Sellery. University of Wisconsin Press, Madison.

National Oceanic and Atmospheric Administration. 1982. *Climatography of the United States. 60. Climate of Wisconsin.* National Climatic Center, Asheville, North Carolina. 17 pp.

Sellery, G.C. 1956. *E. A. Birge. A Memoir.* University of Wisconsin Press, Madison. 221 pp.

Stauffer, R.E. 1980. Windpower time series above a temperate lake. *Limnology and Oceanography* 15: 513–528.

Trelease, W. 1889. The "working" of the Madison lakes. *Transactions of the Wisconsin Academy of Sciences* 7: 121–129.

2. Physical Limnology

Although the main interest in this book is biological, the physical environment of the lake cannot be ignored. The organisms of the lake, and the ecosystem as a whole, are strongly influenced by physical processes such as vertical mixing, horizontal circulation, and sediment-water interactions. We are fortunate to have a detailed understanding of the physical limnology of Lake Mendota, deriving initially from the classical work of Birge but being elaborated to a great extent by the work of Professor Reid Bryson and his students in the 1950's and early 1960's, by Kenton M. Stewart in the 1960's, and by Robert A. Stauffer in the 1970's. In addition to these kinds of physical studies, many biologically- or chemically-oriented workers have carried out routine physical measurements as part of their special research studies.

Morphometry and Hydrography

The morphometry of Lake Mendota is well known. The lake has been mapped in detail at least three times, in the years 1897–98 by Birge and student helpers (Juday, 1914), in the early 1950's by Civil Engineering students (Murray, 1956), and in the late 1970's by Richard Lathrop and associates at the Wisconsin Department of Natural Resources (see Figure 1.6 and End Papers for a map, an enlarged version of which is available through the Bureau of Engineering, Wisconsin Department of Natural Resources, Madison). The underwater con-

Table 2.1. Characteristics of the Madison lakes

Item	Mendota	Wingra	Monona	Waubesa	Kegonsa
Depth, meters-mean	12.4	2.7	7.7	4.4	4.4
Depth, meters-max.	25.3	4.3	22.6	11.3	9.4
Direct Drainage Area, km^2	55	–	104	114	141
Cumulative Drainage Area, km^2	686	23	793	909	1093
Volume, m^3	48×10^7	0.6	10.6	3.8	5.6
Area, km^2	39.4	1.3	13.8	8.3	12.7
Lake Frontage, km	33.8	–	20.9	13.8	15.1

Data from Dane County Regional Planning Commission and other sources.

tours of the Birge map are not too extensive but the latter two maps provide excellent detail (and actually differ only in minor ways from each other).

General Features of the Lake

Lake Mendota (see Figure 1.5 and End Papers) is one of the larger inland lakes in Wisconsin. The physical dimensions were given by Juday (1914) and Stauffer (1974). Lake Mendota has a surface area of 39.1 km^2, a volume of 48.6×10^7 m^3, and a maximum depth of 25.3 m. Lake Mendota is almost elliptical in shape, with a length of 9.50 km and a breadth of 7.40 km. Its long axis runs SW–NE and its short axis about N–S. The SW–NE orientation of the lake is significant, because the prevailing winds in this area are from the west (SW in summer, NW in winter), the long fetch thus leading to the development of relatively high waves on very windy days. Some further data on the morphometry of Lake Mendota and the other Madison lakes are given in Table 2.1.

Lake Mendota has a fairly regular shoreline, with very few bays and no islands. Because of this regular shoreline, the lake is relatively homogeneous horizontally, and there are few places where unusual concentrations or aggregations of organisms can occur. The only confined feature is University Bay on the southern shore, but this bay is open to the northeast and water circulates well with the rest of the lake (Bryson and Ragotzkie, 1955). The lake is also fairly regular in the angle of its underwater slope, which is about 1.5%. There are no sudden drop-offs or areas of extensive shallows. The western end of the lake is separated in part from the rest of the lake by a bar, the so-called "Second Point Bar". Another bar exists off the peninsula on the north, called Governors Island (no longer an island, since it has been artificially connected with the mainland). The deepest part of the lake, the so-called "Deep Hole" sampling site used by many water chemists, is in the southeast part of the lake.

Although Lake Mendota receives water from several streams, about 70% of the water enters through the Yahara River and Six Mile Creek on the north, with smaller amounts entering at Pheasant Branch Creek on the west through the City of Middleton, at springs on the south and west shore, and at Willow

Creek, a drainage ditch on the University of Wisconsin campus. All of the surface water exits through the Yahara River at Tenney Park locks on the west. The Yahara River connection between Lake Mendota and Lake Monona is an artificial waterway that replaced a winding, marshy stream which connected the two lakes during presettlement time (Kanneberg, 1936). The water budget will be discussed in detail later in this chapter and the composition of the bottom sediments in Chapter 3.

Hypsometry of Lake Mendota

It is important to know the underwater contours of the lake, not only in the calculation of lake volume, but in the study of sediment-water interactions, in calculating lake areas at various depths, and in weighting the quantitative significance of water samples taken at various depths. Because of the shapes of lake basins, a water sample taken deep in the lake is representative of a smaller volume of water then a sample taken near the surface. In order to calculate whole lake concentrations, proper weighting of samples for hypsometry must be carried out.

The detailed morphometry of Lake Mendota is given in the map in the End Papers. Hypsometric data based on contour maps were presented by Juday (1914) and Stauffer (1974). The latter, based on a more detailed hydrographic map, is presented in Table 2.2. Such a table can be used in the weighting needed to calculate whole lake data (see later chapters). It is also possible to estimate parameters for a polynomial equation that will permit calculation of either volume or area for any depth or depth interval. These polynomial equations are given at the bottom of Table 2.2.

Climate of the Lake Mendota area

Lakes are strongly affected by meterological conditions, and a knowledge of local weather is essential for the understanding of the physical limnology of a lake. Weather data are obtained on a routine basis by the U.S. Weather Bureau, and Madison, Wisconsin has a first-class weather station at Truax Air Field, which is less than 1 km from the eastern end of the lake. Although weather parameters may differ slightly between the lake and the air field (Stauffer, 1980), it seems likely that the Weather Bureau observations will be close to those that would be made on the lake. In the early years, the U.S. weather station was located on the University of Wisconsin campus, about 200 m from the lake and 40 m above the elevation of the lake (Neess and Bunge, 1956).

Madison has a typically continental climate with cold winters and warm summers. A summary of normal, mean, and extreme values for a variety of parameters is given in Table 2.3. One characteristic of Madison climate is extreme variability. Although there has been little change in the long-term averages over many years, the year-to-year variability is great. A record of

Table 2.2. Hypsometric data for Lake Mendota

Depth	Area, m^2	Volume, m^3
0.5	39.1×10^6	18.8×10^6
1.0	36.0	54.8
2.0	34.0	88.8
3.0	32.1	120.8
4.0	30.3	151.1
5.0	28.7	179.8
6.0	27.8	207.6
7.0	27.0	234.6
8.0	26.0	260.6
9.0	25.0	285.6
10.0	24.0	309.6
11.0	22.7	332.3
12.0	21.1	353.4
13.0	19.9	373.3
14.0	18.4	391.7
15.0	17.6	409.3
16.0	16.4	425.7
17.0	15.2	440.9
18.0	13.7	454.6
19.0	11.3	465.9
20.0	8.6	474.5
21.0	5.9	480.4
22.0	3.5	483.9
23.0	1.6	485.5
24.0	0.1	485.6

Depth for 1/2 lake volume = approximately 8 m.
Total volume of lake 485.6×10^6 m^3.
Polynomial equations for calculating volume and area at any depth:

Vol (V)$=10.8+37.4\ Z-0.722\ Z^2$ $r^2=0.990$.
Area (A)$=36.1-1.01\ Z-0.0188\ Z^2$ $r^2=0.986$.

To calculate V or A for a given depth interval (Z_2-Z_1), calculate for Z_2 and Z_1 and subtract.

temperature measurements for over 100 years has been summarized from U. S. Weather Bureau records and is presented in Table 2.4. Although the mean annual temperature for the 107 years is 7.44°C, the coldest year, 1885, had a mean temperature of 5.4°C and the warmest years, 1878 and 1921, had mean temperatures of 9.7°C. Even more striking are the extreme seasons and months. The warmest winter season (Dec.–Feb. 1877–78) had a mean of −0.11°C whereas the coldest winter season, 1874–75, had a mean of −12.39°C. The largest variation between cold and warm seasons was the spring period (March–May) in which a 9°C variation was found from warmest to coldest year. Over half of the annual precipitation occurs in the warm part of the year, May to September, the wettest single month being June. Most of the significant rain that falls in the summer occurs as thunderstorms and tends to be erratic and widely scattered. The driest season of the year is winter, and snow is scattered and unpredictable from year to year.

Table 2.3a. Normals, means, and extremes of temperature and precipitation at Madison, Wisconsin. Data from the Wisconsin State Climatologist

	Temperatures °C							Precipitation in Centimeters Water Equivalent							Precipitation in Centimeters Snow, Ice Pellets			
	Normals			Extremes														
Month	Daily Maximum	Daily Minimum	Monthly	Record Highest	Year	Record Lowest	Year	Normal	Maximum Monthly	Year	Minimum Monthly	Year	Maximum in 24 hr	Year	Maximum Monthly	Year	Maximum in 24 hr	Year
J	−3.7	−13.2	−8.4	13	1947	−38	1951	3.18	6.22	1974	0.48	1961	3.23	1960	55.6	1971	2.5	1971
F	−1.4	−11.6	−6.5	16	1976	−33	1959	2.41	7.04	1953	0.20	1958	3.94	1953	53.1	1975	6.2	1950
M	4.0	−6.0	−1.0	26	1967	−34	1962	4.90	12.80	1973	0.97	1958	6.40	1973	64.5	1959	34.5	1971
A	13.3	1.4	7.4	32	1952	−13	1972	6.76	18.06	1973	2.44	1946	7.19	1975	44.2	1973	32.8	1973
M	19.6	7.0	13.3	34	1975	−7	1966	8.66	15.90	1960	2.49	1971	9.25	1966	1.8	1966	1.8	1966
J	24.9	12.6	18.8	36	1953	−1	1972	11.00	20.70	1963	2.06	1973	9.32	1963	0		0	
J	27.4	14.9	21.2	40	1976	2	1965	9.68	27.76	1950	3.51	1946	13.34	1950	0		0	
A	26.7	14.1	20.4	38	1947	2	1968	7.75	18.97	1972	1.78	1948	7.37	1965	0		0	
S	21.6	9.2	15.4	37	1953	−4	1974	8.53	24.16	1941	1.24	1952	9.07	1961	Trace	1965	Trace	1965
O	16.1	3.8	9.9	32	1976	−10	1952	5.49	14.10	1959	0.15	1952	5.11	1960	2.3	1967	2.3	1967
N	6.1	−3.1	1.5	24	1964	−24	1947	4.75	10.01	1961	0.28	1976	5.89	1971	22.6	1954	17.3	1954
D	−1.2	10.0	−5.6	17	1970	−30	1962	3.73	9.25	1971	0.64	1960	4.22	1971	52.8	1970	0.6	1970
Yr	12.8	1.6	7.2	40	JUL 1976	−38	JAN 1951	76.84	27.76	JUL 1950	0.15	OCT 1952	13.34	JUL 1950	64.5	MAR 1959	40.6	DEC 1970
Number of years in record (through 1976)				37		37			37		17		28		28		28	

Table 2.3b. Normals, means and extremes of relative humidity, wind, and sunshine at Madison, Wisconsin. Data from the Wisconsin State Climatologist

	Relative humidity pct.				Wind					Mean sky		Mean number of days Sunrise to sunset		
							Fastest km							
Month	Hour 00	Hour 06	Hour 12	Hour 18	Mean speed kpm	Prevailing direction	Speed kpm	Direction	Year	% of possible sunshine	cover, tenths sunrise to sunset	Clear	Partly cloudy	Cloudy
J	77	78	69	73	16.9	NNW	109	E	1947	48	6.7	8	6	17
F	77	79	65	69	17.1	NNW	92	W	1948	52	6.6	7	6	15
M	78	81	63	67	18.2	NW	113	SW	1954	54	6.9	6	8	17
A	76	81	56	57	18.5	NW	117	SW	1947	52	6.7	6	8	16
M	77	80	55	55	16.7	S	124	SW	1950	58	6.5	7	9	15
J	81	83	56	56	14.8	S	95	W	1947	64	6.1	8	9	13
J	84	86	56	57	13.2	S	116	NW	1951	69	5.6	10	11	10
A	86	90	58	61	13.0	S	76	W	1955	69	5.5	10	11	10
S	87	86	60	68	14.2	S	84	W	1948	57	5.6	10	8	12
O	80	86	58	67	15.4	S	117	SW	1951	57	5.7	10	8	13
N	82	86	67	75	17.4	S	90	SE	1947	41	7.1	6	6	18
D	82	83	73	77	16.4	W	105	SW	1949	40	7.2	6	6	19
YR	81	84	61	65	15.9	S	124	SW	MAY 1950	57	6.4	94	96	175
Number of years in record (through 1976)	17	17	17	17	30	14	30	30		30	28	30	30	30

Table 2.4. Warmest and coldest seasons at Madison, Wisconsin, over the 107 year period from 1820–1926. Temperatures in °C

	Year	Mean temperature
Winter (Dec–Feb)		
Coldest	1877–78	−12.4
Warmest	1874–75	−0.11
Spring (March–May)		
Coldest	1843	2.0
Warmest	1830	11.9
Summer (June–August)		
Coldest	1915	17.9
Warmest	1830	22.9
Autumn (Sept–Nov)		
Coldest	1838	6.3
Warmest	1830	11.9
Year		
Coldest	1885	5.4°C
Warmest	1878	9.7°C
	1921	9.7°C

Data from an unpublished paper by Eric R. Miller, A Century of Temperature in Wisconsin, available from The National Climate Center, Asheville, North Carolina.

The precipitation data summarized in Table 2.3 reflect only approximately the wide year-to-year variability in the Madison area. Mitchell (1979) has summarized the precipitation data for Madison and has indicated the frequent occurrence of drought. Annual precipitation at Madison has varied from a high of over 100 cm/yr to a low of 35 cm/yr. Hydrologic drought, defined by Mitchell as a condition in which the annual precipitation is one standard deviation less than the mean, has occurred in Madison in the years 1895, 1902, 1938, 1955, 1958, 1963, 1965, and 1976. This marked variation in precipitation makes difficult the determination of a water budget for the lake (see below). Water is lost to the atmosphere by evapotranspiration, a process that occurs predominantly in the warm period of the year. Water gain by precipitation is counterbalanced by water loss from evapotranspiration, the two processes being approximately equal in the Madison area (Figure 2.1).

An important meteorological factor influencing the physical processes on the lake is wind. Stauffer (1980) has summarized wind data (converted to wind power above the lake surface) obtained from the National Climatic Center, Asheville, North Carolina. He recognized three types of windpower cycles, diel, weekly, and annual. Windpower is highest in mid-day and lowest at night, the diel cycle of wind being related to the diel cycle of atmospheric stability. This diel cycle is superimposed on an annual cycle characterized by a winter-early spring maximum and a late summer minimum (Figure 2.2). The high winter-early spring windpower is related to the incidence of high-powered storms during this period of the year. However, as with temperature and precipitation, there is pronounced year-to-year variation in windpower. In years of warm spring temperatures, windpower drops to the low summer values

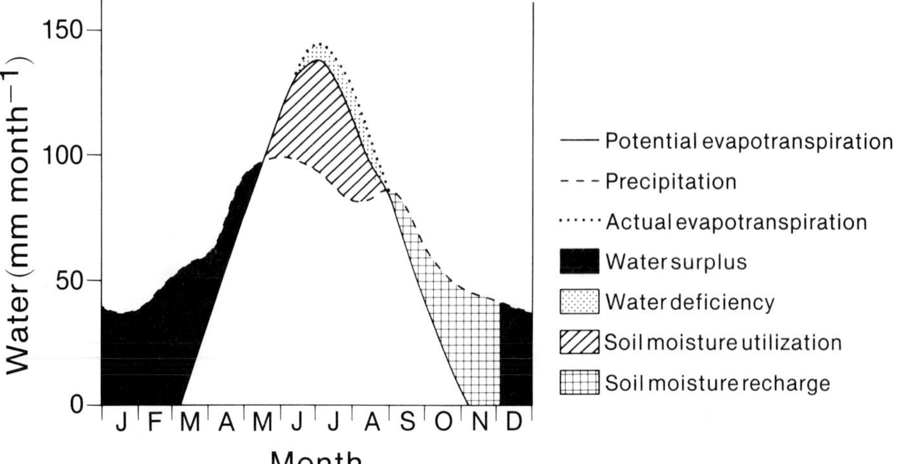

Figure 2.1. The water balance at Madison, Wisconsin. From Thornthwaite et al. (1958).

earlier, whereas cold springs tend to be unusually windy. Since the amount of windpower during spring is a major factor influencing the way in which Lake Mendota stratifies, these year-to-year variations have important consequences for the physical structure of the lake.

Water budget of the lake

Water enters Lake Mendota in a number of sources:

1. Direct precipitation (rain and snow on the lake), P

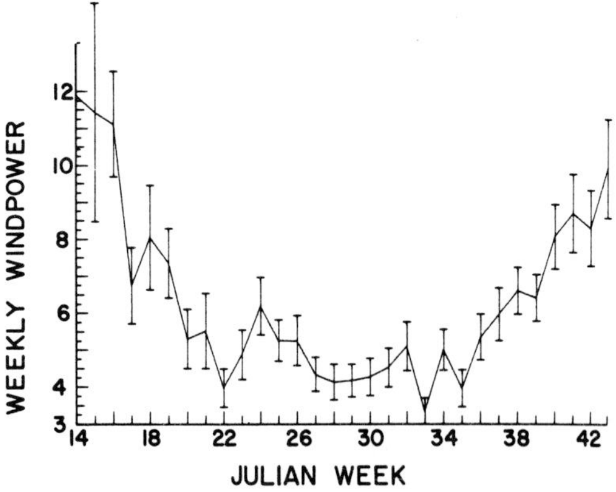

Figure 2.2. Average windpower by Julian week, 1966–1975, over Lake Mendota. Units are power, 10^5 ergs $\cdot$ cm^{-2} $\cdot$ day^{-1}. Bars are ± standard error of the mean. From Stauffer (1980).

Table 2.5. Annual water budgets for Lake Mendota

Year	Precipi-tation	Stream Input	Evaporation	Stream Output	Storage	Ground-water (by difference)
1951	32.2	52.1	30.7	96.5	−7.0	+35.7
1952	40.8	99.5	37.3	126.6	−10.0	+14.1
1953	30.2	72.7	39.2	102.2	+1.0	+39.0
1954	34.1	54.7	36.4	82.9	+5.0	+35.5
1976	21.9	88.6	37.3	80.3	−1.2	+5.9
1977	31.7	49.9	33.2	54.6	+1.9	+8.1

Data for 1950's from McCaskey (1955); for 1970's from Lathrop (1979).
All values to be multiplied by 10^6.
All units, m^3.
Direct groundwater measurements of Brock et al. (1982) in 1977–79: 25.96×10^6 m^3 yr^{-1}.

2. Overland runoff directly into the lake, R
3. Input from streams and rivers, I
4. Ground water seepage, GI

Water leaves the lake from the following sources:

1. Evaporation, E
2. Output through the Yahara River, O
3. Groundwater seepage, GO

If S is the lake storage, that is, the change in lake water volume, + equals an increase, − equals a decrease in lake volume, then the water budget of the lake can be expressed as: $S = P + R + I + GI - E - O - GO$.

For Lake Mendota, all components of this equation have been measured or can be calculated from other data. Since R is small, it is generally ignored. Because of the wide variation from year to year in precipitation in the Lake Mendota basin (see above), the measured values for any one year will not provide a good estimate of the average water budget.

The water budgets presented by McCaskey (1955) and Lathrop (1979) are based on measurements of stream and river input, on the lake storage, S, and outlet, O, from stream gauge data. Precipitation, P, was obtained from the U.S. Weather Bureau. Evaporation, E, was obtained from evaporation pan measurements made at Arlington, Wisconsin (University of Wisconsin Experimental Farms), multiplying the evaporation pan values by 0.7 to convert pan evaporation data to lake evaporation data. Neither McCaskey nor Lathrop measured groundwater seepage, but calculated it by difference, using lake level data by the U.S. Geological Survey to compute lake volume (a lake level variation of 0.00035 m (0.01 feet) corresponds to a volume change of about 120,000 m^2. The summary of these data is given in Table 2.5.

Although McCaskey and Lathrop estimated groundwater by difference,

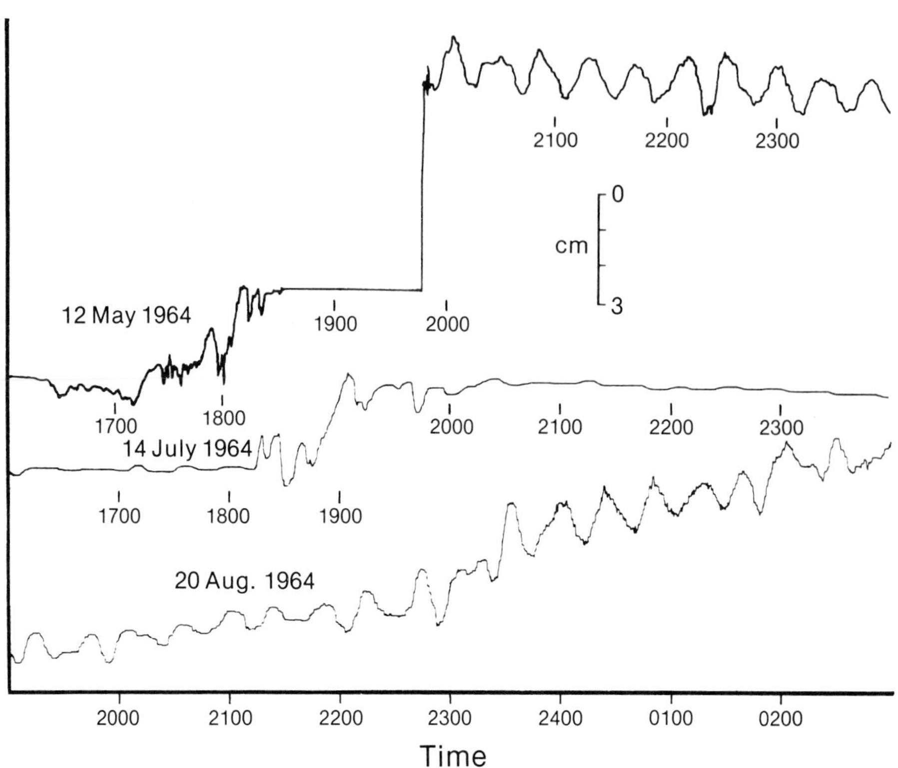

Figure 2.3. Change in water level of Lake Mendota as a result of rainfall events, as measured by a sensitive water-level recorder. From Stewart (1965).

groundwater seepage was measured directly by Brock et al. (1982) using seepage meters of the type described by Lee (1977). These direct measurements, made during the years 1977–79, agree fairly well with the McCaskey data, and with estimates Sonzogni and Lee (1974), but the Lathrop (1979) estimate is considerably lower. It should be noted that one of the years in which Lathrop did his study, 1976, was a drought year (Mitchell, 1979), whereas the year when most of the Brock et al. (1982) data were obtained (1979) involved a very wet summer. Groundwater seepage should also be affected by pumping from water supply wells (Cline, 1965; Lathrop, 1979). Although the mean annual precipitation in the Madison, Wisconsin area is 77 cm (31 inches), there has been wide variation over the past 100 years, from a low of 35 cm (14 inches) to a high of 132 cm (53 inches) (McCaskey, 1955).

Stewart (1965) used a sensitive water-level recorder to measure the response of Lake Mendota to rain fall events, as well as occurrences of surface seiche. As seen in Figure 2.3, an event on 12 May 1964 caused the lake to rise 8–9 cm in less than 3 hours. The annual variation in the water level of Lake Mendota for the years 1963 and 1964 is seen in Figure 2.4. Over this two year

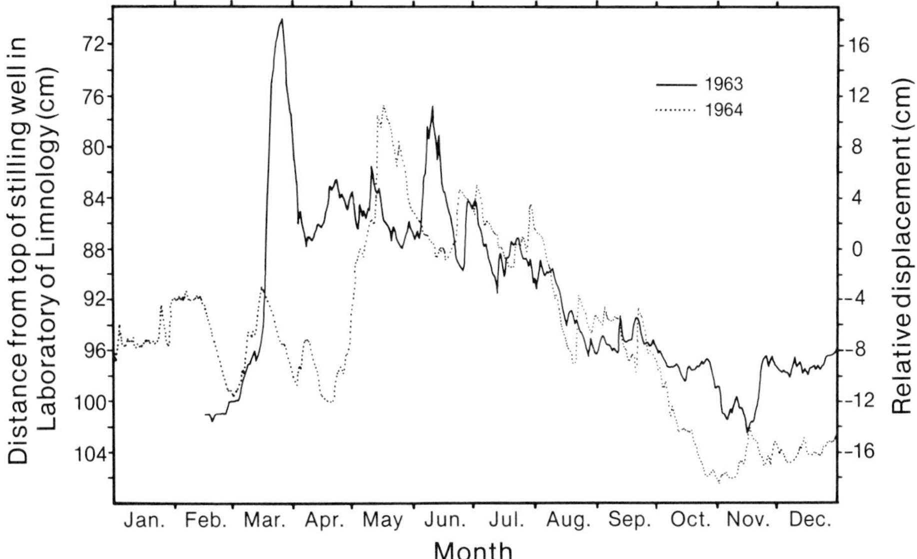

Figure 2.4. Variation in water level of Lake Mendota, 1963 and 1964. The high peak in March 1963 was due to rapid runoff during a warming period. From Stewart (1965).

period, the lake fluctuated about 0.31 meters. The large increase in water level in March 1963 was the result of rapid runoff from the drainage basin during an early spring thaw. Such early spring thaws may be responsible for large increases in nutrient loading to the lake (Lathrop, 1979).

A water-level gauge for Lake Mendota is maintained at the Tenney Park locks. For many years, responsibility for this gauge and for controlling the lake level was assumed by the City of Madison, but the respnsibility was transferred to the County of Dane on January 1, 1980. The target elevations for the lake are: summer maximum, 259.08 meters (850.10 feet); summer minimum, 258.96 meters (849.60 feet); winter minimum, 258.53 meters (848.20 feet). These target elevations are established to provide satisfactory water depth for summer boating, to avoid flooding of property, and to avoid ice damage to property in winter. Weather conditions sometimes make it difficult to maintain these target elevations. For instance, during the very dry summer of 1976, the lake elevation decreased to 258.60 meters (848.43 feet) whereas during the very wet summer of 1978 the elevation increased to 259.34 meters (850.86 feet). Heavy storms in the late summer of 1981 brought the lake level to an even higher elevation, 259.52 meters (851.44 feet). Over a five year period (from 1976–1981) during which the present field work was carried out, the total fluctuation in lake level was 0.96 meters (3.16 feet). Although these deviations from the target elevations do not seem excessive, they are often sufficient to cause considerable shore damage. For instance, flooding occurred of roads in low-lying areas during the summer of 1978. The key point, however, is that the level of Lake Mendota is maintained remarkably constant over long periods of time.

Table 2.6. Mean annual discharge and residence time for the Madison lakes

Item	Mendota	Monona	Waubesa	Kegonsa
Mean discharge $m^3 \times 10^6$ yr^{-1}	77.5	102	126	140
Lake volume $m^3 \times 10^6$	481	106	37.6	56.4
Residence time, yr	6.21	1.04	0.30	0.40
Flushing rate, yr^{-1}	0.16	0.96	3.33	2.63
Half-time, yr	8.96	1.50	0.43	0.55

Data of Lathrop (1979).

From the water budget data obtained, an important lake parameter, the hydraulic residence time, and its reciprocal, the flushing rate, can be obtained. The residence time is calculated as:

residence time, *years* = volume (m^3)/flow rate (m^3 yr^{-1})

This calculation assumes a steady state, which is approximately true for a large lake such as Lake Mendota. For periods of calculations, the rate of discharge through the outlet was used as a measure of the flow rate, ignoring storage (which is generally low enough to be negligible). Data for the four Madison lakes are given in Table 2.6.

From the residence time, a half-life can be calculated. The half-life, $t_{1/2}$, is the time at which some conservative element such as chloride dissolved in the water would be reduced in concentration to one-half its initial value, assuming all inputs of this element were stopped. The formula for calculating one-half life from residence time is:

$t_{1/2}$ = residence time/ln 2 = residence time/0.693

As seen in Table 2.6, the half-time of Lake Mendota is almost nine years. Since this is the time for a dilution to one-half of the original concentration, the length of time required to dilute a substance such as a pollutant to background levels (assuming it is conservative) would be many years.

Also noteworthy in Table 2.6 is the significantly shorter residence times of the Yahara lakes other than Lake Mendota.

Thermal structure and heat budget

The classical studies on the thermal structure and heat budget of lakes were done by Birge using Lake Mendota as a study site (Hutchinson, 1957; Stewart, 1973). A lake gains heat by radiation from the sun and by convection through the mixing of the water by the wind. A lake loses heat by radiation to the sky and by convection to the air, the latter again influenced by the wind. The density of water is strongly temperature-dependent, and the mixing of water layers is influenced by their temperatures. The heat content of the lake is not just a function of temperature but of its size and morphometry. A small deep lake is less readily warmed than a large shallow lake. A lake's heat content is estimated by measuring the temperature at a series of depths and multiplying this value by the volume of water represented by that layer (obtained from hypsometry, see Table 2.2). If this total heat content (degrees · m^3) is divided by the total volume of the lake (m^3), the mean temperature of the lake (degrees) can be calculated. Heat content (enthalpy, cal/cm^3) is equivalent to mean temperature, if it is assumed that water has a specific heat of unity (approximately true for lake water). The total heat content of the lake (cal/cm^2) can then be calculated by multiplying the heat content (cal/cm^3) by the mean depth (cm).

Heat budget

Stewart (1973) has presented detailed data for mean temperature and heat content for three of the Madison lakes, and his data for 1963 are given in Figure 2.5. As seen, there is an inverse relationship between mean temperature and heat content, Lake Waubesa having the highest mean temperature and the lowest heat content, Lake Mendota have the lowest mean temperature and the highest heat content. The higher heat content of Lake Mendota is due to the fact that it has a considerably greater mean depth than the other lakes, making the cooling of the deeper waters more difficult.

The annual heat budget of the lake (cal/cm^2) is the total heat income from the time of minimum heat content until the summer maximum. The mean annual heat budget for Lake Mendota, estimated from 5 years of measurements (1960–1964 and 1966) was 24,073 cal/cm^2 (Stewart, 1973). Corresponding data for Lakes Monona and Waubesa (3 year averages) are 17,559 and 11,362 cal/cm^2 respectively. Year-to-year variation was relatively small, and there was little different between Stewart's data in the 1960's and the data presented by Birge (1915).

Thermal stratification

Of great biological importance is the thermal stratification that occurs in many lakes in the summer. This phenomenon was first studied in detail by Birge on Lake Mendota, but has been widely studied throughout the world (Hutchinson, 1957). Stratification occurs because warm water is less dense than cold water and because lakes are warmed from the top. If the temperature difference between the top and bottom is great enough, then the two layers of water resist mixing by the wind and stratification ensues. If the oxygen demand of the bottom waters is high enough, the bottom layer of water, no longer aerated from the surface, may become anaerobic, with important biological and geochemical consequences. Lakes Mendota and Monona are deep enough to stratify through the summer months, but Lakes Waubesa and Kegonsa only stratify for short, unpredictable periods.

Time of stratification and bottom water temperature

Of major significance is the time at which stratification occurs, since this influences the length of the spring mixing period as well as the length of the

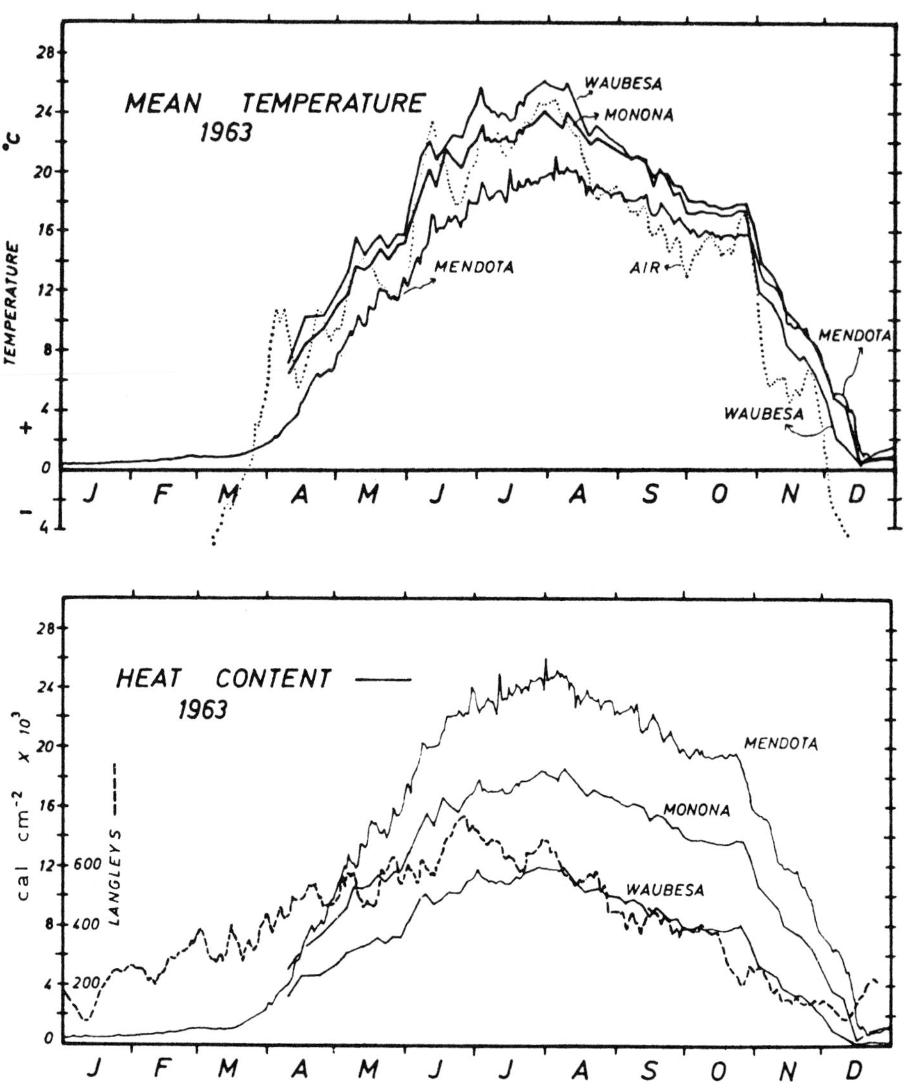

Figure 2.5. Mean temperatures and heat contents (from 0°C) of Lakes Mendota, Monona, and Waubesa during 1963. The dotted and dashed lines represent moving 10-day averages plotted on the tenth day. From Stewart (1973).

summer stratification. The continental weather pattern of the Madison area (see earlier in this chapter) has a considerable influence on when stratification occurs in Lake Mendota. Spring is generally cold and windy (Stauffer, 1980) followed by a rapid warming and calming. With the wind speed low and the temperature high at the beginning of summer, Lake Mendota generally stratifies quickly and firmly. The period during which stratification occurs is usually between mid-May and early June, although in an atypically warm spring, strat-

Table 2.7. Bottom water temperature (24 meter depth) for Lake Mendota in midsummer, Julian day approximately 200 (July 19)

Year	Temperature	Source of data
1895	12.5	Neess and Bunge, 1956
1896	14.5	"
1897	11.5	"
1898	10.0	"
1906	11.2	"
1909	9.8	"
1910	12.5	"
1911	10.5	"
1912	12.0	"
1913	12.5	"
1914	10.8	"
1915	12.8	"
1916	12.8	"
1960	9.5 (20 m)	Stewart, 1965
1961	11.5 (18 m)	"
1962	11.0 (18 m)	"
1963	12.0 (18 m)	"
1971	11.2	Stauffer, 1974
1972	7.65	"
1976	9.0	Present work
1977	11.0	"
1978	8.5	"
1979	8.9	"
1980	11.2	"
1981	14.0	"
1982	9.0	"
1983	9.0	"
1984	11.0	"

ification may come a week or so earlier. On the other hand, if summer is late, stratification may not occur until mid-June.

The temperature of the bottom water at the time of stratification will depend upon when and how stratification occurs. If stratification occurs early, due to an early onset of warm, windless weather, then the bottom water temperature will be lower than if stratification occurs late after a long cold windy spring. The marked year-to-year variation in the stratification process can be seen from the data presented in Table 2.7, which give the bottom water temperature at Julian day 200 (mid July) for the 24 meter depth. As seen, over a seven year period from 1972 through 1981, the 24 meter water temperature ranged from a low of 7.65 to a high of 14.0. The data from 1895–1916 also show considerable variation, from a low of 9.8 to a high of 14.5. Most of this variation in time of stratification does not show up in heat budget calculations, but should have important biological consequences, especially for biogeochemical processes in the sediments.

Detailed temperature profiles of Lake Mendota have been provided from time to time since Birge's first studies in 1894–1896. Aside from Birge himself (see Neess and Bunge, 1956, 1957), the most extensive data have been provided by Stewart (1965), Stauffer, (1974), and the present writer (data for 1976–81 are presented in the Appendix to this book). Data for 1906 (from Birge and Juday, 1911) and from 1976 (present work), which are actually quite similar, are presented in Figure 2.6.

With the kind of contour graph shown in Figure 2.6, it is relatively easy, by inspection, to observe the time at which stratification occurs (shown as the time that the isopleths become horizontal). Also easily obtained from the graph is the depth of the thermocline and the temperature at the thermocline depth. With a lake as large as Lake Mendota, the thermocline is fairly deep (10–12 m throughout most of the summer) and extends over a fairly broad depth (generally at least two meters).

Fall overturn, the period when stratification breaks down, is also a gradual process, the thermocline slowly becoming deeper until, some time around early October and usually after a heavy wind storm, the lake completely mixes to the bottom.

It is important to note that profiles such as those shown in Figure 2.6 have been taken in the center of the lake. By an examination of the hypsometric data (Table 2.2), it can be seen that about half of the area of the lake is shallower than 10 m and hence is essentially never stratified.

We will refer back to Figure 2.6 during subsequent discussions on chemical and biological processes in the lake in later chapters.

Ice and freeze-up

Ice cover has a profound effect on a lake, greatly reducing light penetration and abolishing wind mixing. Although ice formation cannot begin until water reaches the freezing point of 0 °C, the time of ice-in and ice-out are markedly influenced by the depth, size, and fetch of the lake. The larger the lake and the longer the fetch, the later freeze-up occurs. The freezing process, with examples from Wisconsin lakes, has been discussed in detail by Ragotzkie (1978).

Freeze-up of Lake Mendota usually occurs during a period of calm following the arrival of a strong cold front. Although ice formation may occur in University Bay ahead of the main body of the lake, the total lake may freeze up suddenly during a period of a few hours (generally during the night). The date of freeze-up of Lake Mendota can vary by over a month, from early December to mid January, although the mean date of freezing is 19 December (Stewart and Hasler, 1972). In some years, the fast freezing is not permanent, the lake reopening partially before complete closure.

Water temperature at which lakes freeze over is closely related to the fetch. Small lakes freeze when the water temperature is between 2 and 3 °C, while lakes as large as Lake Mendota generally cool to less than 1 °C before freezing. Only in a rare year would the temperature of Lake Mendota after freezing be 1 °C or above.

Reduction of light penetration occurs partly due to the ice cover itself, but

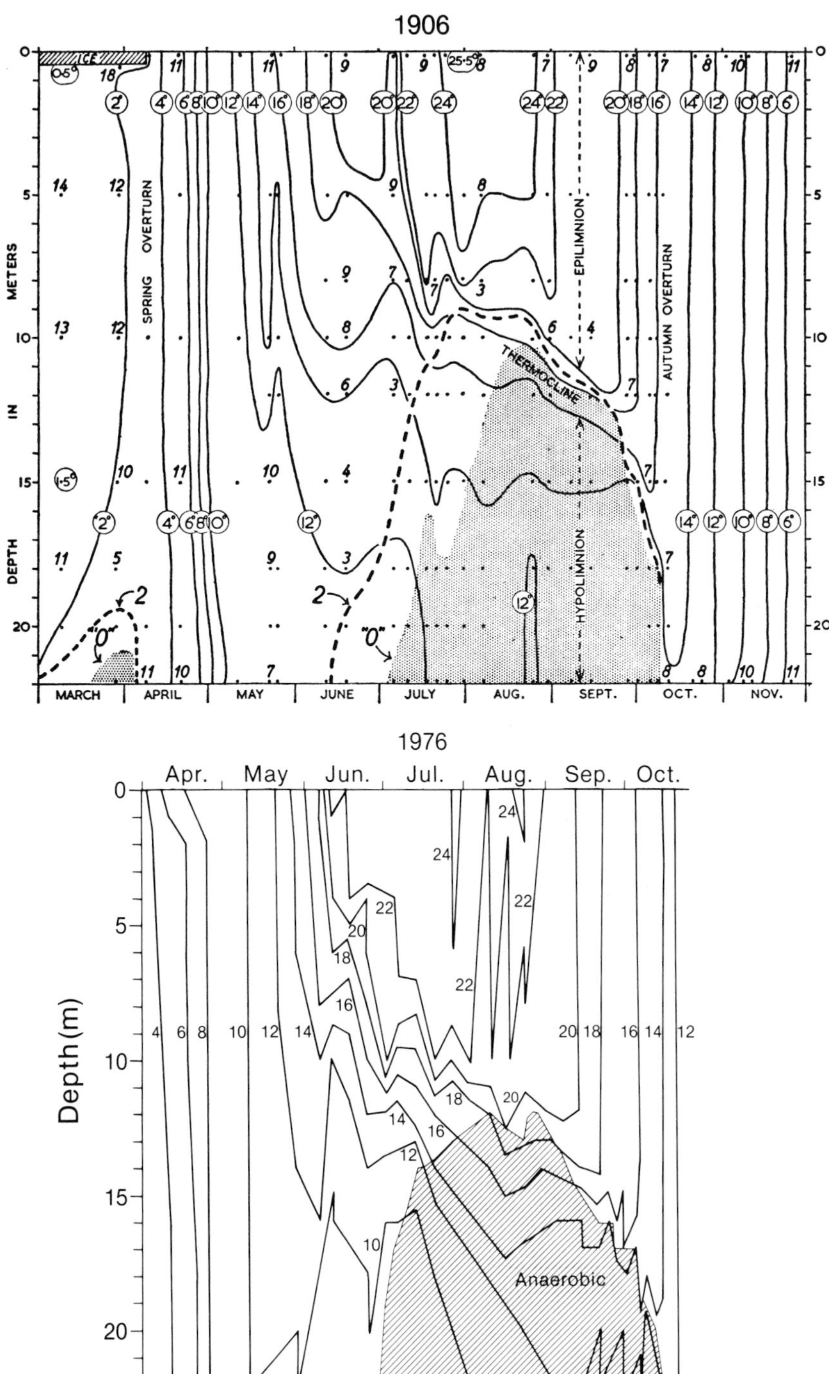

Figure 2.6. Seasonal changes in water temperature for the years 1906 and 1976. The figure for 1906 also presents data for dissolved oxygen concentration (ppm, numbers in italics). Data for 1906 from Birge and Juday (1911) and for 1976 from the present work. The figure for 1906 is from Mortimer (1956).

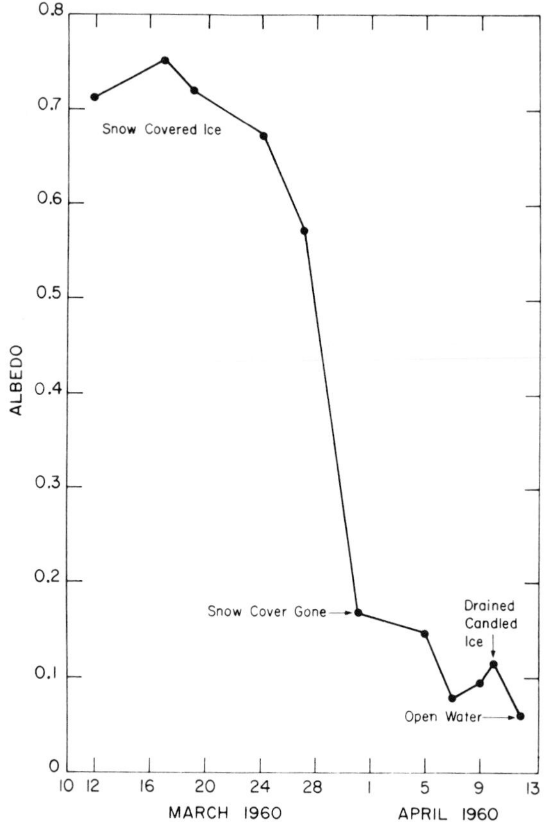

Figure 2.7. Albedo of Lake Mendota preceding ice breakup. From Ragotzkie (1978).

more importantly due to the present of snow (Ragotzkie, 1978). In a heavy snow year, light penetration into Lake Mendota may be very low, thus reducing the development of under-water algal blooms. If snow is minimal or absent, light penetration may be very good and because of the absence of vertical mixing, significant underwater phytoplankton populations may develop (see Chapter 4).

With the onset of the spring season, the snow cover often disappears rapidly and light penetration dramatically increases. Some results of the change in light penetration for Lake Mendota are shown in Figure 2.7. Disappearance of the snow cover is a necessary prelude to the break-up of the ice cover. The melting of the ice occurs mostly by conduction from warming water underneath the ice, which is itself heated by solar radiation. Just before break-up, the ice generally becomes altered into long, vertically-oriented crystals, a process called "candling" (Ragotzkie, 1978). After candling, the physical integrity of the ice sheet in the horizontal dimension is essentially destroyed, and a strong wind can then cause the ice to break up. Once the ice sheet is broken into large

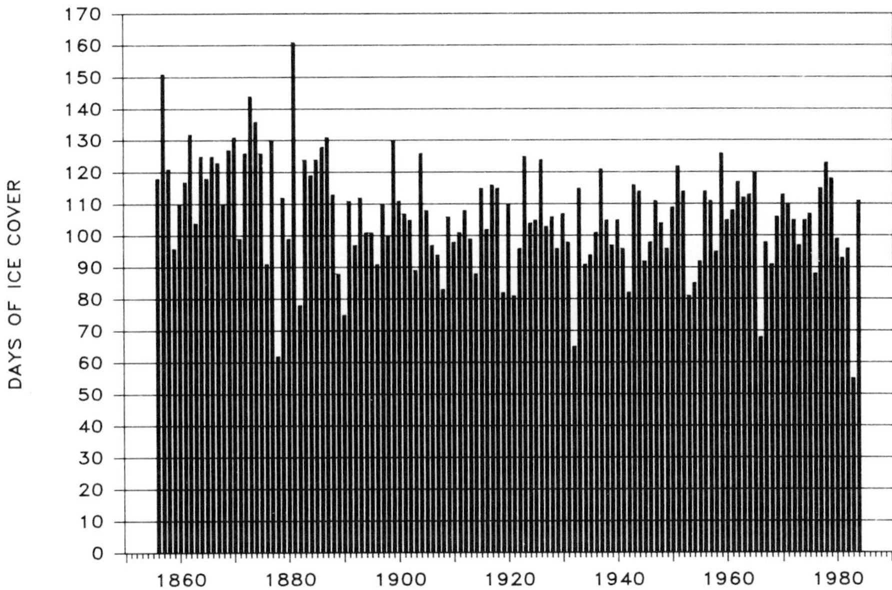

Figure 2.8. Length of ice cover on Lake Mendota for different winter seasons over a 130-year period. Data of the Wisconsin State Climatologist, Madison, Wisconsin.

pieces, the increased ice surface exposed to the water results in very rapid melting. The entire process of break-up and complete melting of the ice can occur in a single day.

In Lake Mendota, break-up can occur at any time from early March until late April depending on the year. The mean date of lake opening is 6 April (Stewart and Hasler, 1972). Because of the wide variability of the weather in the Madison area (see above), the length of time ice is on the lake varies greatly, as shown in Figure 2.8. The duration of ice cover has important effects on mixing and solar radiation penetration, and hence on the biology of the lake.

Water movement

Lake Mendota is large enough to exhibit pronounced water movements. Water movement in lakes can be of several types: 1) Mass flow, due to movement of water into and out of the lake through rivers and streams; 2) Surface wave movement (progressive waves), induced by the wind; 3) Surface currents, primarily wind-induced but also caused by geostrophic effects (the Coriolis force), by changes in atmospheric pressure, and by horizontal density differences; 4) Vertical movement of the whole lake, leading to long standing waves and seiches; 5) Vertical movement of parcels of water within the lake, resulting in either internal waves or large-scale vertical currents such as the Langmuir circulation (Hutchinson, 1957).

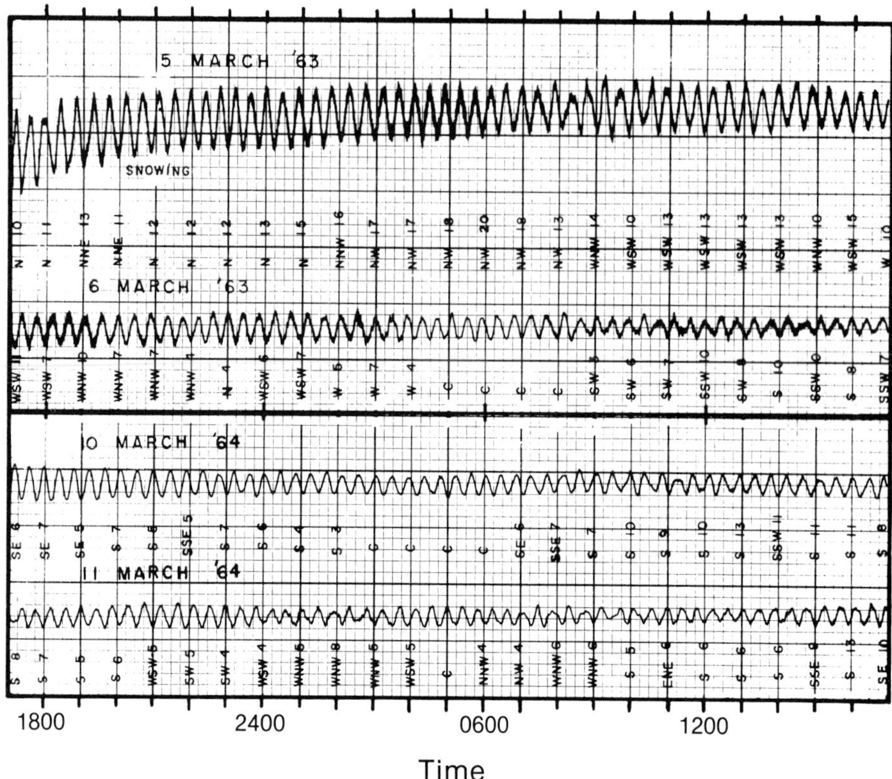

Figure 2.9. Records of uninodal oscillations in Lake Mendota. The mean of 111 periods (trough to trough) was 25.9 minutes. From Stewart (1965).

In general, water movement in the horizontal direction is much more rapid than water movement in the vertical direction. Horizontal water movement in lakes the size of Lake Mendota can be expected to have a strong influence on the distribution of small organisms of the lake, but has little effect on large organisms such as fish. Patchiness of relatively passive organisms such as phytoplankton is due primarily to horizontal water movement. Vertical water movement, although less rapid than horizontal, should have significant effects on photosynthetic organisms, since it influences their distribution in the water column and hence their light regime. The relationship of phytoplankton photosynthesis to light is discussed in Chapter 5.

Although the theoretical and quantitative descriptions of water movements in lakes are well developed (Hutchinson, 1957), these descriptions deal only with water itself or with things dissolved in the water. Movement of organisms the size of phytoplankton or larger cannot be described by the conventional equations of physical limnology. Consequently, the movements of organisms can generally not be predicted mathematically, but must be determined by

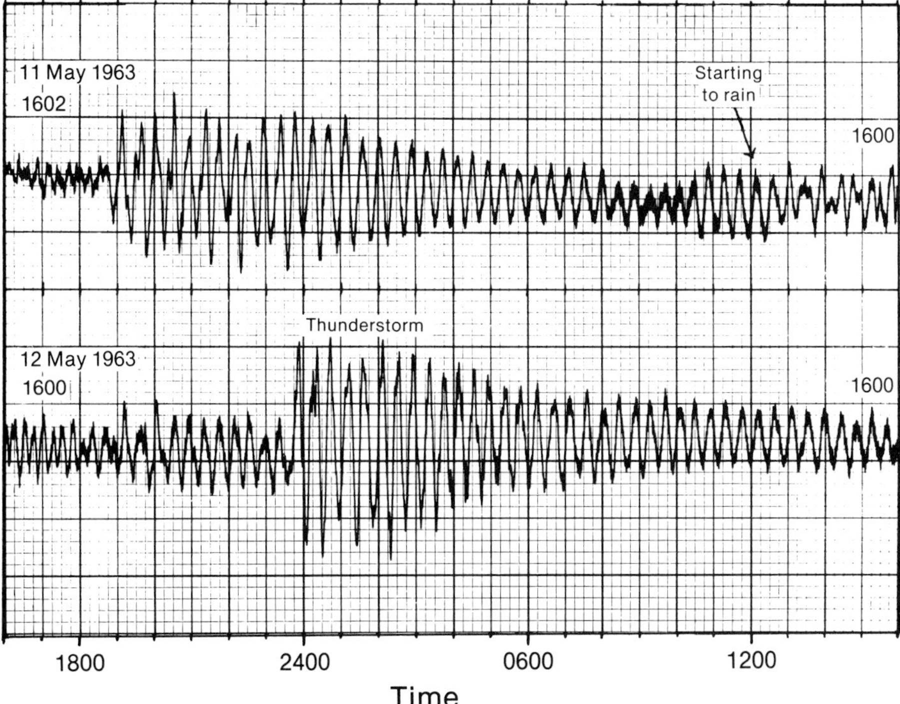

Figure 2.10. Lake Mendota surface seiche and the effect of two thunderstorms. From Stewart (1965).

direct measurement in the lake. The distribution of phytoplankton will be described in detail in Chapter 4.

Water movement in Lake Mendota has been extensively studied by physical limnologists, most prominently by Professor Reid Bryson and collaborators in the 1950's and early 1960's, and Kenton M. Stewart in the mid 1960's.

Stewart (1965) operated a water-level recorder on Lake Mendota and recorded surface standing waves (surface seiche). From the recordings, the period of the standing waves could be calculated. According to theory, the period of a wave is directly correlated with the length of the basin and inversely correlated with the depth, and the theoretical equation fits well with the actual data for Lake Mendota. For a transect across the lake from SW to NE (approximately from the City of Middleton to the Village of Maple Bluff), the data for two persistent surface seiches are shown in Figure 2.9. The records shown are of the uninodal oscillation, which has a period of about 26 minutes. Only twice during two years of continuous recording did Stewart find a regular uninodal oscillation to persist on Lake Mendota for as long as two days. Stewart also calculated the periods of the higher order of nodes, which were: binodal 13.5 minutes; trinodal 9.15; quartanodal, 8.23; pentanodal, 6.41 minutes. The uninodal periods were more frequently recorded than the multinodal frequen-

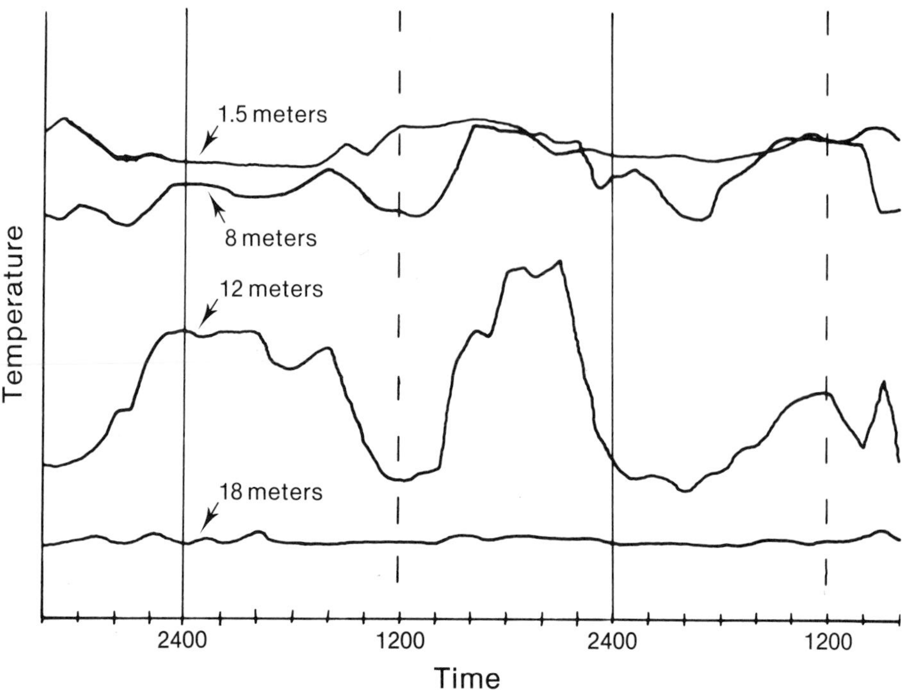

Figure 2.11. Evidence of standing internal waves in Lake Mendota from measurements of temperature fluctuations. From Stewart (1965).

cies, although even with the uninodal oscillations, good records lasting for a full day were not common. Seiches are induced by both atmospheric pressure changes and by persistent strong winds, and both phenomona have been observed to induce seiches on Lake Mendota. Figure 2.10 illustrates the dramatic effect of thunderstorms on two consecutive days. Stewart also observed on 27 March 1964 a change in the water level of Lake Mendota due to a powerful earthquake centered in Alaska.

In addition to oscillation of the lake as a whole, as shown by surface seiches (above), the various water layers of different density can oscillate in relation to one another, leading to the formation of internal waves or seiches. For an internal seiche to occur, a vertical density gradient must exist, the latter ordinarily resulting from thermal stratification. The period of an internal seiche is influenced not only by the length and depth of the basin, but by the vertical density distribution.

Stewart (1965) made detailed measurements with automatic recording devices of temperature at various depths in the lake. One short interval of 48 hours is illustrated in Figure 2.11. As would be predicted by theory, the temperature excursion is much greater at the thermocline than in either the epilimnion or hypolimnion.

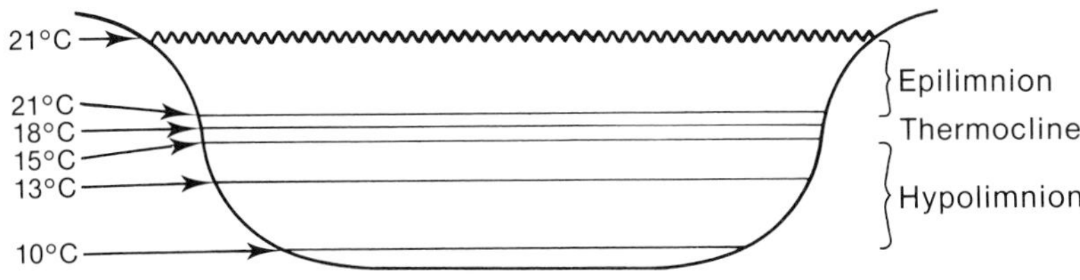

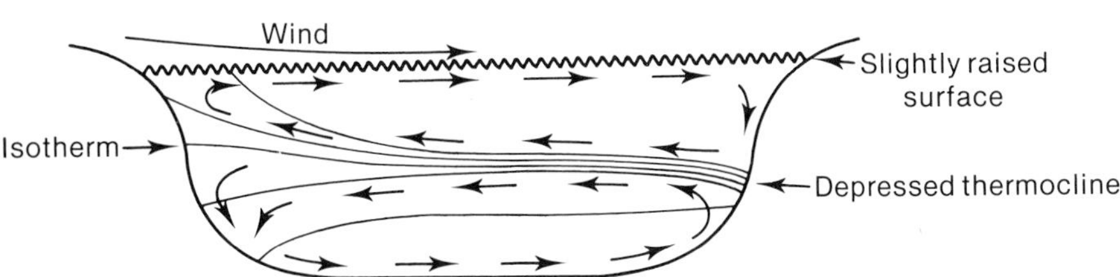

Figure 2.12. Deformation of water masses and the tilting of the thermoclime as a result of wind stress. Upper) Simple two layered lake at rest; lower) after deformation due to wind stress. The arrows show the direction of the underwater currents. From Bryson and Suomi (1952).

Horizontal and vertical circulation

Wind stress is primarily responsible for water movement in a lake the size of Lake Mendota. Studies of the water movement have been presented in several papers by Bryson and associates (Bryson and Suomi, 1952; Bryson and Bunge, 1956; Clarke and Bryson, 1959; Shulman and Bryson, 1961; Lathbury, Bryson and Lettau, 1960; Ragotzkie and Bryson, 1953).

A general discussion of the circulation of Lake Mendota has been given by Bryson and Suomi (1952). Consider a simple two-layered lake (Figure 2.12, upper). Since the wind exerts a horizontal stress on the lake surface, it drives the water downwind, resulting in tilting of the layers as seen in the lower part of Figure 2.12. This results in a lowering of the thermocline on the downwind side of the lake with a convergence or compression of the isotherms, and a raising of the thermocline of the upwind side of the lake with a divergence of the isotherms. In the epilimnion, water moves downwind at the surface and upwind at the top of the thermocline, whereas in the hypolimnion the water movements are just the opposite. An example of actual data from Lake Mendota is given in Figure 2.13. A summary of horizontal water velocities for Lake Mendota is given in Table 2.8, where it can be seen that the currents are much higher at the surface than at depth.

On a theoretical basis, the vertical current speed should be much lower than

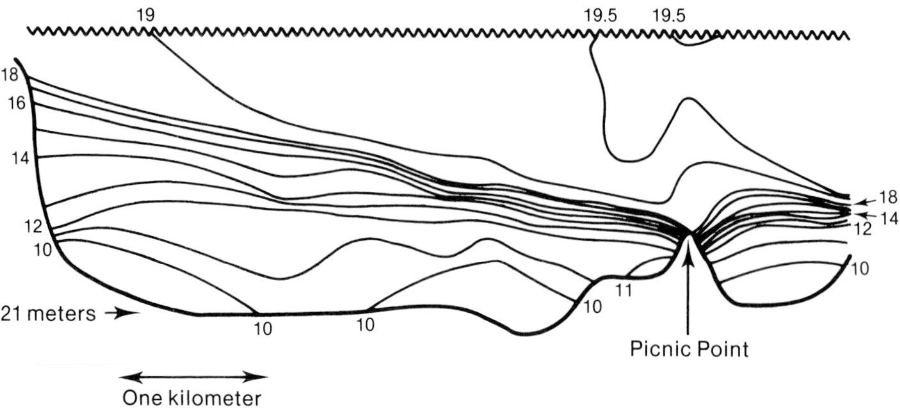

Figure 2.13. Typical tilted thermocline on Lake Mendota shown by temperature measurements across a northwest-southeast section. From Bryson and Suomi (1952).

the horizontal speed, except perhaps where large density differences in water occur, such as at the thermocline (Hutchinson, 1957). Direct measurements of both horizontal and vertical current speed by Bryson and Suomi (1952) confirm this, as seen in Figure 2.14. The rapid vertical current across the thermocline is of considerable importance for understanding transfer of chemical components from the hypolimnion into the epilimnion, as discussed in Chapter 3.

Horizontal and vertical currents influence the concentrations of chemical elements in the lake water, especially for those elements which are derived from or consumed by the sediment. Although a single set of water samples taken in the deepest part of the lake may serve to obtain an estimate of whole-lake concentrations, lateral transport and interaction between water and shallow sediment shelves may render the estimate imprecise. Stauffer (1985) has shown that lateral gradients over sediment shelves of Lake Mendota can cause a slight bias in estimates of whole-lake phosphorus and oxygen concentrations.

Table 2.8. Mean and modal velocities for horizontal currents in Lake Mendota

Depth	Mean cm/sec	Mode cm/sec	Number of cases
0–0.25 m	9.36	6–7	34
0–0.5 m	5.93	4–5	107
0.75–1.25 m	4.48	1–2	583
1.75–2.25 m	4.71	3–4	367
4.75–5.25 m	3.23	1–2	255
9.75–10.25 m	2.78	2–3	248
14.75–15.25 m	1.67	0–1	26
19.75–20.25 m	7.7	not defined	10

From Lathbury et al. (1960)

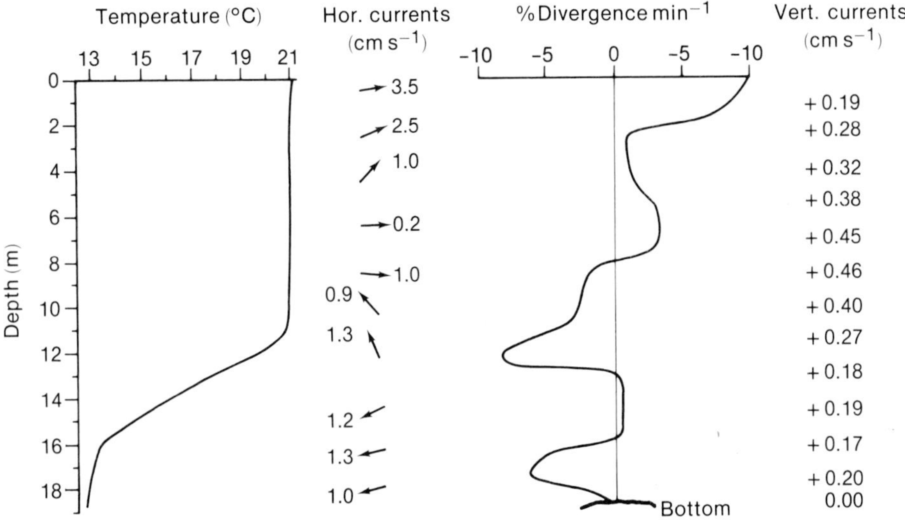

Figure 2.14. Vertical distribution of temperature, horizontal and vertical currents, and divergence, Lake Mendota, August 31, 1950. From Bryson and Suomi (1952).

Oxygen is consumed at the sediment-water interface as the water passes over a shallow shelf, causing an overestimation of total lake oxygen if water is obtained from a deep hole station. Phosphorus, on the other hand, is released from the sediment into the water passing over a shelf, causing an underestimation of total-lake phosphorus concentration. For phosphorus, Stauffer concluded that the error is only 2.5% of the total lake phosphorus, but at periods of the year when prolonged stagnation of water movement occurs (mid-summer or under ice cover), the bias may be greater.

Summary. The four seasons in Lake Mendota

As a lake of moderate size in a temperate climate, Lake Mendota responds to seasonal changes in weather in a typical fashion. It is most convenient to initiate a discussion of the four seasons with the period beginning after fall overturn (Figure 2.15). From early October until ice cover forms (ice-up), the lake is isothermal with depth, gradually cooling from the temperature at fall overturn (12–19°C) to a temperature of 0°C. The date of ice-up is quite variable from year to year, and depends not only on the air temperature but the wind. Ice-up frequently occurs during one of the windless periods that accompanies the establishment over the Madison area of a very cold polar air mass.

Although mixing occurs in both spring and fall, the fall mixing period is the time when the most extensive resuspension of bottom sediments occurs, partly because of the strong winds which occur at that time of the year and

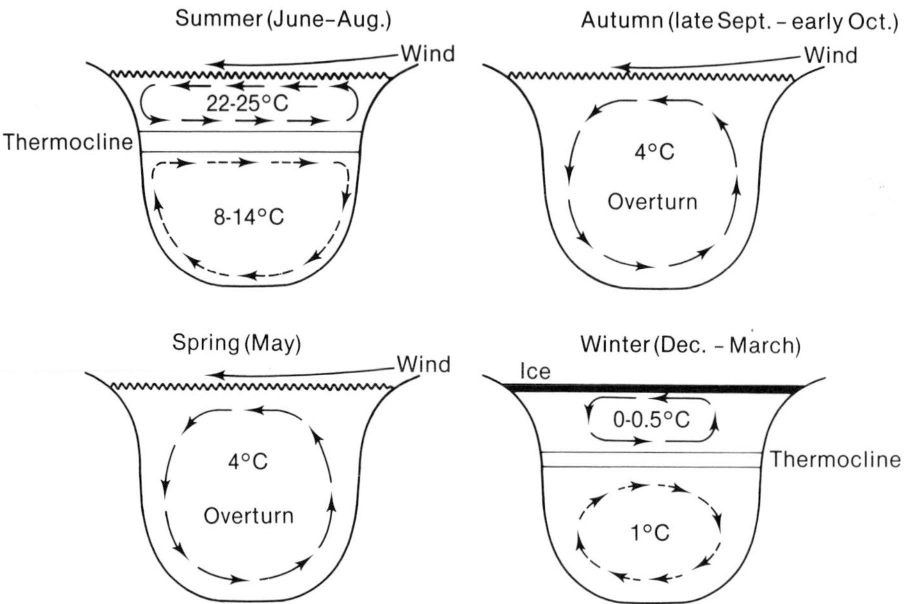

Figure 2.15. The four seasons on Lake Mendota.

partly because the lake tends to mix more extensively when the surface water becomes colder (and hence denser) than the water beneath.

Once ice-up has occurred, mixing processes virtually cease. During the dark days of late December through early March, biological processes, especially those driven by photosynthesis, are subdued. Depending on the snowfall, little or no light may penetrate the ice, and photosynthetic processes almost cease. During this period, materials that had become suspended in the water during the fall mixing period settle out, and the water becomes the clearest it is all year.

By early March, light is again passing through the ice to the lake surface. This is partly because the day length is longer and the sun higher in the sky, and partly because the early spring thaws have melted the snow. In some winters, there is little or any snow, so that photosynthetic processes may occur even in January and February, but generally there is sufficient snow so that a March thaw is required to restore the clarity of the ice.

Depending on the weather, the ice may disappear completely in early to mid-March, but the average date of ice-out is early April and in some years ice-out may not occur until mid-April (see Figure 2.8). Paradoxically, a late ice-out may actually *promote* phytoplankton growth, since the ice layer prevents vertical mixing by the wind, permitting the photosynthetic organisms to develop extensively just below the ice surface, where light conditions are favorable.

Once the ice disappears, the spring mixing period ensues. This is usually not as long or as extensive as the fall mixing period, and because the surface waters are warming up and hence becoming less dense, more energy is required to mix the water than in the fall.

Summer stratification usually occurs suddenly in the period between the end of May and early June, usually brought about by a sudden period of very warm, windless weather. By late June, the lake is firmly stratified, with the surface water about 15°C warmer than the bottom water. Although summer is a period of low wind power (Stauffer, 1980), even if strong winds occur mixing to the bottom cannot occur due to the great density difference between epilimnion and hypolimnion.

References

Birge, E.A. 1915. The heat budgets of American and European lakes. *Transactions Wisconsin Academy of Sciences, Arts, and Letters,* 18: 166–213.

Birge, E.A. and C. Juday. 1911. The inland lakes of Wisconsin. The dissolved gases of the water and their biological significance. *Wisconsin Geological and Natural History Survey* Bulletin 22, Scientific Series No. 7, 259 pp. Madison, Wisconsin.

Brock, T.D., D.R. Lee, D. Janes, and D. Winek. 1982. Groundwater seepage as a nutrient source to a drainage lake; Lake Mendota, Wisconsin. *Water Research,* 16: 1255–1263.

Bryson, R.A. and W.W. Bunge. 1956. The "stress-drop jet" in Lake Mendota. *Limnology and Oceanography,* 1: 42–46.

Bryson, R.A. and R.A. Ragotzkie. 1955. Rate of water replacement in a bay of Lake Mendota, Wisconsin. *American Journal of Science,* 253: 533–539.

Bryson, R.A. and V.E. Suomi. 1952. The circulation of Lake Mendota. *Transactions of American Geophysical Union,* 1: 42–46.

Clarke, D.B. and R.A. Bryson. 1959. An investigation of the circulation over Second Point Bar, Lake Mendota. *Limnology and Oceanography,* 4: 140–144.

Cline, Denzel R. 1965. Geology and ground-water resources of Dane County, Wisconsin. *U.S. Geological Survey Water-Supply Paper 1779-U,* 64 pp. plus maps. U. S. Government Printing Office, Washington, D. C.

Hutchinson, G.E. 1957. *A Treatise on Limnology,* Vol. 1. John Wiley & Sons, New York.

Juday, Chancey. 1914. The Inland Lakes of Wisconsin. The Hydrography and Morphometry of the Lakes. *Wisconsin Geological and Natural History Survey* Bulletin 27, Scientific Series No. 9, Madison, Wisconsin.

Kanneberg, Adolph. 1936. The Dam at the Outlet of Lake Mendota. pp. 17–19 In: *Lake Mendota Origin and History,* The Technical Club of Madison, Madison, Wisconsin.

Lathbury, A., R. Bryson and B. Lettau. 1960. Some observations of currents in the hypolimnion of Lake Mendota. *Limnology and Oceanography,* 5: 409–413.

Lathrop, R.C. 1979. *Dane County Water Quality Plan Appendix B: Water Quality Conditions; Appendix H: Lake Management.* Dane County Regional Planning Commission, Madison, WI.

Lee, D.R. 1977. A device for measuring seepage flux in lakes and esturaries. *Limnology and Oceanography,* 22: 140–147.

McCaskey, A.E. 1955. *Hydrological Characteristics of Lake Mendota Drainage Basin.* Ph.D. Thesis University of Wisconsin, Madison.

Mitchell, V.L. 1979. Drought in Wisconsin. *Transactions of the Wisconsin Academy of Sciences, Arts and Letters,* 67: 130–134.

Mortimer, C.H. 1956. p. 178 In: *E. A. Birge, A Memoir,* by G. C. Sellery. University of Wisconsin Press, Madison.

Murray, R.C. 1956. Recent sediments of three Wisconsin lakes. *Bulletin of Geological Society of America,* 67: 883–910.
Neess, J.C., and W.W. Bunge. 1956. An unpublished manuscript of E.A. Birge on the temperature of Lake Mendota; Part 1. *Transactions of the Wisconsin Academy of Sciences, Arts and Letters,* 45: 193–208.
Neese, J.C., and W.W. Bunge. 1957. An unpublished manuscript of E.A. Birge on the temperature of Lake Mendota; Part 2. *Transactions of the Wisconsin Academy of Sciences, Arts and Letters,* 46: 31–89.
Ragotzkie, R.A. 1978. Heat budgets of lakes. pp. 1–19 In *Lakes Chemistry Geology Physics* (A. Lerman, editor), Springer-Verlag, New York.
Ragotzkie, R.A. and R.A. Bryson. 1953. Correlation of currents with the distribution of adult *Daphnia* in Lake Mendota. *Journal of Marine Research,* 12: 157–172.
Shulman, M.D. and R.A. Bryson. 1961. The vertical variations of wind-driven currents in Lake Mendota. *Limnology and Oceanography,* 6: 347–355.
Sonzogni, W.C., and G.F. Lee. 1974. Nutrient sources for Lake Mendota. 1972. *Transactions of the Wisconsin Academy of Sciences, Arts and Letters,* 62: 133–164.
Stauffer, R.E. 1974. *Thermocline Migration-Algal Bloom Relationships in Stratified Lakes.* Ph.D. Thesis University of Wisconsin, Madison.
Stauffer, R.E. 1980. Windpower time series above a temperate lake. *Limnology and Oceangraphy,* 15: 513–528.
Stauffer, R.E. 1985. Lateral solute concentration gradients in stratified eutrophic lakes. *Water Resources Research.*
Stewart, K.M. 1965. *Physical limnology of some Madison Lakes.* Ph.D. Thesis University of Wisconsin, Madison.
Stewart, K.M. 1973. Detailed time variations in mean temperature and heat content of some Madison lakes. *Limnology and Oceanography,* 218–226.
Stewart, K.M. and A.D. Hasler. 1972. Limnology of some Madison lakes: annual cycles. *Transactions of the Wisconsin Academy of Sciences, Arts and Letters,* 40: 87–123.
Thornthwaite, C.W., J.R. Mather and D.B. Carter. 1958. *Water balance maps of eastern North America.* Resources for the Future, Washington, D.C.

3. Chemistry and nutrient loading

Lake Mendota is a typical hard-water lake whose chemistry has been significantly modified by eutrophication. A sequence of events can be recognized in Lake Mendota which leads to an alteration in the chemistry of the lake water. This sequence will be described briefly here and will be elaborated in the rest of this chapter.

1. As a result of nutrient enrichment by the eutrophication process, marked increases in phytoplankton production have occurred (see Chapters 4 and 5).
2. Some of the organic carbon produced by the phytoplankton has sedimented to the bottom of the lake, where it undergoes decomposition processes.
3. Decomposition results in depletion of oxygen from the bottom waters, leading to anaerobiosis.
4. Anaerobiosis results in the development of new biogeochemical processes, such as sulfate-reduction and methanogenesis (see Chapter 7), and also results in modification of iron and manganese redox states. Among other things, iron sulfides form in the sediments and soluble iron is transferred to the water column.
5. Changes in iron/sulfur oxidation states cause phosphate to be released from iron-phosphate complexes, resulting in transfer of phosphate to the

water column. This leads to further acceleration of the eutrophication process.

The sequence of events described above has been well documented in many lakes and is not unique to Lake Mendota. Lake Mendota provides, however, an excellent example of the biogeochemical changes that a lake undergoes as a result of the eutrophication processes.

Chemistry of the lake water

A lake receives its chemical inputs from three major sources: 1) atmosphere; 2) stream flow and runoff; 3) groundwater seepage. In the case of Lake Mendota, category 2 is by far the most significant source of chemical input (Lathrop, 1979; Brock, et al., 1982).

Because a lake is a dynamic entity, it can never be said to have a fixed chemical composition. Some average values and ranges, however, have been assembled for Lake Mendota and are presented in Table 3.1. Some of the elements listed are relatively conservative and vary little, whereas others are strongly influenced by sediment-water interactions and by biological processes. Those elements or chemical species which are most strongly influenced by biological processes generally show seasonal variation or variation with depth through the water column.

Nutrient loading studies

A number of studies on nutrient loading into Lake Mendota have been carried out, beginning in the 1940's. The Madison lakes have been under study for a long time and are frequently considered a prime example of cultural eutrophication (Hasler, 1947; Edmondson, 1969). However, it is clear from an examination of the literature that there has been much confusion about the timing of eutrophication in Lake Mendota. Edmondson (1969) and others (unpublished documents of the Lake Mendota Problems Committee, 1966) considered Lake Mendota to have deteriorated during the twentieth century, whereas there is no real evidence that this is the case. Historical records and the evidence of sediment cores (discussed below) suggest that the eutrophication of Lake Mendota occurred in the mid-1800's, and the transformation was virtually complete by the time of Trelease's paper in 1889. This conclusion is supported by the analysis of historical data by Stewart (1976) which will be discussed here and in Chapter 9.

Part of the confusion regarding the timing of events in Lake Mendota arises because drastic deterioration did occur in the twentieth century in Lakes Monona and Waubesa as a result of disposal of treated sewage effluents into these lakes (Edmondson, 1969). This situation was well described by Sawyer

Table 3.1. Typical chemical analyses for Lake Mendota

Constituent	Range
Specific conductance μmhos/cm	250–390
pH	6.5–9.2
Turbidity ppm SiO_2	10–50
Color, chloroplatinate unit, mg/l	5–15
Sodium mg/l	4.5–8.0
Potassium mg/l	3.5–4.0
Magnesium mg/l	23–33
Calcium mg/l	26–30
Nitrate mg N/l	0–0.77
Nitrite mg N/1	0.0025–0.02
Ammonia mg N/l	0.04–3.52
Organic nitrogen mg N/l	0.5–5.0
Total Phosphate mg P/1	0.05–0.65
Orthophosphate mg P/1	0.02–0.4
Dissolved Solids mg/l	200 ± 20
Filterable Solids mg/l	10–60
Silicon Dioxide mg SiO_2/1	0.1–1.5
Chloride mg/l	6.2–19
Iron mg/l	0.02–0.2
Manganese mg/l	0.005–0.5
Dissolved Oxygen mg/l	0–15
Chemical Oxygen Demand mg/l	7.0–43
Suspended sediment mg/l	0.5–6.8
Dissolved Organic Carbon mg/l	10 ± 1
Fluoride mg/l	0.09–0.25
Alkalinity total mg/l of $CaCO_3$	96–193
Sulfate mg/l	6–10

Concentrations, given as weight of the significant element, are dependent on sampling location and date.
Data from Lathrop (1979), Hawley (1967), and the University of Wisconsin Water Chemistry Program (personal communication).

(1947,1954) and Lackey and Sawyer (1945). For instance, the classic paper of Sawyer (1947) opens with the statement:

> The lakes at Madison, Wisconsin, are notorious for several reasons: first, because Henry Wadsworth Longfellow once wrote a complimentary poem about them; second, because of the undesirable algal blooms, which of recent years have infested, in particular, the lakes in the chain below the city and which have driven lake-shore cottagers away and resulted in a general depreciation of lake-shore property values; third, because of the large-scale use of copper sulfate in an attempt to control the algal blooms in Lakes Monona, Waubesa, and Kegonsa, having a combined area of 13.5 square miles.

There is no doubt that the lower lakes deteriorated radically in the 20th century. They also recovered rapidly when sewage effluents were ultimately removed (Sonzogni, et al., 1975; Sonzogni and Lee, 1975). However, in contrast to the lower lakes, Lake Mendota has never received significant amounts of sewage effluent in the 20th century (Sawyer, 1947; Lathrop, 1979). Since the

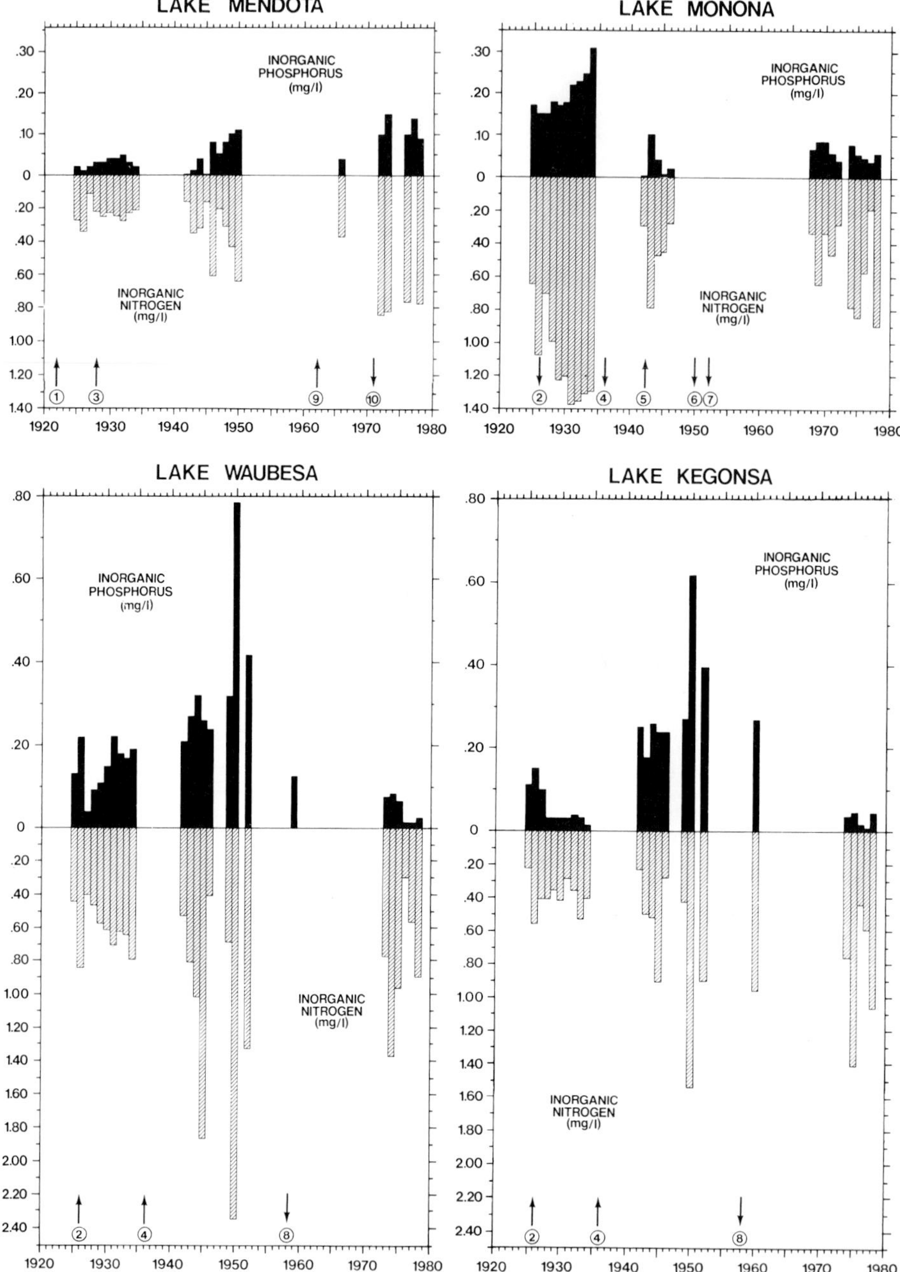

Figure 3.1. Phosphorus and nitrogen analyses for the Madison lakes over a 60-year period. Left page, spring surface concentrations; right page, summer surface concentrations.

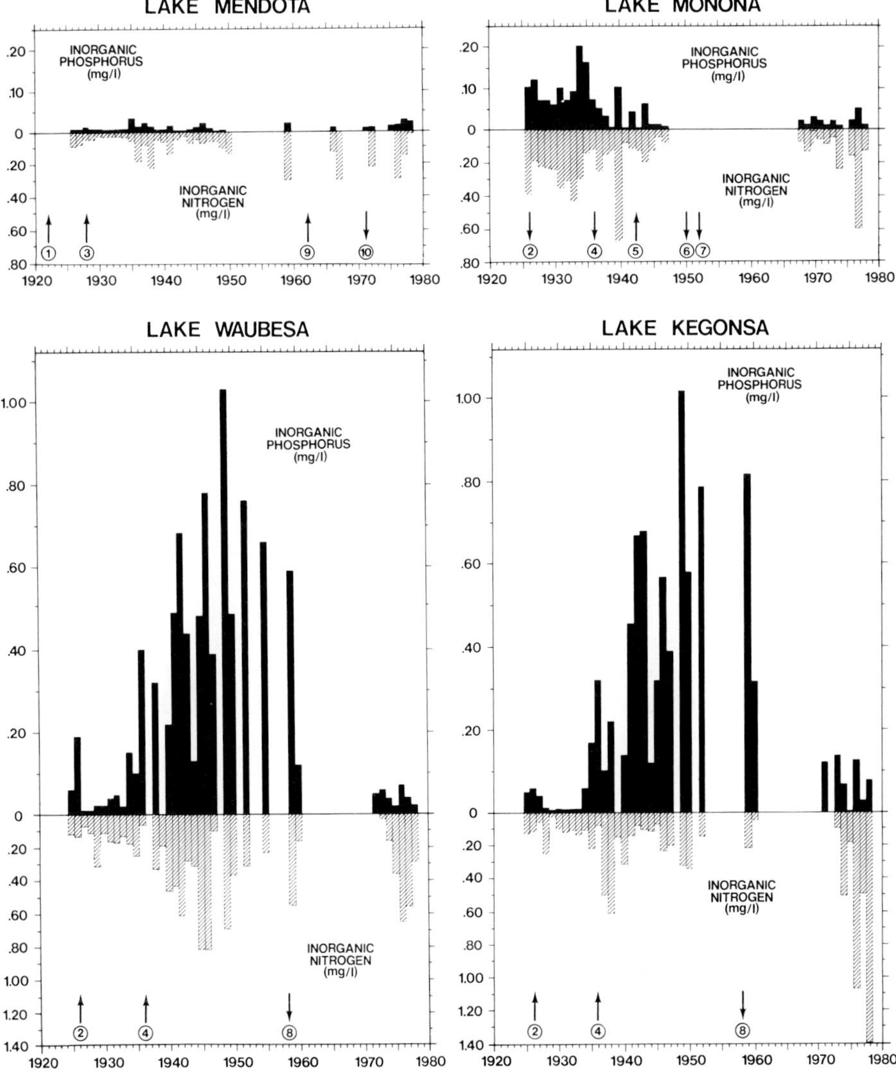

The circled numbers on the graphs refer to the following historic events:

1) 1922, DeForest discharge begins, discharged to Upper Yahara;
2) 1926, First unit operational, Nine Springs Treatment Plant, most sewage still treated at Burke Plant, discharged to Monona;
3) 1928, Waunakee discharge begins, discharged to Six Mile Creek;
4) 1936, Burke Plant closed, all sewage to Nine Springs, discharged to Waubesa;
5) 1942, Burke Plant reopened for Army, discharged to Monona;
6) 1950, Burke Plant closed, all sewage to Nine Springs, discharged to Waubesa;
7) 1952, Brewery discharge to Monona ends;
8) 1958, Nine Springs discharge diverted to Badfish Creek;
9) 1962, Windsor discharge begins, discharged to Upper Yahara;
10) 1971, DeForest, Waunakee, Windsor discharges in Mendota Watershed end. From Lathrop (1979).

mid-1970's, all of the Yahara river basin has been well sewered and essentially no urban effluents, treated or untreated, currently enter the Madison lakes (personal communication from the Madison Metropolitan Sewerage District).

Lathrop (1979) has presented historical data for actual concentrations of inorganic nitrogen and phosphorus on all four of the Madison lakes. As can be seen in Figure 3.1, Lake Mendota had markedly lower concentrations of phosphorus and nitrogen than the lower lakes during the 1930's, when treated sewage was being discharged into Lake Monona. The response of Lake Monona to sewage diversion in the 1940's is also seen in Figure 3.1, as is the response of Lakes Waubesa and Kegonsa to sewage diversion in the 1950's. As described in detail by Lathrop (1979), only minor discharges into any of the Madison lakes now occur, and these are mostly from storm sewer (rather than sanitary sewer) effluents.

Sources of nutrients

What are the sources of nutrients to Lake Mendota? Loading studies were initiated by G.A. Rohlich and associates in the 1950's (see Hasler, 1963) and the data obtained were analyzed by Sonzogni and Lee (1974). However, the drainage basin of Lake Mendota is huge, over 600 km² (see Chapter 2), and stream flow varies sharply as a result of storm events. Thus, accurate loading estimates require an extensive and continuous monitoring system which was only carried out in the mid to late 1970's by the Dane County Regional Planning Commission (Lathrop, 1979). A major monitoring program initiated in 1976 involved measurement of the flow and nutrient concentrations of the major streams entering the lake. A streamflow gauging network was installed to obtain continuous data on high-flow and base-flow discharges. The results of this hydrologic work were discussed in Chapter 2. In concert with gauging, stream water samples were taken for chemical analysis. Certain chemical constituents were measured only under low-flow conditions, either from monthly or annual samples (Table 3.2). Much greater frequency of sampling was carried out for nitrogen and phosphorus, the important phytoplankton nutrients which were expected to vary in concentration during storm events. These "storm-event" samples were collected during high-flow events at various intervals on the hydrograph to provide information on the mass loading of nitrogen and phosphorus.

During a typical storm event, there is usually a rapid rise in stream flow on the hydrograph after the start of the storm, followed by a gradual decrease in flow rate until the base-flow value is reached (Figure 3.2a). To calculate a nutrient loading rate, the concentrations of nutrient at different stages of the hydrograph must be multiplied by the flow rate (Figure 3.2b).

Phosphorus loading rates

Hydrological data for Lake Mendota obtained by the Dane County Regional Planning Commission were given in Table 2.6. Nutrient loading rates for ni-

Table 3.2. Chemical constitution under base flow conditions of streams entering Lake Mendota

Input stream	Conductivity μmhos/cm	Cl^- mg/l	pH	Alkalinity mg/l	Total P mg/l	Soluble P mg/l	Total N mg/l	Organic N mg/l	Nitrate mg/l	Ammonia mg/l	Suspended sediment mg/l	Sulfate mg/l
Token Creek	555	9	8.0	258	0.08	0.04	4.04	0.42	3.45	0.10	57	8
Yahara River (at Windsor)	604	18	8.1	263	0.06	0.04	3.17	0.31	2.80	0.05	28	13
Yahara River (at Highway 113)	466	12	8.1	219	0.28	0.12	2.84	1.25	1.24	0.30		
Six Mile Creek (at Waunakee)	594	12	8.1	298	0.09	0.06	2.80	0.56	2.10	0.08	17	
Six Mile Creek (at Mills Rd.)	623	16	8.0	288	0.13	0.09	2.10	0.53	1.45	0.10	35	15
Spring Creek	579	12	8.0	277	0.13	0.10	3.09	0.43	2.42	0.21	37	
Pheasant Branch (at Highway 12)	862	20	7.6	294	0.09	0.03	2.71	0.69	3.0	0.21	43	49
Pheasant Branch (at Highway M)	678	16	7.9	296	0.09				2.2		30	24
Lake Mendota Outlet	384	14	8.1	157	0.13	0.08	1.61	0.93	0.29	0.37	2.9	

Data of Lathrop (1979). Values are averages from a number of analyses. Analyses for specific chemical compouds are expressed as weight of the element.

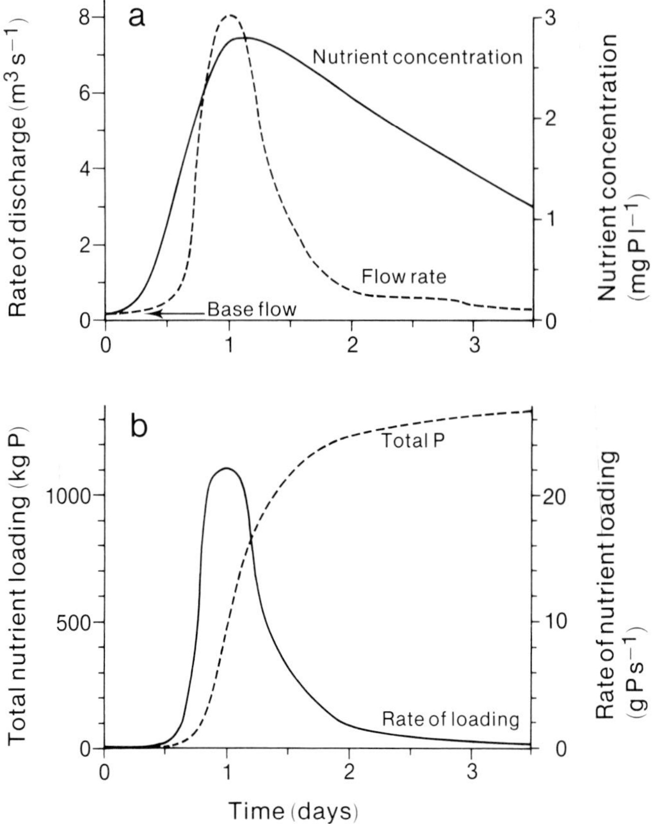

Figure 3.2. Idealized graph showing a) the change in flow rate and phosphorus concentration and (b) the total and rate of phosphorus loading of a stream before, during, and after a storm event.

trogen and phosphorus, calculated as described above, are given in Figure 3.3 and are summarized in Tables 3.3 and 3.4. As described in detail by Lathrop (1979), most of the phosphorus enters the lake in a few major storm events, especially during early spring or late winter before vegetation covers the agricultural fields.

The large drainage basin of Lake Mendota has been mentioned in Chapter 2 and is summarized in Table 3.5. The bulk of the watershed is rural, predominantly agricultural. The significance of rural runoff for loading of both phosphorus and nitrogen to Lake Mendota is emphasized in these three tables.

Groundwater is an insignificant source of either phosphorus or nitrogen (Lathrop, 1979; Brock, et al., 1982). Despite the large amount of water (1/3 of the total) which enters the lake as groundwater, this groundwater is much lower in nutrients than the surface runoff or baseflow. Thus, from a nutrient loading standpoint Lake Mendota is primarily a drainage lake.

Annual loading data can be obtained by averaging the detailed measure-

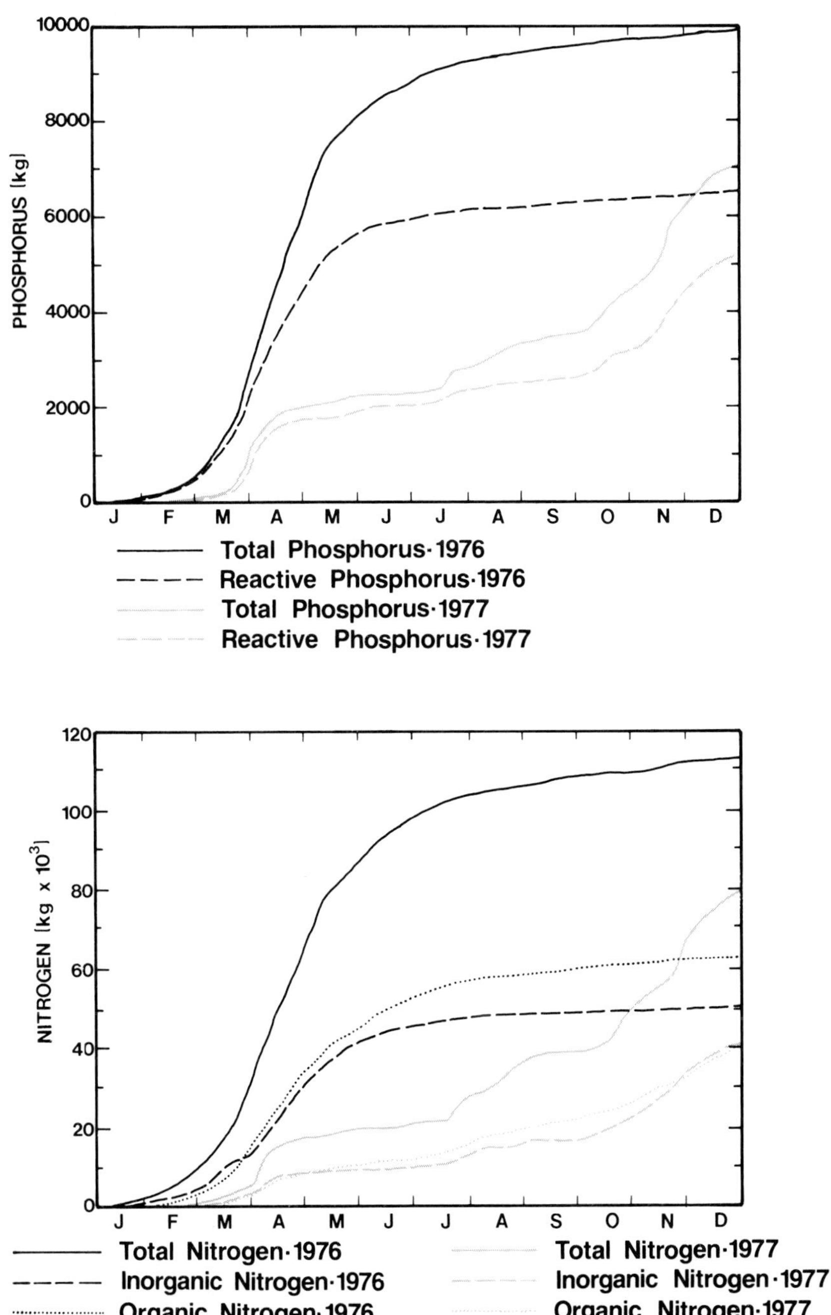

Figure 3.3. Nutrient loading in Lake Mendota. From Lathrop (1979).

Table 3.3. Lake Mendota phosphorus loading

Source	1976 Total P	1976 Soluble P	1977 Total P	1977 Soluble P
Baseflow	4,500	2,600	3,700	2,300
Precipitation	700	550	1,000	790
Dry fallout	2,400	1,000	2,400	1,000
Groundwater seepage	120	120	160	160
Rural runoff				
Monitored (397 km^2)	24,200	11,700	8,200	4,900
Unmonitored (113 km^2)	6,900	3,300	2,300	1,400
Urban (38.6 km^2)	4,200	2,100	5,100	2,500
Total				
kg	43,000	21,000	23,000	13,000
g/m^2 of lake surface/yr	1.1	0.53	.58	.33
g/m^3 of lake volume/yr	0.090	0.043	0.048	0.027
Outlet	9,900	6,400	7,000	5,100

From Lathrop (1979). Values are in kilograms for the whole lake.
The urban loading does not include about 21.5 km^2 of developed area in villages and cities in the monitored rural area.

ments, but the large year-to-year variation makes these averages of doubtful long-term significance. In order to calculate a more precise value for average nutrient loading, data from many more years would be needed. The relationship of nutrient loading to phytoplankton growth and production will be discussed in Chapters 4 and 5. Internal cycling of phosphorus and nitrogen in Lake Mendota is discussed later in this chapter.

Sediment composition and chemistry

Sediment-water interactions are of great importance for lake ecosystem function, and a knowledge of sediment characteristics and chemistry is therefore

Table 3.4. Lake Mendota nitrogen loading

Source	1976 Total Nitrogen	1976 Inorganic Nitrogen	1977 Total Nitrogen	1977 Inorganic Nitrogen
Baseflow	180,000	150,000	140,000	120,000
Rural runoff	160,000	86,000	48,000	23,000
Urban runoff	20,000	5,000	26,000	6,000
(Other)*	(210,000)	(130,000)	(210,000)	(130,000)
Total	(570,000)	(370,000)	(420,000)	(280,000)
Outlet	113,000	50,000	78,000	40,000

From Lathrop (1979). Values are in kilograms for the whole lake. Values in parentheses are estimates.
*From Sonzogni and Lee (1974); includes precipitation, dry fallout, groundwater and nitrogen fixation estimates.

Table 3.5. Lake Mendota drainage area

	Area (km^2)
Lake Mendota watershed	549
Monitored Rural*	397
Pheasant Branch Creek at U.S. 12	49.1
Six Mile Creek at Mills Road	93.6
Token Creek at U.S. 51	63.1
Yahara River at Windsor	191
Monitored Urban	16.7
Spring Harbor Storm Sewer at University Avenue	8.52
Willow Creek Storm Sewer above Observatory Drive	8.18
Unmonitored	135
Rural	113
Urban	21.9

*Includes 21.5 km^2 of developed land in the monitored rural area.
Prepared by Dane County Regional Planning Commission, 4/79.

essential for an understanding of lake chemistry. Sediment itself consists of a viscous mixture of fine particles and water, with chemical constituents present both in solution in the interstitial water and adsorbed to or an integral part of the sediment particles themselves. Sediment also has large numbers of living microorganisms. The sediment-water interface is the boundary across which the movement of materials occurs (Lerman, 1979):

> The sediment-water interface separates a mixture of solid sediment in interstitial water from an overlying body of water. Wherever sediment accumulates, growth of the sediment pile is achieved by sedimentation of solid particles and inclusion of water in the pore spaces among the particles. An observer stationed on shore would observe the sediment-water interface rising as sediments accumulate. But an imaginary observer who is balanced on the interface as sediment particles continue to arrive from above and pile up, will see the particles and pore water flow by in the downward direction. In this sense one can always speak of the fluxes of solids, water, and solutes as moving up or down, *across* the sediment-water interface.

As far as nutrients and other chemical constituents are concerned, sedıments can serve either as sources or sinks. Even further complication arises because a single sediment can serve as a source at one time and a sink at another, or as a source at all times for one nutrient and a sink at all times for another.

As sinks, sediments are the recipients of detrital material settling from the water column. This detrital material can be either exogenous (produced outside the lake) or endogenous (produced in the lake). The sediment itself is made up of materials that had sedimented at some earlier time. A lake can be said to age, and a major part of this aging process is the accumulation of sediments in the lake basin (Hutchinson, 1957). Ultimately, sedimentation leads to the complete filling in of the lake.

Sediments serve as sources of chemical constituents in two ways: 1) mo-

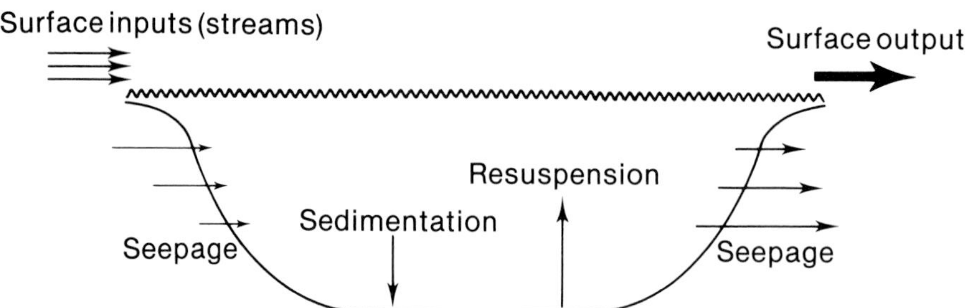

Figure 3.4. Processes affecting nutrient chemistry of a lake.

lecular diffusion out of the sediment into the water, a process which is driven by the concentration difference that exists between sediment and water Lerman, (1979); 2) advective transport (resuspension), a process occurring as a result of bioturbation and vertical and horizontal currents. An excellent overview of the physical aspects of fluxes at the sediment-water interface can be found in Lerman (1979).

As far as the lake ecosystem is concerned, fluxes of nutrients in and out of the sediment modify the chemistry of the overlying water. In the same way that the process of nutrient additions to the lake from the surrounding land is called *external loading,* nutrient additions from the underlying sediment are called *internal loading* (Stauffer, 1981). Thus sedimentation and resuspension are major processes influencing the biological and geochemical events in a lake (Figure 3.4).

Three types of studies on Lake Mendota sediments have been carried out: 1) general descriptions of sediment types and the distribution of various types of sediment throughout the lake basin; 2) chemical and physical analyses on segments of sediment cores taken at different locations in the lake; 3) measurement of the sedimentation process itself, by means of sediment traps.

As emphasized by the quotation from Lerman above, the sediment-water interface is not constant, but is in a continual state of change. Both short-term and long-term change in the sediment-water interface occur. Short-term change occurs as a result of the mixing processes that occur in the water column, primarily during the spring and fall isothermal periods. Because of these short-term changes, the sediment-water interface during one period of the year is different in physical and chemical characteristics from that during another period. Long-term change occurs as a result of the gradual accumulation of sediment as the lake ages. Although resuspension during spring or fall mixing may return sediment to the water, and hence decrease the thickness of the sediment pile, not *all* of the material which settles is returned to the water, so that, over time, there is a gradual accumulation of sediment.

Extensive changes can occur in sediments after deposition, and these are of great concern to the geochemist. The gradual change in sediment chemistry and physics with time is called *diagenesis* (Lerman, 1979), and is of major

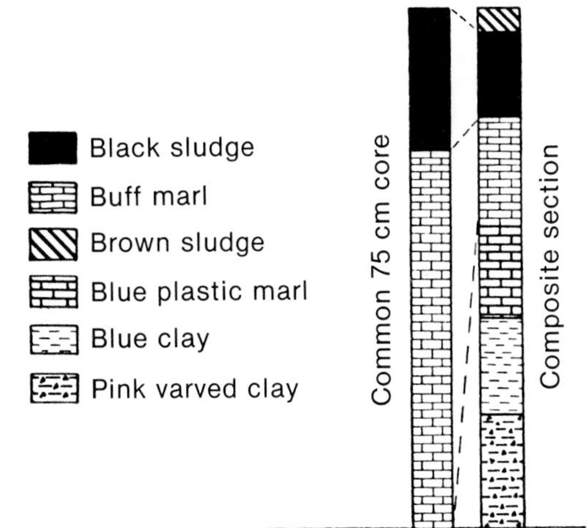

Figure 3.5. An idealized stratigraphic section through the Lake Mendota sediments (right side, labeled "composite") and a typical 75 cm (30 inch) core. From Murray (1956).

concern to those studying sediment cores. Although of considerable geochemical interest, sediment diagenesis is beyond the scope of this book.

From the viewpoint of eutrophication, the most critical knowledge is that related to how sediments bind or release nutrients such as phosphorus. Adherence of phosphorus to sediment particles is influenced by the chemistry and ionic charge of the particles, the pH, and redox conditions.

Recent sediments in Lake Mendota

Early work on Lake Mendota sediments was carried out by associates of Birge and Juday (mostly unpublished) and by Twenhofel (1933) and a more detailed study was performed by Murray (1956) under the auspices of the University of Wisconsin Committee on Lakes and Streams. Using a coring apparatus, Murray (1956) took over 50 samples on a 2000-foot grid pattern throughout the main body of the lake. Each core was broken into 15–20 cm (6–8 inch) sections which were bottled separately and analyzed chemically and physically. Chemical analyses were performed for calcium carbonate, organic carbon, phosphate, sulfur, and iron. Mineralogical examination of the sediments was also carried out.

Murray found that the common deep-water sediment of Lake Mendota was a shiny black sludge (about the consistency of mayonnaise), underlain by a buff marl layer. The contact between the sludge and the marl was very precise ("knife sharp" in Murray's words), as illustrated in Figure 3.5. The black sludge has been called "gyttja" by later workers. The thickness of the black sludge

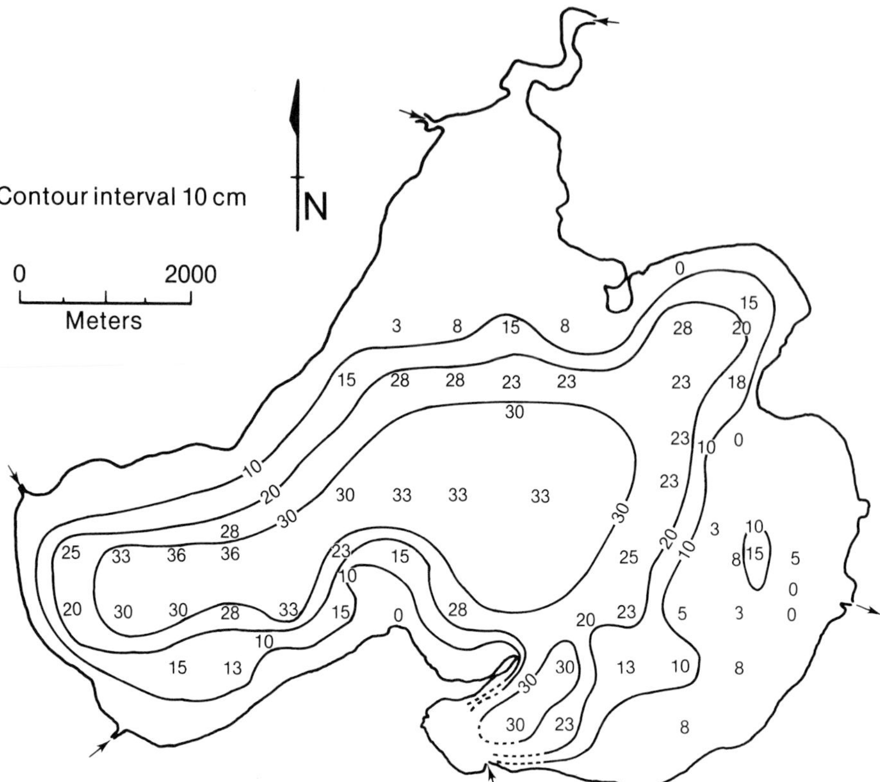

Figure 3.6. Isopleth map of the black sludge layer of Lake Mendota. The numbers indicate the measured thickness in centimeters of the black sludge at each coring site. From Murray (1956).

layer varies throughout the lake, being deeper in the center and west end and relatively shallow in the east end (Figure 3.6). The black sludge layer was low in carbonate and high in organic matter, whereas the marl layer was high in carbonate and low in organic matter (Figure 3.7). Murray hypothesized that the black color of the sludge was due to the presence of iron sulfide and this hypothesis was later confirmed by Doyle (1968) and Nriagu (1968). These latter workers showed that the sulfide of the sludge existed as hydrotroilite ($FeS \cdot nH_2O$), greigite (cubic Fe_3S_4), tetragonal FeS (mackinawite), or a mixture of these. As will be discussed later (Chapter 7), the sulfide is formed by bacterial sulfate reduction and by bacterial protein decomposition. Murray developed a model to explain his results which hypothesized an increase in deposition of clastic materials (predominantly soil mineral particles), resulting in a decrease in carbonate concentration of the sediments. Because of the higher organic content of the black sludge as compared to the marl, this hypothesis would require an increase in productivity of the lake at the time that clastic deposition increased.

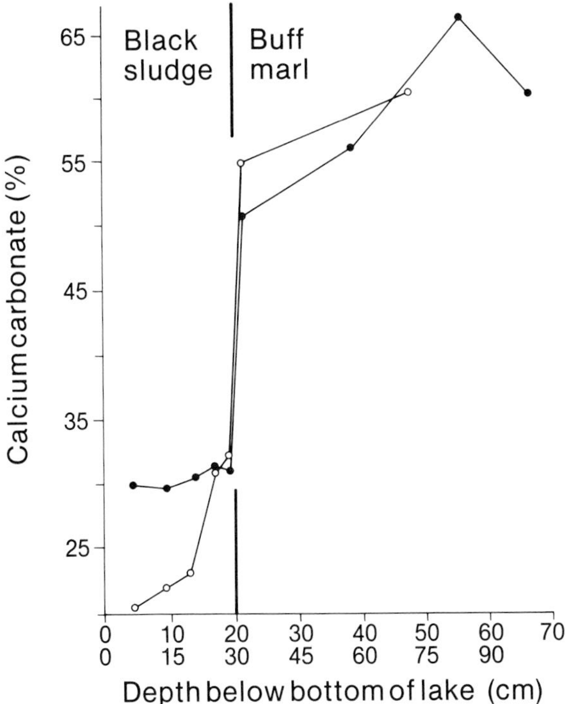

Figure 3.7. Distribution of carbonate with depth in the sediment of Lake Mendota, showing the sharp difference between the black sludge and the marl layer. The two lines refer to separate cores taken at different locations in the lake, both at about 20 m depth. From Murray (1956).

Murray suggested:

> The increase in clastic deposition in Lake Mendota seems best explained by the mid-nineteenth century development of Madison and surrounding farm community. This development brought about the clearing of the land for farming and the clearing of the shore with shore-line development for homes. The building of the dam raised the level of the lake approximately 5 feet. This exposed a new previously undisturbed shoreline to erosion and buried the white cobble beaches referred to by early travelers.

Murray's hypothesis that agricultural impact has been primarily responsible for the massive changes in Lake Mendota has been generally confirmed by later work.

Bortleson (1970) did a much more extensive chemical analysis of sediment cores than did Murray, and also took much longer cores. In addition, Bortleson performed counts for *Ambrosia* (ragweed) pollen as a means of dating the segments of the cores. The ragweed plant is known to have developed extensively only after the land in the Madison area was cleared for agriculture, hence the location of the "ragweed rise" in the cores indicates the time of agricultural

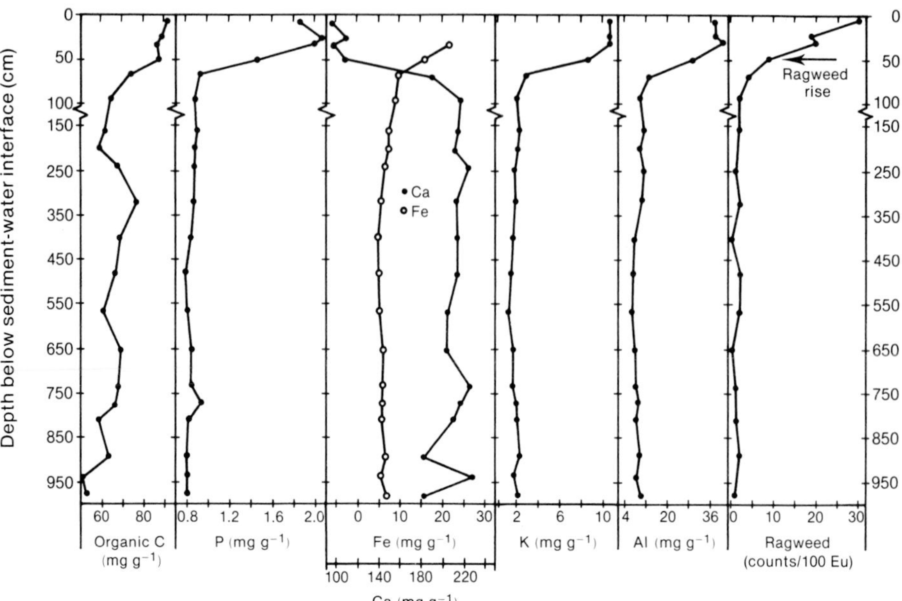

Figure 3.8. Chemical analyses of segments from a long core taken in the center of Lake Mendota. The ragweed rise (expressed in pollen counts relative to an added marker of *Eucalyptus* pollen) is shown by the arrow. From Bortleson (1970).

development. Data from the longest core taken by Bortleson in the deep hole of Lake Mendota are presented in Figure 3.8 and Table 3.6. This almost 10 m (990 cm) core was fractionated into 10 or 20 cm segments. The top 62 cm of sediment consisted of black gyttja (Murray's black sludge), low in carbonate and high in organic matter, and the rest of the core consisted of fine-grained, buff-colored marl deposits high in calcium carbonate. Although the core reached almost 10 m into the sediment, the bottom of the post-glacial marl deposits was not reached. The differences in sediment chemistry above and below the ragweed rise are striking and amply confirm the hypothesis of cultural eutrophication in Lake Mendota as a result of agricultural development. As seen in Figure 3.8, potassium and aluminum are high in the surface layers, consistent with the idea of increased input of clastics (clay minerals are potassium aluminosilicates) as a result of soil erosion. Phosphorus is also high in the surface sediments but markedly lower in the deep (presettlement) sediments. The surface sediments are higher in iron than the deeper sediments, consistent with their black color. Bortleson (1970) concluded that "the chemical stratigraphy of the 9.9 m core indicates that stable conditions existed in Lake Mendota and its watershed for a long historical period before the settlement period in Wisconsin." A corollary of this is that since settlement, these stable conditions have been absent.

Table 3.6. Chemical composition of a long core taken in the center of Lake Mendota

Depth of Sediment (cm)	% Solids	Total C	Org–C	CO_3–C	P	Fe	Mn	Ca	K	Al
0–11	12.4	122	92.4	29.6	1.86	—	1.09	98.5	10.8	38.2
11–22	20.7	122	88.7	33.3	2.06	—	1.12	111	10.8	38.5
22–42	15.1	117	87.4	29.6	1.99	22.0	1.23	98.5	10.8	41.5
42–62	15.8	121	88.0	33.0	1.47	16.3	0.712	110	8.78	30.0
62–82	12.3	133	75.4	57.6	0.940	10.0	0.433	192	3.11	13.0
82–102	12.8	130	65.0	65.0	0.880	9.25	0.400	217	2.23	10.4
157–177	15.6	127	62.1	64.9	0.913	7.75	0.365	216	2.43	11.4
197–217	16.1	123	59.5	63.5	0.878	7.75	0.402	212	2.20	10.8
237–257	17.2	136	68.5	67.5	0.873	6.80	0.370	225	2.01	11.7
312–332	17.5	141	77.2	63.8	0.875	5.73	0.338	213	2.01	10.4
394–414	16.9	133	68.9	64.1	0.834	5.10	0.332	214	1.68	8.00
474–494	16.5	131	66.9	64.1	0.801	5.13	0.330	214	1.55	7.25
554–574	17.4	123	61.5	61.5	0.820	5.20	0.330	205	1.27	6.55
639–659	17.9	131	69.9	61.1	0.866	6.15	0.352	204	1.79	8.05
718–738	20.7	136	68.2	67.8	0.852	5.95	0.395	226	1.69	8.00
758–778	19.9	132	66.9	65.1	0.870	6.00	0.364	217	2.09	8.80
798–818	19.2	122	59.0	63.0	0.825	5.73	0.332	210	2.00	8.55
880–900	17.9	118	63.5	54.5	0.806	6.60	0.328	182	2.30	9.85
935–950	16.3	119	50.6	68.4	0.806	5.45	0.400	228	1.68	8.25
970–990	20.8	108	53.0	55.0	0.820	6.90	0.360	183	2.04	10.2

Data of Bortleson (1970). The core (designated WC–95) was collected 14 November 1968 from a location where the water depth was 23 m. The chemical analyses are expressed as mg of element per gram of sediment.

CAMROSE LUTHERAN COLLEGE LIBRARY

Table 3.7. Chemical composition of surficial sediments of Lake Mendota

Constituent	Bortleson (1970)	Constituent	Konrad et al. (1970)
Sampling depth (m)	23.2	Sampling depth (m)	13
Depth of sediment (cm)	0–10	Depth of sediment (cm)	0–10
Percent solids	13.65	Organic carbon (%)	7.85
Total carbon (mg/g)	122	Total nitrogen (%)	0.92
Organic carbon (mg/g)	85	Organic C:N ratio	8.6
Carbonate carbon (mg/g)	37.5	Fixed ammonium (μg/g)	147
Organic nitrogen (mg/g)	9.67	Exchangeable ammonium in sediment (μg/g)	66
Total phosphorus (mg/g)	1.79	Interstitial water assays	
Acid-soluble phosphorus (mg/g)	0.88	Ammonium nitrogen (μg/ml)	9.4
Iron (mg/g)	18.9	Nitrate nitrogen (μg/ml)	0.21
Manganese (mg/g)	1.39	Nitrite nitrogen (μg/ml)	0.19
Calcium (mg/g)	125	Sampling depth (m)	18–19.5
Magnesium (mg/g)	12.7	Depth of sediment (cm)	0–9
Potassium (mg/g)	9.85	Interstitial phosphorus (μg/ml) (Holdren et al., 1977)	3.13

Chemical composition of surficial sediments

Chemical analyses have been carried out on sediment samples collected with an Eckman dredge (Bortleson, 1970; Delfino, Bortleson, and Lee, 1969). Such dredge samples provide material representative of recently sedimented particles. Average values obtained from a total of 30 samples are given in Table 3.7. Analyses of nitrogen species for one sample have been presented by Konrad et al. (1970) and are also given in Table 3.7.

Phosphorus concentrations of the sediment are of considerable importance from the standpoint of lake eutrophication. The phosphorus data of Bortleson (1970) are for total phosphorus, much of which is present in bound form and probably not available. Holdren et al. (1977) presented data for inorganic phosphorus concentrations of the interstitial water of Lake Mendota sediments and for cores taken at different times of the year. Although there was some variation throughout the year, at the deeper location this variation was not great. The average value for interstitial (soluble) phosphorus calculated from their data, 3.13 μg/ml, is considerably higher at all times than the lake water concentration, showing that the sediments are a source of soluble phosphorus. Phytoplankton and other phosphorus-rich particles settle out of the water column into the surficial sediments and undergo decomposition processes which lead to the release of soluble phosphorus (Fallon, 1978). This soluble phosphorus can then move back into the overlying water by molecular diffusion or advection. Although some phosphorus is adsorbed by Lake Mendota sediments (Shukla, et al., 1971; Williams, et al., 1970), a significant fraction of adsorbed phosphorus can be readily desorbed.

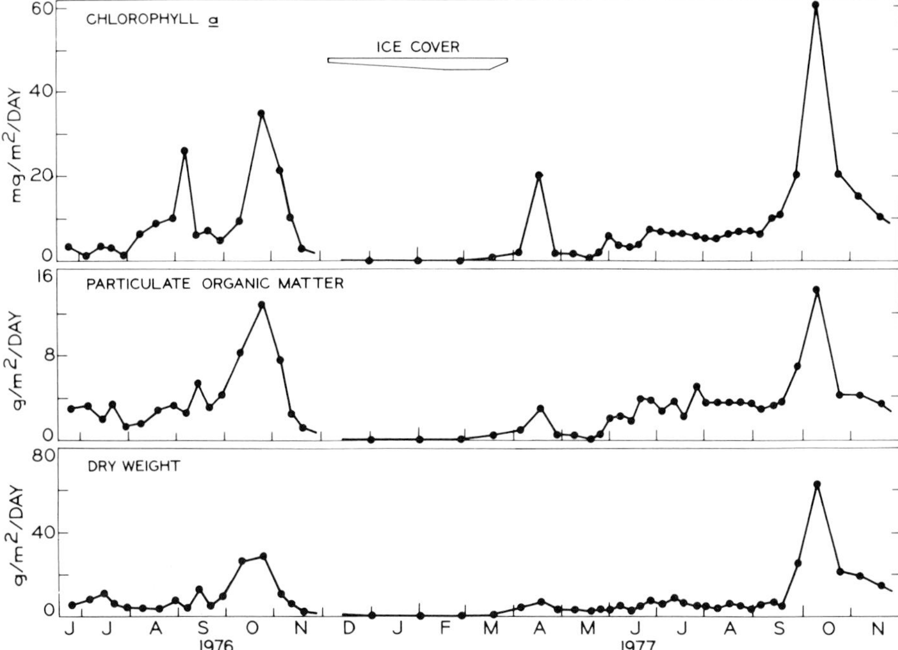

Figure 3.9. Sedimentation rates for chlorophyll, particulate organic matter, and dry weight, as measured with sedimentation traps in Lake Mendota. From Fallon and Brock (1980).

Sedimentation rate

Sedimentation rates have been measured directly by use of sediment traps by Fallon and Brock (1980) in the years 1976 and 1977 (Figure 3.9). A consistent pattern for sedimentation emerged. During summer stratification, the accumulation rate of material in the sediment traps was fairly low and did not show much variation. At the time of fall overturn, there was a sharp rise in accumulation in the sediment traps, probably due to resuspension of bottom sediment. Catches in the sediment traps fell to their lowest recorded values during the ice-covered period and then showed a sharp increase at spring overturn, again partially due to resuspension. However, chemical analyses of the material collected in the traps during the spring and fall mixing periods indicated that resuspension of bottom sediments was less important during the spring mixing period than in the fall (Fallon and Brock, 1980). Sediment resuspension is not as severe during spring mixing due to the less rapid vertical currents that result from wind stress on a warming lake (as in the spring period) than on a cooling lake (as in the fall of the year).

By integrating the daily sedimentation rates for each 4–14-day period over the year, one can estimate the annual sediment accumulation in deep portions

Table 3.8. Estimates for annual sediment input to pelagic sediments of Lake Mendota, based on observed sedimentation rates

Period	Dry wt. (g m^{-2} yr^{-1})	Depth (cm yr^{-1})*	Org. matter (g m^{-2} yr^{-1})	Part. P
17 June 1976	Observed, 1919	1.2	818	5.6
17 June 1977	Corrected†, 1090	0.7	436‡	4.2§
24 Nov. 1976	Observed, 2,662	1.8	668	8.9
24 Nov. 1977	Corrected, 622	0.4	249‡	5.0§
Long-term rate	715–1,100	0.35–0.5	132–206	1.2–1.4

Data for 1976 and 1977 from Fallon (1978). Data for long-term rate from Bortleson (1970).
*Based on water content of 0.16 g dry wt ml^{-1} wet sediment.
†Corrected values based on models for resuspension of sediments.
‡Corrected values based on assumed organic content of new sediment of 40%, a value approximately observed over summer periods when resuspension was least important. Bulk sediment has an organic content of 16% (Bortleson 1970).
§Based on assumption that resuspended sediments contain about 0.002 g P g^{-1} dry sediment (Bortleson 1970).

of the lake. Table 3.8 presents two annual estimates based on integration of data from different times of the year. Corrections for resuspension were made as described by Fallon and Brock (1980).

Sedimentation rate was also estimated by Bortleson (1970) from pollen analyses of sediment cores. For this approach, Bortleson assumed that 100 years had elapsed since deforestation and the initiation of the ragweed rise. Sedimentation rate based on dry weight, obtained from these estimates, must be converted into sediment thickness by adjusting for the density of the sediment.

Fallon's technique gives an estimate of current sedimentation rate but the estimate has a high variance, whereas Bortleson's technique gives a value which is an average over a long period of time but does not permit measurement of short-term changes. When Fallon's values are corrected for resuspension, they are in close agreement with Bortleson's (Table 3.8).

Another approach to the calculation of sedimentation rate was used by Nriagu and Bowser (1969). These workers studied magnetic spherules in Lake Mendota sediments. These spherules were found only in the black sludge, and were shown to be flue products derived from industrial and domestic activity. Since these spherules, like pollen, are not affected by diagenesis, they provide a horizon for estimating sedimentation rate. The ragweed pollen rise corresponds closely to the position in the sediment at which the magnetic spherules first appear, thus providing further confidence in the estimates of sedimentation rate.

For the post-cultural period, Bortleson calculated that the rate of phosphorus sedimentation was 11.9 mg P/cm^2/100 years (deep hole), a value considerably lower than the value of Fallon and Brock (1980), 46 mg P/cm^2/100 years (corrected for resuspension). The difference between sedimentation rate calculated for the sediment as a whole (based on dry weight and adjusted for density) and that for phosphorus alone, is probably due to the fact that considerable decomposition of the recently sedimented material occurs, resulting in a release of phosphorus back to the water.

One problem with all these sedimentation rate measurements must be noted. They involve measurements primarily in the *deep* portions of the lake, with water depths greater than 10 m. There is less mixing of water over the deeper sediments and hence less resuspension. This leads to an overestimation of sedimentation rate for the whole lake. It is uncertain how much bias this introduces, but it is a point of caution that must be applied if whole lake estimates are being made (R.E. Stauffer, personal communication).

Despite this last limitation, an important conclusion can be drawn:

> The primary event leading to the eutrophication of Lake Mendota was the deforestation of land in the drainage basin, an event which occurred rapidly during deforestation in the 1850's.

Oxygen in Lake Mendota

The oxygen cycle for Lake Mendota was described in detail by Birge and Juday (1911) for the year 1906 and has changed little since that time (Stewart, 1976; Figure 2.6; Appendix to this book). Oxygen is an element of central importance in lake ecosystem function. First, many organisms are obligate aerobes and are killed or cease to reproduce when oxygen is reduced or eliminated. Second, presence of oxygen controls the redox state of many other important lake elements, such as C, N, Fe, Mn, and S. As Birge (1910) stated:

> The cycle of (oxygen) changes in a lake illustrates more readily and more conspicuously than perhaps any other facts could do what may be called the 'annual life cycle' of the individual lake.

Oxygen is both added to and removed from the lake water. Oxygen is added to the lake in the first instance by diffusion and advection *from* the atmosphere. Oxygen is also produced in the photosynthetic process ("manufactured oxygen" in Birge's terminology). Oxygen is also removed from the lake by diffusion or advection *to* the atmosphere. Whether oxygen moves into or out of the lake water via the atmosphere is determined by whether the water is undersaturated or supersaturated with oxygen (saturation is affected most strikingly by temperature changes in the water). Oxygen is also removed from the lake by the biological process of respiration, and by chemical reaction with reducing substances such as sulfide and ferrous iron. These reducing substances are produced in the first place by biological processes.

In relatively deep lakes such as Lake Mendota, where the hypolimnion is large, biological and chemical processes are the most important sinks for oxygen, and oxygen evasion to the atmosphere is less important. Oxygen depletion can occur through reactions in either the water column or the sediment, but the sediment reactions are most important.

Oxygen depletion in the hypolimnion is a process which has been extensively studied, and considerable work has been done relating this process to the productivity and trophic status of lakes (Hutchinson, 1938; Cornett and Rigler, 1980). The *oxygen deficit,* more accurately called the *hypolimnetic oxygen deficit,* is an expression of how much oxygen must pass into the hypolimnion of a lake to bring the water of the hypolimnion to saturation. The hypolimnetic oxygen deficit will thus be influenced by the oxygen demand of the hypolimnion (including the sediments). In order to eliminate the effects of lake morphometry, the rate of oxygen depletion is generally expressed as the *areal hypolimnetic oxygen deficit* (AHOD), which is the amount of oxygen which must pass through unit area of lake surface (unit of AHOD is mg $O_2 \cdot m^{-2} \cdot d^{-1}$). The areal dimension chosen is the area of the plane delimiting the upper boundary of the hypolimnion. As Cornett and Rigler (1980) have shown, AHOD does not completely eliminate lake morphometry as a variable, and it is necessary to include the mean depth of the lake in any comparative analysis. Cornett and Rigler (1980) also show that other factors influencing the AHOD include the productivity of the lake (as hypothesized by Hutchinson, 1938) and the mean hypolimnetic temperature.

One important consideration, in assessing oxygen depletion, is the relative importance of sediments and hypolimnetic water as sources of oxygen demand. As shown by Fallon (1978) and Fallon and Brock (1979), decomposition of organic matter was considerably more rapid in hypolimnetic samples containing sediment than in sediment-free water. It was found that hypolimnetic water alone was an insignificant source of oxygen demand. This leads to a focus on sediment-water interactions.

Oxygen profiles in Lake Mendota

There have been a number of measurements of seasonal change in hypolimnetic oxygen concentration at various depths in Lake Mendota, beginning with the work of Birge and Juday (1911). Data for the years 1906 and 1980 are presented in Figure 3.10. Stewart (1976) calculated weighted mean oxygen concentration for the 9–18 m hypolimnetic stratum, using his own data and that of others, and his results are presented in Figure 3.11 together with some results from the present study. From data of this type, areal hypolimnetic oxygen deficits can be calculated and Stewart's results are presented in Table 3.9. As Stewart notes, the data of both Figure 3.11 and Table 3.9 show distinct variations between years, but no prominent long-term trends over the years of measurement: "These variations may reflect differences in mean hypolimnetic temperatures as well as man's 'cultural influences'."

The possible hypolimnetic temperature effects are worth emphasizing. As we noted in Table 2.7, the temperature of the hypolimnion varies considerably from year to year, depending upon weather conditions during the period when the lake is stratifying.

It is possible to calculate an oxygen depletion rate constant, assuming that oxygen depletion is a first order reaction. The value obtained depends on the depth at which the oxygen concentrations were measured. Data from the pres-

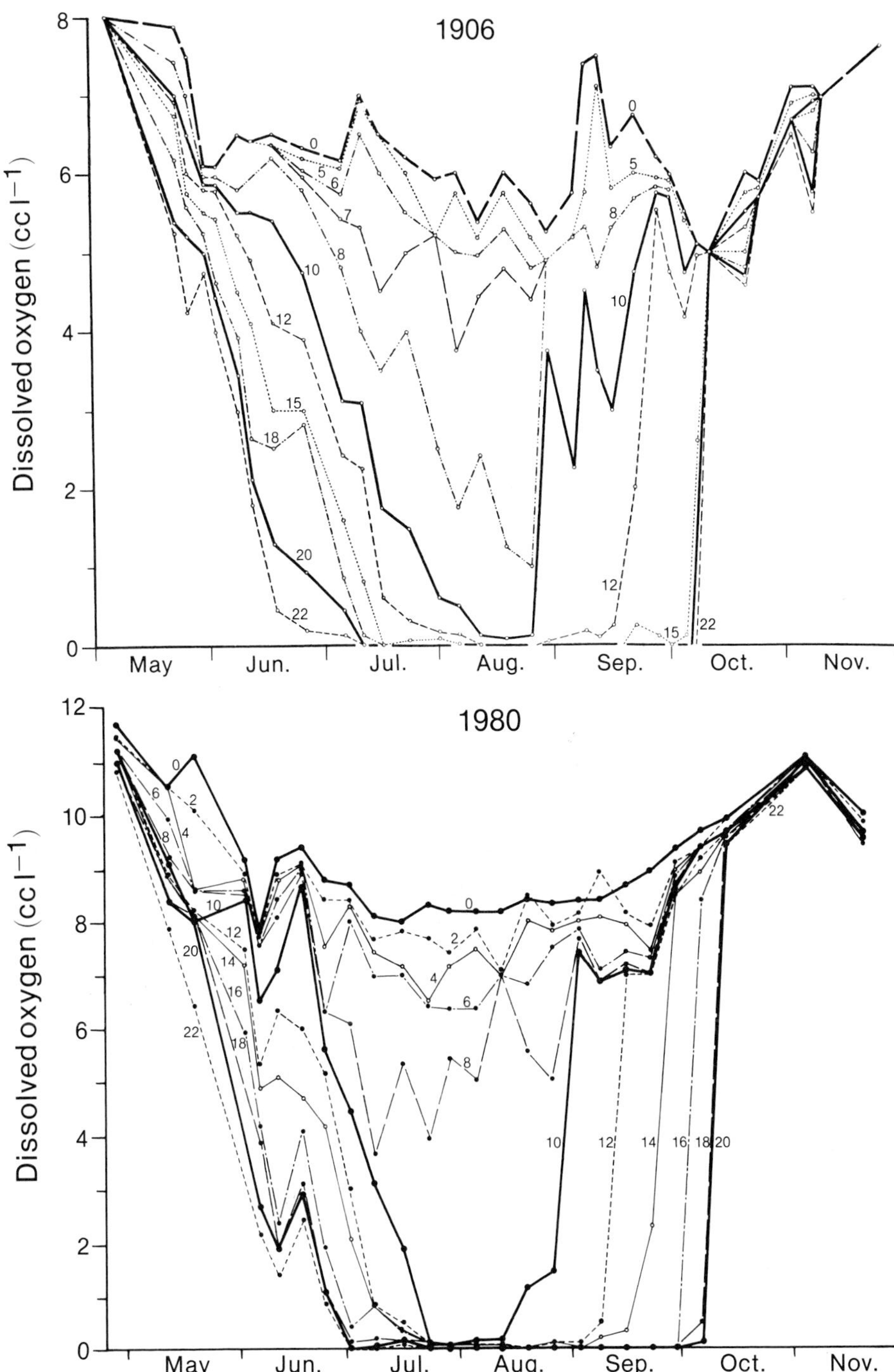

Figure 3.10. Dissolved oxygen at different depths in Lake Mendota in 1906 and 1980. The data for 1906 are from Birge and Juday (1911). The 1980 data are from the present work (Appendix). The numbers on the lines refer to the depth in meters at which the oxygen concentrations were measured.

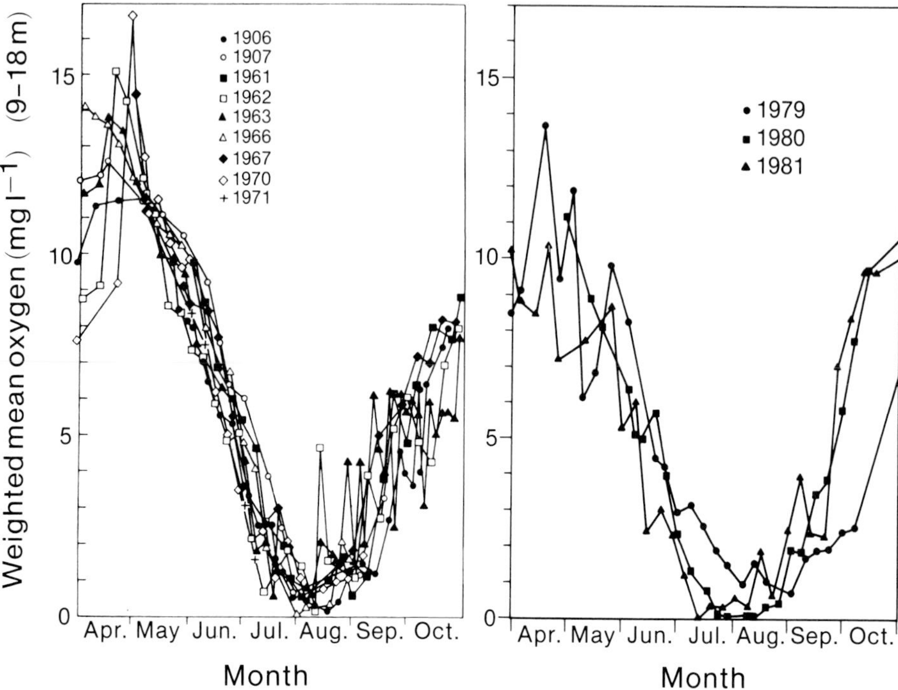

Figure 3.11. Weighted mean oxygen concentration for the 9–18 stratum of Lake Mendota from April to October for different years. From Stewart, 1976, and present study.

ent study for six years for two depths are presented in Table 3.10, together with Stewart's calculations of his and historic data. As seen, considerable year-to-year variability occurs. Also, within a given year the rate constant obtained is influenced by the interval over which the calculation is extended. Although attempts have been made to related oxygen deficits to productivity and trophic status, it is clear from the work of Cornett and Rigler (1980) that variations in oxygen deficit cannot be explained by variations in productivity alone.

Thus, although it is tempting, with the long Lake Mendota data set, to draw conclusions about changes in trophic status, such conclusions can only be made with strong reservations (see Chapter 9). At least as far as the accuracy of the data permit, we can conclude, with Stewart, that Lake Mendota has not undergone any measurable change in oxygen depletion rate since 1906. This conclusion is in keeping with the studies of Bortleson (1970) that cultural eutrophication of Lake Mendota occurred when the land was cleared for agriculture in the mid-1850's (see above).

Nitrogen in Lake Mendota

Nitrogen is one of the key nutrients in lake ecosystems. It is available to organisms in several forms: ammonium (NH_4^+), nitrate (NO_3^-), nitrite (NO_2^-),

Table 3.9. Oxygen deficit calculations for Lake Mendota

Year	O_2 Deficit (mg cm^{-2} month^{-1})	
	A	B
1906	2.80	2.91 2.83
1907	2.00	1.79 2.13
1961	2.32 (2.44)	2.10 2.40
1962	3.08 (3.24)	3.00 2.93
1963	2.72 (2.86)	2.58 2.76
1966	2.43 (2.56)	2.14 2.51
1967	2.56	2.10 2.42
1970	2.93	2.49 2.56
1971	3.58	2.72 2.77

From Stewart (1976).
(A) Results from normal oxygen deficit calculation on 2 dates, 5 May to 5 July (9–18 m). Linear slopes assumed. The upper values for 1961–66 have been reduced by 5% (old values from 18.5 m sampling station beneath in brackets) to correspond more closely to estimated values at the deeper (>22 m) station from which measurements were made in other years. (B) Results from planimetric integration of area under slopes of weighted mean oxygen concentration between sampling dates. Linear slopes not assumed. Upper number 2 months from 5 May to 5 July and lower number 3 months from 5 May to 5 August.

and dinitrogen (N_2). Although all of these forms of N are available to various organisms, ammonium and nitrate are the most important for phytoplankton and bacteria.

As outlined earlier in this chapter (see Tables 3.2 and 3.4), loading of nitrogen into Lake Mendota has been measured directly from watershed studies. Unfortunately, many of the published estimates do not specifically separate ammonium from nitrate, but aggregate nitrogen into the categories of inorganic and organic nitrogen. Detailed nitrogen loading data under base flow conditions for two creeks entering Lake Mendota have been published by Lathrop (1979) and are given in Table 3.11. These data should be compared with the average values given in Table 3.2. As seen, nitrate is the major N species entering Lake Mendota under base flow conditions. Nitrate is also the dominant ion in base flow and groundwater seepage (Sonzogni and Lee, 1974; Brock et al., 1982). About equal amounts of nitrate and ammonia are added from

Table 3.10. Oxygen depletion rates after summer stratification for various years

Year	Depth (meters)	Rate ($mg \cdot l^{-1} \cdot d^{-1}$
1906	18	0.16
1907	18	0.12
1961	22	0.16
1962	22	0.18
1963	22	0.19
1966	22	0.17
1967	18	0.16
1970	18	0.18
1971	18	0.20
1976	16	0.10
1977	16	0.09
1978	16	0.18
1979	16	0.08
1980	16	0.10
1981	16	0.14

Data of 1906–1971 from Stewart (1976). Data of 1976–1981 from present study; calculations by V. Watson.

atmospheric precipitation and dry fallout, and slightly more nitrate than ammonium is added from urban runoff (Sonzogni and Lee, 1974). Organic N always constitutes a significant, albeit minor, part of the N loading to Lake Mendota.

Biological nitrogen fixation

Some nitrogen is also added to the lake as a result of dinitrogen fixation. The use of dinitrogen involves the activity of the nitrogenase enzyme system, a system found only in procaryotes. Among the phytoplankton, only the cyanobacteria (blue-green algae) are capable of fixing dinitrogen. Dinitrogen is the most abundant gas in the atmosphere and is usually present in dissolved form in lakewater at a concentration which is saturating at ambient temperature and pressure. Detailed profiles of dinitrogen concentration in Lake Men-

Table 3.11. Nitrogen analyses of two creeks entering Lake Mendota

	Pheasant Branch		Six Mile Creek	
Component	mean	range	mean	range
Total N (mg/l)	2.71	1.26–4.13	2.10	1.18–3.93
Org. N	0.69	0.56–1.60	0.53	0.22–0.96
NO_3–N	3.0	0.5–7.3	1.45	0.51–3.33
NH_4^+–N	0.21	<0.03–0.25	0.10	<0.03–0.73

Data of Lathrop (1979).

dota, presented by Birge (1910), showed that the concentration was uniform throughout the whole water column and did not change as a function of biological activity. The winter concentration, which should have been the highest of the year because of the low temperature, was about 19 cc/l, equivalent to 23–24 mg/l (uncorrected for temperature and pressure). Relative to biological nitrogen requirements (see Chapter 4), this is a high value.

Nitrogen fixation studies have been carried out on Lake Mendota by Neess et al. (1963), Goering and Neess (1965), Vanderhoef et al. (1975), Vanderhoef (1976), Torrey (1972), Torrey and Lee (1976), Peterson et al. (1977), and Brezonik (1968). Much of this published work is in the nature of physiological ecology or methodology and provides little insight into the role of dinitrogen fixation in the overall N budget of the lake. Only the Torrey (1972) study, using the sensitive acetylene reduction method, provides useful in-lake estimates that can be used to assess the significance of nitrogen fixation for the lake as a whole. Torrey's study was used by Sonzogni and Lee (1974) and Torrey and Lee (1976) to estimate the contribution of nitrogen fixation to the N budget of the lake. It was concluded that nitrogen fixation was only important during the summer, when cyanobacterial blooms were common, but showed a wide variation even then, so that annual estimates had a high degree of uncertainty. It was estimated that about 38,000 kg of N per year was added to Lake Mendota as a result of dinitrogen fixation, which was about 7% of the total input to the lake (see also Table 3.4).

The energy requirement for nitrogen fixation is high, and for cyanobacteria the energy source is light. Since light is often limiting during intense cyanobacterial blooms (Chapter 5), it is understandable that nitrogen fixation would not be a major source of nitrogen input to Lake Mendota.

Annual cycle of inorganic nitrogen

In-lake studies of nitrogen dynamics have been reported by Brezonik (1968), Brezonik and Lee (1968), and Stauffer (1974). Detailed data for various nitrogen species are given in the Appendix to this book.

If dinitrogen gas is ignored, an annual cycle of the dynamics of the inorganic nitrogen species can be presented, based primarily on the work of Brezonik (1968) and our own work (Appendix). Profiles of nitrogen species in the lake are given in Figure 3.12. Because of the long hydraulic residence time of Lake Mendota, external loading does not cause large seasonal or annual changes in inorganic N. On the other hand, internal cycling of N is of major importance.

Beginning in early winter, the ammonium concentration in the lake is high and the nitrate concentration is low (the reasons for this will become apparent below). As the winter progresses, ammonium gradually decreases in amount and nitrate rises, probably due to the process of bacterial nitrification. Nitrification is predominantly a winter process because at this time of the year, N-assimilating phytoplankton are not present in significant amounts. At ice-out, nitrate concentrations are high, generally 3–5 times ammonium concentrations. However, as soon as spring mixing occurs, ammonium concentrations

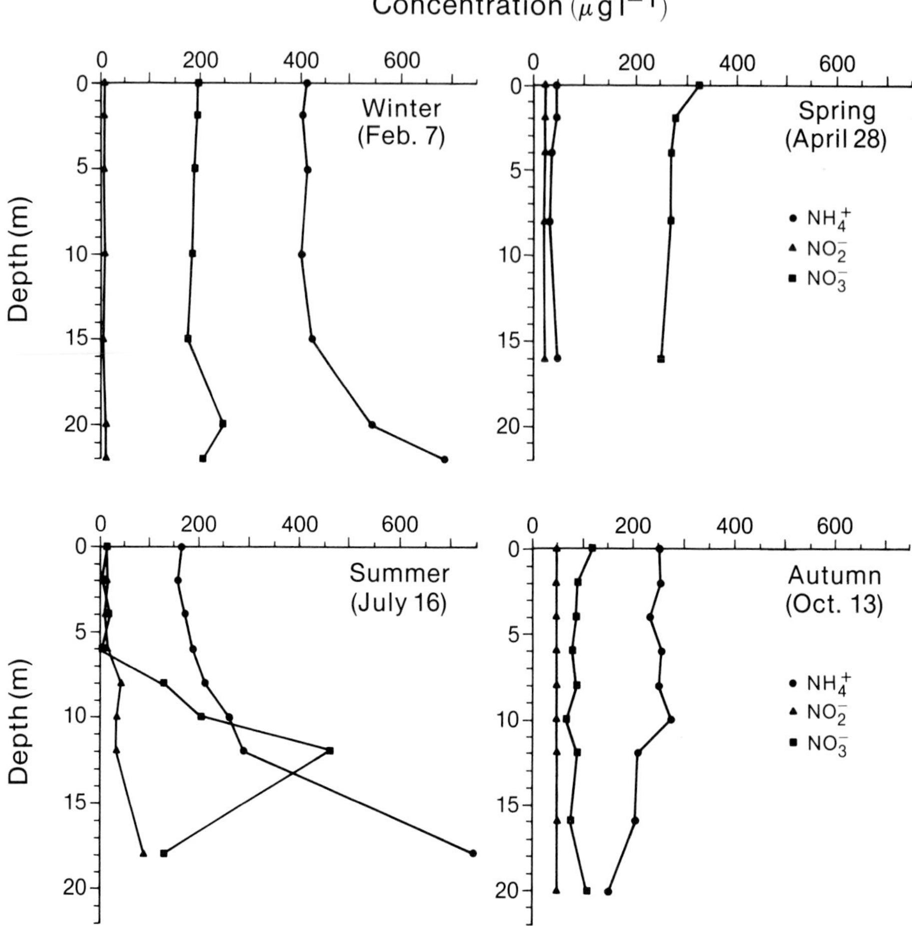

Figure 3.12. Concentrations of ammonium, nitrate, and nitrite in the water column of Lake Mendota at different seasons of the year (1980), from samples collected at the deep hole station. See Appendix for detailed data.

rise, due to the large amount of ammonium in the sediments (Table 3.7) which is added to the water.

During the spring bloom, some depletion of nitrate and ammonium occurs, but at the time of stratification, both species are still present in major amounts. After stratification, the hypolimnion and epilimnion must be treated separately. In the epilimnion, large cyanobacterial blooms often develop (see Chapter 4), causing a major depletion of both ammonium and nitrate. Ammonium disappears first, as would be predicted from the fact that this nitrogen source is used preferentially by phytoplankton over nitrate, but both sources are essentially depleted from the epilimnion. The onset and rate of depletion of nitrogen depends upon the rapidity and extent of development of the cyanobacterial bloom. It is at this time of the year that significant amounts of dinitrogen fixation may occur, although since the cyanobacteria are often light-limited (see Chapter 5), the very energy-demanding nitrogen fixation process is probably never of great quantitative significance. Nitrite is usually only a minor component, although it may development transiently to a higher extent as an intermediate in the nitrification process in winter. During summer, nitrite is occasionally present in significant amounts at the thermocline, probably as a result of either nitrification or denitrification.

In the hypolimnion, the situation is quite different, since phytoplankton assimilation of combined nitrogen does not occur and anoxia changes the redox relationships. As soon as anaerobic conditions develop, there is a marked depletion of nitrate due to the denitrification process, and by the end of the summer nitrate is virtually absent from hypolimnetic waters (Figure 3.12). Ammonium, on the other hand, increases in the hypolimnion. This increase occurs for two reasons: 1) under anaerobic conditions, ammonium is stable; 2) the extensive amounts of ammonium present in the sediment (Table 3.7) are transferred to the water column by advection and diffusion.

Once the fall season arises, the thermocline gradually breaks down. After lake turnover, the ammonium-rich hypolimnetic waters mix with the epilimnion and the ammonium concentration of the epilimnion increases. Nitrate remains low in amount, increasing only to a minor amount as a result of external loading. Finally, once ice forms and phytoplankton activity ceases, nitrate begins to increase again as a result of the nitrification process. The annual cycle is complete.

Phosphorus in Lake Mendota

Phosphorus is the most common cause of excessive algal growth in lakes (Edmondson, 1969) and has been extensively studied in Lake Mendota (Stauffer, 1974; 1985; the present work). Data on phosphorus loading to Lake Mendota have been presented earlier in this chapter. In the present section, we discuss the internal cycling of phosphorus and provide background for discussions of phytoplankton development in Chapters 4 and 5.

Phosphorus is present in lakes at only one oxidation state (PO_4^{3-}), so that the complications regarding oxidation and reduction discussed with N do not arise. However, phosphorus is present in a number of different chemical forms such as soluble orthophosphate, organic phosphates (including those present in organisms), metaphosphate, phosphorus adsorbed to mineral particles, calcium phosphate, etc. The standard chemical assay for phosphorus measures almost solely orthophosphate. Phosphorus in a lake is conveniently separated into two fractions, termed soluble and insoluble phosphorus. Soluble phorphorus is that fraction assayable by orthophosphate assay after filtration of a water sample through a membrane filter (pore size generally 0.45 μm). Total phosphorus is that assayable after digestion of an unfiltered sample with a strong oxidizing agent (perchlorate or persulfate) followed by orthophosphate assay. Insoluble phosphorus is the difference between total phosphorus and soluble phosphorus.

As noted, phosphate is present in the water in both soluble and particulate form. The soluble form is almost certainly available to phytoplankton, whereas the particulate form may or may not be available. Common limnological experience shows that when algae are not present in the water, virtually all of the phosphate is present as soluble orthophosphate, and when the algae develop, soluble orthophosphate concentration decreases and particulate phosphate concentration increases. This increase in particulate phosphorus is, of course, due to the fact that the algae themselves are particulate. It is also well established that the extent of phytoplankton growth in a lake is a function of the phosphate concentration of the water.

The steady-state phosphorus concentration

We have discussed phosphorus loading earlier in this chapter and it is of interest to attempt to predict the phosphate concentration of the lake from loading rates. For this prediction, we begin first by ignoring the biological aspects and simply treat the lake as a large kettle. If we consider phosphate to be completely soluble, then any phosphate entering the lake either remains in the water or flows out through the outlet. Under these conditions, the steady-state concentration of phosphate is expressed as:

phosphate (μg/l) = volume loading (μg/l/year) $\times$ residence time (years)

Thus if we know the residence time of the lake and the loading rate of phosphorus to the lake we can calculate phosphate concentration. The phosphorus loading rate of Lake Mendota has been reasonably well determined and was presented in Table 3.3. From the average loading rate for 1976–1977, 33,000 kg/yr, and the volume of Lake Mendota water, 48×10^7 m^3, we can calculate the volume loading rate to the lake as 69 μg/l/yr. If the residence time of Lake Mendota were exactly one year then the steady-state phosphorus concentration would be 69 μg/l. However, Lake Mendota has a residence time of about six years, so that the steady-state phosphorus concentration, not considering sedimentation, would be 414 μg/l.

However, this calculation assumes that phosphate is completely soluble and remains so. This is not true, because a considerable amount of phosphate is assimilated into phytoplankton at certain times of the year, and a lot of these phytoplankton sediment to the bottom of the lake, carrying phosphate with them (see Table 3.8). In addition, not all the phosphorus entering the lake is soluble. Inorganic particulate phosphate may sediment independently of biological processes. The actual concentration will be lower by an amount which is equivalent to the rate of phosphate sedimentation. In order to complete our calculation, we must therefore have a measure or an estimate of the fraction of phosphate that sediments out of the water.

Although instantaneous sedimentation rates can be measured with sedimentation traps (see earlier in this chapter), what we need is *net* sedimentation for the year. Part of the year, sedimentation occurs at a fairly significant rate, but at other times of the year, during fall and spring turnover, there is massive resuspension of sediment from the water (see Figure 3.9). What we need to know is the net sedimentation rate for the whole year, which is total sedimentation rate minus resuspension rate. One direct way of estimating sedimentation on an annual basis is to measure the amount of phosphate leaving the lake in the outflow. If the loading is known, then the difference between input and output should be sedimentation.

Output values for phosphorus for Lake Mendota were given in Table 3.3 and from these values, sedimentation can be calculated. Lake retention, which is the fraction of phosphorus sedimenting, is 1−(output/input), = 0.75. The mass balance data thus lead to an estimate of phosphorus sedimentation of 24,850 kg/yr, equivalent to 0.63 $g/m^2/yr$. Bortleson's coring data (see earlier in this chapter) permitted an estimate of phosphorus sedimentation rate of 47,000 kg/yr, equivalent to 1.2 $g/m^2/yr$. The value of Fallon and Brock (1980), based on sedimentation trap studies, was 4.6 $g/m^2/yr$, almost four-fold higher than Bortleson's and seven-fold higher than that calculated from output measurements. As Fallon and Brock point out, the higher value for sedimentation trap data is to be expected, since much of the material collected in sedimentation traps undergoes rapid decomposition (measured directly in other studies by Fallon, 1978) when it reaches the surface of the sediment before burial.

The input-output analysis probably provides the best estimate of the *current* sedimentation rate, 24,850 kg/yr. Since the data of Bortleson provide an estimate for the past 100 years of 41,000 kg/yr, these data suggest that phosphorus sedimentation was greater in the past than it is now (perhaps because loading was greater in the past or because the species of phytoplankton present now are less likely to sediment than those of the past).

We can now refine the analyses presented earlier in this section. From the volume loading data, residence time, and phosphorus retention, the steady-state (mean annual) concentration of phosphorus in the lake can be calculated:

mean annual P (μg/l)=volume loading (μg/l/yr)×residence time (years)×(1-fraction sedimenting)

Thus, P=103.5 μg/l

This value is of course different from any measured phosphorus value for any particular water sample of the lake. At times, and at some depths in the lake, the phosphorus concentration may be lower, and at other times, or at other depths in the lake, the concentration may be higher. An average calculation based on the extensive phosphorus assays of Stauffer (1974) is 135 μg/l, which is about 25% higher than the value calculated from the loading data.

The significance of a calculation like this in reference to lake eutrophication has been discussed in detail by Vollenweider (1976), Dillon and Rigler (1974), and Stauffer (1985). We discuss the relationship of phosphorus loading and recycling to phytoplankton growth in Lake Mendota in Chapters 4 and 5.

In addition to phosphorus loss by sedimentation, there is also the possibility of phosphorus *gain* by movement from the sediments. The sediments of Lake Mendota become a net source of phosphorus to the hypolimnetic water when summer anoxia develops. Upward flux out of the hypolimnion into the epilimnion can then occur as a result of entrainment and eddy diffusion (R. E. Stauffer, personal communication). Marked phosphorus gradients develop across the metalimnion during summer stratification, but actual transport of phosphorus into the epilimnion is highly variable, depending strongly on meteorological events. Overall, upward vertical transport should have little effect on *net* phosphorus concentrations, but may influence the extent of duration of specific phytoplankton blooms (see Chapter 4 and 5). Because on an annual basis there is net movement of phosphorus *into* the sediments, phosphorus release from the sediments only serves to modify the value calculated for phosphorus retention. However, because of the way the calculation of phosphorus retention is done (see above), release is already included in the net flux estimation.

Iron and manganese

Iron and manganese are commonly discussed together because of similarities in solubility and redox behavior. Iron has two oxidation states, Fe^{2+} and Fe^{3+}. With manganese, the stable redox states are Mn^{2+} and Mn^{4+}. The reduced forms of both of these elements are more water-soluble than the oxidized forms, although in the presence of sulfide, the reduced forms are taken out of solution as the insoluble sulfides. Extensive analyses of Fe and Mn in Lake Mendota have been carried out by Joseph J. Delfino and collaborators, published in a series of papers (Delfino and Lee, 1968; 1969; Delfino et al., 1969; Delfino and Lee, 1971). Seasonal variation in Fe and Mn occur in the hypolimnion, especially just above the sediments. Delfino et al. (1969) reported that during periods of thermal stratification, Mn is preferentially released from the sediments, even though the ratio of Fe to Mn in the sediments is 19:1.

A typical summer profile of Mn and Fe at the time the hypolimnion is completely anoxic is shown in Figure 3.13. Equilibrium calculations and thermodynamic data were used by Hoffman and Eisenreich (1981) to model the iron-manganese relationships in Lake Mendota sediments. These authors identified a number of solid phases involved in the manganese and iron equilibria, including $MnCO_3$, MnO_2, FeS, and Fe_3O_4. In addition, Mn (II) and Fe (II) were thought to be adsorbed to oxide surfaces (MnO_2, Fe_3O_4, and SiO_2). Hoffman and Eisenreich (1981) showed that both the dissolution of the solid phases and the release of adsorbed metal ion are involved in this solubilization process. These processes are both pH and redox dependent, but they postulate that the appearance of Fe and Mn ions is principally controlled by solid dissolution in high-carbonate, high-phosphate, and high-sulfide waters at the sediment-water interface (Hoffman and Eisenreich, 1981). However, R. E. Stauffer (personal communication) has concluded from field measurements and thermodynamic calculations that manganese (II) accumulation in the hypolimnion is not regulated by mineral equilibrium but by redox shifts in the upper sediments coupled with seasonally-dependent sediment dispersion into the over-

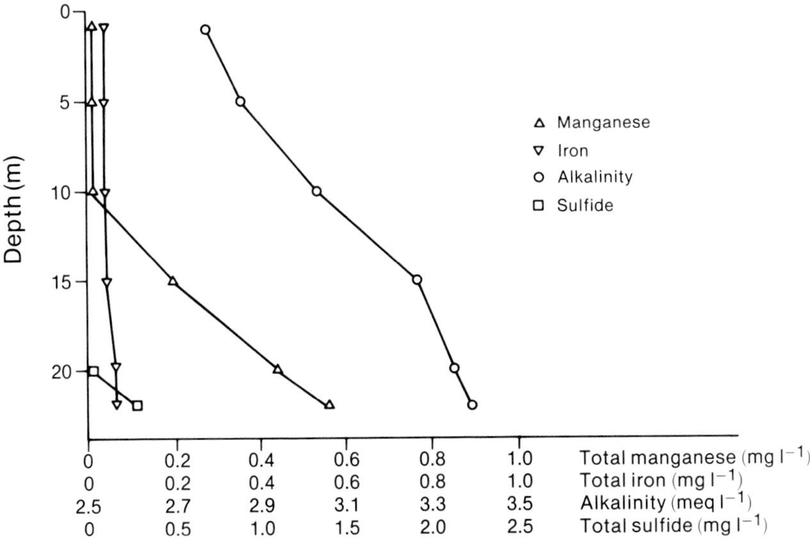

Figure 3.13. Manganese and iron concentrations for water samples collected from Lake Mendota on August 2, 1967. From Delfino and Lee (1971).

lying water column. He has also concluded that the low hypolimnetic iron concentration in the hypolimnion is the result of precipitation as FeS.

Sulfate

Sulfate is one of the major anions of Lake Mendota waters (Table 3.1). Sulfate is important biologically because it is reduced to sulfide by sulfate-reducing bacteria (see Chapter 7) and because the sulfide formed is able to react with and reduce iron, leading to the formation of iron sulfides. Because phosphate is held in sediments at least partly in the form of insoluble iron phosphates, any formation of ferrous sulfide can lead to the release of soluble phosphorus. There are thus complex interactions between sulfate, sulfide, iron, and phosphate.

Lake Mendota is moderately high in sulfate, with values of the order of 6–10 mg/l being found in epilimnetic waters (Table 3.1 and Figure 7.14). The source of sulfate is predominantly the input streams, with only a small amount from direct precipitation on the lake. This conclusion can be shown by the following analysis: If the sulfate concentrations given in Table 3.2 are multiplied by the relative amounts of water that each input stream contributes to the lake, the total sulfate entering the lake from streams can be estimated. For the streams in Table 3.2 (which account for virtually all of the surface water entering Lake Mendota), the following proportions of the total water flow are known (Lathrop, 1979): Token Creek, 41%; Yahara River (at Windsor), 24%;

Six Mile Creek (at Mills Rd.), 12%; Spring Creek, 9%; Pheasant Branch (at Highway M), 13%. The total stream volume flow rate to Lake Mendota is 69.58×10^6 m^3/year. From this volume, the stream flow rates, and sulfate concentrations, it can be calculated that the average sulfate concentration of the entering water is 1.64 mg/l. Since the hydraulic residence time of Lake Mendota is about 6 years, the steady-state concentration of sulfate, if it were all derived from streams, would be 9.84 mg/l, a value quite close to the actual measured concentration.

What is the sulfate contribution to the lake from precipitation? Acidic precipitation in the Madison area does contain sulfate. However, direct measurements and calculations from meteorological data show that precipitation has only a minor effect on the sulfate concentration of the lake. A number of precipitation events were monitored by the present author in 1974 and the pH and sulfate concentration measured. The pH of Madison rain and snow measured between 4.1 and 4.9 and the sulfate concentration averaged 3 mg/l. As shown in Table 2.5, the average input to Lake Mendota from precipitation is 31.82×10^6 m^3/year. With a sulfate concentration of 3 mg/l, the annual input would be 0.1989 mg/l, for a steady-state concentration of 1.2 mg/l. Thus, precipitation contributes about 11% of the total sulfate to the lake.

Local variation in sulfate concentration in the incoming streams is of interest. As shown in Table 3.2, the sulfate levels in the various streams range from 8 to 49 mg/l. The highest values are in Pheasant Branch (see location on Figure 1.6), especially the upper end of this creek. Detailed sulfate analyses in this stream show that the high sulfate levels are derived from a specific region through which Pheasant Branch flows (just above State Highway 12), and geological observations show that in this area a large glacial lake existed before the ice-dam completely disappeared and the present level of Lake Mendota was established. Presumably, the high sulfate levels are due to the mineral assemblage present in this former lake bed. The significance of this region is shown by the fact that although Pheasant Branch contributes only 13% of the stream water to Lake Mendota, it contributes about 28% of the sulfate.

What is the significance of the high sulfate values present in the Lake Mendota waters? Such high sulfate values permit extensive bacterial sulfate reduction, making this process an important part of anaerobic carbon decomposition (see Chapter 7). Further, the sulfide formed from bacterial sulfate reduction is responsible for the precipitation of iron as ferrous sulfide, thus permitting the release of phosphorus from iron phosphates. The extent to which phosphate is released under these conditions is determined initially by the ratio of sulfate to iron. If the sulfate concentration is in excess of the iron concentration, then sulfide formation can bring about the precipitation of all the iron in the system. The relatively high sulfate concentration in Lake Mendota is thus an additional factor which results in a eutrophic condition in the lake.

Chloride

Chloride is a common ion that has little, if any, biological significance in fresh waters, but is of interest because it behaves conservatively in lakes, neither

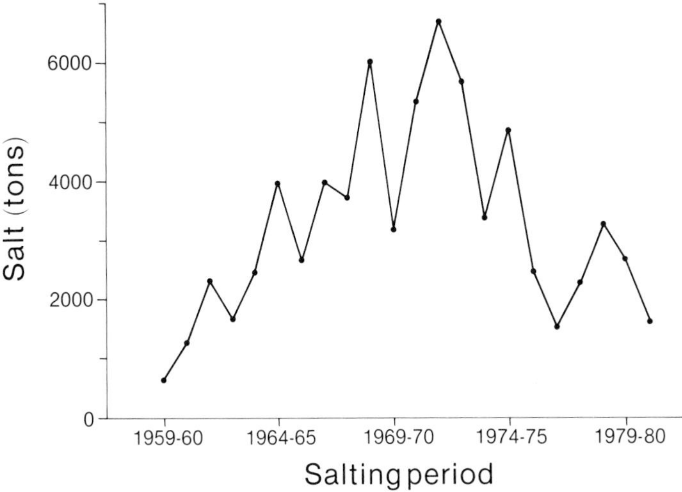

Figure 3.14. Road-salt use by the City of Madison. Data from the Public Health Division, City of Madison. Courtesy of B. Saley.

undergoing precipitation nor redox reactions. As such, chloride serves as a useful marker for water movement and dilution.

In presettlement times, the only sources of chloride would have been natural ones. Because of the continental location of Lake Mendota, over 1000 km from the nearest seawater, chloride levels would be expected to be fairly low. There are, however, two artificial sources of chloride to Lake Mendota, sewage effluents (human wastes are high in chloride) and road salt (used as a deicing agent to improve winter driving conditions). Lake Mendota has never received any significant sewage effluents. Road salting only became a significant activity in the Lake Mendota drainage basin in the 1960's and 1970's, and because of environmental concerns, road salting tapered off in the mid-1970's (Figure 3.14).

The change in chloride concentration of Lake Mendota and three other Madison lakes is shown in Figure 3.15. The high chloride concentrations in Lakes Waubesa and Kegonsa in the 1940's are due to the treated sewage effluents that were placed into Waubesa at this time (as discussed earlier). Lake Mendota and Lake Monona chloride concentrations rose when the road salting program was initiated. Although the salt usage has dropped considerably in the mid 1970s (Figure 3.14), the chloride concentration in Lake Mendota continued high, as would be predicted from the long residence time of the lake. Continued monitoring of chloride in Lake Mendota will permit an independent assessment of the hydraulic residence time of the lake.

Silica

Silica (SiO_2, SiO_3^{2-}) is an interesting element biologically because it is required for growth by those algae which have siliceous frustules. Although siliceous

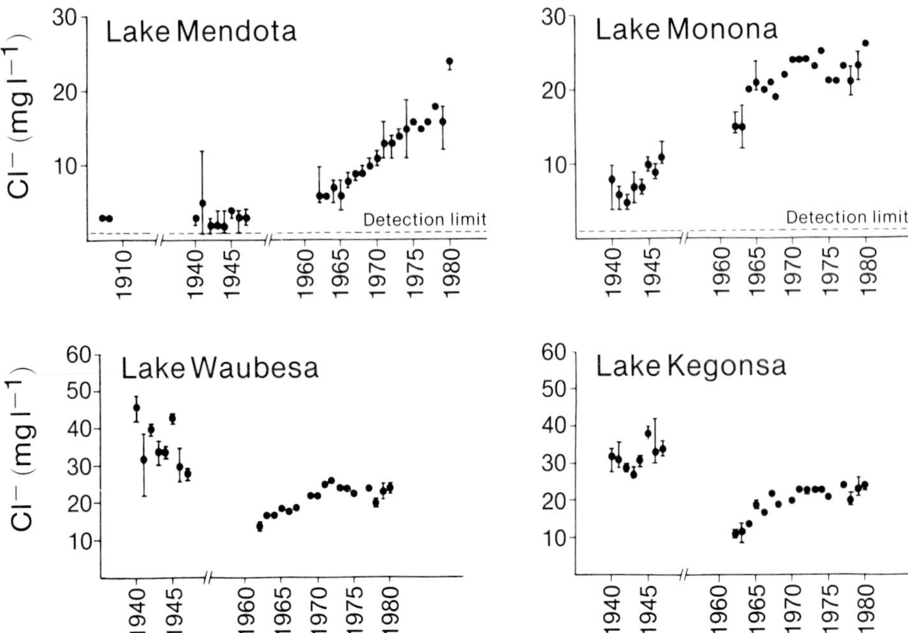

Figure 3.15. Chloride concentrations of the Madison lakes over the period 1900–1980. From Lillie and Mason (1983).

frustules are formed by several classes of algae, the primary siliceous algae in Lake Mendota are the diatoms. Lake Mendota frequently exhibits a spring diatom bloom, as well as occasional other periods of diatom abundance throughout the year. In a study of silica in Lake Mendota, Vigon (1976) showed that silica was often in short supply in relation to phosphorus, and that the spring diatom bloom was probably terminated when the water became depleted of silica. Further, he suggested that sedimentation of diatom cells after the termination of a bloom could lead to the removal of significant amounts of phosphorus from the lake.

Dissolved silica concentrations in the surface waters of Lake Mendota were reported by Vigon (1976) to be around 400–700 μg/l in early spring, but decreased dramatically to 30–50 μg/l at the termination of the diatom bloom. These latter concentrations are just at the analytical detection limit for silica. The phosphorus concentration in Lake Mendota is so high that diatom growth becomes silica-limited. Of course, a large variety of silica-independent algae are found in Lake Mendota (see Appendix), so that silica deficiency does not limit algal growth in general. R. E. Stauffer (personal communication) has estimated that the external loading of phosphorus would have to be reduced by a factor of at least three in order to induce phosphorus rather than a silica limitation for the spring diatom bloom.

Vigon (1976) showed that about 80 percent of the silica input to the water in early spring was from the sediments, the rest coming from the drainage basin. The release of silica from the sediments, including that released from sedimented diatom frustules themselves, was found by Vigon to be a complex process that was influenced by temperature, biological activity, and perhaps the presence of amorphous metal oxides at the sediment surface. The diatom-associated silica could be analyzed independently of the nonbiogenic silica by use of mild alkaline hydrolysis with sodium bicarbonate at autoclave temperature, a treatment which solubilized the diatom-associated silica.

Inorganic carbon in Lake Mendota

Lake Mendota is a typical hard water lake, with pH controlled by a bicarbonate buffer system. The alkalinity is fairly high, around 150 mg/l (3 meq/l), as shown in Table 3.1. The dominant inorganic cations are calcium and magnesium, with lesser amounts of sodium and potassium (Table 3.1).

An extensive study of calcium carbonate equilibria in Lake Mendota was carried out by Hawley (1967). He concluded that Lake Mendota was supersaturated with respect to calcium carbonate and that phytoplankton exerted a strong effect on the carbonate system. Hawley suggested that calcium carbonate precipitation might not occur because of inhibition by the high magnesium ion concentration.

The influence of phytoplankton on the carbonate system is most dramatically shown by the effect on pH, with high pH values developing in the epilimnetic waters during summer blooms.

Profiles of calcium, alkalinity, and pH (the parameters which define the carbonate system) are shown in Figure 3.16. It can be seen that during summer stratification, there is a marked increase in pH in the epilimnetic waters and a decrease in pH in hypolimnetic waters, as would be predicted in a system in which pH was controlled strongly by phytoplankton photosynthesis. More extensive pH data for Lake Mendota are given in the Appendix to this book.

In addition to its inherent interest, a knowledge of inorganic carbon concentrations is necessary for calculations of primary production by the ^{14}C method, as described in Chapter 5.

Summary

The chemistry of Lake Mendota is probably known in as great detail as that of any other lake, thanks to the diligence of generations of students and professors at the University of Wisconsin-Madison. In the present chapter we have presented data on the major chemical elements, and on the sediments. Actual data for the 1976–81 period are given in the Appendix. Not covered in this chapter have been some of the biologically related constituents, such as chlorophyll, organic carbon, methane, sulfate, and sulfide, which will be

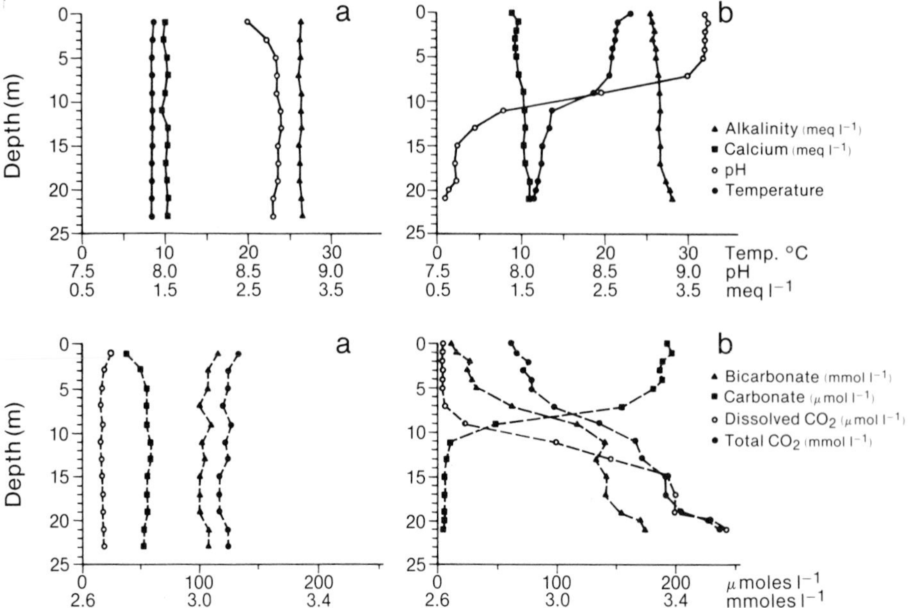

Figure 3.16. Vertical profiles of parameters which define the carbonate equilibrium system for Lake Mendota. a) May 13, 1967, before thermal stratification; b) July 15, 1967, after thermal stratification and during the presence of an *Aphanizomenon* bloom. From Hawley (1967).

discussed in subsequent chapters. Also, seasonal and short-term variations of some of the elements will be discussed later.

A major portion of the discussion in this chapter has related to measurements of external and internal loading of the key phytoplankton nutrients nitrogen and phosphorus. The relationship of nutrient loading to phytoplankton growth will be discussed in Chapters 4 and 5.

The eutrophic nature of Lake Mendota has been stressed and it has been shown that eutrophication of the lake occurred shortly after settlement of the Madison area in the 1850's. One of the strongest pieces of evidence dating the onset of cultural eutrophication has been the chemical analysis of sediments and sediment cores (Bortleson, 1970). Further evidence of change can be obtained from a comparison of historical with contemporary data for such eutrophication-related parameters as hypolimnetic oxygen depletion. It has been shown that the areal hypolimnetic oxygen deficit is the same now as it was in the early part of the century when Birge and Juday made their extensive studies of gases in the lake (see Chapter 9).

It is important to emphasize the lack of any strong evidence for increased cultural eutrophication during the 20th Century. Anecdotal evidence frequently is used to advance the hypothesis that Lake Mendota has become seriously deteriorated since the beginning of the present century. However, anecdotal evidence is notoriously unreliable. If deterioration has occurred since 1900, it is impossible to perceive it by techniques currently available. Lake Mendota has indeed deteriorated, but the deterioration occurred long before 1900 (Trelease, 1889).

References

Birge, E.A. 1910. Gases dissolved in the waters of Wisconsin lakes. *Bulletin of the Bureau of Fisheries,* 28: 1275–1294.

Birge, E.A., and C. Juday. 1911. *The inland lakes of Wisconsin. The dissolved gases of the water and their biological significance.* Wisconsin Geological and Natural History Survey, Bulletin No. 22, Madison.

Bortleson, G.C. 1970. *The chemical investigation of recent lake sediments from Wisconsin lakes and their interpretation.* Ph.D. Thesis, University of Wisconsin, Madison.

Brezonik, P.L. 1968. *The dynamics of the nitrogen cycle in natural waters.* Ph.D. Thesis, University of Wisconsin, Madison.

Brezonik, P.L., and G.F. Lee. 1968. Denitrification as a nitrogen sink in Lake Mendota, Wisconsin. *Environmental Science and Technology,* 2: 120–125.

Brock, T.D., D.R. Lee, D. Janes, and D. Winek. 1982. Groundwater seepage as a nutrient source to a drainage lake: Lake Mendota, Wisconsin. *Water Research,* 16: 1255–1263.

Cornett, R.J. and F.H. Rigler. 1980. The areal hypolimnetic oxygen deficit: an empirical test of the model. *Limnology and Oceanography,* 25: 672–679.

Delfino, J.J., G.C. Bortelson, and G.F. Lee. 1969. Distribution of Mn, Fe, P, Mg, K, Na, and Ca in the surface sediments of Lake Mendota, Wisconsin. *Environmental Science and Technology,* 3: 1189–1192.

Delfino, J.J., and G.F. Lee. 1968. Chemistry of manganese in Lake Mendota, Wisconsin. *Environmental Science and Technology,* 2: 1094–1100.

Delfino, J.J., and G.F. Lee. 1969. Colorimetric determination of manganese in lake waters. *Environmental Science and Technology,* 3: 761–764.

Delfino, J.J., and G.F. Lee. 1971. Variation of manganese, dissolved oxygen and related chemical parameters in the bottom waters of Lake Mendota, Wisconsin. *Water Research,* 5: 1207–1217.

Dillon, T.J., and F.H. Rigler. 1974. The phosphorous-chlorophyll relationship in lakes. *Limnology and Oceanography,* 19: 767–773.

Doyle, R.W. 1968. Identification and solubility of iron sulfide in anaerobic lake sediment. *American Journal of Science,* 266: 980–994.

Edmondson, W.T. 1969. Eutrophication in North America. pp. 124–149, In: *Eutrophication: Causes, Consequences, Correctives.* National Academy of Sciences, Washington, D.C.

Fallon, R.D. 1978. *The planktonic cyanobacteria: their sedimentation and decomposition in Lake Mendota, Wisconsin.* Ph.D. Thesis, University of Wisconsin, Madison.

Fallon, R.D. and T.D. Brock. 1980. Planktonic blue-green algae: production, sedimentation, and decomposition in Lake Mendota, Wisconsin. *Limnology and Oceanography,* 25: 72–88.

Goering, J.J. and J.C. Neess. 1965. Nitrogen fixation in two Wisconsin lakes. *Limnology and Oceanography,* 9: 530–539.

Hasler, A.D. 1947. Eutrophication of lakes by domestic drainage. *Ecology,* 28: 383–395.

Hasler, A.D. 1963. Wisconsin 1940–1961. pp. 55–93, In: *Limnology in North America.* (Frey, D.G., Editor), University of Wisconsin Press, Madison.

Hawley, J.E. 1967. *Calcium carbonate equilibrium in Lake Mendota.* M. Sc. Thesis, University of Wisconsin, Madison.

Hoffmann, M.R., and S.J. Eisenreich. 1981. Development of a computer-generated equilibrium model for the variation of iron and manganese in the hypolimnion of Lake Mendota. *Environmental Science and Technology,* 15: 339–344.

Holdren, G.C., D.E. Armstrong, and R.F. Harris. 1977. Interstitial inorganic phosphorous concentrations in Lakes Mendota and Wingra. *Water Research,* 11: 1041–1047.

Hutchinson, G.E. 1938. On the relation between the oxygen deficit and the productivity and typology of lakes. *International Revue gesamten Hydrobiologie,* 36: 336–355.

Hutchinson, G.E. 1957. *A treatise on limnology,* Vol. 1. John Wiley and Sons, New York.

Konrad, J.G., D.R. Keeney, G. Chesters, and K.-L. Chen. Nitrogen and carbon distribution in sediment cores of selected Wisconsin lakes. *Journal Water Pollution Control Federation,* 42: 2094–2101.

Lackey, J.V., and C.N. Sawyer. 1945. Plankton productivity of certain southeastern Wisconsin lakes as related to fertilization. I. Surveys. *Sewage Works Journal,* 17: 573–585.

Lathrop, R.C. 1979. Dane County water quality plan. Appendix B: water quality conditions; Appendix H: lake management. Dane County Regional Planning Commission, Madison, Wisconsin.

Lerman, A. 1979. *Geochemical processes: Water and sediment environments.* John Wiley and Sons, New York.

Lillie, R.A., and J.W. Mason. 1983. *Limnological characteristics of Wisconsin lakes.* Technical Bulletin No. 138, Department of Natural Resources, Madison, Wisconsin.

Murray, R.C. 1956. Recent sediments of three Wisconsin lakes. *Bulletin of the Geological Society of America,* 67: 883–910.

Neess, J.C., R.C. Dugdale, V.A. Dugdale, and J.J. Goering. 1963. Nitrogen metabolism in lakes. I. Measurement of nitrogen fixation with N^{15}. *Limnology and Oceanography,* 7: 163–169.

Nriagu, J.O. 1968. Sulfur metabolism and sedimentary environment: Lake Mendota, Wisconsin. *Limnology and Oceanography,* 13: 430–439.

Nriagu, J.O., and C.J. Bowser. 1969. Magnetic spherules in sediments in Lake Mendota, Wisconsin. *Water Research,* 3: 833–842.

Peterson, R.B., E.E. Frieberg, and R.H. Burris. 1977. Diurnal variation in N_2 fixation and photosynthesis by aquatic blue-green algae. *Plant Physiology,* 59: 74–80.

Sawyer, C.N. 1947. Fertilization of lakes by agricultural and urban drainage. *New England Waterworks Association,* 61: 109–127.

Sawyer, C.N. 1954. Factors involved in disposal of sewage effluents to lakes. *Sewage and Industrial Wastes,* 26: 317–325.

Shukla, S.S., J.K. Syers, J.D.H. Williams, D.E. Armstrong, and R.F. Harris. Sorption of inorganic phosphate by lake sediments. *Soil Science Society of America Proceedings,* 35: 244–249.

Sonzogni, W.C., and G.F. Lee. 1974. Nutrient sources for Lake Mendota–1972. *Transactions Wisconsin Academy of Sciences, Arts and Letters,* 62: 133–164.

Sonzogni, W.C., G.P.Fitzgerald, and G.F. Lee. 1975. Effects of wastewater diversion on the lower Madison lakes. *Journal of the Water Pollution Control Federation,* 47: 535–542.

Sonzogni, W.C. and G.F. Lee. 1975. Phosphorus sources for the lower Madison lakes. *Transactions of the Wisconsin Academy of Sciences, Arts, and Letters,* 63: 162–175.

Stauffer, R.E. 1974. Thermocline migration-algal bloom relationships in stratified lakes. Ph.D. Thesis, University of Wisconsin, Madison.

Stauffer, R.E. 1981. *Sampling strategies for estimating the magnitude and importance of internal phosphorus supplies in lakes.* Environmental Protection Agency Publication EPA-600/3-81-015, April 1981, 89 pp. Environmental Research Laboratory, Corvallis, Oregon.

Stauffer, R. E. 1985. Relationships between phosphorus loading and trophic state in calcareous lakes of southeast Wisconsin. *Limnology and Oceanography,* 30: 123–145.
Stewart, K.M. 1976. Oxygen deficits, clarity, and eutrophication in some Madison lakes. *International Revue gesamten Hydrobiologie,* 61: 563–579.
Torrey, M.S. 1972. *Biological nitrogen fixation in Lake Mendota.* Ph.D. Thesis, University of Wisconsin, Madison.
Torrey, M.S. and G.F. Lee. 1976. Nitrogen fixation in Lake Mendota, Madison. *Limnology and Oceanography,* 21: 365–378.
Trelease, W. 1889. The "working" of the Madison lakes. *Transactions of the Wisconsin Academy of Sciences,* 7: 121–129.
Twenhofel, W.H. 1933. The physical and chemical characteristics of the sediments of Lake Mendota, a freshwater lake of Wisconsin. *Journal of Sedimentary Petrology,* 3: 68–76.
Vanderhoef, L.N. 1976. Nitrogen fixation in Lake Mendota, 1972–1973. *Hydrobiologia,* 49: 53–57.
Vanderhoef, L.N., P.J. Leibson, R.J. Musil, C.-Y. Huang, R.E. Fiehweg, J.W. Williams, D.L. Wackwitz, and K.T. Mason. 1975. Diurnal variation in algal acetylene reduction (nitrogen fixation) *in situ. Plant Physiology,* 55: 273–276.
Vigon, Bruce, W. 1976. *The role of silica and the vernal diatom bloom in controlling the growth of nuisance algal populations in lakes.* M.Sc. Thesis (Water Chemistry), University of Wisconsin, Madison.
Vollenweider, R.A. 1976. Advances in defining critical loading levels for phosphorous in lake eutrophication. *Memorie dell' Istituto Italiano di Idrobiologia,* 33: 53–83.
Williams, J.D.H., J.K. Syers, and R.F. Harris. 1970. Adsorption and desorption of inorganic phosphorous by lake sediments in a 0.1 *M* NaCl system. *Environmental Science and Technology,* 4: 517–519.

4. Phytoplankton

Introduction

The base of the food pyramid consists of photosynthetic organisms that carry out the synthesis of organic carbon from inorganic carbon using solar radiation as the energy source. One of the main aspects of the present work on Lake Mendota has been a quantitative assessment of phytoplankton changes over the annual cycle and from year to year. A knowledge of phytoplankton dynamics is of interest from both a basic and applied standpoint:

1. Organic carbon derived from phytoplankton drives the biogeochemical cycles of the lake.
2. Phytoplankton provide the main food for the higher trophic levels.
3. The phytoplankton constitute the most numerous readily assessable microbial populations, and as such provide excellent populations for studies on the physiological ecology of microorganisms.
4. Many times of the year, there is more biomass present in phytoplankton than in any other group of organisms.
5. Lake Mendota has long been recognized as a lake with extensive phytoplankton blooms (Trelease, 1889), which have been considered an esthetic nuisance.
6. One group of phytoplankton, the cyanobacteria (blue-green algae), has members which produce animal and human toxins.

7. The cyanobacteria tend to float, contributing to the unsightly state of the lake at certain times of the year.
8. Some cyanobacteria are capable of nitrogen fixation, and may thus be responsible for some nutrient enrichment of the lake.

Phytoplankton ecology has been reviewed many times (Fogg, 1975; Morris, 1980; Reynolds, 1984) and it is not the purpose of the present chapter to cover this same material. Rather, it is intended to place the studies of the dynamics of Lake Mendota phytoplankton within the existing broad theoretical framework. The rest of this introduction is intended merely to set the stage for the material in this chapter.

Factors influencing phytoplankton development in lakes

Phytoplankton are predominantly autotrophic organisms, growing on inorganic substrates and using light as their primary source of energy. Most phytoplankton are aerobes, and only rarely would they show extensive development under anaerobic conditions. Factors necessary for phytoplankton growth in lakes are: solar radiation, suitable inorganic nutrients (of which the most likely to be limiting are nitrogen, phosphorus, silicon, and rarely carbon and iron), adequate temperature (generally, below 30°C), and suitable pH (in the neutral range).

Under various conditions, the most likely factors limiting phytoplankton growth are solar radiation, nitrogen, and phosphorus, or (with siliceous organisms such as diatoms) silicon (see Chapter 3). One significant feature of phytoplankton growth is that it is often self-limiting. Thus, the development of a dense phytoplankton population can lead to self-shading, reducing the availability of light. Also, if a dense phytoplankton population develops, most of the assimilable nitrogen or phosphorus of the lake may become tied up in phytoplankton biomass, leading to a self-imposed nutrient deficiency. Heavy phytoplankton blooms in very eutrophic carbonate-buffered lakes can also lead to a rise in pH, shifting the bicarbonate equilibrium in the direction of carbonate and thus leading to a carbon (that is CO_2) deficiency.

Since the classical work of Talling (1957), a phytoplankton population has been viewed as a compound photosynthetic system in which individual species distinctions can often be ignored. Our work in Lake Mendota is based to a great extent on the work of Talling and its subsequent enhancements (Yentsch, 1980; Harris, 1980; Walsby and Reynolds, 1980; Reynolds, 1984).

Chemical composition of phytoplankton and nutrient-limited growth

The idea of nutrient-limited growth has lead to extensive consideration of the chemical composition of phytoplankton cells. Clearly, if solar radiation is in excess, then the maximum biomass that can develop in a water body will be determined by the total amount of some limiting nutrient present in the water. This makes it of interest to compare the nutrient concentrations on a mass

Table 4.1. Chemical mass balance relationships between algae and Lake Mendota water

Algal chemical composition		Concentration (mg/l) of the element in lake algae, assuming standing crop of 8 mg/l algae	Average element concentration in Lake Mendota water (mg/l)†		Ratio of water/algae (values <1.0 equal deficiency)
Element	g/100 g dry cells*				
Carbon	49	3.9	Inorganic	36	9.23
			Organic	10	
Nitrogen	6.2	0.50	Inorganic	0.6	1.2
			Organic	0.93	
			N_2	24	48
Phosphorus	1.2	0.096	Soluble	0.08	0.83
			Total	0.13	
Sulfur	0.67	0.054		7	129
Potassium	1.9	0.15		3	20
Calcium	0.98	0.078		27	346
Magnesium	0.6	0.048		31	646
Sodium	0.6	0.048		13	271
Iron	0.67	0.54	Total	0.25	4.6
			Fe^{2+}	0.18	3.33

*Algal chemical data of Healey (1973).
†Lake Mendota data, this work (see Chapter 3 and Appendix).

basis in the lake with the nutrient atoms tied up within algal cells. The first analysis of this sort was carried out by Redfield (1958), and the "Redfield ratio" for phytoplankton is a widely discussed parameter. Although the Redfield ratio is often stated to be N:P of 16:1, the specific ratio must vary with the species of phytoplankton. The Redfield ratio reflects the mass balance relationship between phytoplankton and chemicals in the water. Note that nutrient recycling, important for understanding primary production, is not being discussed here. We are considering the simple mass balance between nutrient atoms and algal cells.

We have discussed the chemical composition of Lake Mendota water in Chapter 3 and have shown that there is considerable seasonal variability as well as variability with depth. However, because of vertical mixing (Chapter 2), phytoplankton grow throughout the whole photic zone, and are capable of extracting nutrients from much of the water column. Thus, it is permissible, when considering annual averages, to use an average lake value when considering these mass balance relationships, as listed in Table 4.1.

A large number of measurements of the chemical composition of algae have been made, and a number of these are summarized by Healey (1973). Although algal composition varies markedly as a result of species differences and differences in the nutrient concentration of the growth medium, it is permissible in an overall analysis such as the present one to use an average value for algal composition. In Table 4.1, the element composition of an average algal cell is given and compared to the concentrations of various nutrients in Lake Mendota water. The comparisons in this table provide support for the common assertion that only nitrogen and phosphorus are likely to be limiting for phytoplankton growth in aquatic systems. As seen in this table, the only other element that might potentially be limiting would be iron. Nitrogen would also cease to be limiting if nitrogen fixation were taking place, although in Lake Mendota nitrogen fixation is rarely a quantitatively significant process (Torrey and Lee, 1976), probably due to light limitation (see Chapter 5). Note, however, that it is assumed in Table 4.1 that the Lake Mendota algal concentration is 8 mg/l, a common but somewhat high value. At many times of the year, lower algal densities occur.

Although the above discussion has concerned factors influencing the *growth* of phytoplankton, it is evident that the amount of phytoplankton in the lake, that is, the *standing crop,* is determined also by processes which *remove* phytoplankton. The most important removal processes are grazing (by zooplankton and other small animals), sedimentation, and decomposition. At some times of the year, loss factors may be of relatively minor importance so that standing crop is controlled primarily by growth processes, whereas at other times of the year loss processes may play a major role. In this chapter, growth processes are discussed first, and loss factors are considered later.

Methodology for studying phytoplankton populations in Lake Mendota

Phytoplankton density in Lake Mendota was assessed in the present work by taking discrete water samples at various depths and assaying these water samples chemically for chlorophyll and microscopically for total algal volume. Although there were general correlations between chlorophyll and biovolume, at many times of the year there were marked differences, so that each method provides a different view of phytoplankton change.

Sampling

Intensive sampling for phytoplankton was carried out during the years 1976–81. Samples were also taken in parts of 1983 and 84. For most years, samples were collected at a single mid-lake station, generally called the "deep hole" station (see Chapter 2). In 1976, samples were collected at eight stations located around the lake, in order to determine how much horizontal variability existed (see later). In 1977, samples were collected at four stations around the lake using an integrated sampling system. Because this integrated sampling procedure did not permit analysis of vertical distribution, the 1977 data are not given with most of the data presented in this book.

When large volume samples were needed, a five liter Van Dorn sampler was used, but for most purposes a two liter sampler was employed. The Van Dorn samplers were constructed of opaque polyvinylchloride. At the same time that water samples were taken for phytoplankton, standard limnological parameters were measured such as temperature, dissolved oxygen, and light.

The same water samples that were used for phytoplankton were also used for chemical analyses, as described in the Appendix to this book.

Assessment of algal standing crop

For most purposes, the algal material suspended in the water was removed by filtration through a Whatman GF/C glass fiber filter with a nominal pore size of 1 μm. Experiments that were conducted periodically during the course of these studies indicated that there were virtually no algal cells that were not retained by these glass fiber filters. When chlorophyll concentrations in the water were low, 1000 or 2000 ml of water was filtered, whereas during periods of dense blooms, smaller volumes of water were sufficient for accurate assays. The chlorophyll procedure followed Strickland and Parsons (1972) using the extraction solvent of Shoaf and Lium (1976).

The concentrations of the various algal species were assessed using epifluorescence microscopy following the method described by Brock (1978). In this procedure, algae are stained with the fluorochrome dye primuline, which stains polysaccharide material, and each water sample was filtered through a black membrane filter, 0.45 μm (Sartorius black membrane filters were used for most studies). For most microscopic counts a water volume of 15–30 ml was filtered. The algae were observed with a Carl Zeiss epifluorescence microscope equipped with a vertical illuminator using 10X and 40X bright field objectives (the 40X objective was an oil immersion objective with a numerical aperture of 1.0). The epifluorescence technique permitted the assessment of phytoplankton numbers and volumes. Numbers were obtained by counting cells on a randomly selected group of microscope fields. To obtain statistically valid counts, it is necessary to count at least 30 cells of each species. However, when a given species was rare, it was often not cost-effective to count this many cells. The raw data, given in the Appendix, permit the reader to assess the statistical validity of the counts, and it can be ascertained that for the *dominant* species, the error of counting is satisfactorily low. To estimate biovolume, average dimensions of each species were obtained by measuring around 10 cells to the nearest micrometer using an ocular micrometer. The particular dimension(s) measured depended on the morphology of the organism. A geometrical formula suitable for each shape was then used to calculate the volume of a cell. Similar geometric formulas were also used to calculate surface areas. From the counts and from the calculated volumes, the total volume of each species in the water sample could be calculated. The raw data in the Appendix give the measured dimensions, raw counts, biovolume, and surface area.

Statistical analysis of horizontal and vertical distribution

In the summer and fall of 1976, a detailed study was carried out on the horizontal distribution of phytoplankton across the lake. This was done by sampling for chlorophyll at eight stations distributed throughout the main part of the lake. At each station, samples were taken at 2 m depth intervals through

Table 4.2. Horizontal and vertical variability of chlorophyll-containing particulate material in Lake Mendota, summer-fall 1976, as shown by a Friedman Rank Test

	Station								
Depth	1	2	3	4	5	6	7	8	Sum
0	78	80.5	83.5	76	91	89.5	74.5	43	616
2	43.5	41.5	64.5	68	65	66.5	48.5	34.5	432
4	65	73.5	87.5	73	87	93	79	53	611
6	59.5	59.5	76.5	61	62	67	56	26.5	468
8	74	86	57.5	67.5	75.5	82	57.5	40	540
10	45	56	46	59	53	49.5	48.5	39	396
Sum	365	397	415.5	404.5	433.5	447.5	364	236	

Numbers represent sums of ranks for each station and sampling depth for the period June 8–October 18. For each date, the chlorophyll values were ranked from lowest to highest (rank 1 through 8). The sums of ranks were then made for each station and depth. The data from the rankings show that station 8 was frequently lower in chlorophyll than the other stations. The locations of the stations are shown in the map below. The actual data are in the Appendix.

the epilimnion. Sampling was carried out at weekly intervals over a period of 17 weeks from June 8, 1976 until October 18, 1976 (when the lake turned over). Over 800 separate samples were analyzed for chlorophyll content. Because the chlorophyll values were not normally distributed, a Friedman rank test was carried out. Results comparing the eight stations and the six depths over the whole sampling period are summarized in Table 4.2*. One conclusion from this study is that station 8, the station located in the western part of the lake, is frequently low in chlorophyll when compared to the other stations. Stations 1 and 7 from the center of the lake represent an intermediate level, whereas stations 2, 3, 4, 5, and 6 from the eastern side represent the area of the lake with the highest chlorophyll values. These results are consistent with the fact that the prevailing winds in the Madison area in the summer are from

* These analyses were carried out by Kathie Olson.

the west, leading to a moderate concentration of phytoplankton on the east side of the lake.

Depth distribution is also of interest, and it can be seen in Table 4.2 that the 2 m, 6 m, and 10 m depths are frequently low in chlorophyll, compared to the other depths. Later in this chapter there is more detailed discussion of chlorophyll distribution with depth.

Although there were definitely higher chlorophyll values on the east side of the lake, the horizontal variation in chlorophyll across the lake was not nearly as high as the temporal variation across the summer season. Therefore it was determined that it was not cost-effective, when measuring daily and weekly changes in phytoplankton density, to assay samples from so many stations. In order to be able to sample at frequent time intervals, it was decided in subsequent years to concentrate sampling on a single station in the center of the lake. This strategy was consistent with the conclusion of Stauffer (1985) that a chlorophyll profile taken near the center of Lake Mendota provides an unbiased estimate of pelagic mean chlorophyll concentration with a coefficient of variation of 14%.

Calculations of total lake values

The data for chlorophyll and biovolume data at discrete depths have been integrated over depth to provide an estimate of total lake concentrations. Calculation of total lake chlorophyll involves a simple multiplication of the chlorophyll assays at each depth times the volume of water represented by that depth, as indicated by the hypsometric table (Table 2.2). Biovolume data are summed over species and integrated over depth to provide an estimate of the total lake biovolume. The total lake biovolume (in m^3) has also been included in some of the summaries of the chemical analyses.

Annual cycles of phytoplankton

Seasonal means and variation in chlorophyll a and in phytoplankton biovolume are presented in Figure 4.1. These means are based on the data for 1976 and 1978 through 1981. As seen, chlorophyll and biovolume are highly variable quantities but there is a distinct annual pattern. After the winter low, there is a pronounced rise in both chlorophyll and biovolume at the onset of the spring bloom. The duration of the spring bloom is variable but is always followed by a sharp decrease in phytoplankton at about the time thermal stratification of the lake begins. It is at this time of the year, late May to early June, when the lake has the greatest water clarity. Soon after thermal stratification is complete, the summer bloom ensues, generally due to the development of a large population of blue-green algae (cyanobacteria). From mid-September, phytoplankton biomass gradually disappears through the fall period and is generally at a low value by the end of the year.

From the data presented in Figure 4.1, the ratio of chlorophyll/biovolume

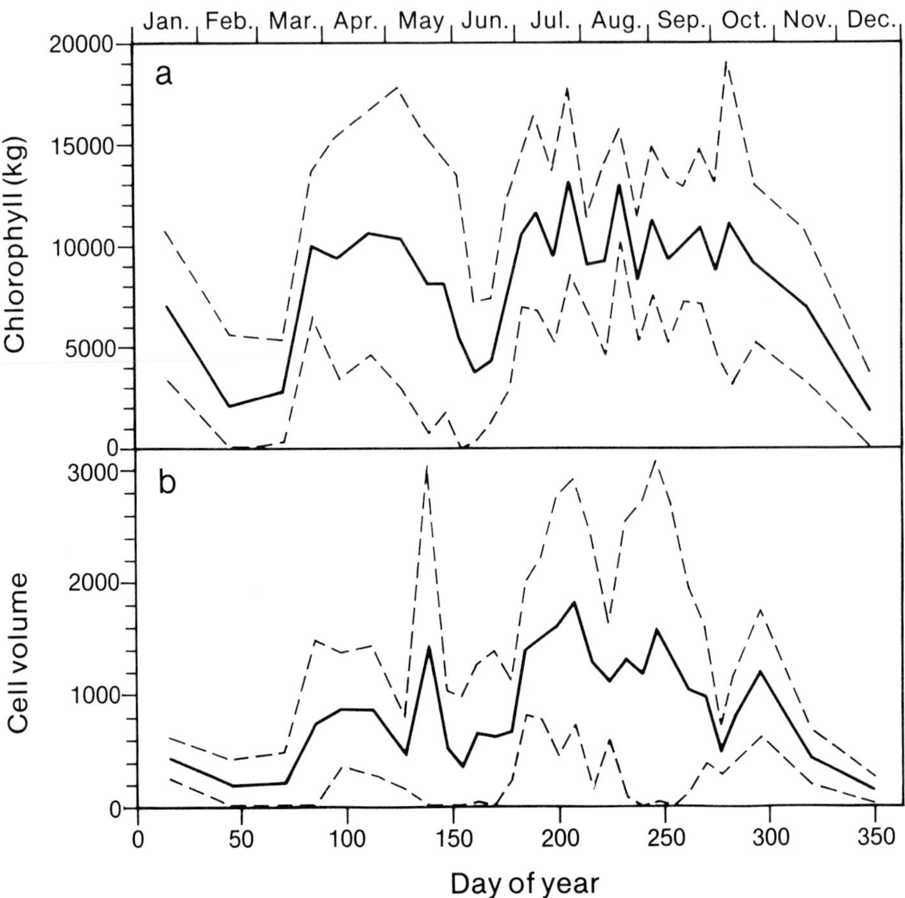

Figure 4.1. Phytoplankton densities in Lake Mendota for the period 1976–1981, as measured by chlorophyll and cell volume. a) chlorophyll, b) cell volume. The values are averages at each sampling date for each year. Data for 1977 are missing. The dashed lines represent ± one standard deviation. Analyses of Dr. Vicki Watson.

can be calculated. This ratio is an estimate of the pigment concentration in the algae and as such provides one assessment of the health of the algal population. Data on annual variation in chlorophyll/biovolume ratio are presented in Figure 4.2. This ratio is quite consistent over the year and from year to year (as shown by the rather narrow standard deviation), except in early June when the mean ratio was found to be high. However, since the year-to-year variation in early June was also very high, the high mean at this time of year is really the result of just a single high value. One might expect the chlorophyll/biovolume ratio to be positively correlated with the concentration of nutrients in the water column, since severe nutrient limitation might be expected to result in decline and senescence of the algal population. However, an assess-

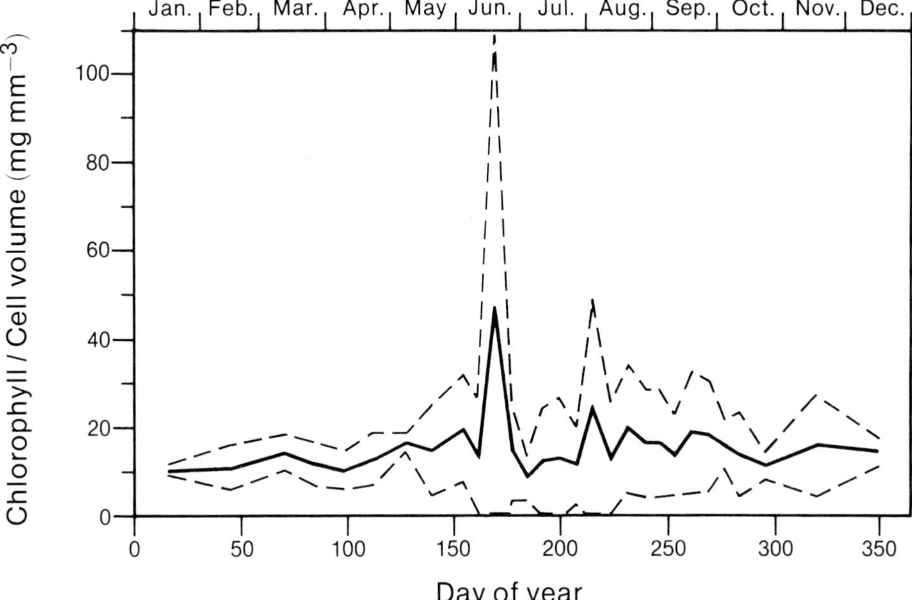

Figure 4.2. Mean and variation in chlorophyll/cell volume ratio, Lake Mendota, 1976–81 (except 1977). The dashed lines represent ± one standard deviation. Analyses of Dr. Vicki Watson.

ment of the 1980 data for both chlorophyll/biovolume ratio and nutrient concentration showed actually a slight negative correlation. (A positive correlation might be found if a time lag were used in the comparison, but such a refinement of the analysis was not carried out.)

The means and standard deviations in Figures 4.1 and 4.2 can be used to determine the relative variability of the phytoplankton parameters over the seasons of the year. Figure 4.3 shows the coefficient of variation (standard deviation as percent of the mean) of chlorophyll and biovolume throughout the year. As seen, the variability of all three parameters are similar throughout most of the year. In analyzing Figure 4.3, the following points can be made:

1. All parameters were quite variable in winter, probably because the algal concentrations at this time of year were low, making the analysis less accurate. Also, fewer samples were taken in winter.
2. Variability is consistently low during the spring bloom. This is probably because the timing and height of the spring peaks are associated with increasing day length and spring mixing, phenomena that are relatively consistent from year to year.
3. The high variability which follows the spring bloom is probably because the phytoplankton crash following the spring bloom is associated with

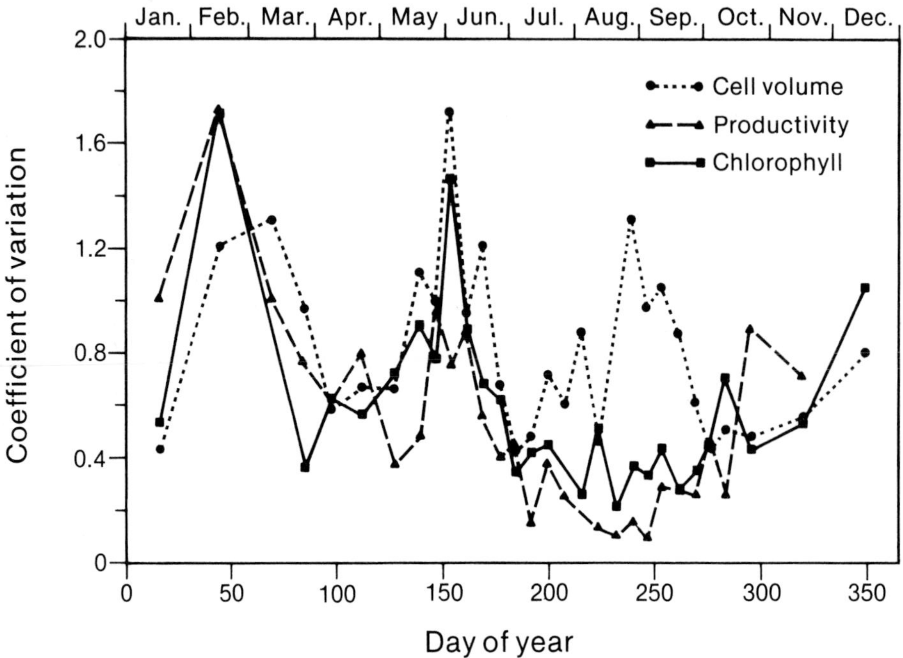

Figure 4.3. Coefficients of variations for phytoplankton parameters, Lake Mendota, 1976–81 (except 1977). The productivity assays are described in Chapter 5. Analyses of Dr. Vicki Watson.

the onset of thermal stratification, the exact date of which is a function of wind patterns which may vary from year to year (see Chapter 2).

4. The low variability for chlorophyll and productivity in the summer are probably because conditions affecting phytoplankton growth and carbon fixation are fairly consistent from year to year. Cell volume, on the other hand, is influenced by nutrient concentration (see later in this chapter), which may vary in the summer due to variable vertical transfer across the thermocline from the nutrient-rich hypolimnion.

Algal species composition

Detailed data on algal species distribution with depth and over the annual cycle are given in the Appendix to this book and some comparisons of historical with modern data are given in Chapter 9. Some phytoplankton and chlorophyll data for 1976 have been presented by Konopka and Brock (1978), for 1976 and 1977 by Fallon and Brock (1980), and for 1979 and 1980 by Pedrós-Alió and Brock (1982). Although in most years a pronounced summer blue-green algal bloom developed (see references just cited and Appendix), as seen in Figure 4.4, the summer of 1981 was different in that little or no blue-green

Table 4.3. Procedure for calculation of mean depth of chlorophyll

Depth m	Chlorophyll g/l or mg/m³	Volume of water in interval $m^3 \times 10^6$	Chlorophyll in interval kg	Depth-weighted chlorophyll m·kg
0.25	32.4	18.8	608	152
1.5	33.0	70.0	2310	3465
3.5	29.5	62.3	1838	6432
5.5	22.6	56.5	1277	7023
7.5	1.7	53.0	90.1	676
9.5	2.3	49.0	113	1071
11.5	0.9	45.1	27.1	312
18	0.0	153.3	0	0
		Sums	6264.2	19131

Mean depth of chlorophyll = 19131 m·kg/6264.2 kg = 3.05 m.
Water samples taken Day 198 Year 1980.

algae developed. (See also the discussion of the summer of 1983, later in this chapter.)

In most years, the spring period is dominated by eucaryotic algae: diatoms, cryptomonads, and green algae. Dinoflagellates (primarily *Ceratium*) generally are present but do not make up a major part of the phytoplankton population, although in the summers of 1978 and 1981, *Ceratium* dominated the summer algal flora. Further discussion of diatom and blue-green algal blooms can be found later in this chapter.

Mean depth of chlorophyll

Because nutrient concentrations in the water column vary with depth, a parameter was needed which could express the average nutrient concentration available to the phytoplankton population at any sampling time. For this purpose, a parameter called the *mean depth of chlorophyll* was calculated. This parameter expresses the average depth in the lake at which a phytoplankton cell is found. As illustrated in Table 4.3, the mean depth of chlorophyll was obtained by multiplying the chlorophyll concentration at a depth (unit, mg/m³) by the volume of water at that depth (unit, m³, obtained from a hypsometric curve) to obtain the total chlorophyll (unit, kg) at that volume interval. Each value was then multiplied by the sample depth it represented, and these weighted depth values (unit, m·kg) were summed over depth. This sum was then divided by the total chlorophyll in the lake (unit, kg) to obtain a mean depth of chlorophyll (unit, m). Generally, the concentration of nutrients at the mean depth of chlorophyll was similar to the average concentration in the mixed zone of the lake but gives a more precise measure of the nutrient concentration to which the average algal cell would be subjected.

The mean depth of chlorophyll over the years 1976–81 of the present study is illustrated in Figure 4.5. Despite the great year-to-year variation in the chlo-

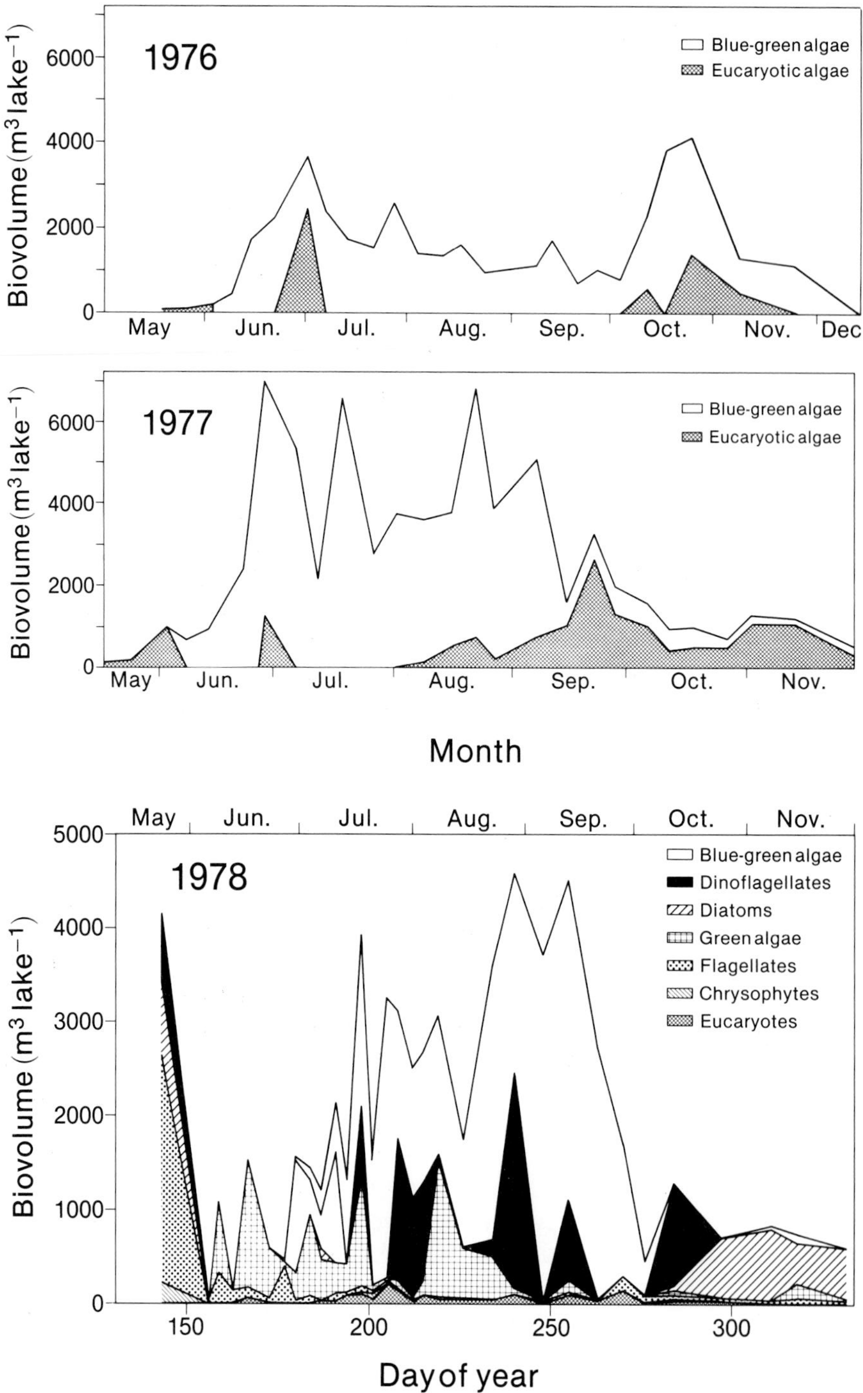

Figure 4.4. Cell volumes of various algal groups (whole lake values), Lake Mendota. Data for 1976 and 1977 from Fallon and Brock (1980).

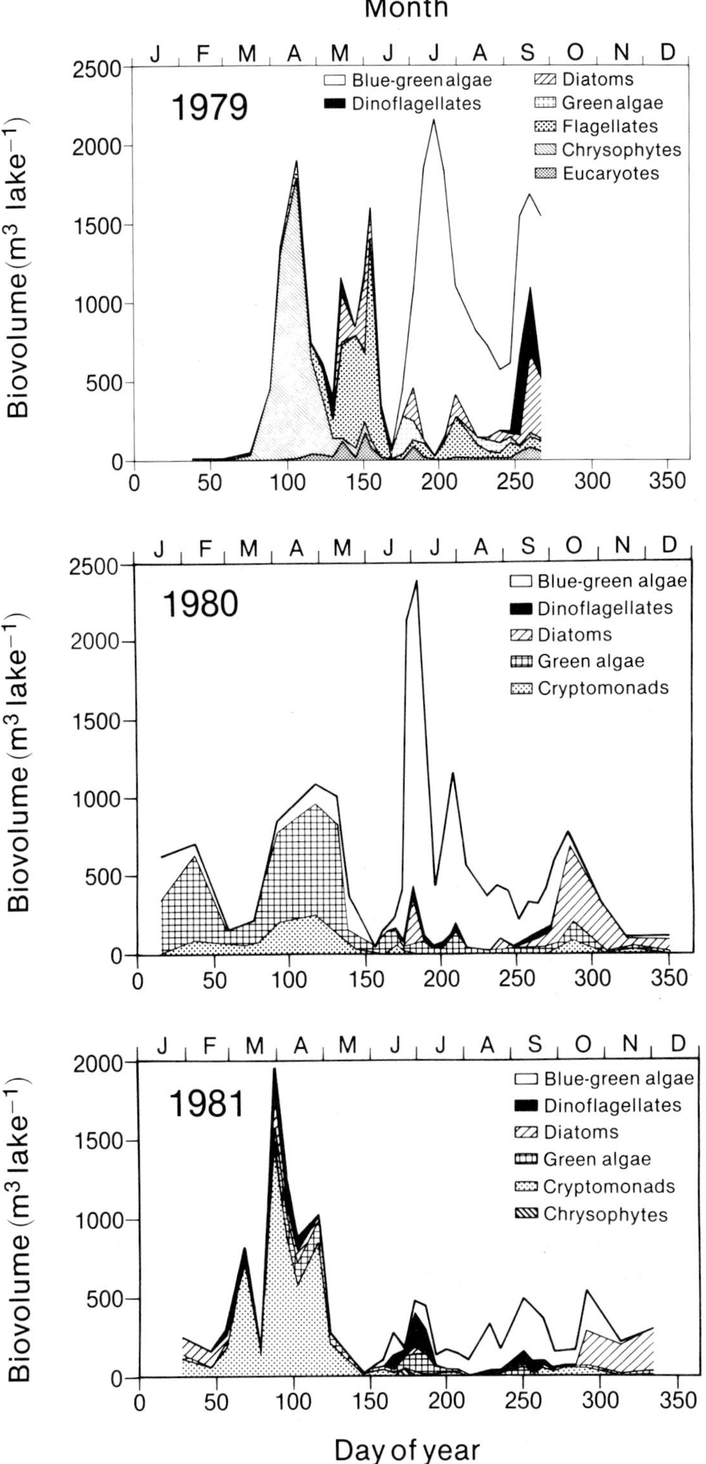
Month
J F M A M J J A S O N D
2500
2000
1500
1000
500
0
1979
Blue-green algae
Dinoflagellates
Diatoms
Green algae
Flagellates
Chrysophytes
Eucaryotes
Biovolume (m3 lake-1)
0 50 100 150 200 250 300 350
1980
Blue-green algae
Dinoflagellates
Diatoms
Green algae
Cryptomonads
1981
Blue-green algae
Dinoflagellates
Diatoms
Green algae
Cryptomonads
Chrysophytes
Day of year

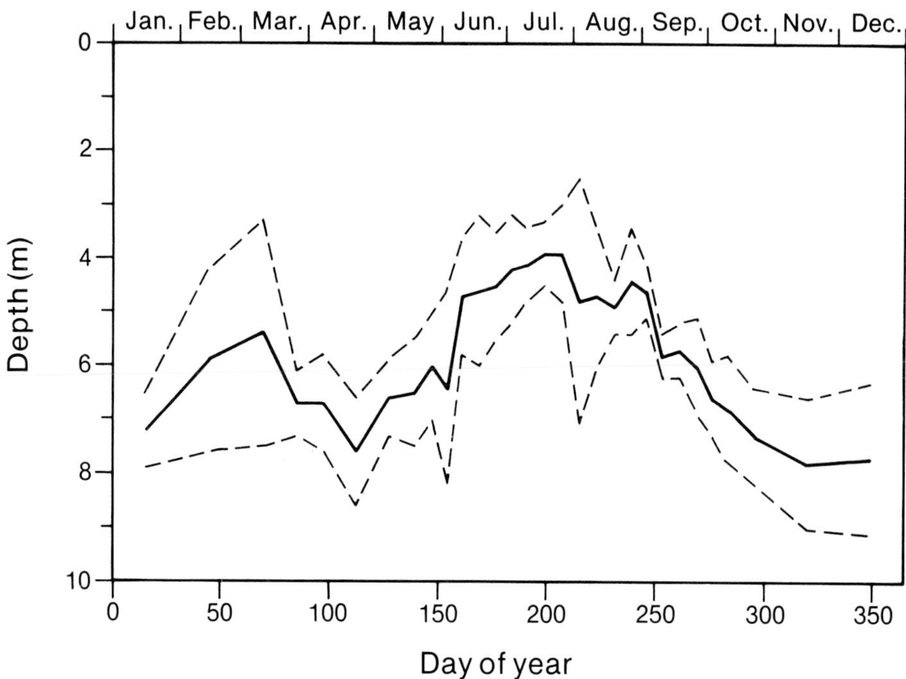

Figure 4.5. Mean depth of chlorophyll-containing particles in Lake Mendota, 1976–81 (except 1977). The dashed lines represent ± one standard deviation. Analyses of Dr. Vicki Watson.

rophyll concentration, the vertical distribution of chlorophyll in the water column seemed to follow a relatively consistent pattern, particularly in the summer. The mean depth of chlorophyll illustrated in Figure 4.5 is generally around 5 to 8 m during spring and fall turnover and at the relatively shallow value of 3 to 5 m during summer stratification. Additionally, the dispersion of chlorophyll around the mean depth is greater in spring and fall than in summer, when it becomes fairly concentrated at the mean depth. This narrower distribution of chlorophyll in summer reflects the fact that phytoplankton, predominantly buoyant blue-green algae, are confined to the relatively stable epilimnion.

A common limnological parameter is the mean depth of the lake, calculated by dividing the lake volume by the area. The mean depth of Lake Mendota is about 13 m. Note that two lakes of different shapes, but with the same area, would have quite different mean depths, since the mean depth calculation does not involve the hypsometry of the lake. A more useful parameter for analyzing algal distribution is the depth at which the total lake volume is divided into two halves. This depth, which I have called the *volumetric mean depth,* is always shallower than the mean depth (because lakes are shaped more like inverted cones than cylinders). The volumetric mean depth for Lake Mendota is 6.7 m, which is similar to the mean depth of chlorophyll during the time of year when the lake is fully mixed. Thus, during periods of complete lake mixing the algae are distributed more or less evenly throughout the water column. In the summer, however, the mean depth of chlorophyll is shallower than the volumetric mean depth, showing that the algae are concentrated in the upper part of the epilimnion. The property of algal buoyancy will be discussed later in this chapter.

Surface/volume ratios of phytoplankton cells

From measurements of phytoplankton volume and surface area, the surface/volume ratio for an algal community can be calculated (Fogg, 1975). Data for 1980 are given in Figure 4.6a. The data for 1979 and 1981 were similar. As seen, the surface/volume ratio is low in spring and goes up sharply in the summer, decreasing again in the fall. Change in surface/volume ratio is correlated with change in the composition of the phytoplankton population. Spring and fall tend to be dominated by large algal forms (diatoms, cryptomonads, or large chlorophytes), while summer is dominated by filamentous or small-coccoid blue-green algae (Figure 4.4). Much of the change in the surface/volume ratio can be attributed to the seasonal change in average cell size (as opposed to cell shape); that is, cell size is much smaller in summer than in spring and fall. This is illustrated by Figure 4.6b.

Correlation of surface/volume ratio with nutrient availability

When seasonal changes in the surface/volume ratio are compared to seasonal changes in inorganic nutrient concentrations (nutrient concentrations measured at the mean depth of chlorophyll), a strong inverse relationship is observed (Table 4.4). The relationship is stronger and more consistent for soluble inorganic nitrogen than for soluble inorganic phosphorus. As discussed by Fogg (1975) and others, small-celled forms, which have large surface-to-volume ratios, should be more efficient at removing nutrients from waters in which the concentration is low. The inverse relationship appears to support the hypothesis that small algae are favored during summer stratification when nutrient concentrations are low, while larger-sized algae are favored during the spring and fall when nutrient concentrations are higher.

Other factors may also influence the surface/volume ratio, such as the vertical mixing regime of the lake or grazing pressure by zooplankton. Unfortunately, although grazing on bacteria has been measured (see Chapter 7), grazing on phytoplankton has not been measured.

Diatom blooms in Lake Mendota

In some years, the spring bloom is dominated by diatoms. The work of Vigon (1976) on silica was discussed in Chapter 3 and it was noted that the dissolved

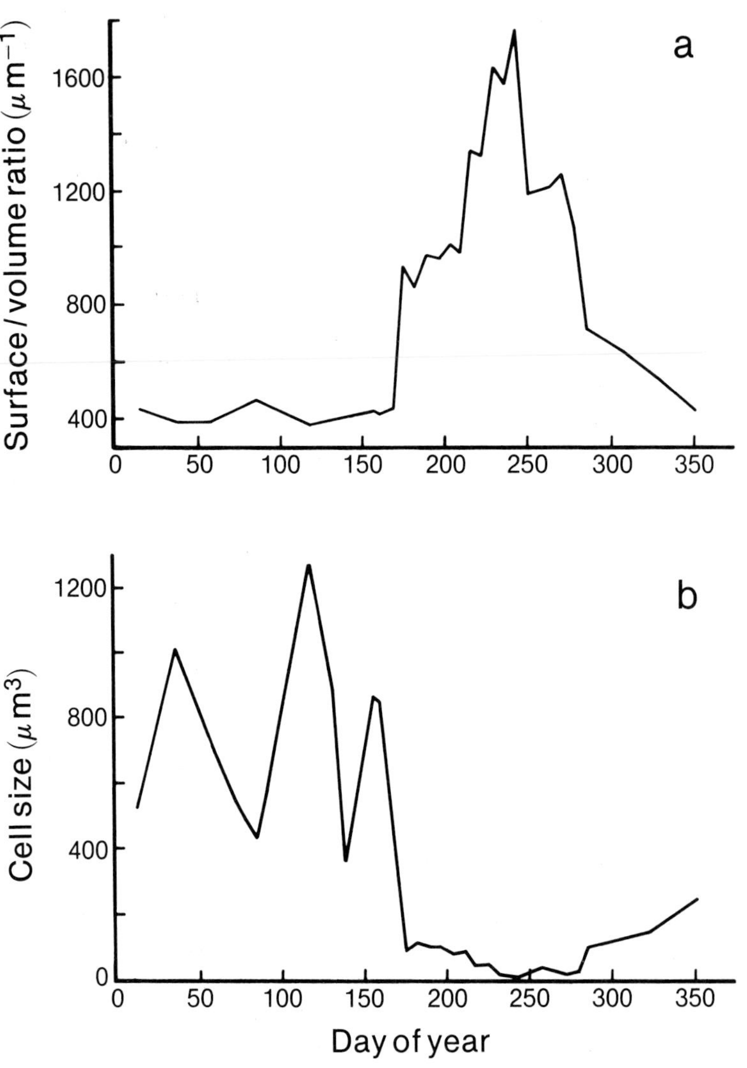

Figure 4.6. Surface/volume ratio and cell size for algae in Lake Mendota during the year 1980. a) Surface/volume ratio. b) Cell size. Analyses of Dr. Vicki Watson.

silica concentrations in the spring could be reduced to almost analytically undetectable levels by the development of a diatom bloom. In the year studied by Vigon, 1975, the diatom bloom was pronounced, but as the data in Figure 4.4 and the Appendix show, a spring diatom bloom does not occur every year. Because of the high phosphorus to silica ratio in Lake Mendota, much phosphorus is left in the lake even if all the diatoms sediment after silica depletion, so that other phytoplankton can develop. Vigon hypothesized that addition of silica to the lake could lead to an increase in the spring diatom bloom and

Table 4.4. Correlation of surface/volume ratio with phosphorous and nitrogen levels*

Constituent	1979	1980	1981
Soluble phosphate	$<-.1$	$-.6$	$-.4$
Total inorganic nitrogen	$-.6$	$-.7$	$-.65$

*Nutrient concentrations measured at a depth corresponding to the mean depth of the chlorophyll concentration were correlated with the surface/volume ratio. The values given are correlation coefficients, r^2.

perhaps to a greater reduction of phosphorus. However, phytoplankton blooms are influenced by other factors than nutrients, especially available light (see Chapter 5), so that it would not automatically follow that an increase in the limiting silica concentration would cause an increase in the diatom bloom.

Blue-green algae of Lake Mendota

The term "water-bloom" is widely used to refer to an accumulation of planktonic (usually blue-green) algae at the surface of a lake or reservoir (Reynolds and Walsby, 1975). Although the algae themselves are only microscopic in size, the accumulation at the surface is so extensive and develops so rapidly that it impresses even the most casual observer. We mentioned in Chapter 1 the paper of Trelease (1889), which described the development of a blue-green algal bloom on Lake Mendota. Trelease referred to this phenomenon as the "working of the lake". Reynolds and Walsby (1975) have discussed in detail the phenomenon of the water-bloom and have noted that it is of world-wide occurrence and has been described by various terms such "flowering of the waters", "wasserblüthe", and "flos-aquae". Bloom formation occurs primarily in eutrophic waters and is the result of the formation of large colonies or aggregates of blue-green algal cells. The most common genera responsible for bloom formation are: *Aphanizomenon, Anabaena,* and *Microcystis* (Figure 4.7). Another blue-green alga occasionally found is *Gloeotrichia* (Figure 4.7). Some of the most massive blooms are formed by *Aphanizomenon,* whose trichomes typically form bundles or rafts of a hundred or more filaments aggregated side by side, often measuring 50 μm across by 500 μm long. Another bloom-forming alga, *Oscillatoria,* does not occur in the blue-green algal blooms of Lake Mendota, but is common in many other lakes.

Although there is considerable year-to-year variability, the blue-green algae generally dominate the summer phytoplankton of Lake Mendota. A summary of the data for the years 1976–81 have been presented in Figure 4.4. Details of the data for this period are given in the Appendix.

The periods of highest chlorophyll values in the lake are generally those during which the phytoplankton is dominated by blue-green algae, although occasionally the chlorophyll values are higher during the spring bloom. In 1976, the genus *Aphanizomenon* was the most common blue-green alga during

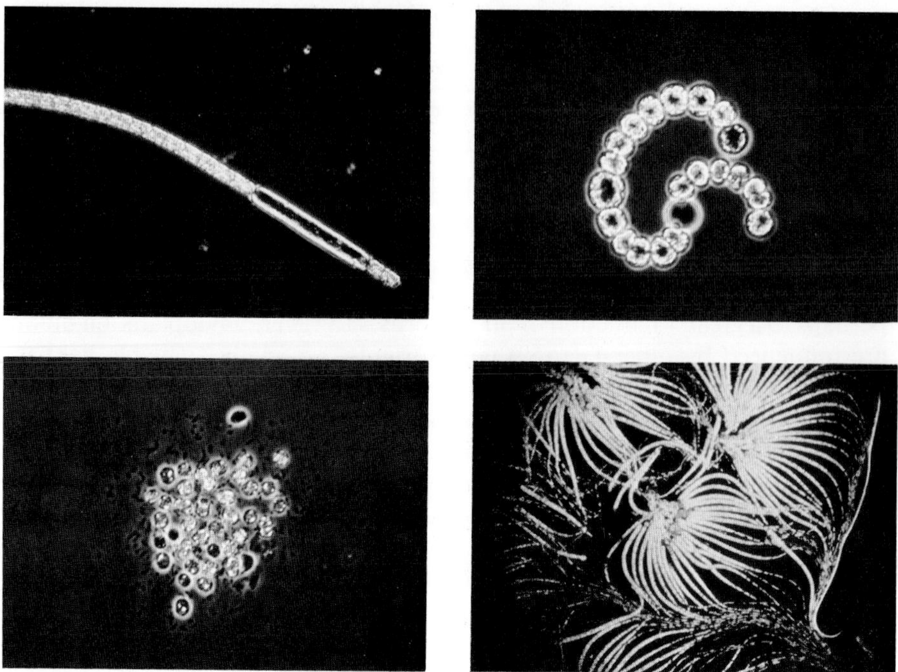

Figure 4.7. Photomicrographs of blue-green algae (cyanobacteria) commonly seen in Lake Mendota. Upper left) *Aphanizomenon* filament and spore; upper right) *Anabaena* trichome and heterocyst; lower left) *Microcystis* colony; lower right) *Gloeotrichia* trichomes. Gas vesicles can be seen in all photomicrographs. Photos taken by Dr. Alan Konopka.

both summer and fall, and eucaryotic algae constituted only a small part of the total. In 1977, the blue-green algal flora was more complex: *Microcystis* constituted a larger portion of the population during late summer and fall, but during the fall the dominant algal species were not blue-green algae but eucaryotic algae (primarily the diatom *Stephanodiscus*).

In 1978, the summer bloom was dominated by eucaryotes (desmids and *Ceratium*), with the blue-green algae *Microcystis* and *Aphanizomenon* not increasing until the middle of July. By late August, *Anabaena* and *Aphanizomenon* began to dominate with a diatom, the large *Stephanodiscus,* rising after fall turnover. In 1979, *Aphanizomenon* dominated throughout the whole summer and *Ceratium* and desmids were virtually nonexistent.

The peak algal biovolume in 1980 (which also corresponded to peak chlorophyll) occurred during the extensive summer blue-green algal bloom, although over the whole year there was more algal biomass in eucaryotic algae than there was in blue-green algae. In 1981, on the other hand, the blue-green algal bloom was very weakly developed and the dominant phytoplankton were eucaryotic algae. In this year, there was a much larger algal development during

Table 4.5. Genus and species list for Lake Mendota phytoplankton

Name	Group
Actinastrum Hantzschii	green
Actinastrum gracilinum	green
Actinastrum sp.	green
Anabaena sp.	blue-green
Ankistrodesmus fractus	green
Ankistrodesmus sp.	green
Aphanizomenon flos-aquae	blue-green
Aphanocapsa sp.	blue-green
Asterionella sp.	diatom
Ceratium hirundinella	dinoflagellate
Ceratium sp.	dinoflagellate
Characium sp.	green
Chlamydomonas sp. 1	green
Chlamydomonas sp. 2	green
Chlorella ellipsoidea	green
Chlorella sp.	green
Chlorochromonas minuta	chrysophyte
Chlorochromonas sp.	chrysophyte
Chlorococcum sp.	green
Chroococcus sp.	blue-green
Closterium sp.	desmid
Coelosphaerium sp.	blue-green
Cosmarium sp.	desmid
Crucigenia Lauterbornii	green
Crucigenia sp.	green
Cryptomonas erosa	cryptophyte
Cryptomonas ovata	cryptophyte
Cryptomonas sp.	cryptophyte
Cyclotella sp.	diatom
Dactylococcopsis sp.	blue-green
Dictyosphaerium pulchellum	green
Dictyosphaerium sp.	green
Dinobryon sp.	chrysophyte
Eudorina sp.	green

(continued)

the spring bloom than there was during the summer, and in this year the spring bloom was dominated not by diatoms but by cryptomonads.

These data show that there is considerable variability from year to year in the nature and extent of the blue-green algal blooms. Although buoyancy is probably one factor which determines the development of blue-green algal blooms (see below), other factors such as nutrient concentration, availability of light, and grazing by larger organisms may also influence the algal bloom development.

A complete genus and species list for the phytoplankton of Lake Mendota is given in Table 4.5 and the detailed data of the occurrence and distribution of these various genera is given in the Appendix to this book. Some discussion of historical data on Lake Mendota phytoplankton is given in Chapter 9.

Table 4.5. Continued

Name	Group
Fragillaria sp.	diatom
Glenodinium sp.	dinoflagellate
Gloebotrys sp.	crysophyte
Gloeotrichia sp.	blue-green
Gomphosphaeria sp.	blue-green
Gymnodinium sp.	dinoflagellate
Kirchneriella sp.	green
Lyngbya limnetica	blue-green
Lyngbya sp.	blue-green
Melosira sp.	diatom
Merismopedia sp.	blue-green
Microcystis aeruginosa	blue-green
Navicula sp.	diatom
Oocystis sp.	green
Oscillatoria rubescens	blue-green
Oscillatoria sp.	blue-green
Pandorina sp.	green
Pediastrum sp.	green
Peridinium sp.	dinoflagellate
Planktosphaeria gelatinosa	green
Planktosphaeria sp.	green
Quadrigula sp.	green
Scenedesmus sp.	green
Schroederia setigera	green
Schroederia sp.	green
Sphaerocystis sp.	green
Staurastrum sp.	green
Stephanodiscus astrea	diatom
Stephanodiscus sp.	diatom
Synechocystis sp.	blue-green
Synura sp.	chrysophyte
Ulothrix sp.	green
Westella sp.	green

Because of taxonomic uncertainties, some forms could not be identified to species, but because they were consistently recognizable they have been kept separate in the data base. See the Appendix for quantitative data.

An unusual summer, 1983

Massive blue-green algal blooms do not develop every summer in Lake Mendota, nor do high chlorophyll values always occur. During the summer of 1983, after the present study was completed, a dramatically low chlorophyll concentration occurred in Lake Mendota. Because of the interest in quantifying this unusual event, some chemical and biological assays were made. It was found that in the middle of July the whole-lake chlorophyll value was 2200 kg/lake, and in early August the chlorophyll value was even lower, 1700 kg/lake. These values are almost *10 times lower* than peak values, and 5 times lower than the mean values, shown in Figure 4.1. The Secchi disk values also indicated an unusual degree of water clarity. During this same period of low chlorophyll, nutrient levels *remained high,* total phosphorus being about 50,000 kg and total inorganic nitrogen about 140,000 kg/lake. The dominant alga of this greatly diminished bloom was *Aphanizomenon,* with minor amounts of *Anabaena,* and *Microcystis,* yet none of these algae reached their normal summer highs, despite the high nutrient levels in the lake. Why?

Unfortunately, no definitive explanation for this atypical situation is available. We note, however, that the weather the previous winter and spring were very unusual. The winter season ended early with a dramatic warming in late February and early March which resulted in the earliest ice-off on record and the shortest ice cover for the 125-year record of ice on Lake Mendota (see Figure 2.8). This early ice-off was then followed by an extremely cold and wet spring, so that the spring mixing period lasted from early March until early to mid May. Such a long mixing period should have caused deep and prolonged resuspension of the sediments, with major inputs of phosphorus and nitrogen to the water. During the summer of 1983, phosphorus and nitrogen values were high, yet they did not lead to the conventional summer bloom. The most interesting hypothesis to account for these facts is that the extensive sediment resuspension somehow *prevented* algal bloom development.

How might extensive sediment resuspension *inhibit* algal bloom development? One possibility is that there is some sort of inhibitory agent in Lake Mendota sediments that became added to the water in high amounts. Another possibility is that something in the sediment added to the water made a key nutrient such as phosphate biologically unavailable even though it was chemically assayable. One candidate here might be ferrous sulfide, a major constituent of the black muds that make up much of the bottom of Lake Mendota (see Chapter 3). If ferrous sulfide were brought into the oxic lake waters it would become oxidized to ferric hydroxide, a substance known to bind phosphate firmly. Unfortunately, no data are available to assess either of these hypotheses, but the existence of this unusual year emphasizes the dangers of generalizing about long-term trends in lake ecosystem biology from too few years of study.

That 1983 *was* an atypical year is shown by the fact that in the summer of 1984 the chlorophyll values were back to their normal highs, with the species list dominated as usual by the blue-green algae *Aphanizomenon, Anabaena,* and *Microcystis.*

Buoyancy of the blue-green algae of Lake Mendota

As discussed in detail by Reynolds & Walsby (1975), most planktonic species of blue-green algae are buoyant at some stage in their development, and evidence is strong that buoyancy is a major factor permitting the blue-green algae to dominate the summer phytoplankton flora of eutrophic lakes. When considering water blooms, Reynolds and Walsby (1975) have shown that lakes can be divided into two major classes with respect to light penetration and vertical mixing of the epilimnion. One group consists of lakes which are generally fairly small, deep, and not especially nutrient-rich, in which the euphotic depth is greater than the mixed depth of the lake. Because in this group of lakes light penetrates to and below the thermocline, blue-green algal blooms are generally found concentrated deep in the water, at or below the thermocline, where nutrients occur in higher concentrations as a result of exchange between sediments and hypolimnion. In these lakes, buoyancy regulation results in a precise positioning of the algae at the optimal deep site. In the other class of lakes, which are generally large, if nutrient enrichment results in high phytoplankton standing crop, self-shading by the high phytoplankton populations results in light penetration to depths less than the mixed zone. Wind-driven mixing in such large lakes is extensive, and frequently brings the phytoplankton into darkness. In such lakes there is strong selection pressure toward buoyant phytoplankton species that are capable of remaining near the surface of the lake. In these lakes, less precise buoyancy regulation is needed, since the algae merely rise to and concentrate near the surface. It is in this latter class of lakes that Lake Mendota belongs.

Buoyancy regulation by planktonic blue-green algae in Lake Mendota was studied in some detail by Dr. A. Konopka in collaboration with Dr. A. E. Walsby (Konopka et al., 1978). In this study, the buoyancy was assessed by laboratory measurements of samples taken directly from the lake. In order to carry out these buoyancy measurements, 1 liter samples of water were placed in glass cylinders (35×6.5 cm) for two hours at 22°C. After this period to allow the algae to assume their position in the water, the top, middle, and bottom thirds of the water were separated by aspiration and the chlorophyll content of each third determined. This method gives a measure of the proportions of algal colonies which were positively or negatively buoyant. Buoyancy distribution was expressed as a buoyancy index, which equaled the percentage of chlorophyll in the top third minus the percentage in the bottom. The chlorophyll in the middle third of the column would be due primarily to algae which were neutrally buoyant. Positive values of the buoyancy index would indicate that more colonies were floating than sinking, and the magnitude of the index would increase as the velocity of movement and percentage of buoyant colonies in the population increased. Negative values would indicate that more colonies were sinking than floating. Details of the method used can be found in the paper by Konopka et al. (1978). In some studies, buoyancy measurements were made of samples that had been irradiated for a period of time with artificial light of various intensities. For these latter buoyancy measurements, after irradiation the cylinders were incubated for 16–18 hours at 4°C before separation of the fractions for assay. This time interval was sufficient to allow equilibrium, and the low incubation temperature retarded any changes in buoyancy that might have occurred due to the lengthy incubation period alone.

The buoyancy study of Konopka et al. (1978) was done during the summer of 1976, when the summer phytoplankton population was dominated by the species *Aphanizomenon flos-aquae*. Buoyancy indices of phytoplankton populations collected at various times of day throughout the summer indicated that the algal colonies collected in the morning were *always* buoyant, although their rate of flotation varied during the summer. Thus, the organisms were able to float to the surface during a period of calm weather, and if the calm weather occurred when the algae were extremely buoyant, a surface accumulation would quickly develop. Whether or not this surface accumulation occurred depended to an important degree on the extent of vertical mixing in the water. For example, if there had been a windy period, so that the epilimnion was well mixed, the algae would be distributed throughout the whole epilimnetic layer. If, after such mixing, a calm period ensued, the algal colonies would quickly rise to the surface. In one series of observations during a windless day nearly 40% of the chlorophyll in the water column accumulated at the surface in a time period between 0850 and 1330 hours.

If the calm conditions persisted for a day or more, *Aphanizomenon* was able to regulate its buoyancy during the day. The population would be concentrated in the surface waters during early morning, but as the day progressed and solar radiation became more intense, a subsurface population maximum would develop, being at its peak by afternoon. Buoyancy indices of algae that were collected during the afternoon showed that there was a decrease in buoyancy concomitant with the downward movement.

Studies were initiated to determine the effect of light intensity upon buoyancy of the *Aphanizomenon* population. These studies were carried out by placing samples of water in clear glass containers at different depths in the lake or by exposing samples to artificial light at different intensities. When the algae were exposed to high light intensities they became nonbuoyant, and it was shown that this was due to a collapse of gas vesicles that were responsible for buoyancy.

The results obtained in Lake Mendota fitted very well with the model of the role of buoyancy in water blooms proposed by Reynolds and Walsby (1975). According to these authors, two advantages could be envisioned for buoyant blue-green algae during calm periods in the lake: 1)The organisms trapped below the euphotic zone by a previous mixing event could rapidly migrate to the more productive surface waters, and 2) the algal colonies would not sink out of the epilimnion into the hypolimnion, from which mixing would never bring them back into the photic zone until fall turnover.

The velocity of upward migration by buoyant algae is determined by two factors: 1) the buoyancy of the organism, and 2) the radius of the colony. The radius of the colony influences buoyancy as described by Stokes' equation, in which upward migration increases as the square of the radius. Species of blue-green algae inhabiting Lake Mendota can form large colonies, and are thus capable of rapid vertical movement. For example, the rate of increase in surface chlorophyll concentration on August 5, 1976 led to an estimate that the algae rose at a rate of 277 μm/sec. The large colony size is most advantageous in unstable water masses, as it allows organisms to migrate quickly both upwards and downwards over a large distance to a favorable depth, with only a small change in density. However, large-colony size is disadvantageous for organisms that stratify in the deep waters of small lakes, in which bloom formation occurs primarily at the thermocline or hypolimnion, since it would make precise depth control difficult. Under these latter conditions, accurate stratification can be accomplished only by filaments which, by virtue of their small size, move slowly through the water.

It can thus be concluded that the function of gas vesicles in the blue-green algae inhabiting Lake Mendota is to keep the algae suspended in the epilimnion during calm periods, and to provide a means of ascending rapidly to the euphotic zone after turbulence has carried the algae into relative darkness. It is quite clear that under conditions when the lake is stratified there is a strong selective pressure in favor of blue-green algae as compared to nonbuoyant eucaryotic algae. During the spring and fall of the year, however, when complete mixing to the bottom of the lake occurs, it is less obvious that the blue-green algae would be at a selective advantage. Indeed, blue-green algal blooms never occur in Lake Mendota in the spring. Although blue-green algae are found in the fall in Lake Mendota, very frequently the populations found at this time developed during the previous summer and have been able to maintain continued populations through the fall period. Conceivably, summer blue-green algal blooms would be low or absent if excessive mixing due to high winds occurred. It is possible that those years in which blue-green algae were not common (1978 and 1981— see Figure 4.4) were unusually windy years.

Where do the algae go?

Figures 4.1 and 4.4 show that large increases in phytoplankton standing crop are generally followed by rapid decreases. Such decreases are often of the order of 50% over a period of 10 days to 2 weeks. It is of interest to determine the causes of such declines. Possible causes for the declines include sinking, decomposition, autolysis, wash-out through the lake outlet, consumption by herbivores, pathogenic infection, and photooxidative lysis. Zooplankton grazing is probably the most significant factor in loss of the small-celled eucaryotic algae that dominate the spring bloom. However, little work has been done on the causes of the decline of the blue-green algae which often dominate the summer phytoplankton flora. A knowledge of the factors influencing such declines is of value for lake management studies, because the decline of massive blooms is often accompanied by detrimental changes in lake chemistry, with impact on the whole ecosystem. Also, from the standpoint of production ecology, the blue-green algae are often the dominant primary producers in Lake Mendota, and it would be desirable to know how the organic matter fixed by these organisms during photosynthesis is transferred to other components of the food chain.

An extensive study of the factors influencing the decline of blue-green algal populations in Lake Mendota was carried out by Fallon (1978). The periodic declines in blue-green algal standing crop could be accounted for primarily by two factors: epilimnetic decomposition, and sedimentation. For *Aphanizomenon* and *Anabaena,* decomposition appeared to be primarily important, whereas sedimentation accounted for more of the decline of *Microcystis.* Note that although decomposition could include zooplankton feeding as one component, there is no evidence that zooplankton eat the very large colonies of the Lake Mendota blue-green algae.

Fallon and Brock (1979) measured decomposition of natural populations of the blue-green algae of Lake Mendota by chemical and microscopic assessment of algal standing crop and by mineralization of ^{14}C-labeled axenic cultures of *Anabaena.* Of the various genera present in the lake, *Aphanizomenon* and *Anabaena* were more sensitive to decomposition than was *Microcystis.* It was concluded that decomposition of blue-green algal material was a rapid process that could occur under both aerobic and anaerobic conditions. Mineralization of the blue-green algal material resulted in the liberation of soluble phosphate and nitrogen. The nitrogen species first liberated was ammonia, but nitrification could occur, resulting in the conversion of some or most of the ammonia to nitrate. Decomposition and mineralization were similar under aerobic and anaerobic conditions, provided that a population of bacteria adapted to the particular conditions was available. However, in hypolimnetic waters, where conditions were anaerobic, high population densities of anaerobic bacteria were not generally present so that the rate of anaerobic decomposition was quite slow. If, however, a sediment-water mixture was used which contained a large anaerobic population, then anaerobic decomposition rates were quite similar to aerobic rates, even though the temperature of the anaerobic system was lower than that of the aerobic system. In Fallon's work, first-order decay constants were estimated for blue-green algal populations collected at different times of the year and an average decomposition rate could be calculated which was only a function of the temperature of the water. Such first-order decomposition rates were then used in making estimates of the importance of decomposition in the lake water itself.

Concerning the mechanism of blue-green algal destruction, little work has been carried out. Fallon and Brock (1979) made a study of the effect of lytic organisms (bacteria and protozoa capable of attacking and lysing blue-green algae) and photooxidative effects. Photooxidation has been considered to be a significant factor in the destruction of blue-green algae because these organisms are often exposed to high light intensities at the surface of the lake (see the previous section on buoyancy). Although Fallon and Brock (1979) were able to show that the blue-green algae were susceptible to photooxidative effects, these effects generally influenced only a small percentage of the lake population and thus were probably not important in causing major declines in algal biomass. Photooxidation was not significant because the blue-green algae were generally well mixed throughout the epilimnetic portion of the lake under most conditions. In lakes with a smaller fetch than Lake Mendota, surface accumulations might be more frequent and photooxidative effects more significant.

Lytic organisms were assessed by a colony counting method which probably underestimated the significance of these organisms. Lytic organisms were shown to increase in numbers in the lake in response to the seasonal development of blue-green algae, but it could not be determined whether these lytic organisms had a major role in algal biomass declines. Microscopic examination of *Aphanizomenon* filaments throughout the rise and crash of an *Aphanizomenon* bloom (Brock, unpublished observations) have shown that bacterial colonization of *Aphanizomenon* filaments only occurs when the *Aphanizomenon* filaments already appear senescent (as indicated by the short lengths of the filaments). Thus, it would seem that the development of lytic organisms may be more a *result* of senescence than a *cause* of this process.

Fallon made extensive measurements of the sedimentation of blue-green algae from the water column. In this work, integrated lake sampling techniques and sedimentation traps were used to evaluate the relative importance of two factors: sedimentation, and epilimnetic decomposition. The details of this work are described in Fallon and Brock (1980). The material caught in the sediment traps was subjected to both chemical analysis and microscopic evaluation, the latter used to determine the amount of blue-green algae present. A special technique was developed to permit microscopic assessment of the material caught in the sedimentation traps. This required the removal of interfering inorganic material by sucrose-gradient centrifugation. The blue-green algal material recovered in this way was quantified using the epifluorescence microscopic technique of Brock (1978). In addition to microscopic quantification, the uncentrifuged sediment trap material was analyzed for organic carbon, phosphorus, protein, carbohydrate, and chlorophyll by standard procedures. Some of the sedimentation rates determined in this work have already been discussed in Chapter 3 and Figure 3.8 presented data on the chemical constituents found in the sedimentation traps over two annual cycles. Figure 4.8 presents data on the sedimentation of blue-green algal material during these same two years. The peak of sedimentation intensity of blue-green algae was during the summer, which was also the period when blue-green algal standing crop was the highest in the epilimnion. However, there are differences in the genera of blue-green algae found in the sedimentation traps and those found in the epilimnion. As seen in Figure 4.8, *Microcystis* seems to preferentially collect in the sedimentation traps, whereas *Aphanizomenon* and *Anabaena*, often major components of the blue-green algal standing crop (Figure 4.4), do not appear to such a great extent in the traps. Reynolds (1976), in a shallow British lake, also found *Microcystis* to have a higher sedimentation rate than any of the other genera that he observed. However, although *Aphanizomenon* and *Anabaena* may decompose more readily in the water column than *Microcystis*, another factor may be that they retain their buoyancy longer during senescence.

Relative importance of decomposition and sedimentation

The rapid declines in algal standing crop illustrated in Figures 4.1 and 4.4 could be due to either sedimentation or decomposition in the epilimnion.

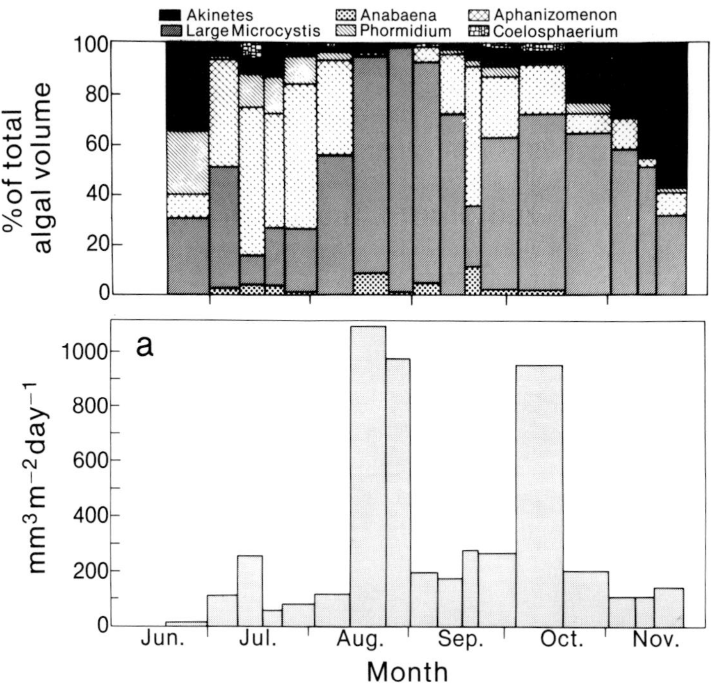

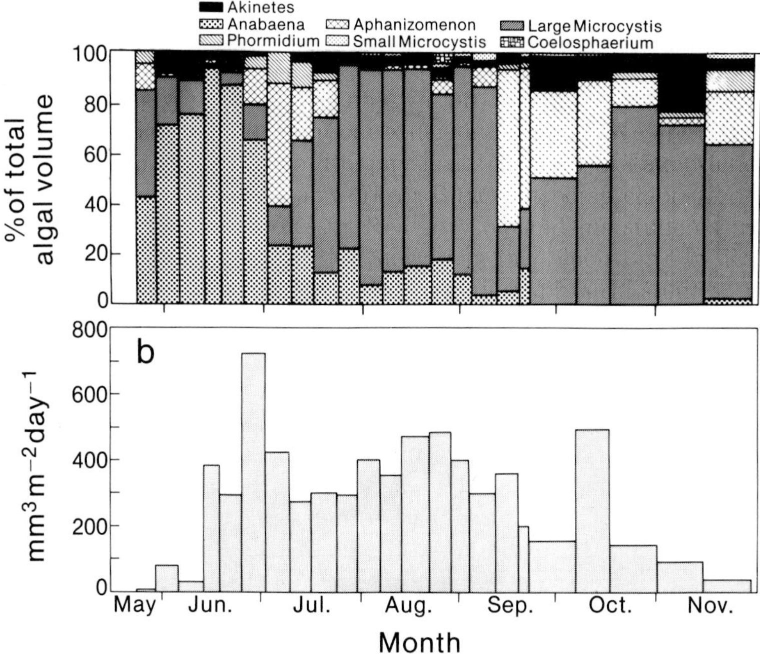

Figure 4.8. Sedimentation of blue-green algae and relative proportions of various genera in sedimentation trap collections. a) 1976; b) 1977.

(Wash-out of the lake is not a factor because of the long residence time.) In the study of Fallon and Brock (1980), an attempt was made to estimate the relative importance of these two processes during the decline phases of blue-green algal blooms. By comparing the chemical and biological composition of material in sediment traps and that in the water column during periods when blue-green algal standing crop was declining rapidly, it could be shown that sedimentation could account for a large portion of the decreases. Based on an analysis of the sedimented material and estimates of primary production in the water column (see Chapter 5), it could be estimated that approximately 30% of the organic material produced in the photic zone during the summer period became sedimented to the mud surface. Most of this material was still readily degradable at the mud surface so that it was not completely removed from the ecosystem. Studies on recycling of this material as a result of sulfate reduction and methanogenesis will be discussed in Chapter 7. Before permanent burial, another 20–30% of the material originating in the photic zone was remineralized, leaving only about 11% of the original organic material to accumulate in the lake sediments. The sediment trap studies thus showed that about 70% of the organic material produced by the blue-green algae in the epilimnion was decomposed in the epilimnion and presumably became transferred to other members of the food chain.

Summary

In this chapter we have described the dynamics of phytoplankton in Lake Mendota and have shown that there are marked year-to-year variations in density and species composition. Detailed data are given in the Appendix. In each year, a spring bloom occurs just about the time of ice-out, which is usually dominated by small flagellates or diatoms. Silica depletion causes a cessation of the spring diatom bloom. After a clear-water period in May, a summer bloom follows which is usually, but not always, composed of blue-green algae. We will make some comparisons between the 1976–81 data and earlier phytoplankton data sets in Chapter 9, but we note here that short-term variability is so marked that it is difficult to discern any long-term trend over the 6 years of the present study. We also noted the unusual year of 1983, in which chlorophyll values were 5–10 times lower than normal.

One striking observation was that the surface/volume ratio of the phytoplankton was low in spring and increased during the summer. This increase in summer was due primarily to the increase in small diameter filamentous blue-green algae. It is suggested that the high surface/volume ratio may increase the rate of nutrient transport, and it may thus be significant that the surface/volume ratio increased at a time that the inorganic nutrient concentration in the lake decreased.

Assay of Lake Mendota waters for key algal nutrients (Table 4.1 and Chapter 3) shows that phosphorus is always in excess of nitrogen if reckoned from N:P ratios found in algal cells. Under some conditions, phosphorus may be limiting for algal growth. Under other conditions, nitrogen may be limiting, whereas under still other conditions, light may be the limiting factor. We discuss the relationship of Lake Mendota phytoplankton to light in the next chapter, which also discusses primary production.

References

Brock, T.D. 1978. Use of fluorescence microscopy for quantifying phytoplankton, especially blue-green algae. *Limnology and Oceanography*, 23: 158–161.

Fallon, R.D. 1978. *The planktonic cyanobacteria: their sedimentation and decomposition in Lake Mendota, Wisconsin.* Ph.D. Thesis, University of Wisconsin-Madison.

Fallon, R.D. and T.D. Brock. 1979. Decomposition of blue-green algal (cyanobacterial) blooms in Lake Mendota, Wisconsin. *Applied and Environmental Microbiology*, 37: 820–830.

Fallon, R.D. and T.D. Brock. 1980. Planktonic blue-green algae: production, sedimentation, and decomposition in Lake Mendota, Wisconsin. *Limnology and Oceanography*, 25: 72–88.

Fogg, G.E. 1975. *Algal cultures and phytoplankton ecology, 2nd Edition.* University of Wisconsin Press, Madison.

Harris, G.P. 1980. The measurement of photosynthesis in natural populations of phytoplankton. pp. 129–187 In *The Physiological Ecology of Phytoplankton* (I. Morris, editor). Blackwell Scientific Publications, Oxford.

Healey, F.P. 1973. Inorganic nutrient uptake and deficiency in algae. *Critical Reviews in Microbiology* 3: 69–113.

Konopka, A. and T.D. Brock. 1978. Changes in photosynthetic rate and pigment content of blue-green algae in Lake Mendota. *Applied and Environmental Microbiology* 35: 527–532.

Konopka, A., T.D. Brock, and A.E. Walsby. 1978. Buoyancy regulation by planktonic blue-green algae in Lake Mendota, Wisconsin. *Archive für Hydrobiologie* 83: 524–537.

Morris, I. 1980. *The physiological ecology of phytoplankton.* Blackwell Scientific Publications, Oxford.

Pedrós-Alió, C. and T.D. Brock. 1982. Assessing biomass and production of bacteria in eutrophic Lake Mendota, Wisconsin. *Applied and Environmental Microbiology* 44: 230–218.

Redfield, A.C. 1958. The biological control of chemical factors in the environment. *American Scientist* 46: 205–221.

Reynolds, C.S. 1976. Sinking movements of phytoplankton indicated by a simple trapping method. 2. Vertical activity ranges in a stratified lake. *British Phycological Journal* 11: 293–303.

Reynolds, C.S. 1984. *The ecology of freshwater phytoplankton.* Cambridge University Press, Cambridge.

Reynolds, C.S. and A.E. Walsby. 1975. Water blooms. *Biological Reviews* 50: 437–481.

Shoaf, W.T. and B.W. Lium. 1976. Improved extraction of chlorophyll A and B from algae using dimethyl sulfoxide. *Limnology and Oceanography* 21: 926–928.

Stauffer, R.E. 1985. Sampling strategies and associated errors in estimating epilimnetic chlorophyll in eutrophic lakes. *Water Resources Research.*

Strickland, J.D. and T.R. Parsons. 1972. A practical handbook of seawater analysis. *Bulletin of the Fisheries Research Board of Canada* No. 167.

Talling, J.F. 1957. The phytoplankton population as a compound photosynthetic system. *New Phytologist* 56: 133–149.

Torrey, M.S. and G.F. Lee. 1976. Nitrogen fixation in Lake Mendota, Madison, Wisconsin. *Limnology and Oceanography* 21: 367–378.

Trelease, W. 1889. The "working" of the Madison lakes. *Transactions of the Wisconsin Academy of Sciences* 7: 121–129.

Vigon, Bruce W. 1976. *The role of silica and the vernal diatom bloom in controlling the growth of nuisance algal populations in lakes.* M.Sc. Thesis (Water Chemistry), University of Wisconsin, Madison.

Walsby, A.F. and C.S. Reynolds. 1980. Sinking and floating. pp. 371–412 In *The physiological ecology of phytoplankton* (I. Morris, editor). Blackwell Scientific Publications, Oxford.

Yentsch, C.S. 1980. Light attenuation and phytoplankton photosynthesis. pp. 95–127 In *The physiological ecology of phytoplankton* (I. Morris, editor). Blackwell Scientific Publications, Oxford.

5. Phytoplankton photosynthesis and primary production*

We discussed in the previous chapter the nature, diversity, and cycles of development of the phytoplankton of Lake Mendota. In the present chapter, we continue a discussion of the phytoplankton of Lake Mendota, with emphasis on the photosynthetic process and on primary production. In this chapter we will consider the factors controlling primary production and photosynthetic efficiency, the manner in which available light is converted into organic matter, and the role of primary production and photosynthetic efficiency in the development of algal blooms.

The efficiency with which algae convert light energy to chemical energy may be assessed under optimum light conditions (physiological maximum efficiency) and under natural conditions (ecological efficiency). The former parameter provides an assessment of the physiological state or "physiological health" of the population. The latter parameter provides an estimate of how efficiently the lake ecosystem as a whole utilizes the primary energy source, light. At the end of this chapter a conceptual model will be presented which attempts to assess the relative importance of various factors that might affect primary production and photosynthetic efficiency.

* Material for this chapter was prepared in collaboration with Dr. Vicki Watson.

Methods

Methods for sampling and assessment of phytoplankton biovolume and chlorophyll have been described in the previous chapter. For most of the work on photosynthesis and primary production, the ^{14}C method was used. Although both *in situ* and laboratory incubation methods were used, the incubator method was used most extensively because it was more reproducible and provided a more detailed assessment of the activity of the phytoplankton population. The incubator method used was based on that described by Fee (1977) and Shearer and Fee (1974).

Sampling

In a well-mixed water body such as Lake Mendota, a single sample from one depth is generally representative of phytoplankton throughout the water column. For convenience, a 2 m water sample was taken. This depth was chosen because it was shallow enough to sample easily, and it was deep enough to avoid unusual surface accumulations or films. (In a lake where considerable vertical inhomogeneity exists, a series of water samples at representative depths would have to be taken.) As shown in the previous chapter and in the detailed data presented in the Appendix, although there is some vertical stratification of phytoplankton from time to time, this vertical stratification never remains for any length of time so that a 2 m sample should be representative of the *average* phytoplankton population of the epilimnion. For the actual ^{14}C procedure, only 400 ml of water was needed, but a two liter sample was generally taken because a chlorophyll measurement was also necessary in order to perform the final calculations. The sample was taken with an opaque Van Dorn water sampler and was immediately transferred to a clean plastic bottle and placed in the dark in an ice chest. Care was taken to avoid exposing the sample to bright light and to avoid changing the temperature significantly. The water sample was returned to the laboratory immediately and the incubations begun within one hour of sampling.

At each sampling time, the light intensity at various depths in the lake was measured. The light meter used was a Li-Cor Quantum Sensor, which provided a measure of photosynthetically active radiation over the range 400–700 nm. Measurements were made at 1 m intervals until a depth was reached at which the intensity was very low. This light profile was used to calculate the extinction coefficient, a parameter needed in the calculation of primary production.

Incubation procedure

Two large aquaria, about 15 gallon capacity, were used for the incubation. A copper coil was inserted into each aquarium to permit circulation of cold tap water. The water circulation was started well in advance of the time of use and ice was used if necessary to reduce the temperature. In all cases the temperature used was within a degree of the temperature of the lake at the time the sample was taken. Because of the short incubation period (one hour), it was not difficult to maintain the desired temperature. Banks of fluorescent and tungsten lights were used to control light intensity. A translucent glass filter was used to reduce the light intensity in one of the aquariums. Positions were established within the aquaria for a variety of light intensities and incubation was in screw-capped test tubes 16×120 mm. The caps used contained Teflon liners, and the tubes used were carefully cleaned (acid-washed) and segregated from general laboratory glassware to avoid any contamination with detergents or laboratory chemicals. Light intensities in the aquaria were also measured with the Li-Cor Quantum Sensor. Light quality was not considered to be a major factor in Lake Mendota, because the lake has little water color, but in lakes with humic acids, a match of the light quality of the incubator to that of the lake would be necessary (Parkin and Brock, 1980; Fee, 1978).

Using a measuring flask containing a 20 ml dispenser, 20 ml of lake water was placed in each tube and three tubes were used for each light intensity. The tubes were kept on ice in the dark until the incubation was begun. Controls were run with each set of incubations; one control tube contained 2 ml of formalin and one tube was wrapped in black plastic tape to provide a darkened sample. The final concentration of ^{14}C used was 1 μCi/ml, provided as sodium bicarbonate. The stock radioactive bicarbonate solution contained 20 μCi/ml, and 0.1 ml of this solution was placed into each tube using a 1 cc syringe. The ^{14}C used was placed into each tube while it was in the ice bath and the tubes were then placed in the water bath at the appropriate temperature for five minutes for equilibration, after which the lights were turned on. The incubation was for exactly one hour, after which the tubes were removed and placed on ice in the dark for immediate processing.

For processing, the liquid in each tube was filtered through a 0.45 μm membrane filter and the retained particulate material washed with filtered lake water. The filters were placed in a desiccator with concentrated hydrochloric acid for one hour to remove any inorganic radioactive carbonate, after which the radioactivity on each filter was counted using a liquid scintillation counter, correcting for quenching by the channels-ratio method.

The light intensities used in the incubators ranged from about 3–50% of summer surface light (40 to 780 μEinsteins/$m^2 \cdot s$). Some representative data using this method are given in Table 5.1. The formula used to calculate carbon fixed was the following:

$$\text{mmoles carbon/l/hr} = (1.06) \times (\text{DPM}) \times (\text{mmoles C/l}) / (\mu\text{Ci/vial}) \times (2.2 \times 10^6) \times (\text{hrs})$$

where *1.06* is a factor to correct for isotope discrimination between ^{14}C and ^{12}C, *DPM* is the disintegrations per minute as measured above, *mmoles C/l* is the inorganic carbon content of the lake, as determined by chemical analysis, *μCi/vial* is the amount of isotope added to each vial (1 or 2 μCi in most cases), *2.2×10^6* is the number of DPM per μCi, and *hrs* is the incubation time in hours. If inorganic carbon in a water body varies considerably, then the amount of inorganic carbon must be determined each time the measurement is made,

Table 5.1. Representative data for ^{14}C fixation of phytoplankton in Lake Mendota and procedure for calculation of carbon fixed

Light intensity Einsteins/m^2/hr	DPM	Avg. DPM (dark-corrected)	Photosynthesis (P) mMC/1/hr
0.14	778	434	2.51×10^{-4}
	906		
	721		
0.21	1200	773	4.47
	1084		
	1136		
0.32	1789	1439	8.32
	1693		
	1936		
0.70	2282	1610	9.31
	1752		
	1896		
1.45	2673	2246	12.9
	2669		
	2494		
2.81	3066	2806	16.2
	2929		
	3524		
Dark	367		
Formalin	376		

Time of sampling: June 18, 1980.
DPM, disintegrations per minute: the value from each incubated vial is listed in this column.
Avg. DPM: the dark value is subtracted from each light value and the three numbers averaged.
Photosynthesis (P): the carbon fixed is calculated from the average DPM as described in the text.
Polynomial equation for the above data:

$$P = 0.000277 + 0.00104\ I - 0.000203\ I^2 \quad (r^2 = 92.7\%)$$

but in many systems, inorganic carbon pool sizes are so large that they show little variation. It has been found in Lake Mendota that two values for inorganic carbon can be used, one of 3.1 mmoles/l which is used from mid-October until the end of May, and one of 2.4 mmoles/l which is used from June through mid-October. Calculations of the carbon fixed using the raw data in Table 5.1 and the formula just given are also presented in Table 5.1. Note that in these calculations the carbon fixed by the dark control has been subtracted from the other values in order to obtain a calculation of the light-stimulated carbon fixed.

Calculating whole lake productivity from incubator data

Incubator data such as that presented in Table 5.1 can be used to calculate carbon fixed for the whole lake (primary production). The original procedure for productivity calculation was described by Fee (1969) and an example of a detailed Fortran computer program is given in Fee (1977). However, the procedure used for Lake Mendota was considerably simplified over that of Fee, and is suitable for use on a microcomputer, so that data can be calculated at remote sites. The principle of the method is as follows:

1) The solar radiation falling on the surface of the lake was either measured (1976 and 1977, U.S. Weather Bureau data) or calculated (1978–81) for the day and hour in question (Brock, 1981). 2) The extinction coefficient of the water was calculated from the underwater light meter measurements. 3) Using 1 and 2, the light at each depth was calculated, assuming that light extinction was an exponential function (which it nearly always was). 4) Using a polynomial equation generated from the photosynthesis/light intensity (P/I) curve (see below), the amount of carbon fixed at each depth selected in 3 is calculated (a simple rectangular integration is of sufficient accuracy). 5) The calculated values are summed over the whole water column, to give whole lake production for that hour. 6) The calculated productivity is corrected for deviation of the phytoplankton density of the incubated 2 m sample from the chlorophyll at each depth or the whole lake average by determining the ratio of chlorophyll in the 2 m sample to the chlorophyll content at each depth or the average chlorophyll content for the photic zone. 7) Values for each hour are summed to provide a daily value. The logic of the procedure is illustrated in Figure 5.1, and the detailed procedures for carrying out these calculations are described below.

The polynomial relating light intensity (I) to carbon fixed (P)

The first step in the calculation of primary production is the conversion of the P/I data into a polynomial equation, using a standard computer curve fitting program such as Minitab. A second-order polynomial equation was suitable for this work and was of the following form:

$$P = a + bI + cI^2$$

where P is productivity (mmoles C fixed/l·hr), I is light intensity (Einsteins/m^2·hr), and a, b, and c are constants for a given P/I curve. Polynomial curves were used rather than some of the other formulations that appear in the literature (Jassby and Platt, 1976; Chalker, 1980) because they give a simple but accurate representation of the photosynthesis-light relationship observed in Lake Mendota at the light levels of interest. Such a polynomial fits the data well at moderate and high light levels, but not at very low light levels. However the amount of photosynthesis at low light levels (when P approaches a in the equation above) is so low that little error is introduced into the estimate of total annual production. The polynomial equation for a typical data set is given in Table 5.1.

Calculating light intensity at any depth

To calculate productivity through the water column, it is necessary to determine the light intensity at a series of depths. Light intensity at any depth is a function of the surface light intensity and the extinction coefficient of the water.

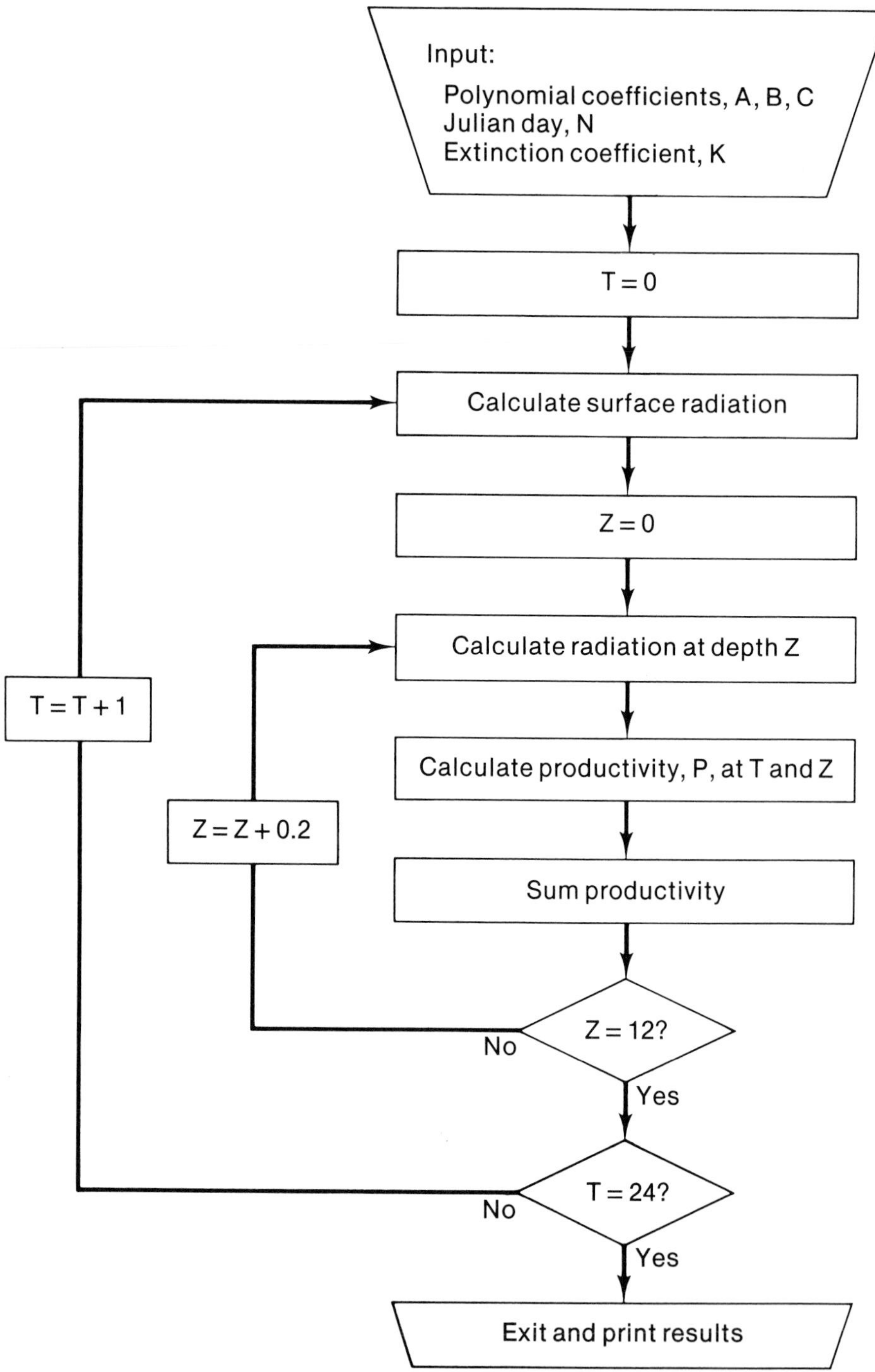

Figure 5.1. Flow sheet for primary production calculations using the incubator method.

The extinction coefficient, k, of the lake water at the time the sample was taken was calculated by fitting the in-lake light measurements to a negative exponential function of the form:

$$I = I_o e^{-kz}$$

where I is the light intensity at any depth, I_o is the light intensity at the surface, and z is the depth in meters. To use actual field data to calculate k, a set of light measurements at a series of depths is transformed by taking the natural logarithms and a least squares approximation is used relating lnI to z. In most cases, the correlation coefficient for the linear regression of the transformed data is very close to a perfect −1.00.

For data from 1978 through 1981, surface light intensity was computed hourly for each sampling day using the method of Brock (1981), yielding photosynthetically active radiation in Einsteins/m²·hr. Using this surface light intensity and the extinction coefficient for that day, light was calculated for every 0.2 m depth through the water column. Light at a given depth was then substituted into the polynomial P/I equation to obtain an estimate of carbon fixed at that depth.

The next step (not shown on Figure 5.1) was to correct for deviation of chlorophyll in the incubated sample from the lake average. The estimate of carbon fixed was multiplied by the ratio of chlorophyll concentration at the depth in question to the concentration at 2 m (from where the sample for incubation was taken). A chlorophyll correction was necessary because it was rare that the lake was so completely mixed that the 2 m chlorophyll would be equivalent to the chlorophyll at other depths. Because chlorophyll content was usually measured at a series of sampling depths each time the primary production measurements were made (see Appendix), it was possible to use these values in the chlorophyll correction. If the lake was stratified, then the chlorophyll content of just the epilimnion was used, whereas if the lake was completely mixed the total lake chlorophyll value was used. (A chlorophyll correction that is almost as accurate and is simpler to program is to determine the deviation of chlorophyll of the incubated 2 m sample from the average chlorophyll concentration of the photic zone. The photic zone is the epilimnion during stratified periods or the whole lake during the rest of the year.) Carbon fixed was integrated over time of day and depth, without correcting for lake morphometry to obtain areal production (in g C/m²·day), and with morphometry correction to obtain whole lake production (in kg C/day). A computer program, based on the logic in Figure 5.1 was used for most of these calculations.

The important assumptions of this method are:

1. Conditions in the laboratory simulate those in the lake for factors relevant to carbon fixation.
2. Algae taken in mid-lake at 2 m depth have a P/I curve fairly typical of algae throughout the mixed zone of the lake; also the P/I curve does not change significantly over the day. Both of these assumptions have been

shown to be reasonable for Lake Mendota from more extensive measurements made during the summer period.

3. Most production occurs at the mean temperature of the mixed zone. This would be a poor assumption for a lake in which the photic zone extended into the metalimnion or hypolimnion, but it is reasonable for Lake Mendota.
4. The chlorophyll profile and extinction coefficient at the sampling site are representative of the lake as a whole. An assessment of data from a number of sites in 1976 and 1977 (see Chapter 4) suggests that this is a reasonable assumption.
5. The extinction coefficient and chlorophyll profile do not change significantly over the day.
6. The method of estimating solar radiation includes an estimate for an average day near the time that the estimate was made but does not provide a measure of actual light intensity on the day of sampling. Thus, it is assumed that over the seven-day period around the sampling date, average cloud cover occurs—a reasonable assumption for the Madison area (Brock, 1981).

Photosynthesis/light (P/I) relationship

The photosynthesis-light relationship based on ^{14}C data of the type illustrated in Table 5.1 can be fitted to second-order polynomial curves with a high degree of correlation ($r^2>0.9$) during the ice-free season. In order to permit a comparison of P/I curves obtained at different times of the year, curves normalized for chlorophyll were prepared and some representative results are given in Figure 5.2. As seen in this figure, the P/I relationship shows considerable variability over the year and cannot be reduced to a single curve that could be used for a long period of time without introducing large errors. Although each P/I curve fits a polynomial with a high degree of correlation, the degree of similarity between curves obtained for different sampling dates was low. When all the P/I data for the ice-free season were lumped together for a given year, the data exhibited a very poor fit to a second-order polynomial curve ($r^2<0.5$). Since chlorophyll concentration varies greatly over the year, the P/I relationship was normalized for chlorophyll but this improved the fit only slightly. A few periods of time in the summers of 1980 and 1981 were found to have P/I data that were more internally consistent (i.e., when sampling days were lumped, the curve fitted to the data had an r^2 between 0.65 and 0.8).

Although satisfactory fitting of the photosynthesis data to polynomial curves could be obtained throughout most of the year, the winter data provided an exception. There was often no obvious relationship between light intensity and carbon fixed during the winter period. Photosynthesis was frequently low throughout the whole winter period relative to the ice-free season and showed neither an increase nor a decrease as the light intensity was altered to the lowest intensity used in the incubator. Presumably temperature rather than

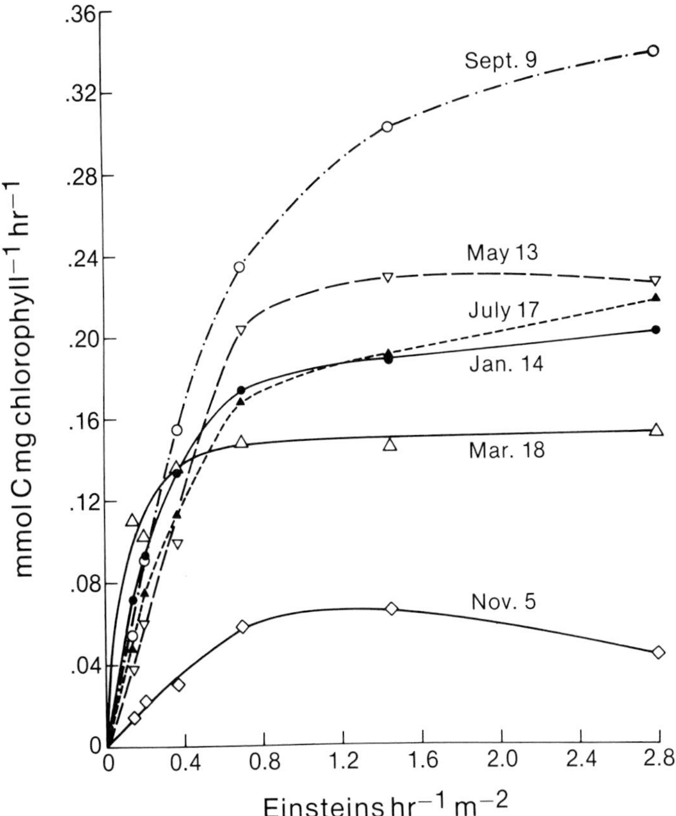

Figure 5.2. Photosynthesis/light intensity curves from incubator data for several dates in Lake Mendota, 1980.

light was limiting for photosynthesis. Because of the very low photosynthetic activity of the phytoplankton in the winter time, and the low solar radiation, little error was introduced into the annual primary production calculations by not having a usable P/I curve for the winter data.

The physiological status of the phytoplankton

The chlorophyll-normalized P/I curves provide some insight into the physiological status of the phytoplankton populations in the lake. Especially valuable are comparisons of chlorophyll-normalized carbon fixation values at saturating light. Konopka and Brock (1978) showed that the photosynthesis at saturating light increased just before an increase in the phytoplankton population and decreased just before an algal bloom crash (Figure 5.3). Similar data using more detailed P/I measurements have been obtained with 1978–81 data. This result shows that the "physiological health" decreases just before the crash occurs.

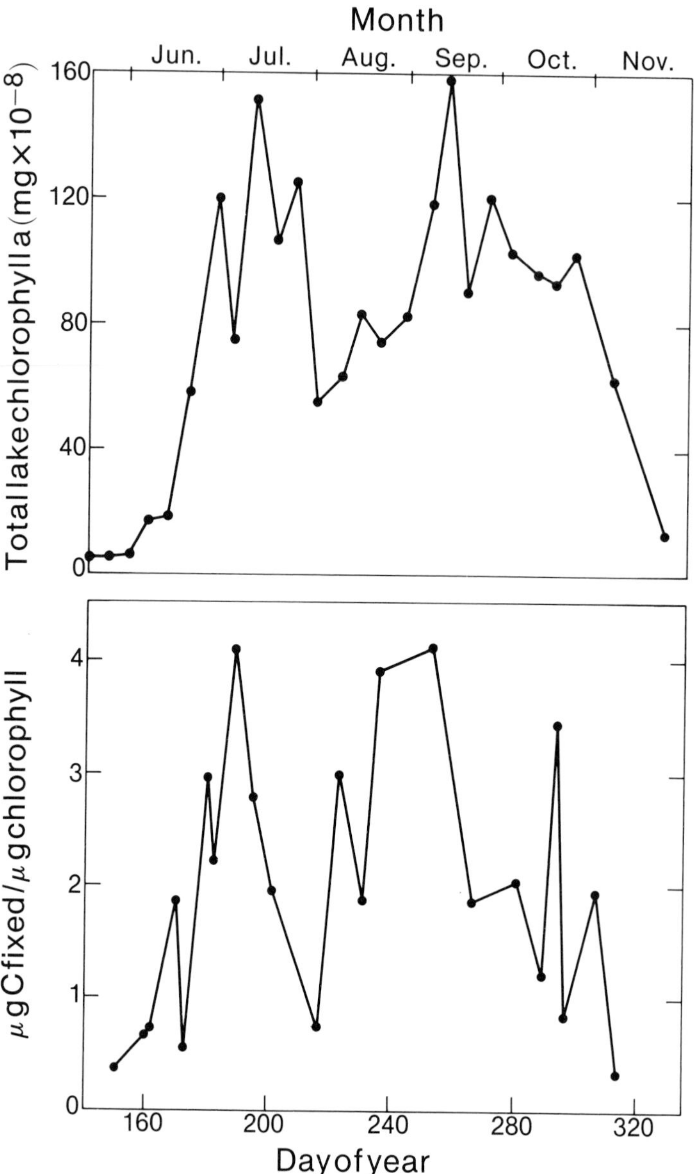

Figure 5.3. Relationship of photosynthesis rate at saturating light intensity to changes in phytoplankton population density as measured by chlorophyll. a) Chlorophyll a concentration; b) photosynthesis rate. Data for 1976. (From Konopka and Brock, 1978).

One explanation for the reduction in light-saturated photosynthesis just before a crash is that the phytoplankton become adapted to low light intensity during times when the population densities are so high that self-shading occurs. However, in the analysis of Konopka and Brock (1978) it appeared that the

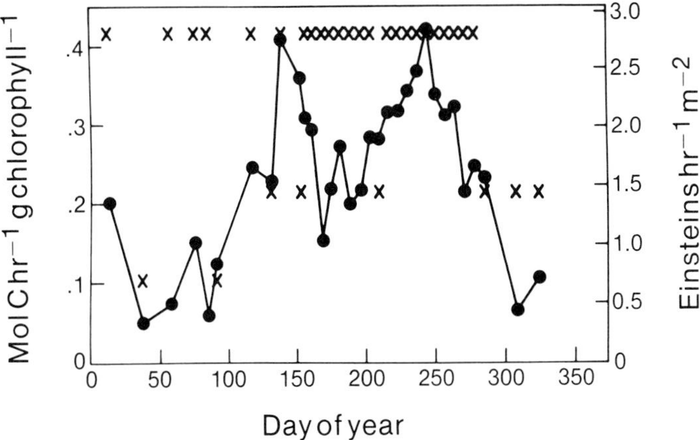

Figure 5.4. Algal production per gram chlorophyll (closed circles) measured at optimally productive light, showing the marked seasonal variation in physiological status of the algae. The light intensity which provided optimal primary production is shown by the crosses. Data for 1980. (From Watson and Brock, unpublished.)

changes in photosynthetic efficiency of the blue-green algal blooms were more related to physiological stress than to light adaptation. One possibility is that the physiological stress is brought about by nutrient deficiency during the extensive summer bloom. That nutrient deficiency was a factor in the work of Konopka and Brock (1978) was indicated by the fact that the phycocyanin content of the algae decreased during the period of extensive bloom formation, a phenomenon which is known to be related to inorganic nitrogen deficiency of blue-green algae.

More precise measurements of this change in physiological status were obtained in later years and the 1980 data, which show algal production per unit of chlorophyll at optimally productive light, are given in Figure 5.4. On this graph, the light intensity which was optimal for photosynthesis on each sampling date is also given.

Photosynthetic efficiency of Lake Mendota phytoplankton

The efficiency of conversion of light energy to chemical energy in aquatic systems has been addressed in a number of ways (Bannister, 1974; Tilzer, et al., 1975; Hickman, 1976; Dubinsky and Berman, 1976, 1981; Morel, 1978; and Morel and Bricaud, 1981). Photosynthetic efficiency has been expressed as carbon fixed per unit chlorophyll (the photosynthetic index of Hickman, 1976, the assimilation number of Bannister, 1974, and the P_B of Morel, 1978), carbon fixation per unit incoming light (Becacos-Kontos and Svansson, 1969); the photosynthetic efficiency of Tilzer, et al. (1975), carbon fixation per unit light absorbed by the water column (the E_z of Dubinsky and Berman, 1976) or absorbed by chlorophyll (the quantum yield of Bannister, 1974, and of Dubinsky and Berman, 1976). In the present work, photosynthetic efficiency was assessed in two ways. A potential or *physiological efficiency* (PE) has been calculated for a specific set of conditions, based on carbon fixed per unit volume in the incubation chamber, normalized for chlorophyll concentration (PEC) and/or for incident light (PELI) or for light absorbed by the water or the chlorophyll-containing particles. From such a physiological efficiency, one may find which conditions lead to the growth of algae with the highest efficiency and give the maximum carbon fixed per unit light or per unit chlorophyll.

Alternatively, the efficiency can be expressed as a realized or *ecological efficiency* (EE) based on the carbon fixed per unit area of lake water column. Carbon fixation of the water column integrated over depth was converted to an ecological efficiency in a number of ways: by dividing by 1) the chlorophyll in the water column; 2) the incident light per square meter; 3) the light absorbed per square meter of the water column; or 4) the light absorbed by some constituent of the water, such as the phytoplankton cells. Such ecological efficiencies are always less than their analogous physiological efficiencies, since light conditions in the lake generally differ from the optimum.

To determine efficiencies based on the light absorbed by the water column as a whole or by various constituents in the water, the total extinction coefficient, k, and extinction coefficients for various constituents have been used in connection with Beer's law to determine the amount of light absorbed by a layer of water or by its constituents (Verduin, 1982). To determine an efficiency based on the light absorbed by algae, one may assess the light absorbed by particulate material or by chlorophyll-containing particles in the water column. In Lake Mendota, particulates are predominantly living or dead algal cells (Fallon and Brock, 1980). Since the percent transmission of a mixture of two solutions is the product of the percent transmissions of each of the solutions evaluated separately, the percent transmission of the unfiltered lake water may be calculated if the extinction coefficients of the various components are known. The average extinction of filtered Lake Mendota water is about 0.3 (for photosynthetically active light), based on data given by James and Birge (1938). To estimate extinction by algae, the procedure of Bannister (1974) has been used, in which the amount of light attenuated by algae is calculated by relating the measured extinction of light in the water column to the concentration of chlorophyll-containing particles. When the extinction coefficient for Lake Mendota was regressed on chlorophyll concentration for data from 1979 through 1981, the regression yielded the following equation:

$$k=0.67+0.016\times \text{grams chlorophyll/m}^3$$

The value of 0.016 in this equation, calculated from Lake Mendota data, is the same as that calculated by Bannister from his own data. Since the intercept value of 0.67 is considerably greater than the extinction coefficient of 0.3 found for Lake Mendota water, these results suggest that frequently there is considerable particulate matter in the water column of Lake Mendota that does not contain chlorophyll, probably dead or senescent algal cells.

For the purpose of estimating efficiencies, production was expressed both in moles (or grams) of carbon fixed and in calorie-equivalents of carbon fixed. The caloric equivalents per gram dry weight of a number of algae are reported in Cummins and Wuycheck (1971). A typical value for mixed algae is around 3300 calories/gram dry weight; for a population dominated by blue-green algae, the value given is 1400 calories/gram dry weight. Based on seventeen samples analyzed over a 1.5 year period by Fallon (1978), the average amount of carbon in particulate organic matter in Lake Mendota was 8% by weight. Taking the value of 1400 calories/gram dry weight and 8% carbon, the Lake Mendota particulate matter can be calculated to have a caloric equivalent of 17.5 Kcal/gram carbon. This compares to 9.3 Kcal released per gram of carbon in the oxidation of hexose and 10 Kcal per gram carbon fixed assumed by Tilzer et al. (1975). The wide range of possible values for this conversion factor add substantial error to the estimate of the caloric efficiency.

Whole lake production normalized for whole lake chlorophyll or for absorbed light is the basis for calculation of ecological photosynthetic efficiencies. When not normalized for incident light, daily areal carbon fixation per unit chlorophyll in the water column generally follows the seasonal cycle of light and temperature (Figure 5.5), being highest in spring and early summer when light is increasing and nutrient concentrations are still high. This quantity drops throughout the late summer and fall as nutrients and then light become less available. When correlation analyses were run of carbon fixed per unit chlorophyll, the correlation was stronger with solar radiation ($r^2=0.79$) than with temperature ($r^2=0.5$). These data suggest that available solar radiation is more important than water temperature as an influence on whole-lake production.

Our data permit us to make an estimate of how much of the solar radiation falling on the lake is used productively for algal photosynthesis. Generally in Lake Mendota, all the light reaching the lake surface is absorbed by the water column in the first 12 m (the mean depth of the lake). During periods of high chlorophyll concentration, much of this light is absorbed by particulates (mainly algae). As seen in Figure 5.6, around 1 to 2% of the absorbed solar radiation is used by the algae, although there are marked variations throughout the year. Efficiency of utilization of solar radiation is highest in spring and mid- to late-summer. If this efficiency is corrected for the amount of chlorophyll present (not shown in the figure), efficiency is highest during the periods of low chlorophyll in early summer and late summer. Perhaps nutrients are less limiting at these times than in mid-summer, or perhaps the phytoplankton distribution in the water column allows more advantageous use of light. Chlorophyll-containing particles are concentrated in a thin layer in mid-summer due to buoyancy (see Chapter 4) and are more evenly distributed in early and late summer, though not mixed to such a great depth as in spring and fall. When the efficiency of light utilization is averaged through the whole year, a value of 1.5% is obtained. Thus, the lake phytoplankton population as a whole is able to capture, in ecologically usable form, about 1.5% of the solar radiation impinging on the lake surface during the whole year.

Estimates of ecological efficiency for various aquatic systems are given in

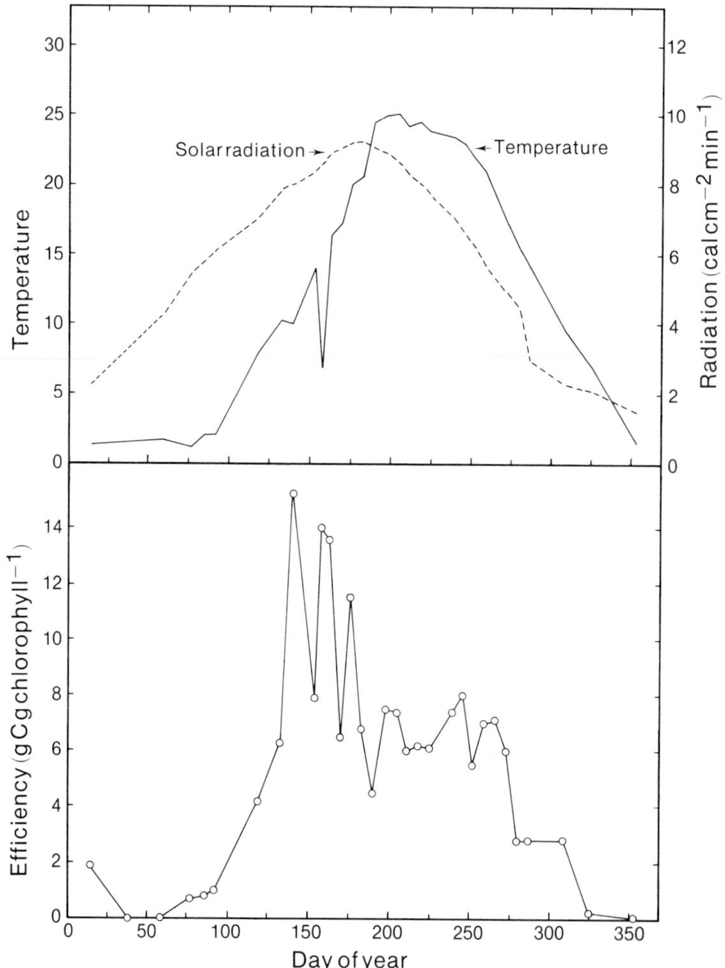

Figure 5.5. Relationship of photosynthetic efficiency to temperature and solar radiation. Upper) Temperature of the water at the mean depth of chlorophyll and solar radiation measured at the surface of the lake. Lower) Photosynthetic efficiency. Data of 1980. (From Watson and Brock, unpublished.)

Table 5.2. Most estimates in Table 5.2 taken from the literature represent efficiencies observed at a specific light and depth, while others are for the whole water column. However, few estimates are integrated over a day or year; most represent hourly efficiencies evaluated around midday. Thus, the Lake Mendota study provides the most extensive data available on efficiency of a single lake. Lake Mendota's average annual integral value (1.2% when converted to calorie equivalents) is similar to the annual integral of Loch Levin and is higher than those of other lakes assessed by Tilzer et al. (1975). Ecological efficiency for Lake Kinneret (Dubinsky and Berman, 1976) was lower (0.35%) on the

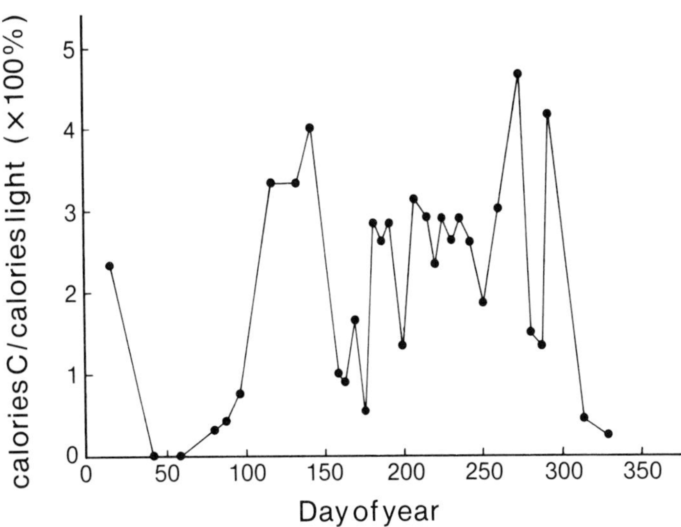

Figure 5.6. Ecological efficiency of Lake Mendota phytoplankton, expressed as calorie equivalents of carbon fixed per m^2 divided by calories of light absorbed by particulates per m^2. Data of 1980. (From Watson and Brock, unpublished.)

midsummer day when it was assessed than was true of Lake Mendota on most midsummer days (1.9%). The Ethiopian lakes investigated by Talling et al. (1973) exhibited a range of efficiencies similar to those in Lake Mendota. Talling et al. claimed that the Ethiopian lakes exhibited efficiencies near the theoretical maximum, given certain assumptions. However, when comparing these values to those of Lake Mendota, it should be kept in mind that the factor used in the present work to convert carbon fixed to calories is almost twice the factor assumed by Talling et al. and by Dubinsky and Berman.

Only a few values are available for photosynthesis normalized for light absorbed by algae, and these generally are normalized for light absorbed by chlorophyll-containing particles rather than all particulates. The annual integral of photosynthesis per unit light absorbed by particulates in the water column is about 1.5% in Lake Mendota; similar estimates are apparently not available for other lakes. In Lake Kinneret, 0.001 to 0.07 moles of carbon were fixed per Einstein of light absorbed by chlorophyll-containing particles at different depths. Although in Lake Mendota we do not have a similar in-lake value, we can use our annual value based on chlorophyll at saturating light to obtain an analogous value. Using this calculation, Lake Mendota's 1980 average value is 0.033 moles per Einstein of light absorbed. Ecological efficiencies are probably always a small fraction of physiological efficiencies, because the light available to the average algal cell in the lake is much lower than that available to the average cell in the incubation chamber. As a result, the ratio of ecological to physiological efficiency provides an assessment of the extent to which actual conditions are lower than optimum conditions.

Table 5.2. Photosynthetic efficiencies of aquatic systems

System	Efficiency	Assumptions[1]	Reference
Saronicos Gulf of Aegean Sea July & Dec.	0.08%	based on incoming radiation	Becacos-Kontos & Svansson 1969
Lock Levin	1.76%	based on incoming PHAR[2]; values are annual means of daily values; gram of C fixed=10 Kcal; values for last 3 lakes based on ice-free season only	Tilzer et al. 1975
L. Sammamish	0.42		
L. Chad	0.26		
Clear L.	0.12		
L. Tahoe	0.035		
L. Wingra	0.45		
Castle L.	0.04		
Finstertaler	0.07		
L. Kinneret[3] July 11	0.35%	based on light absorbed in epilimnion	Dubinsky & Berman 1976
	0.07–7.4	based on light at several depths in water column	
	0.32–15	based on light absorbed by chlorophyll for all of above, measurements made in late morning	
Ethiopian soda lakes[3]	0.51–3.3%	based on PHAR just below surface of water	Talling et al. 1973
Lake Mendota			
annual integral	1.2(1.5)%	based on incident PHAR (based on PHAR absorbed by particulates), gram C fixed=17.5 Kcal integrated over water column	This study
winter average	0.3(0.5)		
spring average	1.6(2.2)		
summer average	1.9(2.5)		
fall average	1.4(1.7)		
annual range	0–7(0–9)[4]		

The efficiency is calculated as calorie equivalents of carbon fixed per calorie equivalents of incident light, expressed as a percent.
[1] Indicates quality of light and whether light used in calculation was incident light or light absorbed by chlorophyll or particulates.
[2] PHAR, photosynthetically active radiation.
[3] Assumes one gram C fixed = 9.3 Kcalories.
[4] The value zero is actually a number less than 0.1%.

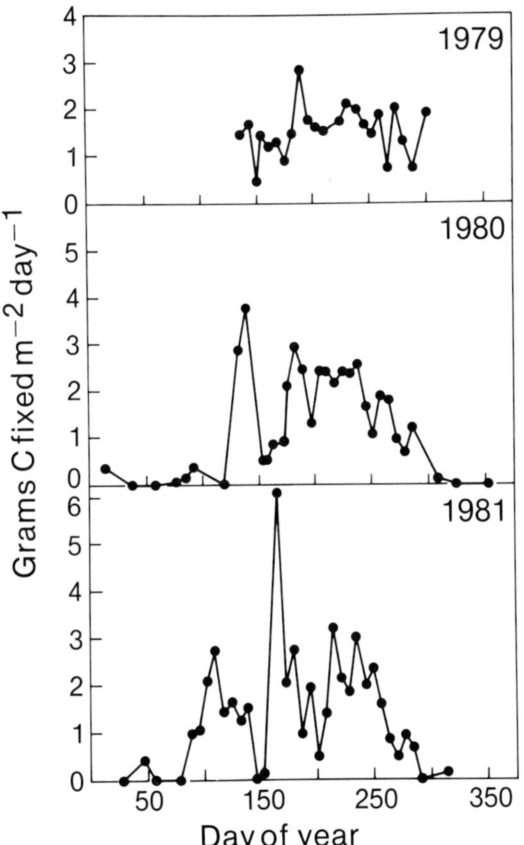

Figure 5.7. Whole lake productivity for several years, illustrating the considerable yearly and seasonal variability. (From Watson and Brock, unpublished.)

Primary production in Lake Mendota

Data on primary production in Lake Mendota have been reported earlier by Fallon and Brock (1980) for the year 1977, Konopka and Brock (1978) for the year 1976, and Pedrós-Alió and Brock (1982) for the years 1979 and 1980. Primary production data for the years 1979 through 1981 are summarized in Figure 5.7. Note that the values presented in Figure 5.7 have been corrected for lake morphometry. Fee (1980) had concluded that production estimates not corrected for morphometry *overestimated* areal production in lakes in the Canadian Experimental Lakes Area by 20%. However, in Lake Mendota, production estimates not corrected for lake morphometry are generally about 50% *lower* than these corrected values because the contribution of the highly productive surface waters is *underestimated.* This is due to the fact that most production occurs in surface waters in eutrophic Lake Mendota but in deeper waters in the more oligotrophic Canadian lakes.

Table 5.3. Primary production of lake ecosystems

Lake	Production	Assumptions	Reference
Experimental Lakes Area	6–172	Integrated over the year and over the euphotic zone, based on actual weather, corrected for morphometry	Fee 1980
Ethiopian soda lakes	0.18–0.96	Hourly values at midday, mole O_2 released = mole C fixed	Talling et al. 1973
	1.35–7.1	Daily values based on O_2 change over 2 days, based on actual weather	
Lake Erie	1.8	Daily value based on mean hourly fixation in summer euphotic zone	Verduin & Munawar 1981
Lake Wingra	345	Mean of 4 annual integrals	Prentki et al. 1977
Dec–Feb	0.13	Daily mean of each 3 month period, based on 4 years data, uses actual weather, integrated over water column, not corrected for morphometry	
Mar–May	0.58		
Jun–Aug	2.3		
Sept–Oct	0.77		
Lake Mendota	295–390	Range of 3 annual integrals	This study
Winter	0.1	Daily mean for each season based on 3 years data, uses average weather, integrated over water column, corrected for morphometry.	
Spring	1.8		
Summer	1.8		
Fall	0.6		
Range	0.6		

Production is expressed as grams of carbon fixed per square meter of water column for the time periods indicated.

The primary production data for Lake Mendota indicate that there is marked year-to-year variability in production; the only typical seasonal patterns are winter lows, summer highs, and an extreme low around day 150 when phytoplankton population densities are also low. Day 150 is usually near the end of the spring mixing period, at which time stratification first occurs (see Chapter 2). A comparison of primary production data for Lake Mendota with those of some other lakes is given in Table 5.3. Annual production for Lake Mendota (295 to 390 grams C/m^2) is similar to estimates made for Lake Wingra, Wisconsin (Prentki et al., 1977), a small eutrophic lake near Lake Mendota. Both of these lakes are much more productive than the most productive lakes in the Experimental Lakes Area (Fee, 1980). As for daily production values, the range of values and the pattern over the year for Lake Mendota were similar to those for Lake Wingra, but did not include the highest values reported for Ethiopian lakes by Talling et al. (1973). The year-to-year variability in primary production in Lake Mendota can be explained partly by the year-to-year var-

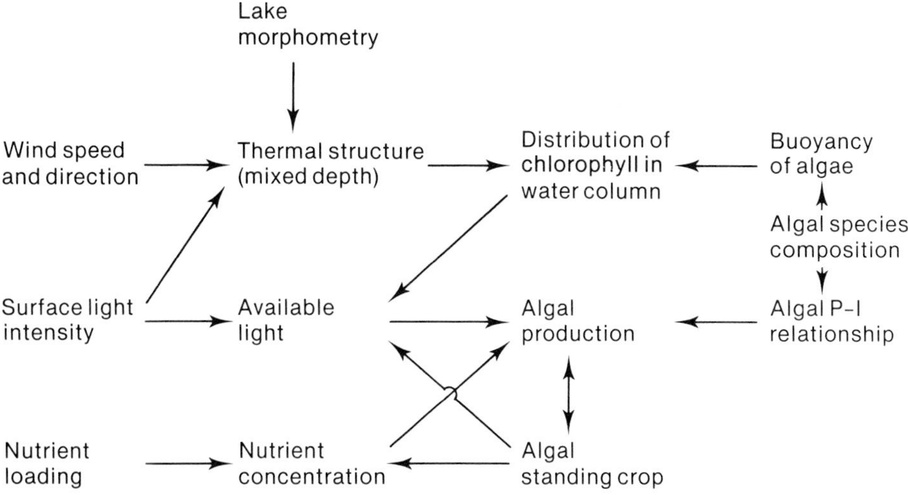

Figure 5.8. Conceptual model describing the effect of various factors on algal production. (From Watson and Brock, unpublished.)

iability in phytoplankton standing crop, and partly by the variability in efficiency of the algae.

A model for Lake Mendota production

An analysis of the Lake Mendota data, including the physical, chemical, and biological parameters described in earlier chapters, suggests that the model depicted in Figure 5.8 might be applicable. In this model, lake morphometry, wind, and solar radiation determine thermal structure (including mixed zone depth). The depth distribution of algal chlorophyll is a function of vertical mixing and species composition, since some algae can regulate their depth. Chlorophyll depth distribution, water clarity, and surface light determine total light available to the algae. Available light along with chlorophyll and nutrient concentrations and the photosynthesis/light intensity (P/I) relationship are the major factors assumed to control production.

In this chapter we have presented data which permit an assessment of the validity of this model. Specifically, it is hypothesized that increases in algal standing crop (biomass and chlorophyll) are preceded by periods of relatively high production and/or photosynthetic efficiency. Production and photosynthetic efficiency are expected to be closely related to day of the year (a predictor of surface light), to chlorophyll concentration and depth distribution (both affect available light), and to the P/I relationship. The relative importance of these factors would be expected to differ when assessed for different time periods within the year.

While the data in this chapter indicate that production and efficiency are

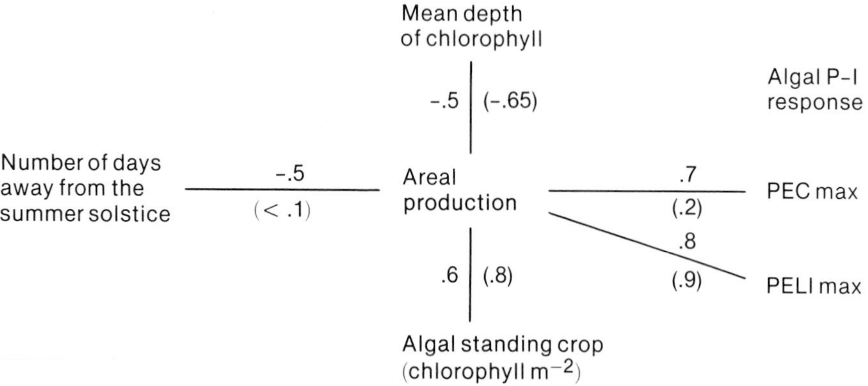

Figure 5.9. Relationship of algal production to other variables based on correlation coefficients for 1980 data (values in parentheses are based on summer data only). PEC max, maximum photosynthetic efficiency per unit chlorophyll; PELI max, maximum photosynthetic efficiency per unit incident light.

not related in a simple way to changes in biovolume and chlorophyll, a number of other relationships do emerge from an analysis of the parameters outlined in Figure 5.8. Figure 5.9 shows correlation coefficients calculated for a number of variables for the 1980 whole year and for the summer. As predicted from the model, production is closely correlated with algal standing crop (chlorophyll concentration) and depth distribution (the mean depth of chlorophyll, which was defined in Chapter 4), with physiological efficiency of the phytoplankton population ("physiological health") and with surface light intensity (as predicted by deviation in time from summer solstice). However, chlorophyll concentration and distribution and carbon fixed per unit light are most important in the summer, while carbon fixed per unit chlorophyll and surface light become important when whole year data are considered. This result is probably best explained by the fairly uniform values of surface light and carbon fixed per unit chlorophyll over the summer, a period in which the phytoplankton consists predominantly of blue-green algae. Mean depth of chlorophyll (see Chapter 4) is negatively correlated with production if summer data are considered, but is less so if the entire year is analyzed. This suggests that in the summer, algal buoyancy, as reflected in the mean depth of chlorophyll, may largely determine available light, while surface light values may be more important when the entire year is considered. Analysis of the chlorophyll-depth distribution showed that algal populations concentrated in a thin layer of the water column (as occurs in midsummer blooms of buoyant algae) had lower efficiencies relative to their potential maximum efficiencies than did populations more uniformly distributed over the water column. Thus, it would appear that these concentrations of algae achieve a smaller fraction of the production of which they are capable than do algae mixed through the water column. The reason for the lower efficiency of these surface algae is not clear but could be due to some inhibitory phenomenon, such as an allelopathic effect.

References

Bannister, T.T. 1974. Production equations in terms of chlorophyll concentration, quantum yield and upper limit to production. *Limnology and Oceanography,* 19:1–12.

Becacos-Kontos, T. and A. Svansson. 1969. Relation between primary production and irradiance. *Marine Biology,* 2:140–144.

Brock, T.D. 1981. Calculating solar radiation for ecological studies. *Ecological Modelling,* 14:1–19.

Chalker, B.E. 1980. Modelling light saturation curves for photosynthesis: an exponential function. *Journal of Theoretical Biology,* 84:205–215.

Cummins, K.W. and J.C. Wuycheck. 1971. Caloric equivalents for investigations in ecological energetics. *International Association of Theoretical and Applied Limnology,* Mitteilungen Communications No. 18:1–158.

Dubinsky, Z. and T. Berman. 1976. Light utilization efficiencies of phytoplankton in Lake Kinneret (Sea of Galilee). *Limnology and Oceanography,* 21:226–230.

Dubinsky, Z. and T. Berman. 1981. Photosynthetic efficiencies in aquatic ecosystems. *Verhandlungen Internationale Verein Limnologie,* 21:237–243.

Fallon, R.D. 1978. *The planktonic cyanobacteria: their sedimentation and decomposition in Lake Mendota, Wisconsin.* Ph.D. thesis, University of Wisconsin, Madison.

Fallon, R.D. and T.D. Brock. 1980. Planktonic blue-green algae: production, sedimentation, and decomposition in Lake Mendota, Wisconsin. *Limnology and Oceanography,* 25:72–88.

Fee, E.J. 1969. A numerical model for the estimation of photosynthetic production, integrated over time and depth, in natural waters. *Limnology and Oceanography,* 14:906–911.

Fee, E.J. 1977. *A computer program for estimating annual primary production in vertically stratified waterbodies with an incubator technique.* Fisheries and Marine Services Technical Report no. 741; Department of Fisheries and the Environment, Winnipeg, Manitoba, Canada. 38 pp.

Fee, E.J. 1978. A procedure for improving estimates of in situ primary production at low irradiances with an incubator technique. *Verhandlungen Internationale Verein Limnologie,* 20:59–67.

Fee, E.J. 1980. Important factors for estimating annual phytoplankton production in the Experimental Lakes Area. *Canadian Journal of Fisheries and Aquatic Sciences,* 37:513–522.

Hickman, M. 1976. Phytoplankton population efficiency studies. *Internationale Revue gesamten Hydrobiologie,* 61:279–295.

James, H.R. and E.A. Birge. 1938. A laboratory study of the absorption of light by lake waters. *Transactions of the Wisconsin Academy of Science, Arts, and Letters,* 31:1–154.

Jassby, A.D. and T. Platt. 1976. Mathematical formulation of the relationship between photosynthesis and light for phytoplankton. *Limnology and Oceanography,* 21:540–547.

Konopka, A. and T.D. Brock. 1978. Changes in photosynthetic rate and pigment content of blue-green algae in Lake Mendota. *Applied and Environmental Microbiology,* 35:527–532.

Morel, A. 1978. Available, usable, and stored radiant energy in relation to marine photosynthesis. *Deep-Sea Research,* 25:673–688.

Morel, A. and A. Bricaud. 1981. Theoretical results concerning light absorption in a discrete medium, and application to specific absorption of phytoplankton. *Deep-Sea Research,* 28A(11):1375–1393.

Parkin, T.B. and T.D. Brock. 1980. The effects of light quality on the growth of phototrophic bacteria in lakes. *Archives of Microbiology,* 125:19–27.

Pedrós-Alió, C. and T.D. Brock. 1982. Assessing biomass and production of bacteria in eutrophic Lake Mendota, Wisconsin. *Applied and Environmental Microbiology,* 44:203–218.

Prentki, R.T., D.S. Rogers, V.J. Watson, P.R. Weiler, and O.L. Loucks. 1977. *Summary tables of Lake Wingra basin data.* Institute for Environmental Studies, Report 85, University of Wisconsin, Madison.

Shearer, J.A. and E.J. Fee. 1974. *Phytoplankton primary production in the Experimental Lakes area using an incubator technique-1973 data.* Fisheries and Marine Service Technical Report No. 474. Department of the Environment Fisheries and Marine Service, Winnipeg, Manitoba, Canada.

Talling, J.F., R.B. Wood, M.V. Prosser, and R.M. Baxter. 1973. The upper limit of photosynthetic productivity by phytoplankton: evidence from Ethiopian soda lakes. *Freshwater Biology,* 3:53–76.

Tilzer, M.M., C.R. Goldman, and E. DeAmezaga. 1975. The efficiency of photosynthetic light energy utilization of lake phytoplankton. *Verhandlungen Internationale Verein Limnology,* 19:800–807.

Verduin, J. 1982. Components contributing to light extinction in natural waters: method of isolation. *Archiv für Hydrobiologie,* 93:303–312.

6. Zooplankton

Zooplankton are usually defined as small free-swimming animals that can be caught in nets. Zooplankton range in size from about 0.1 mm to 1–3 mm in length. The main components of the zooplankton are protozoa, rotifers, and crustaceans. Although rotifers may be more abundant numerically, the crustaceans generally constitute most of the biomass. They include the cladocerans, cyclopoids, and calanoid copepods. Some zooplankton are herbivorous and feed on phytoplankton or bacteria, whereas others are predatious and feed on other zooplankton. Zooplankton, in turn, are fed upon by small fish.

Zooplankton are an important component of the herbivore food chain and are often considered to be the main agents responsible for cropping the phytoplankton population. Although it is unlikely that zooplankton graze significantly on blue-green algae, they probably play an important role in cropping the eucaryotic algae (see Chapter 4). An understanding of the dynamics of zooplankton is therefore important if we are to understand the food chains of the lake.

Zooplankton are also important as indicators of environmental change. Because the taxonomy of zooplankton is reasonably well understood, it is possible to relate measurements made in recent years with measurements made many years ago. In the case of Lake Mendota, an extensive record of crustacean zooplankton dynamics was provided by Birge in the years 1894–96 (Birge, 1897). Although behavioral and physiological studies were carried out on Lake Mendota zooplankton in the 1950's and 1960's by McNaught and Hasler (1961) and Ragotzkie and Bryson (1953), no further synoptic work was done on Lake Mendota zooplankton until the late 1970's, when E. Woolsey and C. Pedrós-Alió initiated the studies on zooplankton dynamics which are summarized here. The studies of these latter workers provide an excellent data base against which to compare the earlier work of Birge. It is the purpose of the present chapter to provide data on zooplankton dynamics in the period 1976–80, and to compare these data with those of Birge. It is concluded that there has been no significant change in the zooplankton of the lake and that there has been more year-to-year variation within each separate time period than there was between the two time periods separated by over 80 years. These data are thus of interest in the prediction of long-term changes in the lake.

Methods for studying Lake Mendota zooplankton

Most of the samples taken for zooplankton analysis in the 1970's were obtained by Carlos Pedrós-Alió as part of his study of zooplankton feeding on bacterioplankton (see Chapter 7), but samples for 1976 were gathered by Dr. Vicki Watson and winter samples for 1978 were kindly provided by Richard Lathrop of the Wisconsin Department of Natural Resources.

Sampling

In the work of Pedrós-Alió (1981), samples were taken at the deepest part of Lake Mendota (24.2 m depth). Three different techniques were used to collect zooplankton. Vertical tows were taken with a 20 cm diameter plankton net (mesh size 156 μm) during all years of the study, to provide an integrated picture of the zooplankton throughout the water column. During 1979, a discrete sampling technique was also used to obtain information on vertical distribution. Samples were taken at a number of depths with a 5 liter Van Dorn bottle and at least 1 liter of water from each depth was concentrated to a few milliliters on a Whatman GF/C glass fiber filter. This filtration method had the added advantage of retaining the small members of the zooplankton that would be lost with the plankton net, but the small total volume of lake water that could be examined made it unsuitable for the less abundant species. A simplified method, based on that of Likens and Gilbert (1970), was used during 1980. Discrete samples were taken at several depths with a Van Dorn bottle and these samples were filtered through circular pieces of Nitex net (64 μm mesh size). The 64 μm net proved efficient at retaining rotifers and nauplii of copepods, although it would not retain protozoa. The main advantage of this method was its simplicity and speed, while providing the necessary data to calculate biomass. During the period when the lake was stratified, two replicates were filtered from the epilimnion and two from the hypolimnion.

All samples were examined under a dissecting microscope. Except in 1979, the lengths of at least 30 individuals of each major species were measured at each sampling date to the nearest 30 μm and the average size for that date

Table 6.1. Average lengths and dry weights for Lake Mendota zooplankton

Species	Length (mm)*		Dry weight (μg)† 1980
	1979	1980	
Crustacea			
Daphnia galeata	1.01(0.024)	1.14 (0.058)	9.62 (3.96)
Daphnia retrocurva	1.37(1.350)	1.00 (0.086)	7.72 (3.96)
Daphnia pulex	—	1.37 (0.075)	25.96(24.62)
Daphnia parvula	0.98(0.020)	—	—
Diaphanosoma leuchtenbergianum	1.93(2.300)	0.88 (0.035)	2.93 (4.9)
Chydorus sphaericus	0.26(0.070)	0.26 (0.005)	0.45
Bosmina longirostris	0.40	0.34 (0.0002)	0.91
Ceriodaphnia sp.	0.49(0.140)	0.42	1.44
Calanoid copepods	1.02(0.030)	0.93 (0.018)	6.15
Cyclopoid copepods	0.70 (0.001)	0.78 (0.016)	3.15
Nauplii	0.20	0.17 (0.006)	0.19
Rotifera			
Keratella cochlearis	0.10	0.089(0.00001)	—
K. quadrata	0.20	0.158 (0.012)	—
Brachyonus sp.	0.20	0.089 (0.002)	—
		0.425 (0.040)	—
Polyarthra sp.	0.10	0.108 (0.009)	—
Conochiloides sp.	—	0.112 (0.010)	—
Trichocerca sp.	—	0.168 (0.006)	—
Filina sp.	—	0.120 (0.021)	—
Asplanchna sp.	—	0.500 (0.288)	—

*Average, and between parentheses, 95% confidence limits.

†In 1979 weights were calculated for size classes and an average weight was not calculated. For 1980, average weight and, between parentheses, 95% confidence limits.

calculated. In 1979, a lesser number of individuals was measured. The most abundant species were divided into size classes and individuals assigned to them. Species too rare to count accurately were assigned the average value of all measurements for that species during the year. Average lengths and dry weights for the Lake Mendota zooplankton are shown in Table 6.1.

Except in 1979, average dry weights were calculated with the regression equations of dry weight versus length given by Dumont, Van de Velde, and Dumont (1975). In 1979, the most abundant species were divided into size classes and the dry weights calculated for each size class independently with the same equations. A carbon to dry weight ratio of 0.40 was used to convert dry weight to carbon biomass.

Statistics

Statistics of zooplankton counting have been discussed by Ruttner-Kolisko (1977). Fortunately, in the present study, differences between replicates and techniques were not high. Average coefficients of variation (C.V.) for replicates taken with Van Dorn bottles were 19% for cyclopoids, 15% for calanoids, 31% for *Daphnia* species, 17% for *Daphnia galeata mendotae,* 20% for *Chydorus sphaericus*, 29% for other cladocerans, 26% for rotifers, and 15% for the most abundant rotifer *Keratella cochlearis*. When simultaneous samples taken with Van Dorn bottles and vertical tows were compared, average C.V.'s were 22% for cyclopoids, 21% for calanoids, 42% for *Daphnia* species, 38% for other cladocerans, and 26% for *Chydorus*. These C.V.'s are only slightly higher than those for replicates of Van Dorn samples. In addition, two way analysis of variance without replication was performed on abundance from vertical tows and Van Dorn samples for seven sampling dates to compare variability due to different sampling techniques with variability due to the different zooplankton species. As expected, the latter was always highly significant due to the different population dynamics of the different species. On the other hand, variability due to the different sampling techniques was never significant ($P<0.05$), making possible comparisons obtained with all of them. Rotifers could not be counted in vertical tow samples because they were not retained by the zooplankton net, so that equivalent comparisons could not be made for these animals. See Pedrós-Alió (1981) for the raw data.

Species composition

In the present study, fourteen main species of zooplankton were found: 3 cyclopoid copepods (*Diacyclops bicuspidatus thomasi* (S.A. Forbes), *Mesocyclops edax* (E.A. Forbes), and *Acanthocyclops vernalis* (Fischer)), 3 calanoid copepods (*Leptodiaptomus sicilis* (S.A. Forbes), *L. siciloides* (Lilljeborg), and *Aglaodiaptomus clavipes* (Schacht), 4 species of the genus *Daphnia* (*D. galeata mendotae* Birge, *D. pulex* Leydig, *D. parvula* Fordyce, and *D. retrocurva* Forbes), and 4 other cladocerans (*Chydorus sphaericus* (O.F. Müller), *Bosmina longirostris* (O.F. Müller), *Diaphanosoma leuchtenbergianum* Fischer, and *Leptodora kindtii* (Focke)). In addition, one cyclopoid copepod (*Tropocyclops prasinus* (Fischer)), one calanoid copepod (*Skistodiaptomus oregonensis* (Lilljeborg)), *Ceriodaphnia quadrangula* G.O. Sars, and *Ergasilus* sp. were observed occasionally in very small numbers.

In the case of rotifers, animals were only identified to genus, except for the easily distinguished *Keratella cochlearis* (Gosse) and *K. quadrata* Müller). The other genera present were *Polyarthra* Ehrenberg, *Brachionus* Pallas, *Asplanchna* Gosse, *Trichocerca* Lamarck, *Filinia* Bory de St. Vincent, *Conochilus* Hlara, and *Conochiloides* Hlara.

Although there have been a number of revisions of the taxonomy of the crustacean zooplankton, it is relatively easy to ascertain which species present now were found by Birge (1897). For compactness in presenting the data here, certain species have been aggregated into larger groups, but it has been ascertained that the species within these groups are the same as those reported by Birge. Some genera and species currently recognized were not distinguished by Birge, who tended to lump species together. A summary of the names used by Birge and in the present work, including the manner in which these have been aggregated in the summary graphs, is given in Table 6.2.

Table 6.2. Major zooplankton species in Lake Mendota

Designation on Figure 6.1–6.7	Name; this study	Birge's nomenclature
Chydorids	*Chydorus*	*Chydorus*
Cyclopoids	*Diacyclops*	*Cyclops*
	Mesocyclops	
	Acanthocyclops	
Calanoids	*Leptodiaptomas*	*Diaptomus*
	Aglaodiaptomus	
Diaphanosoma	*Diaphanosoma*	*Diaphanosoma*
Daphnia galeata	*D. galeata mendotae*	*D. hyalina*
Daphnia retrocurva	Same	Same
Daphnia pulex	Same	*D. pulicaria*

Zooplankton abundance

Some data on zooplankton abundance was presented by Pedrós-Alió and Brock (1983). An overall comparison of the zooplankton dynamics of the Birge study and that of the present study was given in Pedrós-Alió and Brock (1985). Detailed presentation of all the zooplankton data for the period 1976–81 can be found in Pedrós-Alió et al. (1985). The comparisons are summarized in Figures 6.1 through 6.7. Note that the vertical and horizontal scales of the graphs of the two data sets are the same. In general, it can be stated that the annual cycles of the various zooplankton groups are quite similar in the two data sets. The peaks tend to occur at the same time of year, and the peak heights are also similar. Although there is considerable year-to-year variability in the phasing of the annual cycles, this variability is no greater in the present data set than in Birge's. When it is considered that the sampling methods are different, the correspondences between the two data sets are surprisingly high.

General description of the annual cycle

The annual cycle of zooplankton showed similarities from year to year, even though there were differences between years. The annual cycle of 1979 is dis-

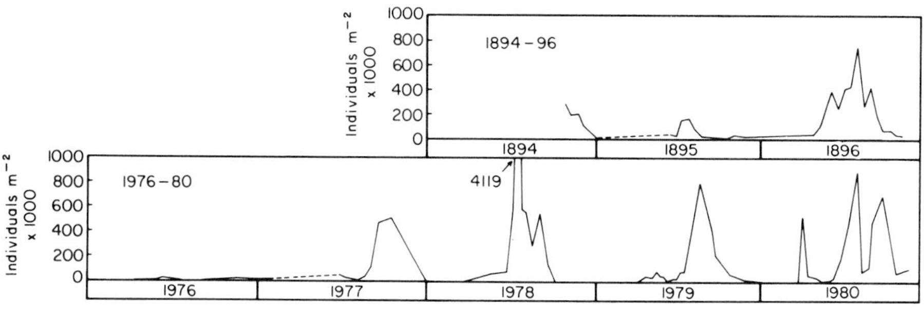

Figure 6.1. Standing crop of chydorids in Lake Mendota. From Pedrós-Alió and Brock (1985).

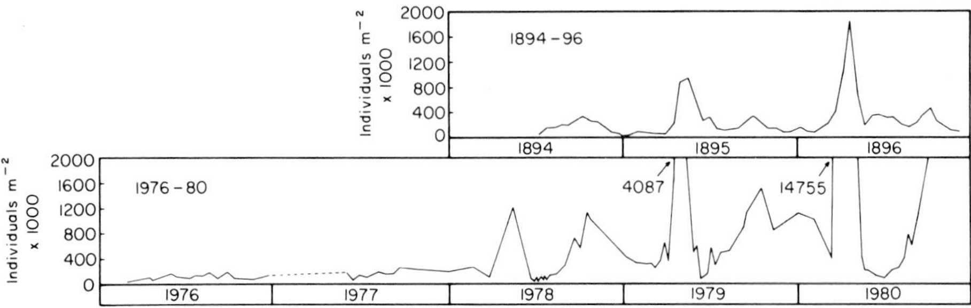

Figure 6.2. Standing crop of cyclopoids in Lake Mendota. From Pedrós-Alió and Brock (1985).

cussed briefly here, with the detailed data being presented in Pedrós-Alió et al. (1985). There were typical winter and summer zooplankton populations with different species composition, a sharp spring transition with its characteristic fauna, and a very smooth and long autumn transition. Winter numbers

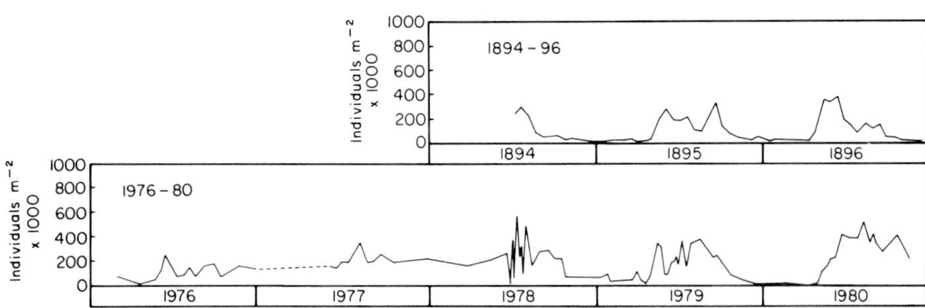

Figure 6.3. Standing crop of calanoids in Lake Mendota. From Pedrós-Alió and Brock (1985).

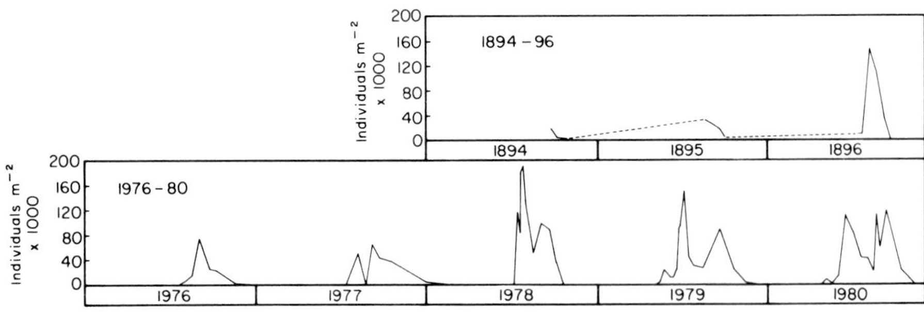

Figure 6.4. Standing crop of *Diaphanosoma* in Lake Mendota. From Pedrós-Alió and Brock (1985).

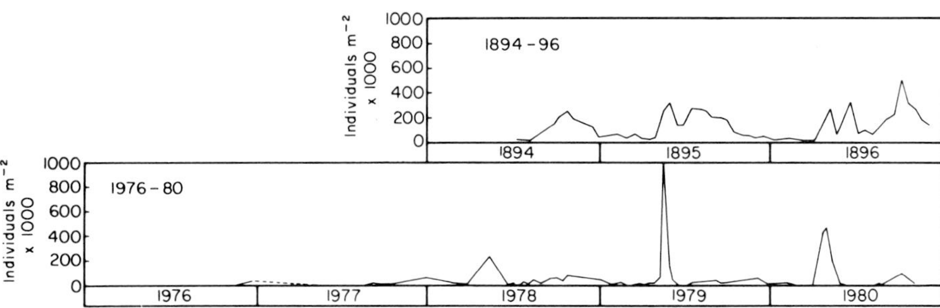

Figure 6.5. Standing crop of *Daphnia galeata* in Lake Mendota. From Pedrós-Alió and Brock (1985).

were low under the ice, mostly composed of *Diacyclops bicuspidata thomasi*, the rest being a few diaptomid copepods and large adults of *Daphnia galeata mendotae*. Vertical distribution indicated that most of these organisms were concentrated toward the bottom of the lake, possibly because of the warmer temperatures found there in the winter.

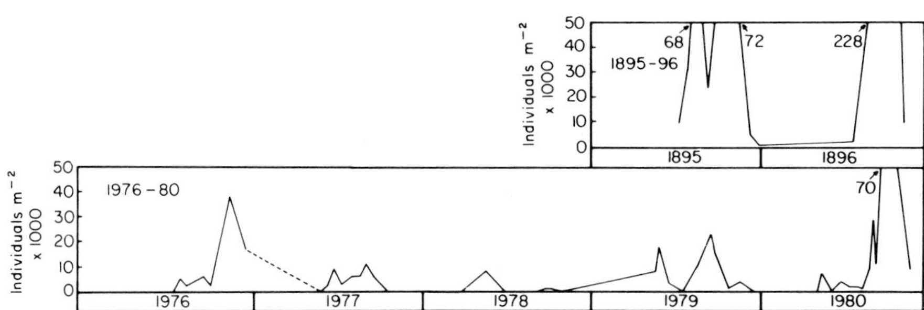

Figure 6.6. Standing crop of *Daphnia retrocurva* in Lake Mendota. From Pedrós-Alió and Brock (1985).

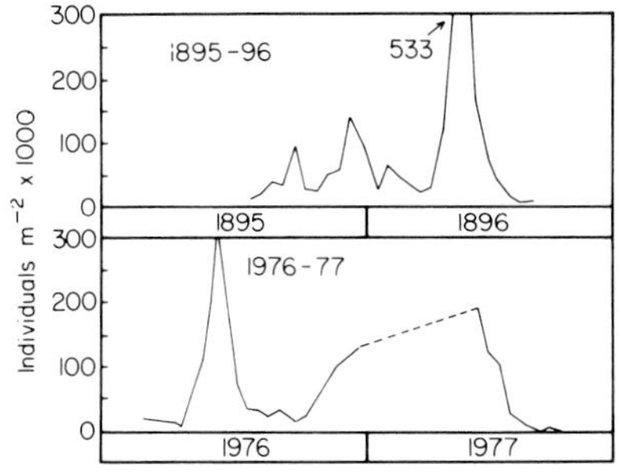

Figure 6.7. Standing crop of *Daphnia pulex* in Lake Mendota. From Pedrós-Alió and Brock (1985).

From late April to late June, a transition from winter to summer species occurred. During this period, first a huge peak of *D. bicuspidatus thomasi* appeared, probably due to the copepodites that had overwintered. These animals developed very slowly at the mud-water interface, but grew to adulthood and reproduced when the ice melted and the temperature started to increase. Maximum zooplankton biomass was attained at this point with *Diacylops* constituting from 80–90% of the biomass. Right after this maximum, a peak of *Daphnia galeata mendotae* (*D. pulex* in other years) developed in which this species constituted from 50–60% of the biomass. However, this part of the annual cycle was characterized by the presence of many other species of cladocerans that did not appear the rest of the year, such as *Bosmina, Ceriodaphnia,* and *Daphnia parvula,* or that were typically summer species, such as *D. retrocurva, Diaphanosoma,* and *Chydorus.* At the same time, diaptomids started to be more abundant and, after the *Diacyclops* population collapsed, summer species of cyclopoids appeared, such as *Mesocyclops* and *Acanthocyclops.* Therefore, the spring seemed to be a period during which winter species disappeared while summer ones got started. During this change, a few species of cladocerans attained differently sized peaks, giving the richest diversity of species of the whole year.

The summer fauna was dominated by calanoid copepods, which presented a series of ups and downs throughout the summer. In addition, typically aestival cladocerans appeared, *Chydorus sphaericus* being the most abundant. *Leptodora kindtii,* a large (up to 1.8 cm long) predacious cladoceran, appeared early in July and kept more or less constant numbers until winter. Throughout the summer, cyclopoids kept increasing in numbers slowly during stratification, and quickly reached their winter levels in October or November when the other summer zooplankton species disappeared and *Diacyclops* came back. There usually was a small peak of *Daphnia retrocurva* toward October, which disappeared from the lake toward the end of November, at which time *Diaphanosoma* and *Chydorus* also disappeared. The number of calanoids went down slowly and with this, the return to winter conditions was complete.

Species changes over the 1976–80 sampling period

The most dramatic change in species composition was the substitution of *Daphnia galeata* in 1979 and 1980 for *D. pulex,* which had been the dominant cladoceran the previous years (compare Figure 6.5 with Figure 6.7). This substitution was almost complete and very few individuals of *D. pulex* could be observed after 1978. The remaining changes in species composition were only due to the presence of certain species during some years but not in others. For instance, *D. parvula* appeared in 1979, presenting high numbers in the metalimnion immediately before and after the main *D. galeata* peak. In 1980, there was again a small peak, but no individuals of this species were observed previous to 1979. Finally, *Ceriodaphnia quadrangula* was observed in significant numbers only in February 1976 and in June of 1979 and 1980.

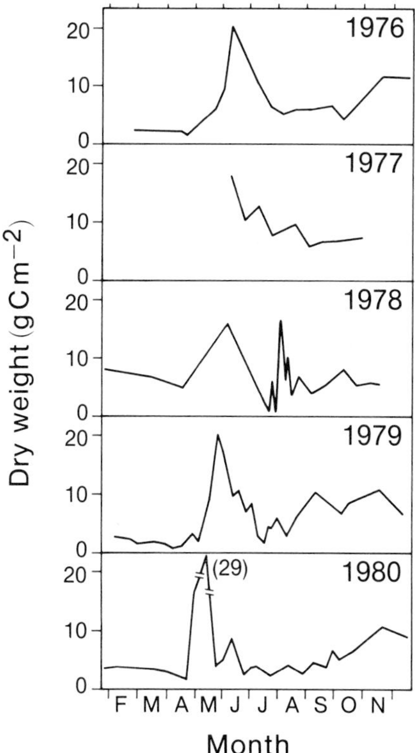

Figure 6.8. Total zooplankton biomass. Dry weight in g carbon/m² for the five years of the present studies. From Pedrós-Alió et al. (1985).

Biomass of zooplankton

Zooplankton biomass has been calculated from average weights and percent composition of the different groups for each sampling date. Data for the 5 years of the study are presented in Figure 6.8. Comparison of zooplankton, phytoplankton, and bacterioplankton abundances for two years is given in Figure 6.9. Comparing algal biomass with zooplankton biomass, it can be seen that zooplankton biomass in general is larger than phytoplankton biomass. Zooplankton biomass was negatively correlated with algal production also, and no relationship was apparent with algal biomass in either year. In 1979, zooplankton biomass was significantly correlated with numbers of bacteria per ml and with bacterial biomass, but these latter correlations were not observed in 1980. For a discussion of bacterial biomass and zooplankton feeding on bacteria, see Chapter 7.

That zooplankton biomass is larger than phytoplankton biomass has been noted several times in planktonic habitats (Nauwerck, 1963; Coveney et al., 1979). This is not unreasonable when it is considered that zooplankton can

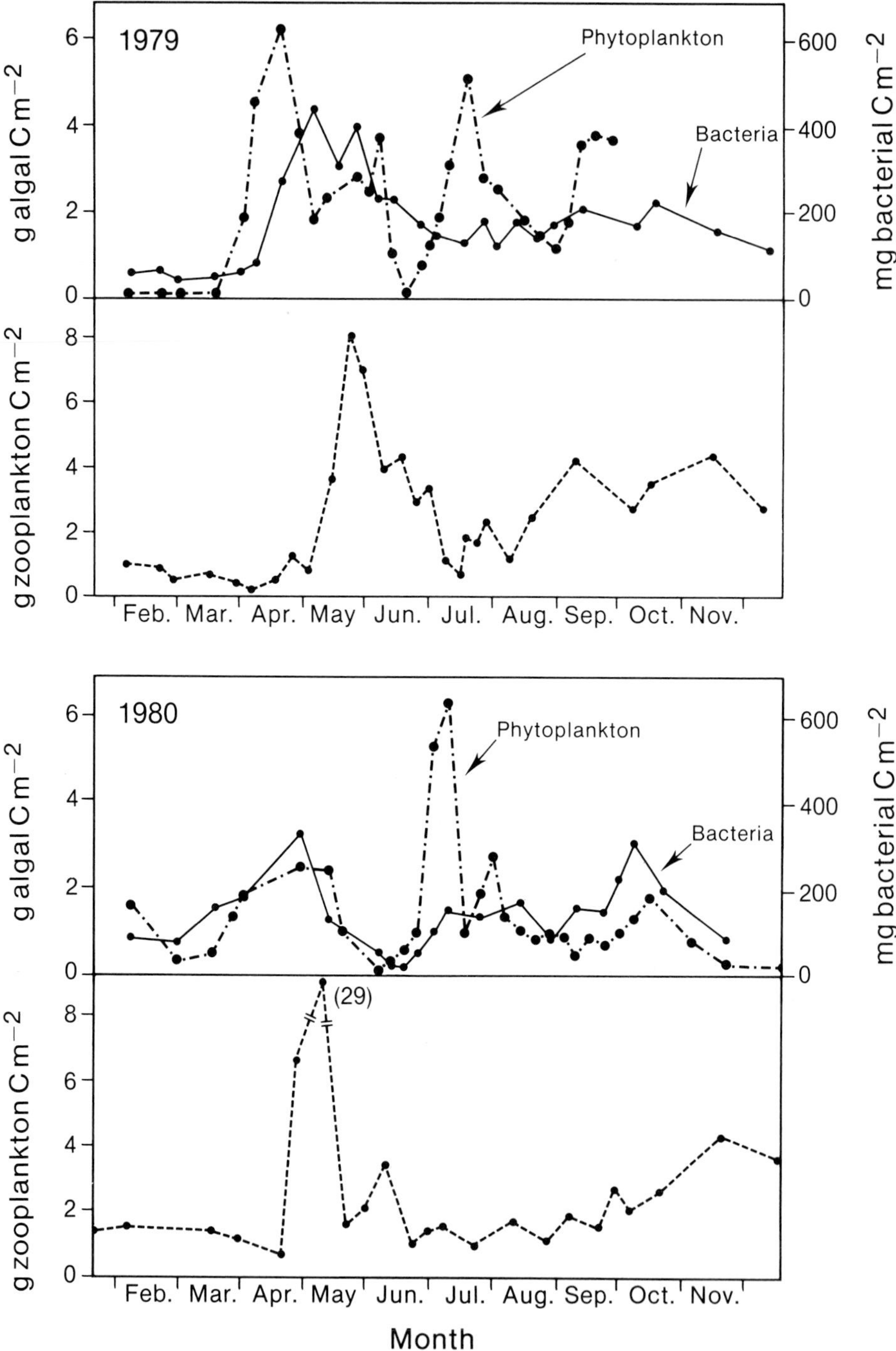

Figure 6.9. Zooplankton, phytoplankton, and bacterioplankton biomasses for 1979 and 1980. From Pedrós-Alió and Brock (1983).

ingest close to 100% of some kinds of phytoplankton and the ecological pyramid of biomass can be inverted as long as turnover rates of the lower trophic levels are considerably faster than turnover rates of the upper levels.

Conclusions

The zooplankton of Lake Mendota are typical of hard-water temperate lakes of a eutrophic character. One of the main values of the zooplankton data from the 1976–80 period is that it permits comparison with Birge's data from 1894–97. Further discussion of these comparisons is reserved for Chapter 9, but it can be stated here that as far as species of crustacean zooplankton are concerned, Lake Mendota is quite similar now to what it was over 80 years ago.

Zooplankton are sensitive indicators of the nature and extent of long-term change in lake ecosystems. Hopefully, the present data set (for details see Pedrós-Alió et al., 1985) and that of Birge's (Birge, 1897) will provide useful material for future workers interested in assessing the extent of change which may be occurring in the Lake Mendota ecosystem.

References

Birge, E. A. 1897. Plankton studies on Lake Mendota. II. The crustacea of the plankton, July 1894–December 1896. *Transactions of the Wisconsin Academy of Sciences, Arts & Letters,* 11: 274–448.

Coveney, M. F., G. Cronberg, M. Enell, K. Larsson and L. Olofsson. 1977. Phytoplankton, zooplankton and bacteria. Standing crop and production relationships in a eutrophic lake. *Oikos,* 29: 5–21.

Dumont, H. J., I. Van de Velde and S. Dumont. 1975. The dry weight estimates of biomass in a selection of cladocera, copepoda and rotifera from the plankton, periphyton and benthos of continental waters. *Oecologia,* 19: 75–97.

Likens, G. E. and J. J. Gilbert. 1970. Notes on quantitative sampling of natural populations of planktonic rotifers. *Limnology and Oceanography,* 15:816–820.

McNaught, D. C. and A. D. Hasler. 1961. Surface schooling and feeding behavior in the White bass *Roccus chrysopus* (Rafinesque), in Lake Mendota. *Limnology and Oceanography,* 6: 53–60.

Nauwerck, A. 1963. Die Beziehungen zwischen Zooplankton und Phytoplankton in See Erken. *Symbolae Botanica Upsaliensis,* 17(5).

Pedrós-Alió, C. 1981. *Ecology of heterotrophic bacteria in the epilimnion of eutrophic Lake Mendota, Wisconsin.* Ph.D. thesis, University of Wisconsin, Madison.

Pedroś-Alió, C. and T. D. Brock. 1983. The impact of zooplankton feeding on the epilimnetic bacteria of a eutrophic lake. *Freshwater Biology,* 13: 227–239.

Pedrós-Alió, C. and T. D. Brock. 1985. Zooplankton dynamics in Lake Mendota: short-term versus long term changes. *Freshwater Biology,* 15: 89–94.

Pedrós-Alió, C., E. Woolsey and T. D. Brock. 1985. Zooplankton dynamics in Lake Mendota: abundance and biomass of the metazooplankton from 1976 to 1980. *Transactions of the Wisconsin Academy of Sciences.*

Ragotzkie, R. A. and R. A. Bryson. 1953. Correlation of currents with the distribution of adult *Daphnia* in Lake Mendota. *Journal of Marine Research,* 12:157–172.

Ruttner-Kolisko, A. 1977. Comparison of various sampling techniques, and results from repeated sampling of planktonic rotifers. *Archiv für Hydrobiologie,* 8: 13–18.

7. Bacteria, carbon cycling, and biogeochemical processes

The study of bacteria in lakes has begun relatively recently because suitable techniques have only been available during the past decade. The qualitative importance of heterotrophic bacteria in lake ecosystems has been recognized for a long time, but their quantitative importance has seldom been determined. It has been commonly accepted that bacteria play an important role in the detrital food chain of lakes, being responsible for the decomposition of both phytoplankton and higher organisms. In addition, there is mounting evidence that bacteria are able to use photosynthetically-derived products excreted by living phytoplankton. If zooplankton then consume some of the bacterial carbon so produced, the bacteria would play a significant role in the well-known grazing food chain. In the present work, a number of techniques have been used in an attempt to define the various roles of planktonic bacteria in the aquatic food chains and carbon cycling of Lake Mendota.

We have already discussed in Chapter 3 some aspects of carbon cycling related to sedimentation of particulate material and sediment-water interactions. In Chapter 4 we discussed the development of phytoplankton and the subsequent loss of phytoplankton-derived particles as a result of sedimentation. In this earlier discussion, the role of bacteria was implicit but no specific studies on the bacterial populations themselves were presented. In the present chapter, we present data on the standing crop and productivity of heterotrophic planktonic bacteria, primarily those living in the epilimnion of the lake. We will show that bacterial heterotrophic production constitutes about 50% of gross primary production from phytoplankton. Thus the bacteria constitute a major component of the Lake Mendota food chains.

The other major habitat in which bacteria play an important role is in the sediment of the lake. As described in Chapter 3, the sediments of Lake Mendota, rich in ferrous sulfide, are predominantly anaerobic except for a narrow layer at the surface which may be oxidizing when the lake is unstratified. During summer stratification, oxygen is rapidly depleted from the bottom waters due to the oxygen demand of the sediments (Chapter 3). This oxygen consumption is the result of two processes, a strictly chemical process involving primarily ferrous sulfide oxidation, and a biological process involving the bacterial decomposition of organic matter. Organic carbon, derived primarily from phytoplankton, settles to the surface of the sediment and undergoes decomposition. This decomposition process is initially aerobic, but once all oxygen has been consumed the process is completed anaerobically.

It is well established that a series of bacteria are involved in the anaerobic decomposition process. The macromolecular constituents which are the main components of the sedimented material are first broken down to monomeric materials by hydrolytic bacteria. These monomeric materials are then fermented to low-molecular weight organic acids, alcohols, and hydrogen gas by an ubiquitous group of heterotrophic fermentative bacteria. The organic acids and alcohols are subsequently converted to carbon dioxide by the sulfate-reducing bacteria, if sulfate ion is present in the sediment, or to carbon dioxide and methane by the methanogenic bacteria, if sulfate is low or absent. Methane can diffuse into the overlying water column and then may be lost from the lake. However, there is also a group of methane-oxidizing bacteria which can develop either at the sediment-water interface or near the thermocline, and reoxidize some of the methane back to CO_2. Thus, we must consider that carbon from organic materials is cycled in the lake in two ways: 1) in the epilimnion, primarily by heterotrophic bacteria which oxidize it fully to CO_2; and 2) in the sediments, by concerted acton of fermentative, sulfate-reducing, and methanogenic bacteria. If methane is produced, some of it can be transferred back into the water column. Both of these carbon cycling processes are important in Lake Mendota. One goal of the present chapter is to assess the relative importance of these two carbon cycling processes.

In addition to the direct interest in carbon cycling in the sediments, there is another interest which involves the interrelationships of the carbon cycle, the sulfur cycle, the iron cycle, and the phosphorus cycle. It is well established that phosphate, an important algal nutrient, can be held in sediments by bonding to oxidized iron components such as ferric hydroxide. However, if sulfide is present, ferric iron will be reduced to the ferrous state and will form an insoluble ferrous sulfide complex, a process discussed in Chapter 3. If ferrous sulfide formation occurs, the phosphorus that had been bound to the iron is solublized and may be released into the water column. Hypolimnetic accumulation of phosphate is an important summer process in Lake Mendota (Stauffer, 1974). Since the sulfide involved in this process is derived from bacterial sulfate reduction (see below), it can be seen that anaerobic bacterial processes in the sediment can play an important role in internal phosphorus cycling in the lake. An additional complication is that the reduced organic carbon for the sulfate-reduction production is derived primarily from primary production by phytoplankton developing in the overlying water. Phytoplankton growth in the first place has been stimulated by the presence of phosphate. In low-sulfate lakes, the importance of sulfate reduction in liberating phosphorus is minor, but in Lake Mendota the concentration of sulfate in the water is high enough (see Chapter 3) so that sulfate reduction can be a significant process in releasing phosphorus from the sediments.

One problem with assessment of bacteria in aquatic systems is that the small size and low numbers make microscopic counting on unconcentrated samples virtually impossible (Brock, 1984). Early workers used viable counting procedures, but it was quickly recognized (Snow and Fred, 1926) that these procedures grossly underestimated bacterial numbers. Direct microscopic counting always provides higher counts (Bere, 1933), but for this latter method the bacteria must be concentrated in some way which avoids cellular damage and which still permits their visualization at high power under a light microscope. The method of choice today is the acridine orange direct counting (AODC) method, which involves staining of the bacteria with the fluorochrome acridine orange, filtration through a black membrane filter, and visualization of the bacteria with epifluorescence microscopy (Hobbie et al., 1977). This method has only been available since the mid-1960's and is the method used in the work described in this chapter. Discussion of some historical work on the bacteria of Lake Mendota can be found in Chapter 9.

Methods

Bacterioplankton

The methods used for bacterioplankton counting and production have been described in detail by Pedrós-Alió (1981) and by Pedrós-Alió and Brock (1982). Water samples were taken at the deepest part of the lake (24.2 m) using a 5-liter Van Dorn bottle at a number of depths. For direct microscopic examination, subsamples were preserved in formalin. For incubation experiments, water samples were placed in polyethylene bottles and brought back to the laboratory in a cooler. Incubation experiments were started less than one hour after sampling.

The acridine orange direct counting (AODC) method was used for direct microscopic counting (Hobbie et al., 1977). Water samples were stained with acridine orange and filtered through 0.45 μm black Sartorius membrane filters or through 0.2 μm Nuclepore polycarbonate filters which had been stained with acid black no. 107. Filters were observed with a vertical illuminator under a Zeiss fluorescence microscope at $\times 1250$. At least 600 bacteria were counted per filter and usually more than ten fields.

For some of the quantitation work, observations with a scanning electron microscope were carried out. For this procedure, 10–40 ml of lake water was filtered onto Nuclepore filters of different pore sizes. After fixing with glutaraldehyde and dehydrating, a small square was cut from each filter and subjected to critical-point drying with CO_2. The filter pieces were then mounted on metal stubs, shadowed with gold-palladium, and observed with a scanning electron microscope. Pictures of both selected and random fields were taken, enlargements made, and the cells in each picture measured with a caliper, and the proportion of cells undergoing division counted. Volumes of rods, cocci, and oval-shaped bacteria were calculated assuming they were cylinders, spheres, or oval solids, respectively. At least 300 bacteria were measured on each sampling date.

Bacterial biomass was calculated by using the average volume for a given date $\times$ the number of cells present per square meter. Conversion to carbon assumed a density of 1.025 mg/mm^3 and a carbon to wet weight ratio of 0.092 (conversion factor 94.3 fg of C per μm^3). These conversion factors were obtained from the bacterial physiology literature (Pedrós-Alió, 1981).

Bacterial production was estimated by the frequency-of-dividing-cells (FDC) technique and by a technique using radioactive sulfate. For the latter procedure, incorporation of radioactive sulfate into bacterial-sized particles was measured in the dark. The assumption with this technique is that all of the sulfur that the bacteria use for production is derived from inorganic sulfate present in the water. If the carbon to sulfur ratio of the bacteria and the sulfate concentration of the water are known, it is then possible to calculate bacterial carbon production from radioactive sulfate uptake measurements. The details of this procedure are presented by Pedrós-Alió and Brock (1982). A carbon to sulfur ratio of 50:1 was used when converting sulfur uptake values to carbon units. Sulfate uptake measurements were done with the three size fractions analyzed, >8, 3–8, and <3 μm. Uptake in the <3 μm fraction was assumed to be exclusively bacterial, since both scanning electron micrographs and fluorescence microscopic observations of samples filtered through 3 μm Nuclepore filters never showed any evidence of algae or protozoa. The interpretation of the other fractions, however, is not as straightforward, since at least some algae do take up sulfate in the dark. The sulfate uptake in the larger size fractions can be corrected by measuring ^{14}C-acetate uptake (see below) in those fractions and using the distribution of that uptake as a measure of the proportion of the bacterial heterotrophic activities in each size fraction (Campbell and Baker, 1978). Note that in this procedure, acetate uptake is used only as a relative measure of bacterial activity in different fractions used for measuring sulfate uptake. The sulfate concentration in the lake was measured and this concentration was used to calculate the total amount of sulfate uptake represented by the radioactive material. The concentration of radioactive sulfate added was negligible compared with the natural concentration, so that the radioactive sulfate was behaving as a true tracer.

Frequency of dividing cells (FDC) was measured in scanning electron micrographs. The regression lines obtained by Hagström et al. (1979), and by Newell and Christian (1981) were used to convert FDC to growth rate μ (hr^{-1}).

From μ and biomass B (cells l^{-1}), calculated from AODC counts, productivity P (cells $l^{-1} \cdot hr^{-1}$), could be calculated according to the formula $P = \mu \times B$.

Assimilation of algal excretory products by epilimnetic bacteria

Differential filtration was used to assess the importance to epilimnetic bacteria of the carbon excreted by algae (Brock and Clyne, 1984). This was done by using the ^{14}C technique to estimate the uptake by the bacteria of algal-excreted photosynthetically-fixed carbon, and relating this uptake to total bacterial production rates. By means of differential filtration through Nuclepore filters of defined pore sizes, it was possible to follow ^{14}C from inorganic bicarbonate through the phytoplankton to the bacterioplankton.

The ^{14}C technique for measuring algal photosynthesis has been described in Chapter 5. The procedure used here was very similar except that a single light intensity of 130 μEinstein m^{-2} sec^{-1} was used (approximately saturating for photosynthesis) and 10 μCi of $NaH^{14}CO_3$ was added to each tube containing 25 ml of lakewater. The samples were incubated for periods from 1–4 hours and the incubation was terminated by placing the samples on ice in the dark until filtration. Filtration was through a series of Nuclepore filters (3 μm, 1 μm, 0.4 μm, and 0.2 μm) and the filtrate from the final filter collected. The filters were rinsed with 10 ml of 0.2 μm-filtered lakewater, dried, and placed in HCl fumes to remove any inorganic carbon that might have formed during the incubation period. The filters were then counted conventionally with a liquid scintillation counter. To measure excretion, 5 ml of the final filtrate was placed in a large-mouthed vial with 5 drops of concentrated HCl and the solution allowed to equilibrate for 24 hours to remove inorganic ^{14}C (predominantly bicarbonate) before addition of liquid scintillation fluid. Tests showed that all radioactive inorganic carbon was removed by this procedure so that only excreted organic carbon should have remained. In each experiment, parallel nonradioactive samples were filtered through the various pore-size filters for determination of bacterial number by the AODC method and analysis of chlorophyll (as a measure of the sizes of the phytoplankton colonies). The chlorophyll assays were performed as described in Chapter 4.

The total light-stimulated carbon fixation was defined as the sum of the radioactivity found in all fractions, particulate and soluble, less uptake by samples incubated in the dark. Radioactivity from samples incubated in the dark was similar to that from samples incubated in the light in the presence of 1% (v/v) formalin. The percent total excretion was calculated by dividing the sum of the particulate fractions less than 3 μm (assumed to be excreted radioactivity incorporated into bacteria) plus the radioactivity in the soluble fraction by the total carbon fixed (all size fractions plus soluble). The percent net excretion (that carbon not taken up by the bacteria) was calculated by dividing the soluble fraction by the total carbon fixed. Assays of dissolved inorganic carbon in the lakewater were used to convert radioactivity to amount of carbon fixed. The uptake rate by bacteria of carbon excreted by algae was calculated from the rate of uptake in the 0.2 μm–3 μm particulate fractions.

An assumption of this calculation is that the specific radioactivity of the excreted material is not diluted by nonradioactive material already present in the water. It is likely that this assumption is sometimes false, so that the bacterial uptake rates should be higher than the calculated rates by some unknown (probably small) factor.

Methods for studying anaerobic bacterial processes

Anaerobic habitats of the lake include both the sediments and the hypolimnetic water, but most anaerobic bacterial activity occurs in the sediments (Fallon, 1978). Activity of anaerobic sediment bacteria was measured both directly on sediment samples and indirectly by measuring chemical changes in the overlying water. The latter involved determination at various times of chemical profiles in the water column for key constituents such as sulfide and methane.

The details of the chemical profile determinations are given in Ingvorsen and Brock (1982). Chemical profiles of the water column were determined every second or third week during stratification at a 24 m site. Samples were collected at 1 or 2 m intervals at depths below 10 m, using a 4 liter opaque plastic Van Dorn water sampler. Upon retrieval, thick-walled (ca. 4 mm inner diameter) amber latex tubing was connected to the outlet port and flushed thoroughly before the sample bottles were filled. Water samples for analysis of sulfate, sulfide, and oxygen were collected in 60 ml glass-stoppered (BOD) bottles, allowing for two to three volumes of overflow. Samples (10 ml) for determination of dissolved methane were removed through the tubing with a 10 ml glass syringe fitted to an 18-gauge hypodermic needle. The samples were injected into 25 ml N_2-gassed anaerobic vials containing 1 ml of 6 N HCl. Methane analysis on head-space gas was done by transferring the gas with a pressure-lock syringe to a gas chromatograph. A peristaltic pump was used to collect water samples within 5 cm of the sediment-water interface. The analytical procedures were the same as for samples collected with the Van Dorn sampler.

The details of sediment sampling procedures are given in Ingvorsen et al. (1981) and Zehnder and Brock (1980). All sediment samples were from the 24 m deep station. Surface sediment samples were collected with an Eckman dredge. Upon opening the top of the dredge, the overlying water was carefully removed and the top 2 to 3 cm of sediment was carefully scooped off with a spatula and transferred to N_2-gassed, 250 ml thick-walled glass bottles. The glass bottles were filled to the top and sealed anaerobically with butyl rubber stoppers. During the closure process, excess sediment was allowed to escape through an 18-gauge needle temporarily inserted through the stopper. Sealed bottles were transported to the laboratory in well-insulated containers and, if not processed immediately, were stored in the dark at *in situ* temperature. Subsequent manipulations of sediment were done by using standard anaerobic techniques.

Quantification of sulfate reduction was done using a radioactive tracer technique. Undiluted homogenized sediment (10 ml) was dispensed anaerobically into a 20 ml serum vial containing five glass beads to facilitate subsequent mixing. The headspace was then flushed with N_2-CO_2 (99:1) for an additional 2 min, stoppered with a black-lip butyl rubber stopper, and crimped with an aluminum seal. The isotope was injected through the stopper with a 1 ml syringe as a mixture of carrier-free $^{35}SO_4^{2-}$ diluted in a solution of Na_2SO_4. The total sulfate concentration of the isotope mixture was adjusted for each experiment to establish an initial sulfate concentration identical to that at the sediment surface when 0.1 ml of the solution was added. Combined addition of tracer and carrier sulfate was important to obtain an even distribution of both compounds within the sample. The sulfate concentration of the tracer solutions varied between 5 and 20 mM and contained 3 to 5 μCi ml^{-1}. The sediment-isotope mixture was blended vigorously on a Vortex mixer and incubated in the dark at *in situ* temperature. After incubation, each vial was injected with 1 ml of 20% zinc acetate to trap sulfide, followed by submersion in an ethanol-dry ice mixture for 15 min and storage at −25C until analysis was performed. Samples could be stored up to 14 days without any loss in sulfide radioactivity, and sulfate reduction did not resume when samples were thawed. After thawing took place, an anoxic sample (usually 5 ml) was removed from the sealed vial with a syringe fitted with an 18-gauge needle and injected anaerobically into a distillation apparatus similar to that described by Jorgensen and Fenchel (1974). The sediment was acidified to pH 0 with 6 N HCl, and the liberated H_2S gas was quantitatively recovered in a trap containing 15 ml of 2% zinc acetate solution. A sample (10 ml) from the zinc acetate trap was mixed with 10 ml of liquid scintillation cocktail for determination of sulfide radioactivity. The slurry remaining in the reaction vessel was diluted to 200 ml with water and centrifuged. A 1 ml sample of the supernatant was used for quantification of sulfate radioactivity. Sulfate reduction was calculated as described by Jorgensen (1978). Most incubations were stopped before 10% of the initial $^{35}SO_4^{2-}$ was converted to $^{35}S^{2-}$ (i.e., normally within 0.5 to 1 h of incubation) to ensure linearity of sulfate reduction. Isotope recovery ($^{35}SO_4^{2-}$ + $^{35}S^{2-}$) from vials was >93% in incubated samples and in controls which had been either zinc acetate treated or autoclaved before incubation.

Measurement of sulfate reduction in the water column was also done using an isotope technique. Rates of sulfate reduction in water samples were measured in 60 ml glass-stoppered (BOD) bottles immediately upon return to the laboratory. The tapered glass-stopper of a sample bottle was removed, and 0.050 ml of carrier-free $^{35}SO_4^{2-}$ solution (about 50 μCi ml^{-1}) was rapidly injected near the bottom of the bottle. Anoxic distilled water (approximately 1.5 ml) was carefully added with a syringe to avoid entrapping air when the stopper was replaced. The bottles were incubated in the dark within ±1°C of *in situ* temperature for 17 hours. Sulfate reduction rate in the water-column samples was found to be linear for at least 36 hours. The procedure for determining sulfate and sulfide radioactivities is described above; details can be found in Jorgensen (1978).

Sediments for measuring methanogenesis were obtained in a manner similar to that described above (see Zehnder and Brock, 1980). In addition to dredge

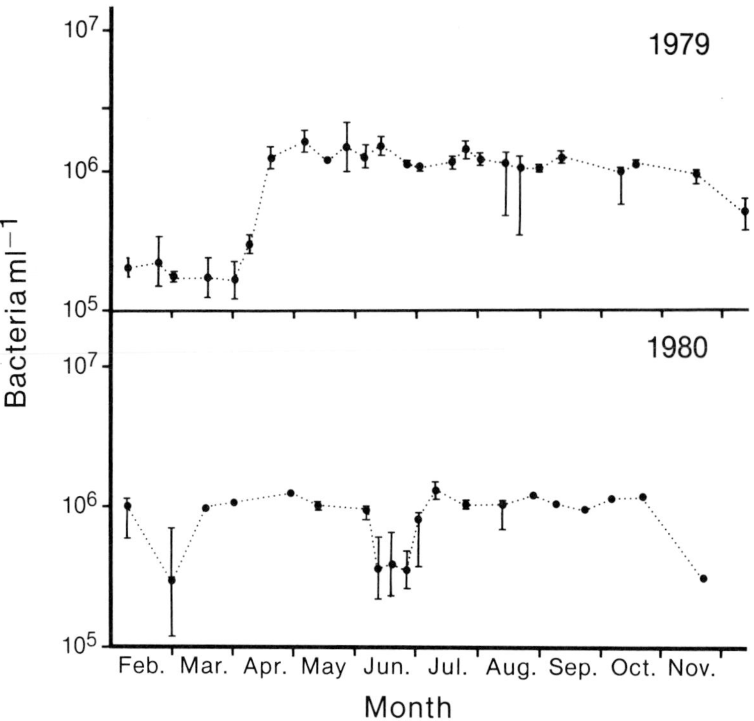

Figure 7.1. Bacterial numbers from epifluorescence counts for 1979 and 1980. Bars indicate 95% confidence limits. (From Pedrós-Alió and Brock, 1982.)

samples, some methanogenesis rates were measured on sediment cores. Cores (6.6 cm diameter) were ordinarily taken by a scuba diver. The sediment was sealed in the coring tube with black rubber stoppers at the lake bottom and the whole apparatus immediately taken to the laboratory in the liner. Sediment samples from the different depths were taken with a syringe through holes which were drilled at 1-cm intervals into the liner before core sampling in the lake (the holes were sealed with adhesive tape to avoid loss of liquid during transport). In some cases a commercial Benthos core sampler was used, taking into account that the uppermost 20 cm of the sediment was considerably compressed by this method. Subsamples from these cores were taken as described above. All incubations were made in the dark in 35 ml serum vials with 20 ml of liquid. The vials were closed with black-lip rubber stoppers, sealed with an aluminum seal, and made anaerobic by evacuating and flushing alternately several times with the gas mixture desired. Methane was assayed by removing gas samples from the headspace using pressure-lock syringes and injecting into a gas chromatograph equipped with a flame ionization detector.

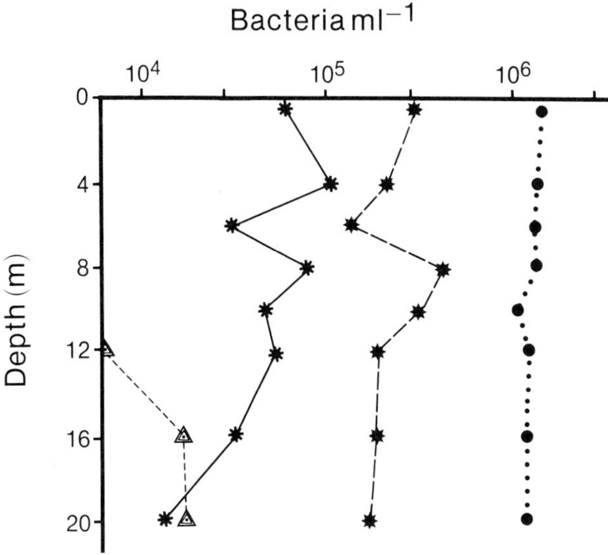

Figure 7.2. Vertical profile for Lake Mendota bacteria for 26 August 1980. Solid circles, total count; stars, bacteria attached to particles; asterisks, *Planctomyces;* open triangles, curved anaerobic rod (procaryote T3 of Caldwell and Tiedje (1975). (From Pedrós-Alió and Brock, 1982.)

Bacterial biomass and production in the water column

Total bacterial numbers for the years 1979 and 1980 are shown in Figure 7.1. Data for years 1976, 1977, and 1978 were similar. The total number of bacteria was different in the winter and spring than during the summer. Low bacterial numbers were found under the ice in the first part of the year. During the spring, bacterial numbers roughly followed phytoplankton abundance but after permanent summer stratification had occurred bacterial numbers attained a more or less constant value (1×10^6 to 2×10^6 cells per ml) and this level was maintained throughout the summer despite wide fluctuations in phytoplankton abundance. Data on phytoplankton abundance and productivity for the years in question have been presented in Chapters 4 and 5.

To determine whether there was any difference in the vertical distribution of bacteria in the lake, vertical profiles were obtained at each sampling date. In general, vertical stratification of bacteria was not found, except for a few times during mid- to late-summer when the lake was firmly stratified and oxygen had been completely depleted from the hypolimnion. A vertical profile for late August is shown in Figure 7.2. As seen, although total bacterial numbers remained constant throughout the water column, differences in recognizable bacterial types were found. For instance, the organism *Planctomyces,* a characteristic bacterium which appears in the epilimnion only in the late summer,

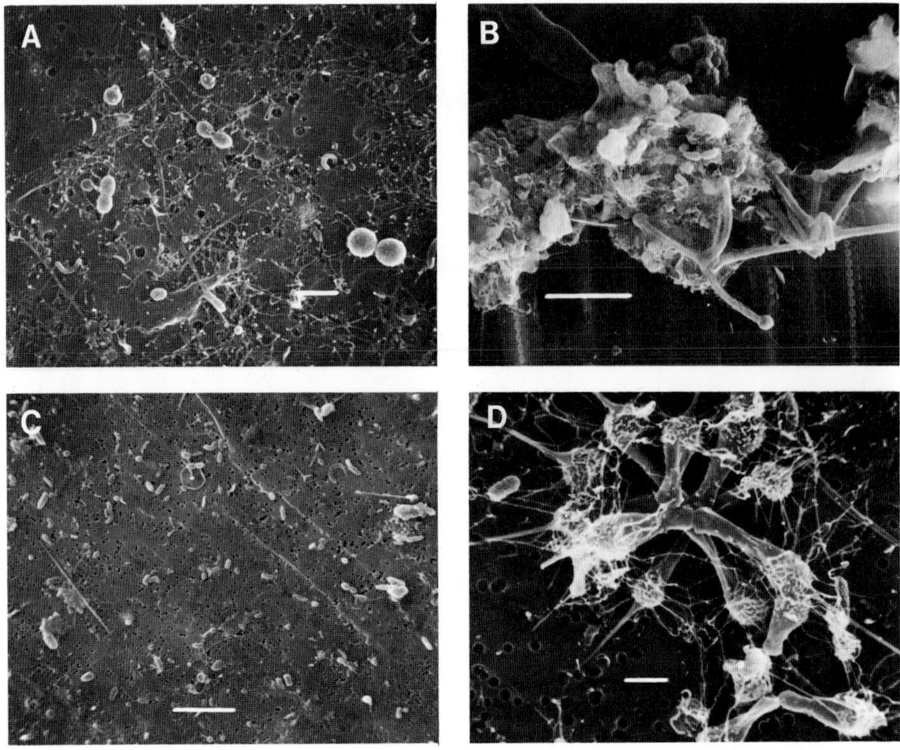

Figure 7.3. Scanning electron micrographs of Lake Mendota bacteria. Dimension of bar: A, 1 μm; B, 5 μm, C, 5 μm; D, 1 μm. (From Pedrós-Alió and Brock, 1982.)

was low or absent in the hypolimnion, whereas another distinctive bacterium, an unidentified procaryote designated T3 by Caldwell and Tiedje (1975), was found only in the hypolimnion. This latter organism was thought to be a strict anaerobe by Caldwell and Tiedje. For the purpose of calculating bacterial biomass and production, the counts from vertical profiles were used.

Bacterial volume and biomass

Figure 7.3 shows examples of scanning electron micrographs of representative bacterial types found in the lake. It was from figures similar to these that bacterial volumes were measured. Bacterial volumes measured during 1980 are shown in Table 7.1 for differently shaped bacteria. The average bacterial volume, calculated by aggregating all measurements done (4654 bacteria), was 0.159 μm^3. This is equivalent to a sphere 0.67 μm in diameter, substantially larger than the volumes reported in the literature for other aquatic systems. By using the conversion factor from volume to carbon (see earlier) and the abundance of bacteria, the biomass in mg of carbon per square meter can be calculated. Bacterial biomass is shown in Figure 7.4 for the year 1980. Data

Table 7.1. Average bacterial volumes in the epilimnion of Lake Mendota in 1980

Shape	Volume (μm^3)		Number of cells measured
	Average	95% CI*	
Free-living			
Rods	0.100	0.022	1,973
Cocci	0.120	0.049	744
Ovals	0.128	0.026	457
Attached			
Rods	0.232	0.069	552
Cocci	0.273	0.123	350
Ovals	0.229	0.088	276
Total			
Rods	0.135	0.036	2,525
Cocci	0.183	0.081	1,094
Ovals	0.160	0.044	733
Summary			
Free-living	0.118	0.009	3,389
Attached	0.271	0.020	1,265
Total	0.159	0.013	4,654

*CI, Confidence interval.

for the year 1979 have been presented in Pedrós-Alió and Brock (1982). In Figure 7.4, phytoplankton and zooplankton biomass values are also given and it can be seen that bacterial biomass is about 10% of either algal or zooplankton biomass, and that the latter two are about the same magnitude. Bacterial biomass changed during the year between approximately 50 and 400 mg of carbon per m^2. The biomass minimum in June of 1980 for both algae and bacteria is probably related to a simultaneous peak in 1980 of the zooplankton *Daphnia galeata mendotae*, an animal which feeds heavily on these organisms.

Bacterial production

Bacterial production calculated both from dark sulfate uptake and from frequency of dividing cells (FDC) is shown in Figure 7.5. Although on any given sampling date these two estimates do not match except in an extremely rough way, they give values within the same order of magnitude. Thus, annual values calculated from these two different methods are comparable, with the FDC values tending to be lower.

Bacterial production ranged from 0.1 to 2.5 g of carbon per m^2 per day (8 to 200 mg of carbon per m^3 per day). Annually, bacteria produced from 100 to 200 g of carbon per m^2. When these values are compared with bacterial production values from the literature, Lake Mendota fits right with the highly eutrophic lakes studied before. The relationship between bacterial production, algal production, and zooplankton feeding will be discussed later in this chapter.

When attempting to relate bacterial productivity to algal productivity, it is

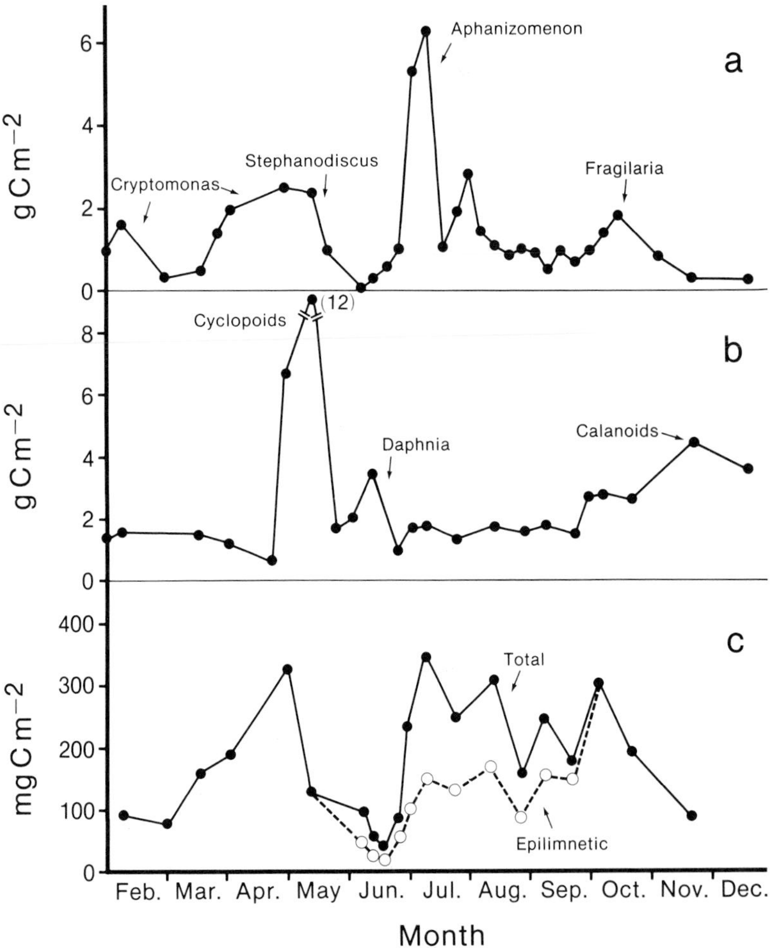

Figure 7.4. Biomass of a) phytoplankton, b) zooplankton, and c) bacterioplankton during 1980. During summer stratification, bacterial biomass in the epilimnion is also presented separately. (From Pedrós-Alió and Brock, 1982.)

important to note that algal production measured with the ^{14}C method measures gross production because of the short incubation periods involved. Bacterial production, on the other hand, is net production. If 50% of the assimilated materials are respired by bacteria, the total carbon flux through the heterotrophic bacteria would be twice the measured bacterial production and would amount to 50–60% of the gross primary production annually. These calculations emphasize how important the bacterial component is for the lake ecosystem as a whole. As will be shown below, feeding by zooplankton removes from 1–10% of the bacterial net production annually and therefore does not seem to impose a heavy pressure on the bacterial population.

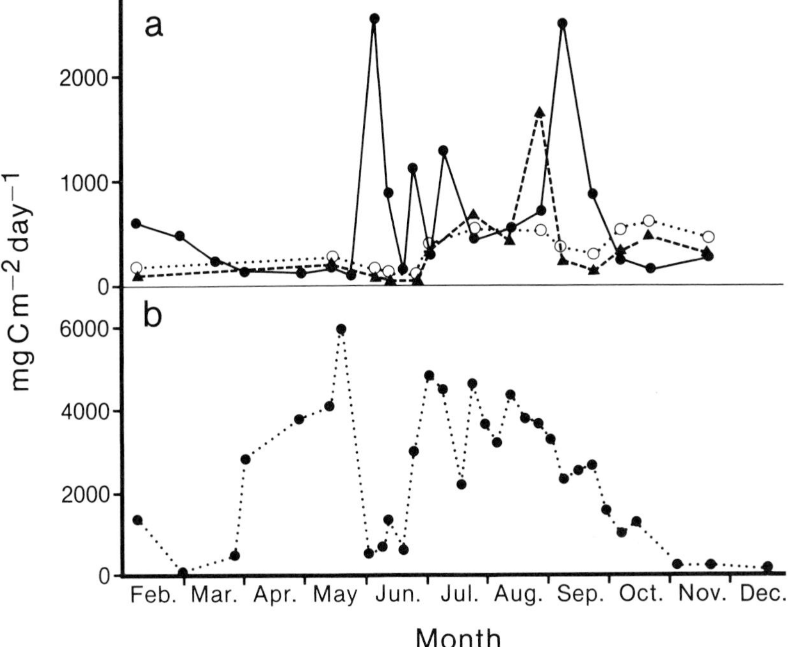

Figure 7.5. a) Bacterial heterotrophic productivity and b) algal primary productivity for 1980. Bacterial productivity is given in two ways, from radioactive sulfate uptake (closed circles) and from frequency of dividing cells, using the equation of Hagstrom et al. (1979) (open circles) or Newell and Christian (1981) (closed triangles). (From Pedrós-Alió and Brock, 1982.)

Correlations

Spearman's rank correlation coefficients were calculated for bacterial biomass and production and for the other biological parameters measured in Lake Mendota. The results are presented in Table 7.2. As seen, bacterial biomass was positively correlated with algal biomass for both years. Additionally, phytoplankton biomass was positively correlated with primary production and with bacterial biomass. Thus it seems that there may be a tight coupling between bacteria and their most probable food source, the algae. This will be discussed further below when we present data on the assimilation by bacteria of algal excretory products.

The impact of zooplankton feeding on bacteria

Data on zooplankton dynamics were presented in Chapter 6. In addition to measuring zooplankton abundance and bacterial abundance, experiments were done to measure directly the feeding of zooplankton on naturally occurring

Table 7.2. Correlation coefficients among environmental parameters in Lake Mendota

	Spearman's rank correlation coefficients							
Parameter[a]	BB	AC	PS	PF	AB	AP	ZB	ZF
BB	X	.251	−.312	.788[b]	.491[c]	.305	.066	.189
AC		X	0.379	.319	.580[c]	.421	−.464	.196
PS	.642	NM[d]	X	−.144	−.263	−.123	−.246	−.105
PF	.049	NM	−0.444	X	.478	.430	−.287	.474
AB	.435[c]	NM	0.219	.094	X	.747[b]	−.233	.208
AP	.083	NM	0.910[d]	.463	.523[c]	X	−.463[c]	.277
ZB	.499[b]	NM	−.328	−.581	−.076	−684[b]	X	.439
ZF	.553[b]	NM	.150	−.696	.012	−.281	.859[c]	X

[a]BB, Bacterial biomass; AC, acetate uptake; PS, bacterial productivity from sulfate; PF, bacterial productivity from frequency of dividing cells; AB, algal biomass; AP, algal productivity; ZB, zooplankton biomass; ZF, zooplankton feeding on bacteria. The upper right hand side of the matrix presents coefficients for 1980, and the lower left hand side presents those for 1979.

[a]NM, not measured.

[b]$0.01 > P > 0.001$.

[c]$0.05 > P > 0.01$.

[d]$P < 0.001$.

bacteria. Two methods were used, one which involved microscopic assessment of bacterial disappearance as a result of zooplankton feeding, the other which measured the incorporation into zooplankton of ^{14}C-labelled bacteria. These methods are described in detail in Pedrós-Alió and Brock (1983a). Briefly, for the microscopic method, animals were removed from the lake and enclosed in vials containing 15–20 ml of lake water which had been filtered through 80 μm Nitex net. Vials were incubated at lake temperature in the dark for 2–4 hours and incubation stopped with formalin. Bacteria were counted in the experimental vials and in controls at zero time and after incubation, using the fluorescence microscope (AODC) technique. Filtering rates were calculated with the formula:

$$\text{filtering rate (ml/animal/day)} = 24/t \times (\ln(C_t/C_0) - \ln C'_t/C'_0)) \times \text{ml/animal}$$

Where t is the incubation time in hours, C_0 and C_t are the numbers of bacteria at the beginning and at the end in the experimental vial, C'_0 and C'_t are the numbers of bacteria at the beginning and at the end in the control vial, and *ml/animal* is the volume of lake water in the vial divided by the number of animals present.

To label the bacteria for these feeding experiments, 250 ml of lake water was filtered through 8 μm Nuclepore filters to remove algae and zooplankton. The filtrate was placed in 250 ml darkened bottles, ^{14}C-acetate added (about 100 μCi), and the bottles incubated at room temperature for about 48 hours. The labelled bacterial suspension was concentrated by filtering through a 0.2 μm Nuclepore filter to reduce the volume to a few milliliters. Lake water, filtered through 0.2 μm Nuclepore filters, was added and the volume reduced again at least twice to remove the soluble radioactivity. Radioactivity in the liquid and in the particulate (retained by a 0.2 μm Nuclepore filter) fractions of this suspension was determined by liquid scintillation counting on 0.1–0.2 ml samples. The bacteria in the suspension were counted by fluorescence microscopy. In this way, the specific activity per bacterium and per milliliter of suspension could be determined. The radioactively-labelled suspension was immediately used in the feeding experiments to minimize loss of radioactivity by bacterial respiration or excretion. Animals for these feeding experiments were collected from 500–1000 ml of lake water on 64 μm Nitex net. A few ml of the radioactive bacterial suspension was added and the bottles were incubated in the dark at *in situ* temperatures for 15–20 minutes. Controls with formalin-killed animals and with live animals feeding on the filtrate of the bacterial suspension that went through 0.2 μm Nuclepore filters (the soluble radioactivity) were run every time. Incubation was stopped with formalin and the bottles kept on ice. The contents were filtered immediately on Nitex net, rinsed with distilled water, and washed into Petri dishes. Individuals of different zooplankton species were then sorted with a pipette and each species put in a separate scintillation vial with 0.5 ml of sodium hypochlorite to dissolve the animals and avoid self-absorption of radioactivity. The radioactivity was then counted in a scintillation counter. Filtering rates were then calculated with the formula:

$$\text{filtering rate (ml/animal/day)} = (I - C)/(\text{DPM added/ml}) \times (1440\ \text{min/day})/(\text{incubation time (min)})$$

Where I equals disintegrations per minute (DPM) incorporated per animal in the experimental and C equals DPM per animal in the control vial for that species.

Measurement of total feeding in the lake

With zooplankton abundance and feeding rates for each species and sampling date, and numbers of epilimnetic bacteria for each date (see earlier in this chapter), total feeding was calculated for each species and expressed in three ways.

Procedure 1: F_1, total feeding as epilimnetic bacteria eaten per m^2 and per day:

$$F_1 = f \times A \times E_{b1}$$

Procedure 2: F_2, total feeding as mg of bacterial carbon eaten per m^2 and per day:

$$F_2 = f \times A \times E_{b1} \times C_b$$

Procedure 3: F_3, total feeding as percent of the epilimnetic bacteria eaten per day:

$$F_3 = (f \times A \times E_{b1})/E_{b2} \times 100$$

Where f = filtering rate in ml/individual/day, A = abundance of the zooplankton species in individuals per m^2, E_{b1} = epilimnetic bacteria per ml, E_{b2} = epilimnetic bacteria per m^2, C_b = carbon weight per bacterium (based on measured bacterial volumes and a carbon to volume ratio of 94.3 fg carbon μm^{-3}). The contributions of each species were then summed for each date to obtain total zooplankton feeding per day.

Feeding rates

Average filtering rates of zooplankton on bacteria measured in 1979 and 1980 are shown in Table 7.3. These rates are considerably lower than those for algae found in the literature and are toward the lower end of the scale for published rates on bacteria (see Pedrós-Alió and Brock, 1983a, for a detailed review of literature data).

Total feeding as percent of the epilimnetic bacteria eaten per day, as bacteria eaten per m^2 per day, and the percent feeding by different zooplankton taxa are compared with algal and bacterial biomasses in Figure 7.6. During the

Table 7.3. Average filtering rates on bacteria (ml/individual/day) for Lake Mendota zooplankton

Species	Filtering rate		Number of experiments
	Average	95% CI	
1980			
Daphnia galeata mendotae	1.626	0.925	8
D. retrocurva and *D. parvula*	0.300	—	1
Diaphanosoma leuchtenbergianum	0.393	0.319	5
Chydorus sphaericus	0.245	0.342	12
Bosmina longirostris	0.365	0.064	2
Cyclopoid copepods			
(adults + copepodites)	0.112	0.088	21
Nauplii of copepods	0.045	0.029	20
Calanoid copepods			
Calanoid adults	1.023	0.387	21
Calanoid copepodites	0.321	0.370	6
Rotifers	0.012	0.011	10
	0.021	0.016	6
1979			
Daphnia galeata mendotae	2.602	1.986	5
Cyclopoid copepods			
(adults + copepodites)	1.085	1.006	4
Calanoid copepods			
(adults + copepodites	3.820	4.526	4

From Pedrós-Alió and Brock (1983a).
C.I., confidence interval.

winter, most of the feeding was by *Diacyclops*, but the absolute amount was extremely low. During June, *Daphnia galeata mendotae* was the main animal feeding on bacteria. Finally, during the summer, calanoids and cladocerans shared the dominance as bacterial eaters. Rotifers contributed only a very small part for most of the year.

In Figure 7.7 feeding by zooplankton and bacterial heterotrophic production are compared for the year 1980. While bacterial production was of the order of 200 to 2500 mg carbon m^{-2} day^{-1}, zooplankton feeding was only 1–20 mg carbon m^{-2} day^{-1}. This indicates that even at peak feeding periods, the influence by zooplankton on the bacterial population could not be very important. Table 7.4 compares annual bacterial production with annual zooplankton feeding. Feeding was only 1–10% of the bacterial heterotrophic production, depending on which measure of production was used. Thus, although feeding of zooplankton could remove up to 60% of the bacterial biomass daily when *Daphnia galeata mendotae* had its June peak, the value was usually 10–20% of the bacterial biomass during the summer and close to nil during the winter. Considering the high bacterial productivity (Figure 7.5) and the fact that bacterial biomass did not change significantly over the year or from year to year, some other factors must be more important than zooplankton feeding in controlling bacterial numbers in the epilimnion of Lake Mendota. This question is discussed in detail later in this chapter.

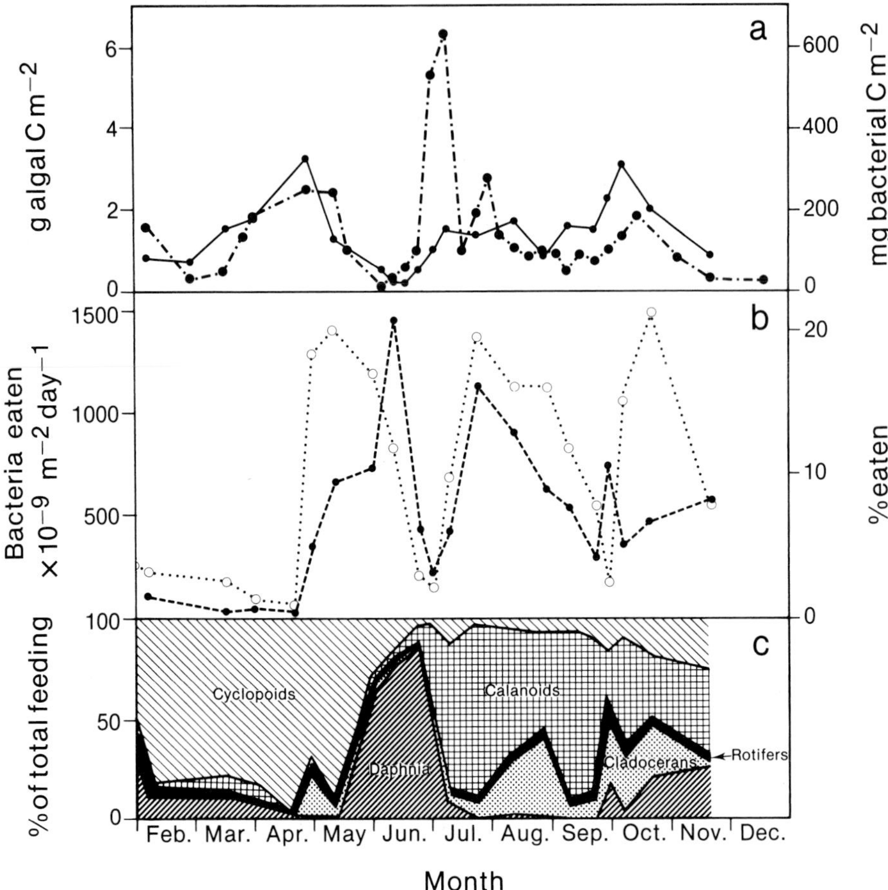

Figure 7.6. Bacterioplankton, phytoplankton, and zooplankton biomass and zooplankton feeding on bacteria for 1980. a) Biomass of phytoplankton (large filled circles) in gC m^{-2} and bacterial epilimnetic biomass in mgC m^{-2} (small filled circles, continuous line). b) Total feeding of zooplankton expressed as F_1: epilimnetic bacteria eaten per m^2 per day (open circles) and as F_3: percent of the epilimnetic bacteria eaten per day (filled circles). c) Percent of total feeding due to different groups of zooplankton. Cladocerans means Cladocera other than *Daphnia*. (From Pedrós-Alió and Brock, 1983a.)

Importance of bacterial attachment to particles

Planktonic bacteria live either in the free-living state or attached to particles. The importance of bacterial attachment to particles has been widely discussed (Floodgate, 1972; Paerl, 1980). Attachment to particles could be useful to bacteria in several different ways. 1) Organic matter adsorbs to particles, creating local accumulations of nutrients which may benefit bacteria. 2) The particles themselves could serve as bacterial food sources, perhaps being derived from algal remains or zooplankton fecal pellets. In this case, the attachment of

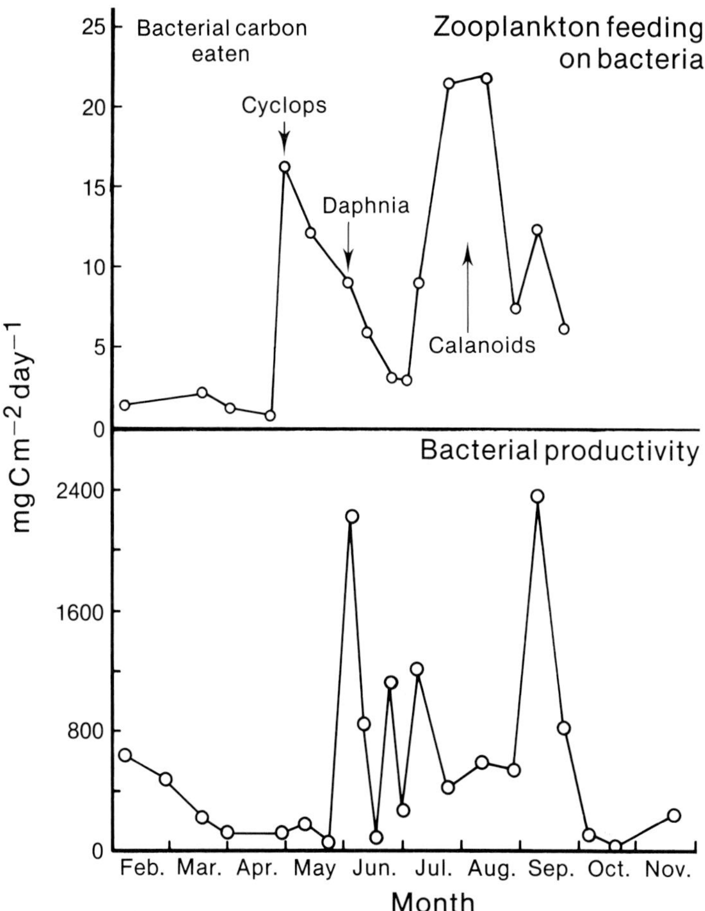

Figure 7.7. Zooplankton feeding on bacteria compared with bacterial production for 1980. (From Pedrós-Alió and Brock, 1983a.)

bacteria to particles is of obvious advantage, since bacteria would be able to use the polymeric particulate matter directly as sources of energy and nutrients. 3) When attached to a particle, a bacterial cell is transported through the water more rapidly than when the bacterium is free-living. This could be an advantage in the following way: A free-living bacterial cell moves with a parcel of water which inevitably becomes depleted of nutrients, whereas an attached cell will constantly be moved to new water and thus will have more nutrients available (Wangersky, 1977).

However, there are disadvantages for bacteria in attachment to particles. Sedimentation out of the water column to the bottom of the lake will occur more readily for attached than free-living bacteria, and during times of the year when the lake is stratified bacteria which settle out of the water column will likely be removed permanently from the ecosystem. Further, bacteria which

Table 7.4. Annual bacterial production and feeding on bacteria for 1979 and 1980

Parameter	1979	1980
Bacterial production (from sulfate uptake)[a]		
Mean daily (mgC m^{-2} day^{-1})	553	561
Range daily (mgC m^{-2} day^{-1}	50–1430	62–2562
Total annual (gC m^{-2})	202	205
Bacterial production (from FDC)[b]		
Mean daily (mgC m^{-2} day^{-1})	292	320
Range daily (mgC m^{-2} day^{-1})	93–1065	80–557
Total annual (gC m^{-2})	107	117
Zooplankton feeding		
Mean daily (mgC m^{-2} day^{-1})	27.3	9.0
Range daily (mgC m^{-2} day^{-1})	1.2–136.5	1.0–22.0
Total annual (gC m^{-2})	10.0	3.0
Percent of production (sulfate)	4.9	1.5
Percent of production (FDC)	9.3	2.6

[a]Bacterial production estimated from dark sulfate uptake.
[b]Bacterial production estimated from frequency of dividing cells.
From Pedrós-Alió and Brock, 1983a.

are attached to particles are more likely to be eaten by zooplankton than free-living bacteria, since both filtering and grabbing, the two feeding mechanisms of zooplankton, more effectively remove particles larger than bacterial size.

In the present section, studies are discussed that were carried out to estimate the relative activity of attached and free-living bacteria in the epilimnion of Lake Mendota (Pedrós-Alió and Brock, 1983b). The procedures are extensions of the procedures used to measure bacterial production and biomass already discussed earlier in this chapter. Differential counts were made of attached and free-living bacteria during annual cycles to determine the relative importance on a biomass basis of the attached bacterial component. Using scanning electron microscopy, the sizes and volumes of attached and free-living bacteria were assessed. Bacterial activity was measured separately for attached and free-living bacteria using three distinct methods: frequency of dividing cells (FDC), uptake of $^{35}SO_4^{2-}$, and uptake of ^{14}C-acetate. To carry out these measurements, differential filtration through Nuclepore filters of defined pore sizes were used to separate the bacteria from algae and attached free-living bacteria.

Abundance and biomass of attached bacteria

The abundance of free-living and attached bacteria is illustrated for the year 1980 in Figure 7.8. As seen, attached bacterial abundance follows the same general pattern as total bacterial abundance: low numbers during the winter and high numbers during the spring and summer. However, the percent of bacteria attached to particles changed substantially, from 1 to 10% in the winter to 15 to 30% in the summer. These values are fairly similar to those reported in a wide variety of other aquatic environments (Pedrós-Alió and Brock, 1983b). The number of bacterial cells per particle varied from 4 to almost 30, being

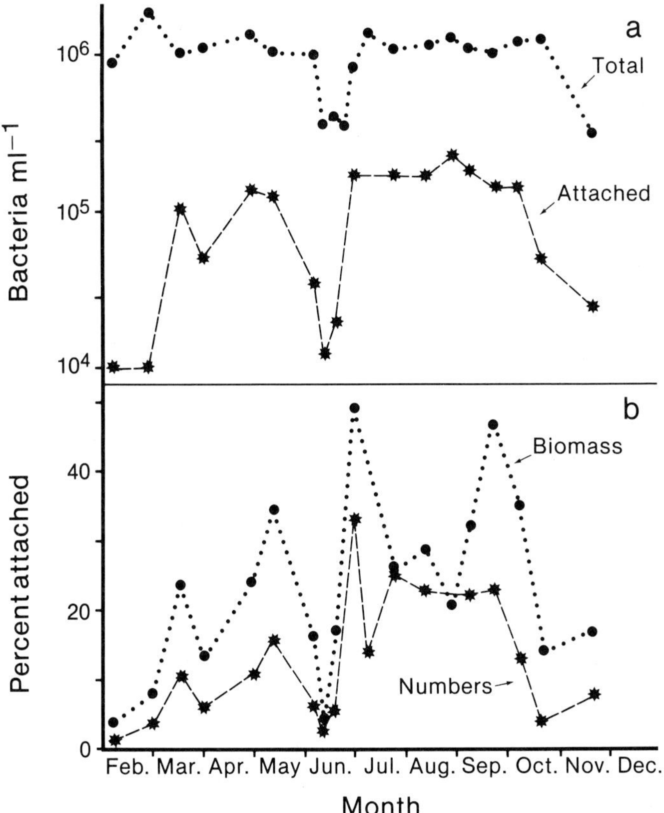

Figure 7.8. Frequency of attached bacteria in 1980. a) Total counts; b) percent of bacterial numbers and biomass due to attached bacteria. (From Pedrós-Alió and Brock, 1983b.)

higher during late summer and fall when the number of attached bacteria was also high.

The volumes of free-living and attached bacteria, measured with the scanning electron microscope (see Figure 7.3), were compared and it was found that the average volumes of attached bacteria were significantly larger than those of free-living bacteria. Because of the larger size of the attached bacteria, the bacterial biomass on particles can be up to 50% of the total biomass, even though the relative fraction of attached bacteria is low.

Activity of attached and free-living bacteria

The growth rate measured by frequency of dividing cells (FDC) of attached bacteria was higher than that of free-living bacteria on 10 of 13 sampling dates in 1980, and the differences were statistically significant on 3 of the 10 occasions. Because of the difficulty of finding particles with more than one dividing cell, the statistics of estimation of FDC on attached bacteria are fairly poor, explaining why significance was found on only three occasions. The two tracer methods for measuring bacterial activity gave different results. Sulfate uptake was greater in the free-living than in the attached populations and ranged throughout the year from 20 to 100%. On the other hand, most of the radioactive acetate was taken up in fractions retained by 3 μm Nuclepore filters, indicating that the attached bacteria were more active in taking up acetate than the free-living bacteria. These differences could be explained if the sulfur of the attached bacteria came from the particulate material that they were decomposing instead of sulfate from the water. If that were the case, sulfate uptake would not be a good measure of bacterial production in environments where attached bacteria constitute a significant portion of the total. Several other possible explanations for the discrepancy between these two procedures are discussed by Pedrós-Alió and Brock (1983b).

Although the mechanism of bacterial attachment to particles was not studied, scanning electron micrographs (Figure 7.3) indicate a large amount of fibrillar material surrounding the attached bacteria. In other work (Corpe, 1980), bacteria capable of attaching to particles have been shown to synthesize complex matrices of extracellular polysaccharides.

Model for bacterial association with particles

A conceptual model which includes the various components thought to be significant in the bacterioplankton dynamics in Lake Mendota is shown in Figure 7.9. Although it is not required by the model, it is conceivable that all bacteria may, at some time, attach to particles. This attachment process has been discussed in detail by many workers (for example, Marshall and Bitton, 1980). The attachment process is found not only in stalked and budding bacteria, but in conventional unicellular bacteria. Even bacteria with no apparent attachment structures are influenced by the presence of increased surface area. In many cases, permanent adhesion to the particles may occur. The relative advantage of particulate attachment for bacterial activity has been shown in some cases and is indicated by the activity measurements described above. It is well established that high-molecular-weight organic matter from aquatic environments adsorbs strongly to inert particulate surfaces, and bacteria which also adsorb to the same surfaces may then be able to grow faster due to increased nutrient availability. Although bacteria may grow better attached to particles, a large population of free-living cells is necessary for two reasons: 1) if no cells were free-floating, colonization of new particles would be impossible; 2) if no cells were liberated from the particles, all the attached bacteria would sink with the particles and disappear from the epilimnion. Thus it is likely that even species of bacteria which are predominantly particle-bound contribute cells to the free-living bacterioplankton.

However, although there are definite advantages of attachment, attached bacteria also suffer negative effects that may counteract the possible benefits of a higher growth rate. In the first place, feeding by zooplankton is more likely

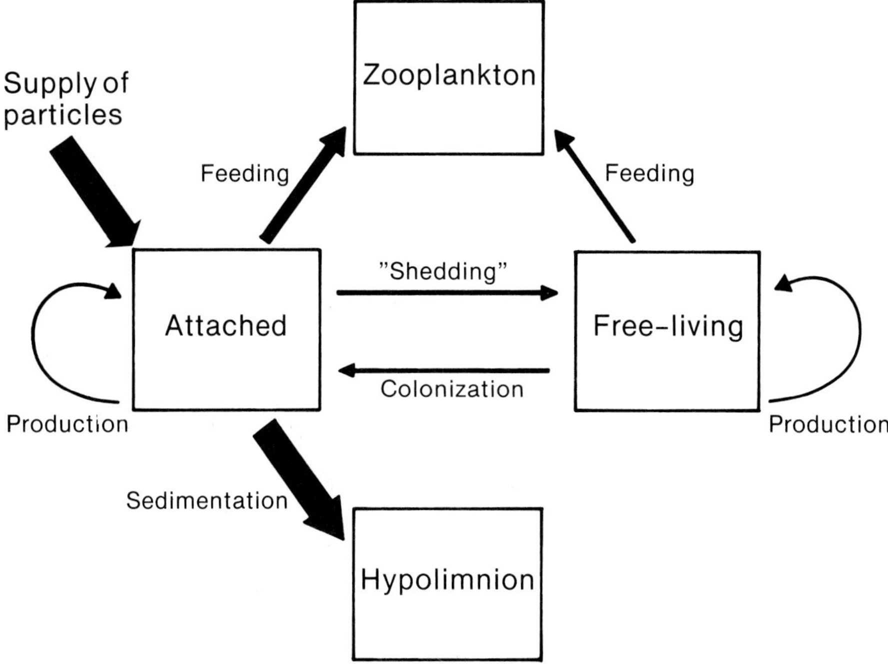

Figure 7.9. Schematic representation of the processes affecting the importance of the attached bacterial population in aquatic systems. (From Pedrós-Alió and Brock, 1983b.)

to affect the attached bacterial population. In the second place, free-living bacteria do not sediment (Jassby, 1975), while attached bacteria do sink with the particles they are attached to and disappear from the plankton.

The following factors probably determine the importance of attached bacteria in aquatic ecosystems: concentration of particles in the system, the extent of flushing of the system (flow rate of water out of the system), sedimentation (influenced by lake stratification), and feeding by zooplankton. If flushing is too high then most bacterial biomass and activity should be found adhered to fixed (nonmoving) surfaces rather than on suspended particles. However, flushing rate is quite low in Lake Mendota (see Chapter 2). The abundance of particles should also be a common determinant of bacterial attachment; if particle concentration is low or lacking then particle attachment may be an insignificant factor. Because of its large size and long residence time, Lake Mendota is relatively free of inorganic (mineral) particles, but has high particle densities during periods of high phytoplankton blooms (see Chapter 4). The importance of zooplankton feeding will vary seasonally within a system, depending upon the abundance and species composition of the zooplankton. We have presented data on zooplankton dynamics in Chapter 6 and discuss the significance of zooplankton for bacterial dynamics below. Finally, sedimentation will cause a more or less constant loss of particles depending on the sizes of the particles and the turbulence of the system that might resuspend particulate materials. In Lake Mendota, permanent bacterial loss because of sedimentation should occur primarily during winter under the ice and during summer stratification.

We have discussed earlier in this chapter the changes in bacterial abundance through the year, and have discussed in Chapter 6 the changes in zooplankton abundance. We have also presented data on direct measurements of zooplankton feeding on bacteria. In Lake Mendota, feeding by zooplankton demonstrated marked seasonal differences. During the winter most animals were cyclopoids, which have very low filtering rates on bacteria. During the spring, bacterial numbers markedly increased, as did zooplankton numbers. Two overwintering animals, *Diacyclops* and *Daphnia*, increased rapidly and reached huge peaks of abundance in succession. In spite of very low feeding rates, the tremendous population sizes resulted in a significant impact on the bacterioplankton. It seems that this was the only period of the year when the zooplankton might have a major impact on bacteria. Note that during this large zooplankton peak, both free-living and attached bacteria were effectively eaten.

In 1979, zooplankton feeding accounted for 5–10% of the annual bacterial production, and in 1980, the value was 1.5–2% (Table 7.4). If zooplankton feeding does not account for all the bacterial loss, where then do the bacteria go? Studies on attachment to particles presented above suggest that sedimentation of bacteria to particles could be the most important factor. Over 60% of the primary production due to phytoplankton photosynthesis undergoes decomposition in the epilimnion, the rest of the primary production sedimenting to the bottom of the lake (Fallon and Brock, 1980). It is likely that bacterioplankton are the main agency involved in this extensive decomposition process in the water column, since, as we have shown earlier in this chapter, bacterial heterotrophic production constitutes about 50% of gross primary production.

Thus it can be concluded that a combination of factors will determine the relative proportion of bacterial biomass and activity associated with particles. Depending on the abundance of suitable particles, on the grazing pressure, and on the degree of turbulence counteracting sedimentation, the proportion of bacteria living attached or free should change from one time of the year to another. In Lake Mendota, on an annual basis, the data presented here show that the predominant activities of the bacterioplankton are found in the free-living population.

A model for bacterioplankton growth and disappearance

To compare in a dynamic way the relative importance of measured biomass changes, bacterial productivity, and losses of bacteria due to zooplankton feeding, a simple exponential growth model, similar to that of Coveney et al. (1977), was used. Bacterial growth was assumed to conform to the equation:

$$C_t = C_o e^{kt}$$

Where C_o and C_t are the bacterial biomasses at two consecutive sampling dates, t is the time between those sampling dates, and k is the exponential growth rate constant describing the net (observed by direct count) biomass change. The growth rate constant k can be subdivided into several components:

$$k=(k_1-k_2)\pm k_3$$

where k_1 is the rate constant for bacterial productivity (always positive), k_2 is the rate constant for zooplankton feeding (always negative; measurement described above), and k_3 is the rate constant corresponding to other, not measured, gains or losses (and therefore is either positive or negative). The value of k can be calculated simply by rearranging the first equation:

$$k=[\ln(C_t/C_o)]/t$$

The other constants can be calculated in the following manner:

$$k_{1'}=P/B$$

$$k_{2'}=F/B$$

Where P is the productivity for a specific date, B is the biomass, and F is the total zooplankton feeding on bacteria per day for the same sampling date. These various k terms are instantaneous rate constants and, to compare the processes they represent through a time period, average constants were calculated for periods between every two sampling dates. Thus,

$$k_1=(k'_{1,0}+k'_{1,t})/2$$

$$k_2=(k'_{2,0}+k'_{2,t})/2$$

and then:

$$k_3=k_1\pm(k-k_2)$$

where the k' values represent instantaneous rates and k is the average rate for the entire period. Subscripts o and t refer to the starting and ending sampling dates for a given time period.

Finally, the change in bacterial biomass that would occur if only one of the factors were operating can be calculated with the formula:

$$\Delta B_a=k_a(C_o+C_t)/t$$

Where ΔB denotes positive or negative change in biomass and the subscript a can be substituted by any of the k terms (k, k_1, k_2, or k_3).

The calculations were carried out with these formulas to determine the relative importance of the various factors that might affect bacterial biomass change. The results are summarized in Table 7.5. As seen, biomass increases due to production were extremely large, whereas biomass decreases due to zooplankton feeding were small, although still larger than actual observed biomass changes. Thus, behind the apparent steadiness of the bacterial population, there is a very fast turnover that feeding by zooplankton could not balance. Therefore, other loss factors had to be operating.

Table 7.5. Changes in bacterial biomass owing to different factors*

Year	Biomass change (k)	Bacterial production (k_1)	Zooplankton feeding (k_2)	Other losses (k_3)
1979	−0.123	131.962	−12.965	−116.779
1980	−0.080	166.997	−2.618	−164.384

From Pedrós-Alió and Brock (1982).
*All values are in gC m^{-2} $year^{-1}$. Negative values indicate decrease in biomass; positive values increase in biomass. See text for calculation procedures.

The possible loss factors for bacterioplankton of Lake Mendota were discussed in some detail by Pedrós-Alió and Brock (1982), and will be discussed only briefly here. One commonly mentioned loss factor is lysis due to viruses or other bacteria, to autolysis, or to the action of antibiotics produced by other bacteria. It seems unlikely in an aquatic system in which the bacterial population is so dilute that lysis due to viruses or lytic bacteria, or antibiotic action could be responsible for the loss of bacteria. In other systems where bacterial lysis has been studied, it has been found that when bacteria are this dilute, the probability that a virus or lysing bacterium could encounter a host bacterium would be very low. Fallon and Brock (1979) studied organisms capable of lysing blue-green algae in Lake Mendota. The organisms they studied were both bacteria and amoeboid protozoa, and the numbers they found were generally low, around 10^3/ml. If these numbers could be used as an estimate of bacteria-lysing organisms, it can be concluded that they would have very little influence on population declines of the bacterioplankton. Autolysis is another possible process that could lead to bacterial loss, but it seems illogical that bacteria which are growing fast, as indicated by the productivity measurements, could be undergoing autolysis and death at the same time.

Since free-living bacteria are too small to sediment (Jassby, 1975), it seems that the only alternative left is that bacteria attach to detritus particles (organic or not) and sink out of the epilimnion, since sedimentation would primarily be a process occurring in attached bacteria. As shown above, attached bacteria grow more rapidly than free-living bacteria and so there is probably a preferential association of bacterioplankton with particles. The end result is that a considerable portion of the bacterial production becomes locked into particles and sediments with them to the hypolimnion and eventually to the bottom of the lake. However, further quantitative studies on bacterial sedimentation are necessary to confirm the importance of this mechanism for the removal of bacterial biomass from the lake. The sedimentation studies by Fallon described in Chapter 3 involved measurements of organic carbon, an undetermined amount of which would be due to bacterial cells. Obviously, the food chain through the bacterial compartment would operate quite differently if the bacteria were primarily sedimenting to the bottom of the lake than if the bacteria were being consumed within the epilimnion itself.

Significance of algal excretory products for epilimnetic bacteria

Three main sources of organic matter can be envisioned for planktonic bacteria: living phytoplankton cells, dead phytoplankton and other nonliving organic particles, and dissolved organic carbon. The dissolved organic carbon (DOC) pool in lakes has several sources: runoff from the land, excretion by aquatic animals, lysis of plankton cells, and excretion by living phytoplankton. Much of the DOC pool may consist of highly polymerized macromolecules such as humic acids derived from terrestrial sources, and thus be partially unavailable for use by epilimnetic bacteria. The portion of the DOC pool consisting of low-molecular-weight compounds such as simple sugars, carboxylic acids, amino acids, and short peptides should originate primarily from the plankton in the water column and is probably turned over rapidly. The low-molecular-weight DOC released by phytoplankton may be the result of cell lysis upon the crash of an algal bloom (Fallon and Brock, 1979) or may be excreted during photosynthesis by a healthy population. Release by a healthy population is usually defined as "excretion."

It is generally accepted that heterotrophic bacteria directly utilize products excreted by algae and that bacterial production and primary production are closely correlated. However, carbon excreted by phytoplankton may be of greater or lesser significance for bacterial growth, depending upon the ecosystem under study. Excretion of dissolved carbon by phytoplankton and subsequent uptake by heterotrophic bacteria can be studied using ^{14}C and differential filtration with natural samples. In the present work, these techniques were combined to permit an assessment of the importance to epilimnetic bacteria of carbon excreted by algae. Because the seasonal cycle of primary production (Chapter 5) and bacterial production (see above) have been measured, the fraction of the total bacterial production that can be supported by excreted carbon can be estimated.

The methods used for ^{14}C-labelling and differential filtration have been presented earlier in this chapter. The uptake rate of excreted carbon by bacteria was calculated from measurements of the rate of uptake in the 0.2 μm–3 μm particulate fraction. Direct microscopic counts and chlorophyll measurements had shown that there were no phytoplankton cells in this fraction and that the majority of the planktonic bacteria were present in this fraction. In this study, bacterial uptake by light-stimulated algal excretion was measured a number of times during the annual cycle to provide a detailed assessment of the significance of this process.

Details on the study of algal excretion for photosynthesis and primary production can be found in Brock and Clyne (1984). There was considerable variation in algal photosynthesis and algal excretion rate throughout the year, with an average excretion rate for the year of 41.4% of the total light-stimulated ^{14}C incorporation. The excretion values obtained were higher than those in earlier work because previously excretion had been estimated only from the fraction of radioactivity that passed a 0.4 μm filter, thus not counting for the excreted material that passed into the bacterial fraction. In the present work, this bacterial uptake has been included in the total excreted material.

The relative importance of algal excretion for bacterial production is presented in Table 7.6. Bacterial assimilation rates in this table have been calculated in two ways: 1) from the radioactivity present in the >0.2 μm to <3 μm fraction; 2) from the total bacterial count in the samples, assuming a carbon content per cell of 15×10^{-15} g and an average production rate of 4×10^3 mg C/g bacterial carbon/day (annual averages from the data of Table 7.4). Also presented in Table 7.6 are calculations intended to express the percentage of bacterial growth (calculated from total numbers and average production rates, that could be accounted for by algal excretion). As seen, throughout most of the year, a significant proportion of bacterial production can be accounted for by algal excretion. Only during the fall of the year, when extensive algal decomposition was occurring, did excretion account for only a small fraction of the bacterial production. It should be noted that in the year in which this work was carried out a marked increase in bacterial numbers took place during the fall. It is likely that the bacteria seen at that time in the water had developed primarily on decomposing algal material, whereas at other times of the year they were primarily consuming algal excretory products.

These results, taken together with those presented above on bacterial production, provide strong support for the idea that epilimnetic bacterial production in Lake Mendota is closely tied to algal production, and that throughout much of the year a significant portion of bacterial carbon is derived from algal excretion. Although it is obvious that other sources of organic carbon for bacteria exist in aquatic systems, such as humic acids, phytoplankton probably constitute the major source in most cases. In aquatic habitats such as deeply stained lakes, where poor light penetration limits phytoplankton production and high humic acid provides a significant source of bacterial nutrients, phytoplankton may be only a minor source of bacterial carbon and energy. Sediment resuspension also may constitute a significant source of bacterial carbon at certain times of the year (Fallon and Brock, 1980). However, in large clear-water lakes such as Lake Mendota, bacteria must be mostly dependent on phytoplankton*.

However, the dependence of bacteria on phytoplankton need not be a direct one via immediate utilization of excreted organic carbon by living phytoplankton. Extensive decomposition of phytoplankton does occur (Fallon and Brock, 1979). In Lake Mendota, it seems that phytoplankton decomposition constitutes the major source of bacterial carbon during the fall of the year, when phytoplankton breakdown occurs at an extensive rate.

The values reported in Table 7.6 show wide seasonal variation. It is clear that at certain times of the year, or with certain phytoplankton populations,

* Note, however, that all measurements of bacterial uptake of algal excretory products were done at light intensities saturating for photosynthesis. At lower light intensities, excretion might be less important.

Table 7.6. Bacterial carbon assimilation of algal excretion products

Date (1981)	Total bacterial count $\times 10^6$	Assimilation calculated from total count (mmol of C/liter/h)	Measured assimilation of algal products (mmol of C/liter/h)	Fraction from algal products (%)	Bacterial carbon uptake (% of total algal excretion)
3/27	4.96	1.03×10^{-3}	1.85×10^{-4}	17.9 (8.9)[a]	28
5/21	0.05	1.02×10^{-5}	7.54×10^{-5}	>100	20
6/2	7.01	1.46×10^{-3}	1.53×10^{-5}	1.0 (0.5)	19
6/26	8.46	1.76×10^{-3}	1.49×10^{-5}	8.5 (4.2)	17
8/4	7.73	1.61×10^{-3}	6.6×10^{-5}	4.1 (2.0)	11
9/3	6.89	1.43×10^{-3}	1.04×10^{-4}	7.3 (3.6)	10
9/10	3,550	7.39×10^{-1}	1.65×10^{-4}	0.17 (0.008)	12
9/29	2,000	4.17×10^{-1}	1.36×10^{-4}	0.033 (0.016)	19
10/6	4.08	8.54×10^{-4}	1.41×10^{-4}	16.6 (8.3)	24
10/13	2,800	5.83×10^{-1}	1.62×10^{-4}	0.028 (0.014)	31
11/3	5.78	1.20×10^{-3}	2.55×10^{-5}	2.1 (1.0)	12
4/14[b]	2.50	5.21×10^{-4}	3.58×10^{-4}	68.6 (34.3)	39
Average				(13.6)	

From Brock and Clyne (1984).

[a]Numbers in parentheses are the values obtained if it is assumed that 50% of the bacterially assimilated carbon is respired rather than incorporated into cell material.

[b]1982.

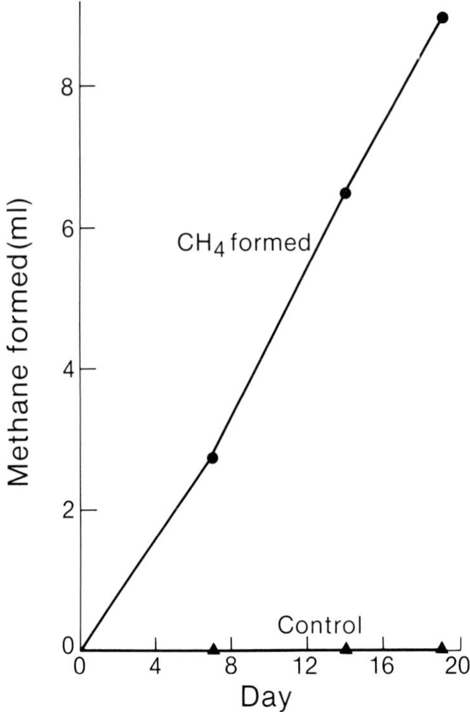

Figure 7.10. Typical time course of methane formation by anaerobic sediment. The controls consisted of heat-inactivated (70°C) and $HgCl_2$-treated sediment. (From Zehnder and Brock, 1980.)

bacterial production is almost totally supported by algal excretion, whereas at other times of the year, when algal decomposition is high, only a very small fraction of the bacterial production is derived from materials excreted by living phytoplankton. Overall, these studies provide strong support for the hypothesis that bacterioplankton constitute a major comsumer of algal production in Lake Mendota.

Bacterial processes in the anaerobic deep-water sediments

Two major bacterial processes were measured in anaerobic sediments: sulfate reduction and methanogenesis. The methods for measuring these processes have been described earlier in this chapter.

Methane production in Lake Mendota sediments

When anaerobic sediment from Lake Mendota was incubated under *in situ* conditions, methane was readily formed (Figure 7.10). Methane formation

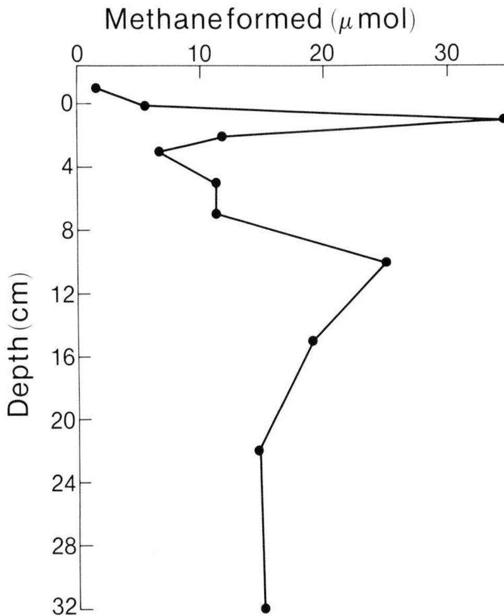

Figure 7.11. Profile of methane formation in a core sampled 26 August 1977 and measured after 6 weeks incubation at 10°C. Values are given for 10 cm^3 wet sediment. The dry weight values, mg/ml, for each core segment are as follows: 0 cm, 0.5; 1 cm, 12.2; 2 cm, 8.0; 3 cm, 15.2; 5 cm, 33.2; 7 cm, 26.8; 10 cm, 134.8; 15 cm, 161.0, 22 cm, 199.2; 32 cm, 286.0.

occurred only under anaerobic conditions: If the sediment was aerated by shaking it vigorously in the air for a short period of time, methane formation was inhibited. Upon aeration, the black mud turned to brown as a result of the oxidation of the ferrous sulfide to ferric hydroxide, but if this brown aerated sediment was incubated in a sealed serum vial with nitrogen gas in the head space, the sediment gradually became black again and methane formation resumed. The addition of a carbon source, such as acetate, resulted in a faster restoration of anaerobiosis and consequently a shorter lag phase before methane formation resumed. However, even in the absence of an additional carbon source, anaerobiosis returned within 2–4 days and the methane formation rate was restored to the normal rate within 1–2 weeks. As shown initially by Zeikus and Winfrey (1976), and confirmed by Zehnder and Brock (1980), the temperature optimum for methane formation in Lake Mendota sediments was about 37°C, well above the temperature of the sediments even during the summer. However, significant methane formation could be measured at ambient temperature (Zehnder and Brock, 1980).

Methane formation varied with depth in the sediment, as shown in Figure 7.11. The core from which the experiment shown in Figure 7.11 was obtained was taken in the late summer in Lake Mendota, a time when the sediments are very soft due to accumulation of a large amount of blue-green algal material as a result of sedimentation. Although this core shows two distinct peaks of active methane formation, if the data are plotted on a sediment dry weight basis, only a single peak of high methanogenesis is found, and this occurs at the sediment surface. This is evidence that methane is derived primarily from the recently sedimented organic material. Extensive studies on factors limiting methanogenesis in Lake Mendota sediments have been carried out by students of J.G. Zeikus (Winfrey, 1978; Phelps, 1984).

These studies, and those reported by Zehnder and Brock (1980), can be used to estimate a methane formation rate for the Lake Mendota sediments. In addition, the rate of methane formation can be estimated during the summer stratification from profiles of methane determined throughout the water column. The rationale and procedures for doing this have been presented by Fallon et al. (1980). This latter study was part of a broader study on the fate of organic carbon in Lake Mendota, which has been discussed in Chapter 3 (see also Fallon, 1978). The methane production rates were estimated by taking water samples at periodic intervals throughout the summer from various depths in the lake and assaying these samples for methane by gas chromatography.

A typical profile for methane distribution and methane oxidation during the summer is shown in Figure 7.12. Methane showed peak concentrations between the 16 and 24 m depths, with a rapid decline in methane concentration at the thermocline (10–14 m). Concentrations in the epilimnion were always <1 nmol·ml^{-1}. The sharp decrease in methane concentration at the thermocline is due partly to vertical mixing into the overlying epilimnion, and partly due to methane oxidation by methane-oxidizing bacteria that develop within the thermocline region (Harrits and Hanson, 1980). Methane evasion rates could be estimated according to a model using Fick's Law for diffusive flux (Fallon et al., 1980).

Table 7.7 shows the calculated values for methane dynamics in Lake Mendota for the 3-month summer period. Methane production was estimated to be 2,210 mmol C·m^{-2}. Total lake methane oxidation equalled 45% of the methane produced and methane evasion to the atmosphere was estimated at 8% of the total produced. Thus, during the stagnation period of the summer, about 47% of the total estimated methane production accumulated in the hypolimnion.

Data for primary production yielded an estimated 250 mmol C·m^{-2}·d^{-1} fixed. From the sedimentation trap data (Chapter 3), the input of particulate material was estimated at 67.5 mmol or about 25–30% of the primary production. If we assume that most of the methane in the pelagic portions of the lake was evolved from the sediments directly below, areal methane production can be estimated to be 35.8 mmol C·m^{-2}·d^{-1}. Methane production rate (Table 7.7) agrees well with production rates measured in the laboratory in sediment cores (Figure 7.11).

Thus, about half of the organic carbon sedimenting to the bottom of the lake is returned to the water column during the summer stratified period as methane. Fallon and Brock (1980) had shown that about two-thirds of the primary production in the epilimnion was decomposed by bacterial action in

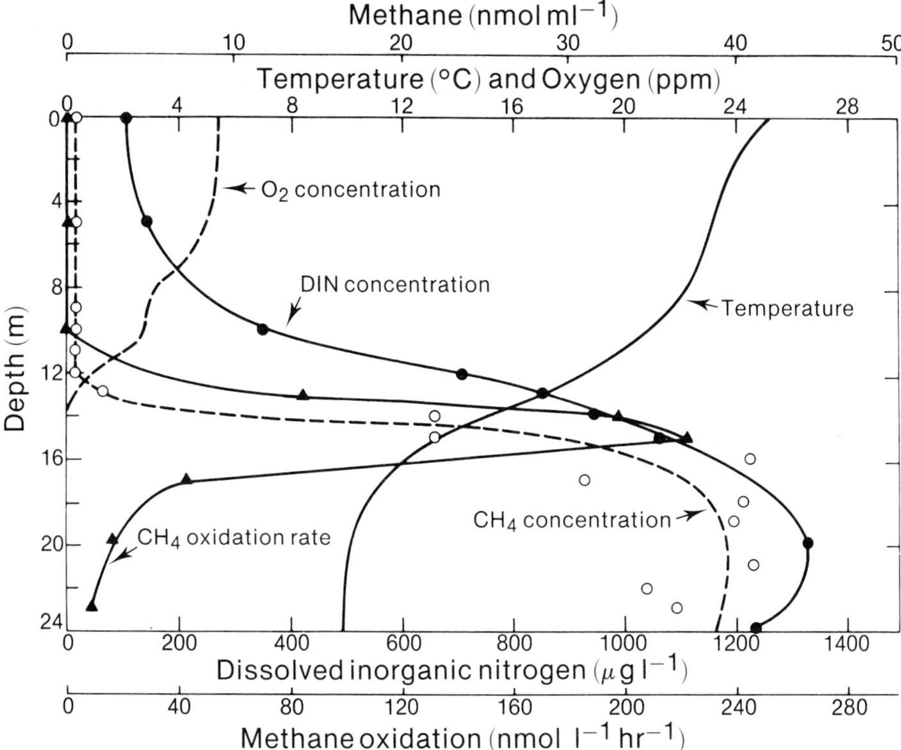

Figure 7.12. Vertical profile of methane, oxygen, and temperature, and methane oxidation rate (measured with ^{14}C-methane), in July 1977. DIN, dissolved inorganic nitrogen. (From Harrits and Hanson, 1980.)

the epilimnion and the other one-third of the primary production sedimented to the bottom of the lake. We thus see that because of the methane formation process, even less than one-third of the primary production becomes incorporated permanently into the sediments. Another process which causes a further loss of organic carbon in the sediments is sulfate reduction, discussed below.

Sulfate reduction in Lake Mendota

It has been well established in many environments that methanogenesis and sulfate reduction are competing processes. In environments where sulfate is high, such as marine environments, a major part of the organic carbon is decomposed via the sulfate reduction process. In freshwater environments, however, sulfate concentrations are considerably lower and little information has been available about the significance of sulfate reduction as a process for the decomposition of organic carbon. In Lake Mendota, inorganic sulfides

Table 7.7. Partial carbon balance for Lake Mendota for stratified period, June 14–September 14, 1977

	Methane cycling	
	mmol $C \cdot m^{-2}$	% CH_4
Process		
Methane oxidized	999.5	45
Methane evaded to atmosphere	181.7	8
Hypolimnetic methane accumulation	1,028.8	47
Total methane produced	2,210	
	Carbon cycling	
	mmol $C \cdot m^{-2} \cdot d^{-1}$	% C sedimented
Process		
Methane produced by deep sediments	35.8	54
Sedimentation of particulate organic carbon	67.5	

From Fallon et al. (1980).

(H_2S,FeS) have been frequently detected in the anaerobic sediments (see Chapter 3). However, the presence of these sulfides is not a direct proof for the sulfate reduction process, since these inorganic sulfide compounds may also arise from organically-bound sulfur (Zinder and Brock, 1978). In an extensive study, K. Ingvorsen studied the sulfate reduction process in Lake Mendota sediments and determined the relative importance of sulfate reduction and methanogenesis in the anaerobic decomposition process. Figure 7.13a shows a profile of sulfide, sulfate, and oxygen at various depths in Lake Mendota toward the end of summer stratification. Hydrogen sulfide was high in the anaerobic hypolimnion, reaching a maximum concentration of 0.14 mM immediately above the sediment surface. Sulfate was depleted in the hypolimnion, and the sulfate profile was almost complementary to that of hydrogen sulfide. Although sulfide may arise from the anaerobic degradation of protein-bound sulfur, the pronounced sulfate depletion in the hypolimnion can only be explained by sulfate reduction.

The rate of bacterial sulfate reduction, measured with radioactive sulfate, is shown in Figure 7.13b. As seen, the rate of sulfate reduction increased by a factor of 10 when a small amount of sediment was incorporated into the hypolimnetic water just at the sediment-water interface. Sulfate reduction was not detected in samples from the aerobic epilimnion (5 and 10 m). Sulfate reduction rates in the top layer of sediment were typically 1,000 times higher than those in the water column just above the sediment.

The optimum temperature for bacterial sulfate reduction in Lake Mendota surface sediment was 37°C, considerably higher than the temperature of the sediment even during the summer maximum. However, even at ambient temperature significant sulfate reduction does occur. It had been shown above that methanogenesis also shows a high temperature optimum in Lake Mendota

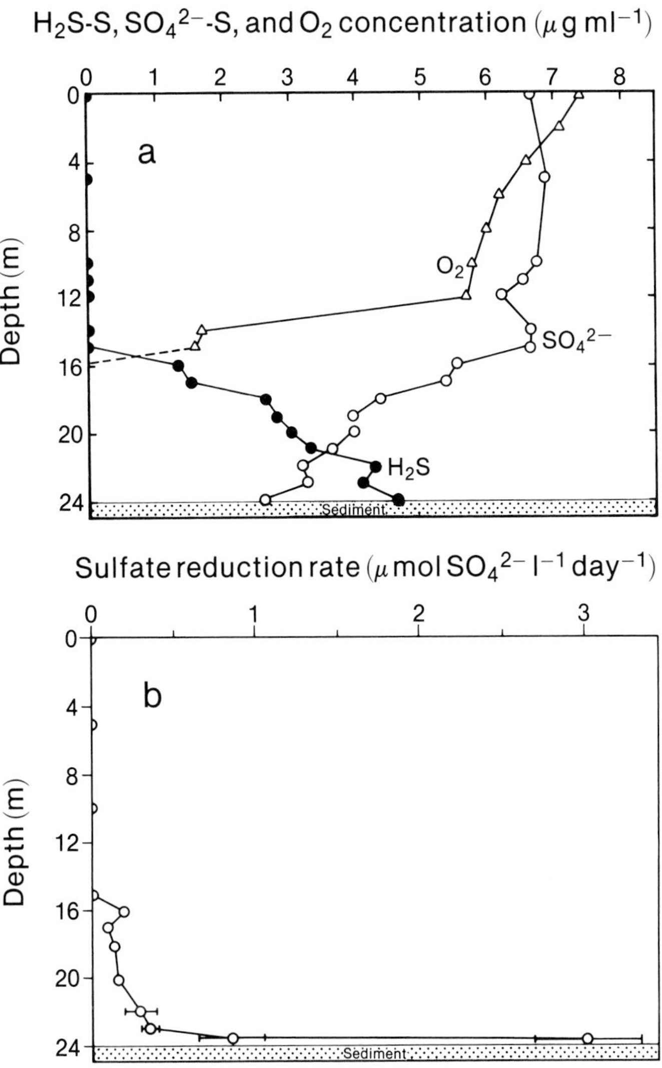

Figure 7.13. a) Distribution of sulfide, sulfate, and oxygen in the water column at the end of summer stratification (18 September 1979). b) Sulfate reduction in the hypolimnion measured with radioactive sulfate. The deepest water sample contained a small amount of resuspended sediment. (Data of K. Ingvorsen.)

sediments. Sulfate reduction rates varied considerably throughout the sampling period, depending upon temperature and sampling date. However, the sulfate reduction rates in Lake Mendota surface sediment, when compared on a dry-weight basis and normalized for temperature, fell in the upper range of rates for this process in marine environments and were of the same order of magnitude as those reported for a productive salt marsh and productive saline

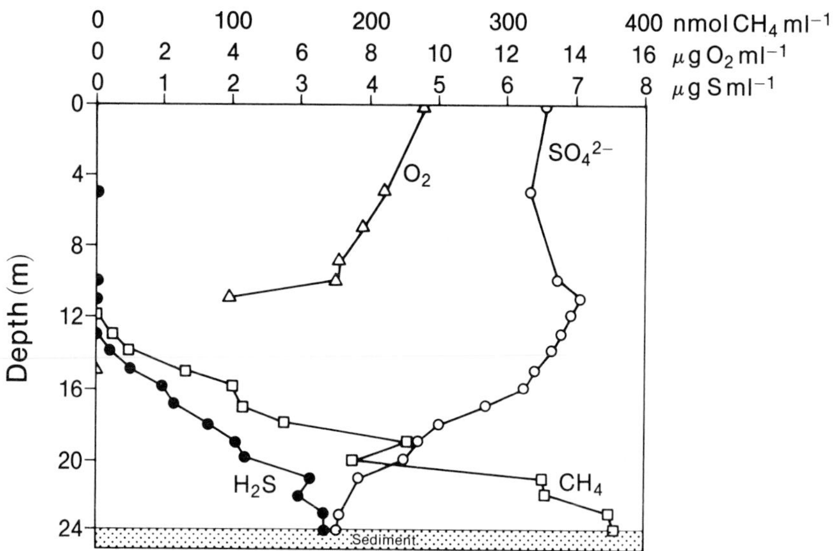

Figure 7.14. Vertical profiles of sulfate, sulfide, oxygen, and methane on 2 September 1979. The thermocline was at 12 m. (From Ingvorsen and Brock, 1982.)

tropical lake. If increased sulfate was added to Lake Mendota sediments, the sulfate reduction process was not stimulated, showing that the process was not limited by the concentration of sulfate. On the other hand, additions of organic compounds strongly stimulated sulfate reduction, suggesting that electron donors were limiting the process.

In order to determine the relative importance of sulfate reduction and methanogenesis in Lake Mendota, a mass balance study similar to that discussed above with methane was carried out. A mass balance model was constructed from changes in the vertical profiles of sulfate, sulfide, and methane during the entire period of summer stratification. A typical profile of the water column, which also shows methane concentrations, is shown in Figure 7.14. Although sulfate is strongly depleted in the hypolimnion, the duration of stratification is too short to allow for complete depletion of hypolimnetic sulfate. Table 7.8 summarizes data collected during stratification in 1979. Data collected in 1976 (see Appendix) and 1977 were similar. The total sulfate consumption in the hypolimnion during summer stratification is the sum of the net sulfate depletion and the diffusion of sulfate from the epilimnion into the hypolimnion. The total sulfate content of the hypolimnion decreased almost linearally during stratification. The net sulfate depletion given in Table 7.8 was calculated as the difference in hypolimnetic sulfate content at the beginning and end of stratification. The diffusion term in Table 7.8 was estimated by applying Fick's first law of diffusion to the slopes of the sulfate profiles determined during stratification, using an average eddy diffusion coefficient supplied by R.E. Stauffer (personal communication). The rate of sulfate consumption in the hypo-

Table 7.8. Partial carbon and sulfur balance for Lake Mendota during summer stratification, 1979

Process	Rate
Methane	
Hypolimnetic methane accumulation	14
Methane oxidation[a]	13
Methane evasion[b]	2
Methane production in sediment	29
Electron eq (mmol $e^- \cdot m^{-2} \cdot d^{-1}$)[c]	232
C mineralization (mmol $C \cdot m^{-2} \cdot d^{-1}$)[d]	58
Sulfate	
Net hypolimnetic sulfate depletion	6
Sulfate diffusion into hypolmnion[e]	1
Total sulfate consumption	7
Electron eq (mmol $e^- \cdot m^{-2} \cdot d^{-1}$)[c]	56
C mineralization (mmol C $m^{-2} \cdot d^{-1}$)[f]	14

From Ingvorsen and Brock (1982).
Data are expressed as nmol$\cdot m^{-2} \cdot d^{-1}$.
[a]Calculation based on data of Table 7.7.
[b]Based on assumption that methane evasion was 8% of total estimated methane production (Table 7.7).
[c]Methane and H_2S (produced by sulfate reduction) both contain 8 electron equivalents.
[d]Assuming that methanogenic processes produce 50% CH_4 and 50% CO_2.
[e]Calculated from slopes of chemical profiles using Fick's first law of diffusion and a mean vertical eddy diffusion coefficient of $2.0 \times 10^2 \cdot cm^2 \cdot s^{-1}$.
[f]Assuming 2 moles of organic carbon oxidized to CO_2 per mole of sulfate reduced to sulfide.

limnion over the entire period of thermal stratification was on the average 7 mmol $SO_4^{2-} \cdot m^{-2} \cdot d^{-1}$. The total rate of sulfide production (net sulfide accumulation rate plus sulfide diffusion rate out of the hypolimnion) during the same period was calculated to be 8 mmol $H_2S \cdot m^{-2} \cdot d^{-1}$. The maximum amount of sulfide that could originate from organic sulfur during summer stratification in Lake Mendota can be estimated on the basis of sedimentation data (Fallon et al., 1980) and the organic content of the sediment, using a commonly accepted carbon to sulfur ratio for the organic material of 100:1. If we assume that all organic material in the upper 1 cm of the sediment and all particulate carbon sedimenting into the hypolimnion were completely mineralized during summer stratification (probably an overestimation), this would produce 0.4 and 0.6 mmol $H_2S \cdot m^{-2} \cdot day^{-1}$. On a molar basis, therefore, organic sulfur would account for $<15\%$ of the total sulfide accumulation.

The data in Table 7.8 thus represent an estimate of the electron flow through sulfate reduction and methanogenesis in the hypolimnion during stratification. It appears that the anaerobic electron flow through dissimilatory sulfate reduction is 25% of that through methanogenesis. Brezonik and Lee (1968) have provided data for Lake Mendota which permit an estimate of electron flow through nitrate reduction (denitrification) of about 14% of the electron flow through bacterial sulfate reduction. Thus, when compared with methanogenesis and sulfate reduction, denitrification is a minor process in anaerobic carbon mineralization in Lake Mendota. Although it is difficult to provide a precise estimate of the amount of organic carbon oxidized to CO_2 by methanogenesis and sulfate reduction, a tentative estimate (Ingvorsen and Brock, 1982) showed that sulfate reduction constituted about 25% of the carbon mineralization via methanogenesis and 20% of the estimated particulate carbon sedimentation. Thus, a significant fraction of the anaerobic organic carbon mineralization is mediated by sulfate reduction, even in a low-sulfate system such as Lake Mendota. Sulfate in Lake Mendota is reduced primarily at the sediment-water interface rather than in the water column. It must also be remembered that higher sulfate concentrations are available for sulfate reduction in sediments covered by oxic water (the epilimnetic sediments during stratification, and whole lake when destratified). Thus, on a whole lake basis, sulfate reduction may be more important for organic matter decomposition than is shown in Table 7.5.

Conclusion

The data in this chapter, taken together with the data in Chapter 5, provide a solid basis for further work on the carbon cycling in Lake Mendota (see also, Chapter 10). It can be seen from the work presented that the microbial components are the most important components in the carbon cycle of the lake. Primary production, which is almost exclusively a process of the phytoplankton, leads to the production of a large amount of organic matter. Some of this organic matter is decomposed in the water column whereas other organic matter is sedimented to the bottom of the lake. The organic matter decomposed in the water column is used primarily in bacterial production and some of these bacteria are consumed by zooplankton for a direct epilimnetic food chain, but most of the bacteria attached to particles settle to the bottom of the lake. Thus, the organic carbon which began as phytoplankton becomes converted into bacterial carbon attached to particles which sediment. However, the carbon which sediments to the bottom of the lake does not remain there. The major part of this carbon is recycled via the two major anaerobic processes of sulfate reduction and methanogenesis. We had earlier (Chapter 3) discussed oxygen depletion (respiration) in the bottom of Lake Mendota and had shown that this process was also driven by organic carbon decomposition. In addition, a small amount of organic carbon breakdown occurs as a result of denitrification. Taken together, these four processes, respiration, denitrification, sulfate reduction, and methanogenesis, recycle up to 90% of the primary production which settles to the bottom of the lake. In the case of methanogenesis, only part of the organic carbon is converted to carbon dioxide and the rest is converted to methane. Some of the methane produced in this process, however, becomes reoxidized by methane-oxidizing bacteria which grow primarily in the thermocline region. A small amount of the methane is lost to the atmosphere by evasion. Thus, at most only 10% of the primary production becomes buried permanently in the sediments. Even this organic matter is not completely fixed because some decomposition continues after the organic matter has been incorporated into the permanent sediments (see Figure 7.11).

In Chapter 5, we showed that only about 1.5% of the solar radiation impinging on the surface of the lake is converted by the photosynthetic organisms into fixed energy in the form of organic carbon. Since about 90% of the carbon which is fixed into photosynthetic organisms is recycled by bacteria either directly by oxidation in the epilimnion or indirectly after sedimentation to the bottom of the lake, the efficiency with which the Lake Mendota ecosystem captures solar energy for use by higher trophic levels in the lake is extremely low, probably less than 0.2%. When it is considered that Lake Mendota is a highly productive lake due to its eutrophic character, it can be appreciated that oligotrophic and mesotrophic lakes must be even less efficient in converting solar energy into biomass.

References

Bere, Ruby. 1933. Numbers of bacteria in inland lakes of Wisconsin as shown by the direct microscopic method. *Internationale Revue der Gesamten Hydrobiologie und Hydrographie,* 29:248–263.

Brezonik, P.L. and G.F. Lee. 1968. Denitrification as a nitrogen sink in Lake Mendota, Wisconsin. *Environmental Science and Technology,* 2: 120–125.

Brock, Thomas D. 1984. How sensitive is the light microscope for observations on microorganisms in natural habitats? *Microbial Ecology,* 10:297–300.

Brock, Thomas D. and Jenny Clyne. 1984. Significance of algal excretory products for growth of epilimnetic bacteria. *Applied and Environmental Microbiology,* 47:731–734.

Caldwell, D.E. and J.M. Tiedje. 1975. A morphological study of anaerobic bacteria from the hypolimnion of two Michigan lakes. *Canadian Journal of Microbiology,* 21:362–376.

Campbell, P.G.C. and J.H. Baker. 1978. Estimation of bacterial production in freshwaters by the simultaneous measurement of (^{35}S)-sulfate and D(^{3}H)-glucose uptake in the dark. *Canadian Journal of Microbiology,* 24:939-946.

Corpe, W.A. 1980. Microbial surface components involved in adsorption of microorganisms onto surfaces. pp. 105–144. In *Bitton, G. and K.C. Marshall (editors), Adsorption of microorganisms to surfaces.* John Wiley, New York.

Coveney, M.F., G. Cronberg, M. Enell, K. Larsson and L. Olofsson. 1977. Phytoplankton, zooplankton and bacteria: standing crop and production relationships in a eutrophic lake. *Oikos,* 29:5–21.

Fallon, R.D. 1978. *The planktonic cyanobacteria: their sedimentation and decomposition in Lake Mendota, WI.* Ph.D. Thesis, University of Wisconsin, Madison.

Fallon, R.D. and T.D. Brock. 1979. Lytic organisms and photooxidative effects: influence on blue-green algae (cyanobacteria) in Lake Mendota, Wisconsin. *Applied and Environmental Microbiology,* 38:499–505.

Fallon, R.D. and T.D. Brock. 1980. Planktonic blue-green algae: Production, sedimentation, and decomposition in Lake Mendota, Wisconsin. *Limnology and Oceanogrraphy,* 25:72–88.

Fallon, R.D., S. Harrits, R.S. Hanson and T.D. Brock. 1980. The role of methane in

internal carbon cycling in Lake Mendota during summer stratification. *Limnology and Oceanography,* 25:357–360.
Floodgate, G.D. 1972. The mechanism of bacterial attachment to detritus in aquatic systems. *Memorie dell'Instituto Italiano di Idrobiologie (Suppl.),* 29:309–323.
Hagström, A., U. Larsson, P. Hörstedt and S. Normark. 1979. Frequency of dividing cells (FDC)—A new approach to the determination of bacterial growth rates in aquatic environments. *Applied and Environmental Microbiology,* 37:805–812.
Harrits, S.M. and R.S. Hanson. 1980. Stratification of aerobic methane-oxidizing organisms in Lake Mendota, Madison, Wisconsin. *Limnology and Oceanography,* 25:412–421.
Hobbie, J.E., R.J. Daley and S. Jasper. 1977. Use of Nuclepore filters for counting bacteria by fluorescence microscopy. *Applied and Environmental Microbiology,* 33:1225–1280.
Ingvorsen, K. and T.D. Brock. 1982. Electron flow via sulfate reduction and methanogenesis in the anaerobic hypolimnion of Lake Mendota. *Limnology and Oceanography,* 27:559–564.
Ingvorsen, K., J.G. Zeikus and T.D. Brock. 1981. Dynamics of bacterial sulfate reduction in a eutrophic lake. *Applied and Environmental Microbiology,* 42:1029–1036.
Jassby, A.D. 1975. The ecological significance of sinking to planktonic bacteria. *Canadian Journal of Microbiology,* 21:270–274.
Jorgensen, B.B. 1978. A comparison of methods for the quantification of bacterial sulfate reduction in coastal marine sediments. I. Measurements with radiotracer techniques. *Geomicrobiology Journal,* 1:11–27.
Jorgensen, B.B. and T. Fenchel. 1974. The sulfur cycle of a marine sediment model system. *Marine Biology,* 24:189–201.
Marshall, K. and G. Bitton. 1980. Microbial adhesion in perspective. pp. 1–6. In *Bitton, G. and K.C. Marshall (editors), Adsorption of microorganisms to surfaces.* John Wiley and Sons, New York.
Newell, S.Y. and R.R. Christian. 1981. Frequency of dividing cells as estimator of bacterial productivity. *Applied and Environmental Microbiology,* 42: 23–31.
Paerl, H.W. 1980. Attachment of microorganisms to living and detrital surfaces in freshwater systems, p. 375–402. In *G. Bitton and K.C. Marshall (eds.), Adsorption of microorganisms to surfaces.* John Wiley and Sons, New York.
Pedrós-Alió, C. 1981. *Ecology of heterotrophic bacteria in the epilimnion of eutrophic Lake Mendota, Wisconsin.* Ph.D. Thesis, University of Wisconsin, Madison.
Pedrós-Alió, C. and T.D. Brock. 1982. Assessing biomass and production of bacteria in eutrophic Lake Mendota, Wisconsin. *Applied and Environmental Microbiology,* 44:203–218.
Pedrós-Alió, C. and T.D. Brock. 1983a. The impact of zooplankton feeding on the epilimnetic bacteria of a eutrophic lake. *Freshwater Biology,* 13:227–239.
Pedrós-Alió, C. and T.D. Brock. 1983b. The importance of attachment to particles for planktonic bacteria. *Archiv für Hydrobiologie,* 98:354–379.
Phelps, Tommy J. 1985. *Ecophysiology of terminal carbon metabolizing bacteria in anoxic sedimentary environments.* Ph.D. Thesis, University of Wisconsin, Madison.
Snow, L.M. and E.B. Fred. 1926. Some characteristics of the bacteria of Lake Mendota. *Transaction of the Wisconsin Academy of Sciences, Arts and Letters,* 22:143–154.
Stauffer, R.E. 1974. *Thermocline migration-algal blooms relationships in stratified lakes.* Ph.D. Thesis, University of Wisconsin, Madison.
Wangersky, P.J. 1977. The role of particulate matter in the productivity of surface waters. *Helgoländer wissenschaftliche Meeresuntersuchungen.* 30:546–564.
Winfrey, M.R. 1978. *Influence of environmental parameters on methanogenesis in freshwater lakes.* Ph.D. Thesis, University of Wisconsin.
Zehnder, A.J.B. and T.D. Brock. 1980. Anaerobic methane oxidation: occurrence and ecology. *Applied and Environmental Microbiology,* 39:194–204.

Zeikus, J.G. and M.R. Winfrey. 1976. Temperature limitation of methanogenesis in aquatic sediments. *Applied and Environmental Microbiology,* 31:99–107.

Zinder, S.H. and T.D. Brock. 1978. Microbial transformations of sulfur in the environment, pp. 446–466. In *J.O. Nriagu (editor), Sulfur in the environment, Part 2.* Wiley-Interscience, New York.

8. Higher Trophic Levels

Although there has been considerable research on animals in Lake Mendota over the past 90 years, there has been little attempt at a quantitative assessment of standing crop, except for the work on zooplankton already discussed in Chapter 6. The present chapter is intended to present a brief overview of previous work on higher trophic levels.

Bottom fauna

Preliminary work on the bottom fauna of Lake Mendota was reported by Muttkowski (1918), and a more detailed study was published by Juday (1921). For the assessment of bottom fauna, Juday separated the lake into three zones: 1) from the shoreline to the 7 meter depth (area of 12 km^2); 2) 7–20 meters depth (area of 21 km^2); 3) greater than 20 meters depth (area of 6.6 km^2). Mud samples taken with an Ekman dredge were diluted with water and washed through gauze net which was fine enough to retain all macroscopic organisms. The various forms extracted were then enumerated in the living state and fresh and oven dry weight values obtained. A summary of the results is presented below. Detailed comparisons of data from a number of stations, which indicate some of the horizontal variability across the lake, can be found in the original paper.

Deep-water bottom fauna

According to Juday (1921), the benthic animal populations in the deep water zone (>20 m) consisted of animals belonging to three separate taxonomic groups: 1) aquatic worms (Oligochaeta) belonging to the genera *Limnodrilus* and *Tubifex*; 2) a bivalve mollusk, *Pisidium idahoense* Roper; 3) larvae of three dipteran insects, *Corethra punctipennis* Say, *Chironomus tentans* Fabricius, and *Protenthes choreus* Meigen. As Juday notes, the bottom-dwelling organisms living in the deep-water area have a fairly stable environment as far as temperature and substratum are concerned. The annual range of temperature at this depth varies from a winter minimum of around 1 °C, to a maximum of approximately 14°, although as we noted in Chapter 2, the bottom temperature of the lake in the summer differs somewhat from one year to the next. In this deep-water area, disturbance by wave action is less and the deposition of material is relatively slow. We discussed the rate of sediment accumulation in Chapter 3.

Although temperature and substratum are relatively constant, dissolved oxygen concentration in this zone varies markedly throughout the year. At depths of 20 meters or more, the dissolved oxygen usually disappears before the middle of July, and is not found again until early October, so that a summer anaerobic period of about two and one-half months occurs. The animals of the genera *Limnodrilus* and *Tubifex* showed little variation in numbers in the bottom sediment throughout the year, being virtually as numerous in the summer during anaerobic conditions as in the winter. They also showed relatively little variation from station to station across the lake. The clam *Pisidium* also showed no significant variation throughout the annual period. Juday concluded that *Pisidium* passed the anaerobic interval in a quiescent or dormant state; specimens which had been kept under observation for a month or so showed no activity whatsoever as long as anaerobic conditions were maintained, but resumed normal activities as soon as they were placed in aerated water. The chironomids were found in a variety of sizes during all seasons of the year. Pupae were most abundant in May and early June and again in the latter half of October. Juday regarded the larvae of *Chironomus tentans* to be an important source of food for fishes and consequently carried out some chemical analyses which showed that these larvae were an excellent source of protein.

The larvae of *Chorethra punctipennis* were in the water column during the night and burrowed into the muddy ooze at the bottom of the lake during the daylight hours. At night they could be found throughout the whole water column, even coming to the water surface. These larvae were also completely absent from the bottom mud during the summer anaerobic period and reached a maximum density in the mud during the winter period from November through March. Juday estimated the total annual production of *Chorethra* larvae in the deep water region of Lake Mendota as about 1200 kilograms of live material per hectare, equivalent to about 100 kilograms of dry matter per hectare. However, since these figures apply only to that portion of Lake Mendota which lies deeper than the 20 meter contour, it cannot be used as an

Table 8.1. Average number of benthic animals per square meter at the various depths in the intermediate zone and at the regular stations

Depth, Meters	Number of Samples	*Oligochaeta*	*Pisidium*	*Chironomus*	*Corethra*	*Protenthes*
5–7		10	19	15	10	24
8	21	311	157	519	747	280
9	24	805	374	291	2,907	196
10	27	1,004	330	557	1,005	370
11	16	1,321	413	953	3,930	200
12	21	1,253	230	716	1,800	340
13	13	1,535	286	1,409	1,470	130
14	5	2,215	502	1,220	3,720	202
15	18	1,750	380	1,815	3,430	350
16	9	2,466	420	1,585	4,830	359
17	5	4,148	295	2,203	10,500	240
18	31	2,370	416	2,300	5,500	195
19	3	2,785	605	1,972	5,850	260
20	17	2,800	506	1,650	7,630	195
8–20	210	1,485	352	1,200	3,460	258
Station I	64	2,320	570	871	5,571	432
Station II	95	3,000	690	549	12,890	191
Five stations	263	3,500	557	593	10,830	185

From Juday (1921).

indication of the rate for the whole lake since *Chorethra* larvae are found in much smaller numbers in shallow waters. Juday noted that the chemical composition of *Chorethra* larvae is such that it should have been a very high-quality food source for fish.

Bottom fauna of the intermediate zone

A summary of the data obtained from sampling of the intermediate zone of the lake, which also includes results of Muttkowski's earlier study, are given in Tables 8.1 and 8.2. As can be seen, the 5–7 meter region is very sparsely populated as compared with the areas a meter or two deeper. Presumably this lower density in the shallow water area is due to the fact that this area is more disturbed by wave action. *Limnodrilus* and *Tubifex* showed a fairly regular increase in numbers with increase in depth, the average between 16 and 20 meters being about 8 times larger than that at 8 meters. There was also a marked increase in *Pisidium* in the intermediate zone, between 8 and 9 meters, the latter depth yielding more than twice as many animals as the former depth. The larvae of *Chironomus* showed a rather irregular increase with depth in the intermediate zone and reached their maximum density at about 18 meters. Beyond this depth, the number decreased, and the average for the whole zone was a little more than twice as large as that of the deep-water zone. *Corethra punctipennis* larvae showed a decided preference for the deeper water, the general average for the intermediate zone being a little less than one-third as

Table 8.2. Live weight and dry weight of the various forms found in the intermediate zone

	Live Weight Kilograms per Hectare	Dry Weight Kilograms per Hectare
Oligochaeta	24.5	4.6
Pisidium		2.7
Sialis		0.8
Chironomus	141.4	22.8
Corethra	186.4	16.0
Protenthes	7.3	1.3
Total	359.6	48.2

From Juday (1921).

large as that of the deep-water zone. The larvae of *Protenthes* were virtually absent from the deep-water zone but were fairly common in the intermediate water zone.

In addition to these common forms, a number of other animals were found in the intermediate zone in relatively small numbers or in isolated samples: leeches, *Hyallella*, aquatic snails, larvae of *Sialis infumata*, and *Palpomyia.*

Standing crop

As part of a comparative study of standing crop in Wisconsin lakes, Juday (1942) presented some data from quantitative studies of the bottom fauna of Lake Mendota. The bottom fauna was stated to have a standing crop of 414 kg/hectare (organic matter on a wet-weight, ash-free basis) which is equivalent to about 2 g organic carbon/m^2. A comparison of these data with those for other components of the food chain is given in Chapter 10.

Fish

Lake Mendota has a relatively rich fauna of fishes, and throughout the period in which the Lake Mendota area has been settled commercial and sport fishing has been a popular activity. A summary of the fishes of Lake Mendota was presented by McNaught (1963) and his species list is given in Table 8.3.

A total of 61 species of fish have been reported for Lake Mendota, 60 of which are among the 173 species found in the Great Lakes drainage. The fishes of Lake Mendota are related primarily to those of the Mississippi and Great Lakes drainages, and the primary fishes in the lake are those which have without a few exceptions been restricted to fresh water throughout their known history. Many of the species of fish listed in Table 8.3 have been stocked at one time or another and McNaught (1963) summarizes these stocking histories.

Certain fish which have been stocked have become a nuisance, of which

Table 8.3. List of fishes found in Lake Mendota

Scientific Name	Common Name	Notes
Petromyzontidae–lampreys		
Lampetra lamottei	American brook lamprey	1
Acipenseridae–sturgeons		
Acipenser fulvescens	Lake sturgeon	2,14
Lepisosteidae–gars		
Lepisosteus osseus	Longnose gar	
Lepisosteus platostomus	Shortnose gar	4
Amiidae–bowfins		
Amia calva	Bowfin	
Salmonidae–trouts, whitefishes, and graylings		
Coregonus artedii	Cisco or lake herring	14
Salmo gairdneri	Rainbow trout	14
Salmo trutta	Brown trout	3,14
Salvelinus fontinalis	Brook trout	14
Umbridae–mudminnows		
Umbra limi	Central mudminnow	
Esocidae–pikes		
Esox americanus vermiculatus	Grass pickerel	15
Esox lucius	Northern pike	14
Esox masquinongy	Muskellunge	5,14
Cyprinidae–minnows and carps		
Carassius auratus	Goldfish	6
Cyprinus carpio	Carp	14
Notemigonus crysoleucas	Golden shiner	15
Notropis anogenus	Pugnose shiner	
Notropis atherinoides	Emerald shiner	
Notropis blennius	River shiner	
Notropis cornutus	Common shiner	
Notropis heterodon	Blackchin shiner	15
Notropis heterolepis	Blacknose shiner	15
Notropis hudsonius	Spottail shiner	15
Notropis spilopterus	Spotfin shiner	1
Notropis umbratilis	Redfin shiner	7
Pimephales notatus	Bluntnose minnow	
Pimephales promelas	Fathead minnow	
Semotilus atromaculatus	Creek chub	1
Catostomidae–suckers		
Catostomus commersoni	White sucker	
Ictiobus cyprinellus	Bigmouth buffalo	15
Moxostoma macrolepidotum	Northern redhorse	1,15
Ictaluridae–freshwater catfishes		
Ictalurus melas	Black bullhead	14
Ictalurus natalis	Yellow bullhead	14
Ictalurus nebulosus	Brown bullhead	14
Ictalurus punctatus	Channel catfish	8
Noturus gyrinus	Tadpole madtom	15

continued

Table 8.3. Continued

Scientific Name	Common Name	Notes
Anguillidea—freshwater eels		
Anguilla rostrata	American eel	9
Cyprinodontidae—killifishes		
Fundulus diaphanus	Banded killifish	
Fundulus notatus	Blackstripe topminnow	
Gadidae—codfishes and hakes		
Lota lota	Burbot	15
Gasterosteidae—sticklebacks		
Eucalia inconstans	Brook stickleback	
Serranidae—sea basses		
Roccus chrysops	White bass	14
Roccus mississippiensis	Yellow bass	
Centrarchidae—sunfishes		
Ambloplites rupestris	Rock bass	
Chaenobryttus gulosus	Warmouth	13
Lepomis cyanellus	Green sunfish	
Lepomis gibbosus	Pumpkinseed	
Lepomis macrochirus	Bluegill	14
Micropterus dolomieui	Smallmouth bass	14
Micropterus salmoides	Largemouth bass	14
Pomoxis annularis	White crappie	14
Pomoxis nigromaculatus	Black crappie	14
Percidae—perches		
Etheostoma exile	Iowa darter	
Etheostoma flabellare	Fantail darter	
Etheostoma nigrum	Johnny darter	
Perca flavescens	Yellow perch	14
Percina caprodes	Logperch	10
Stizostedion vitreum vitreum	Walleye	14
Sciaenidae—drums		
Aplodinotus grunniens	Freshwater drum	
Cottidae—sculpins		
Cottus bairdi	Mottled sculpin	
Atherinidae—silversides		
Labidesthes sicculus	Brook silverside	

From McNaught (1963).

1. Unconfirmed or unpublished but probable, based on reasonably reliable visual reports.
2. A lake sturgeon, *Acipenser fulvescens*, was captured, tagged and released from a fyke-net located off Governor's Island on 22 May 1963. (Length 156.6 cm or 61 3/4 inches, age of 29–31 years.)
3. Brown trout, *Salmo trutta*, captured in a fyke-net at Maple Bluff on 28 May 1963. (Length 51.7 cm.).
4. Shortnose gar, *Lepisosteus platostomus*, examined and verified by Prof. John C. Neess from a sample taken from carp-holding pens in Catfish Bay–Yahara River area. Lake Mendota is close to the northern limit of the shortnose gar.
5. Muskellunge, *Esox masquinongy*, caught by an angler and taken to a local butcher shop, identified there by Prof. John Neess (personal diary, 1946).
6. Goldfish, *Carassius auratus*, stocked in 1855 by Gov. Farwell; captured in a gill-net about 1960.

continued

Table 8.3. Notes—continued

7. Redfin shiner, *Notropis umbratilis*, introduced into one of the University of Wisconsin Arboretum Ponds in 1954. Became established in Lake Monona and Lake Mendota via Murphy Creek.
8. Channel catfish, *Icalurus punctatus*, captured in Mendota in 1862.
9. American eel, *Anguilla rostrata*, captured in Lake Monona in 1880.
10. Log perch, *Percina caprodes*, observed in what was possibly a spawning aggregation of approximately 500 individuals below the Sherman Ave. Bridge at the Yahara River outlet to Lake Mendota.
11. Bluntnose minnow, *Pimephales notatus*, collected along the shore at Maple Bluff with cast-net.
12. The banded killifish, *Fundulus diaphanus*, is especially abundant in the area of Spring Harbor.
13. The warmouth, *Chaenobryttus gulosus*, is infrequently taken by fishermen.
14. Has been stocked in Lake Mendota. See McNaught (1963) for details.
15. Specimens in the fish collection, Museum of the Department of Zoology, University of Wisconsin, Madison.

the carp is the best example. Carp apparently were first introduced into Lake Mendota before 1877. Other species which have been stocked in Lake Mendota include white bass, yellow bass, muskellunge, walleye, northern pike, bluegill, smallmouth bass, largemouth bass, crappie, yellow perch, bullhead, and cisco. The lake sturgeon was likely native to the Madison lakes but it has also been stocked. A sturgeon 156.6 cm long with an estimated age of 29 years was caught in Lake Mendota in 1963 by netting.

The yellow perch, *Perca flavescens,* and the white bass, *Roccus chrysops,* are the most important of the larger pelagic fishes and provide an important sport fishery. Another fish that has been caught commercially is the cisco (also called whitefish), *Coregonus artedii,* a formerly important component of the pelagic fishery which has undergone catastrophic die-offs in recent years (see below). Sport fishing for trout and muskellunge does not exist in Lake Mendota.

Perch

The most important fish in Lake Mendota both quantitatively and in terms of sport fishing is the yellow perch, which has been caught commercially and for sport throughout the length of time that the Lake Mendota area has been settled. There have apparently been large fluctuations in the perch populations over the years, including massive die-offs. Of considerable interest is a massive die-off of perch that was reported in Lake Mendota in 1884 (Dunning and Stevens, 1884; Birge, 1884). Perch died in such large numbers that the phenomenon attracted widespread interest. The die-off began in the middle of July, 1884, and as the fish came to the surface they were driven by waves to the shore. Great numbers of perch accumulated on the south shore adjacent to the city of Madison, decomposing to cause a very offensive odor. An estimated 200 tons of dead fish were found. The fish that died were almost all perch, but a considerable number of cisco and a few suckers and white bass were also found. Interestingly, the perch of Lakes Monona, Waubesa, and Kegonsa were not affected in this 1884 event. According to Dunning and Stevens, fish die-offs had also occurred in earlier years, and every year there were some dead fish. In or about 1844, dead cisco came ashore in quantities as great as the perch did in 1884. Birge (1884) estimated that 200,000 fish had died in two weeks during the month of July. He examined many of these and found no trace of fungus or other evidence of disease.

If it is assumed that fish are 90% water and the ratio of dry weight to organic carbon is 2.5:1, then 200 tons is about 0.9 g C/m^2 of lake surface. These calculations will be considered below when discussing standing crop of perch.

An extensive study of yellow perch in Wisconsin lakes was made by Pearse and Achtenberg (1920). Studies were made of the food eaten by perch and the following results were found: chironomid larvae, 25%; cladocerans, 22%, *Corethra* larvae, 6%; chironomid pupae, 6%; other fish, 5%, amphipods, 3.6%; *Sialis* larvae, 3%; and smaller amounts of many other animals. These workers showed that there were marked seasonal variations in all the types of food that the perch eat and that in general foods were eaten indiscriminately in proportion to their abundance and availability.

An adult perch was found to eat about 7% of its own weight each day. Individuals feeding in shallow water ate a greater variety of types of food than those eating in deeper water. The food eaten by the perch varied with the age of the fish. Young fish ate zooplankton of the genus *Cyclops* and entomostracans, whereas older fish ate *Hyallella* and insect larvae.

The perch were found to become sexually mature in two years. During the spawning season, the males came into shallow water and remained for some time, whereas the females remained only long enough to breed. Except during the spawning season, or when the deeper water was anoxic, most of the perch remained in deep water throughout the year. Perch were found to migrate upward at night and downward during the daytime. The perch swam in schools throughout the year and did not remain in one locality, but moved along the shore.

Hasler and Bardach (1949) studied daily migration of perch in Lake Mendota, using gill nets to trap the fish. They showed that a daily migration of perch took place during the summer months in Lake Mendota at sundown. The fish moved toward the shore in the hours before sunset and, although less intensively, at sunrise. After the fish moved shoreward they cruised along the shore at the 6 meter contour until the sun disappeared.

Hasler (1945) made a series of winter observations on perch in Lake Mendota. These observations were intended to supplement the observations of Pearse and Achtenberg (1920) which had been done primarily during the spring, summer and fall. During the winter time, the perch were found to become concentrated in the deeper waters of the lake and migrated very little into the upper strata. There was a definite correlation between the concentration of perch in deep water and the quantity of *Chironomus* larvae. Hasler also showed that schooling occurred in the winter as well as the summer and that fish of similar size tended to school together.

Bardach (1951) made a study of changes in the yellow perch population in Lake Mendota between the years 1916 and 1948. The 1916 data were those of Pearse and Achtenberg (1920). Bardach concluded that there was a striking

Table 8.4. Average total length and weight of adult yellow perch in Lake Mendota in the years between 1916 and 1948

Year of capture	1916	1931	1932	1939	1943	1946	1948
Number of fish in sample	169	261	51	25	297	375	210
Average total length in mm.	162	198	180	188	214	220	243
Average weight in grams	50	84	76	86	128	137	180

From Bardach (1951).

decrease in the *numbers* of yellow perch in Lake Mendota over the preceding 50 years but that during the same period the *average size* of a yellow perch more than doubled. Although the size increase of yellow perch over that period of time is unquestioned, the data on which the decrease in numbers were calculated were based on fairly small numbers of gill net catches.

Hasler and Wisby (1958) suggest that Bardach's population estimate for 1948 was low. A daily census of fish caught in Lake Mendota during the winter of 1956 by Hasler and Wisby showed nearly 1,500,000 perch were caught. Accordingly, the adult perch population would have had to have been considerably higher than 4,000,000 to allow such a winter catch (4,000,000 was Bardach's estimate). Although standing crop estimates are uncertain it is, however, undisputed that the average weight of the fish increased markedly over the 50 year period between 1916 ad 1948. The data are summarized in Table 8.4.

What might have brought about these dramatic changes in the Lake Mendota perch population? To quote Hasler and Wisby (1958):

> Between 1920 and 1948, large sporadic die-offs of perch were noted. Investigation indicated that a tiny parasite—a small one-celled sporozoan which causes inflamed sores on the sides of perch but which is considered harmless to man—was one of the causes which was killing off large numbers of the fish. Even though some perch die from this disease each year, the mortality rate has decreased greatly from the "mass slaughters" of 1890, 1929, 1946 and 1949. As a result of these declines in Mendota's perch population, remaining fish became longer and heavier. Ten perch weighed a pound in 1916; in 1946, three did. Each acre of lake can support only a certain weight of fish irrespective of numbers if lake conditions remain constant. If the number of individuals is too large, overcrowding results, and each fish remains small. One solution to an over-population problem rests with removing a greater percentage of the population. The remaining fish are then able to grow to a larger size. Another solution to the problem is to feed artificially, but this is impractical. A third solution is to increase the fertility of the water; however, nuisance algae blooms may offset the advantages gained. Since it is not known for certain how much the fish population of Lake Mendota has declined, it would appear that two of these factors—decreasing of the perch population by both disease and harvesting of fish, and an increase fertility of the lake—would raise the amount of food organisms for the remaining fish, thus increasing their length and weight. Land fertilizing practices have increased around the lake during the past decades and, as a consequence, more fertilizer has drained into the lake, enriching the water. Also treated sewage from communities to the north,

Waunakee and DeForest, has contributed much fertility; for example, of all the phosphorous coming into Lake Mendota, 40% entered at Six-Mile creek. This may have contributed to the increase in perch food.

What is the standing crop of perch in Lake Mendota? Hasler and Wisby (1958) suggest that the perch population must be considerably greater than 4 million. Mackenthun and Herman (1949) showed that the average weight of a perch in Lake Mendota was about 0.2 kilograms wet weight. We might thus estimate that the perch population of Lake Mendota amounts to at least 1×10^6 kilograms wet weight. Since the area of Lake Mendota is 39×10^6 m^2 the standing crop of fish would be 25 grams m^{-2}, which would be equivalent to about 1 gram m^{-2} of organic carbon. We noted earlier that a perch die-off occurred in 1884 equivalent to about 0.9 g C/m^2, almost as much as the standing crop in the 1950's!

Cisco

The cisco, *Coregonus artedi* LeSueur, also called lake herring or whitefish, is a species which has been found in Lake Mendota throughout the historical period. The cisco was first studied in Lake Mendota in some detail by Cahn (1927). The cisco is a deep-water fish, spending most of its life in the colder water during the times of the year when the oxygen content is favorable. Breeding only occurs during the early winter when the temperature of the lake has cooled to a mean temperature of less than 5°C. At this time of the year, the breeding fish migrate toward the shallows, the males arriving first and the females later. In the years that Cahn studied, spawning occurred in Lake Mendota between late November and the middle of December.

John and Hasler (1956) carried out a study of factors affecting the hatching of eggs and survival of the young cisco in Lake Mendota. These workers were interested in determining the causes of year class failures of cisco and were trying to explain a marked decline in the populations of cisco in Lake Mendota. They found that the eggs do not hatch under the ice and that hatching begins in middle to late April. The dates of hatching are unrelated to the dates of spawning, the dates of formation or departure of the ice, or the temperatures of the water prior to ice-up. Cisco larvae require illumination in order to feed, and begin to swallow water and feed during the day of hatching. The larvae feed on zooplankton, primarily *Cyclops, Daphnia, Diaptomus, Bosmina, Chydorus,* and rotifers (Cahn, 1927). As the fish reach a larger size, their diet becomes more varied and they have been found to eat mollusks, insect larvae, crustaceans, and small fish.

The cisco is primarily a pelagic fish and is found throughout the deep water of the lake. During that part of the year when the oxygen conditions permit, cisco remain in the deep water, spending most of their time within a meter or two of the bottom. After the formation of the thermocline and the development of anaerobic conditions in the hypolimnion (see Chapter 3), the cisco are forced to leave the bottom waters and move higher in the water column.

As the thermocline climbs upward, the cisco come up with it and remain directly above the thermocline. The massive die-offs which have been reported in various years for cisco are probably due to the fact that in some years the amount of oxic water above the thermocline that is cold enough for the cisco is very limited. Dead fish begin to appear on the surface of the water and if the wind is blowing, the shores become lined with dead fish.

John and Hasler (1956) discussed the cisco fishery in Lake Mendota and the apparent disappearance of this fish from the lake in the 1940's. For almost a century after the settlers arrived, the cisco provided a good fishing. Cisco fishing was done during the spawning season in late November or early December, using gill nets, dip nets, or spears. However, since 1934, the use of gill nets has been forbidden. John and Hasler describe the large cisco catches that were obtained between the years 1890 and the 1930's, but by 1940 the cisco catch began to decrease. From 1942 through 1948, several seasons passed in which the anglers failed to catch any cisco. According to John and Hasler, although cisco fishing became scarce in the early 1940's, the population had started to decline around 1900. However, since actual population assessments were not done except by census of catch, it is uncertain how accurate the data are. One striking thing about cisco is that reproduction is much better in certain years than in others, so that there are strong year classes and weak year classes. Because die-offs have occurred throughout the whole period that Lake Mendota has been under observation, it is more likely that the virtual disappearance of the cisco in the 1940's was not due to a particular die off but to an inability of the fish to reproduce successfully. One striking thing about the cisco die offs in Lake Mendota is that the summer mortalities always comprise only large fish. John and Hasler were unable to provide a satisfactory explanation for the disappearance of cisco in the 1940's. However, it should be noted that cisco did not *completely* disappear from Lake Mendota at this period of time, since cisco are still present and in fact still show certain strong year classes. The cisco year class of 1977 was especially strong (John Magnuson, personal communication), and massive cisco die-offs in 1982 and 1983 were from fish of the 1977 year class.

Summary

Juday (1942) illustrated the weight relationships of the biota in Wisconsin lakes in the form of food pyramids, with dissolved organic matter and phytoplankton at the bottom (large amounts of organic matter) and zooplankton and fish at the top (small amounts of organic matter). However, Juday's data were based almost exclusively on summer observations rather than on annual averages and reasonable measurements for fish were not available.

The standing crops for the various trophic groups are discussed in Chapter 10. We note here that the data for the higher trophic groups, fish and bottom fauna, must be viewed as only approximate, because no detailed studies have been done. Even with this proviso, it is reasonable to conclude that there is no food pyramid, in the usual sense, in Lake Mendota, since the standing crop of fish is almost as high as that of phytoplankton.

The data in this chapter emphasize how little knowledge of higher trophic levels is available. It has occasionally been hypothesized that predation by fish on zooplankton can have significant, even profound, effects in an indirect way on the development of phytoplankton blooms. Although there is no evidence that such indirect effects of fish on phytoplankton do exist in Lake Mendota, further work on fish dynamics would be of great value in helping to understand the interrelationships of the various trophic levels.

References

Bardach, J.E. 1951. Changes in the yellow perch population of Lake Mendota, Wisconsin, between 1916 and 1948. *Ecology,* 32: 719–728.

Birge, E.A. 1884. The dead perch in Wisconsin. *U.S. Fish Commission Bulletin,* 4: 440–441.

Cahn, A.R. 1927. An ecological study of southern Wisconsin fishes. *Illinois Biological Monographs,* 11:94–115.

Dunning, Philo and B.J. Stevens. 1884. Two hundred tons of dead fish, mostly perch, at Lake Mendota, Wisconsin. *U.S. Fish Commission Bulletin,* 4: 439–443.

Hasler, A.D. 1945. Observations on the winter perch population of Lake Mendota. *Ecology,* 26: 90–94.

Hasler, A.D. and J.E. Bardach. 1949. Daily migrations of perch in Lake Mendota, Wisconsin. *Journal of Wildlife Management,* 13: 40–51.

Hasler, A.D. and W.J. Wisby. 1958. Perch and lake research on Lake Mendota. *Wisconsin Conservation Bulletin,* March 1958, 16–20.

John, K.R. and A.D. Hasler. 1956. Observations on some factors affecting the hatching of eggs and the survival of young shallow-water cisco, *Leucichthys artedi* LeSueur in Lake Mendota, Wisconsin. *Limnology and Oceanography,* 1: 176–194.

Juday, C. 1921. Quantitative studies of the bottom fauna in the deeper waters of Lake Mendota. *Transactions of the Wisconsin Academy of Sciences, Arts, and Letters,* 20: 461–493.

Juday, C. 1942. The summer standing crop of plants and animals in four Wisconsin lakes. *Transactions of the Wisconsin Academy of Sciences, Arts, and Letters,* 34: 103–175.

Mackenthun, K.M. and E.F. Herman. 1949. A preliminary creel census of perch fishermen on Lake Mendota, Wisconsin. *Transactions of the Wisconsin Academy of Sciences, Arts, and Letters,* 39: 141–150.

McNaught, D.C. 1963. The fishes of Lake Mendota. *Transactions of the Wisconsin Academy of Sciences, Arts, and Letters,* 52: 37–55.

Muttkowski, R.A. 1918. The fauna of Lake Mendota: A qualitative and quantitative survey with special reference to insects. *Transactions of the Wisconsin Academy of Sciences, Arts, and Letters,* 19: 374–482.

Pearse, A.S. and H. Achtenberg. 1920. Habits of yellow perch in Wisconsin lakes. *Bulletin of the Bureau of Fisheries,* 36: 297–366.

9. Long-term Change in Lake Mendota

In the previous chapters, we have presented considerable data from studies on Lake Mendota which have been carried out across a time period of over ninety years. Lake Mendota is one of the few lakes in the world on which so much systematic work has been done over such a long period. Comparisons of modern with historic data would seem to offer an unparalleled opportunity to assess the changes which take place in a lake ecosystem over time. Although there have been a number of studies of cultural eutrophication (Edmondson, 1966, 1969, 1972), they have dealt mainly with lakes that have undergone recent degradation of a massive character. In the present case, there have been no major human-induced changes in the Lake Mendota basin over the past 90 years, although there have probably been small, gradual changes. Perhaps the most significant change in the Lake Mendota basin over this time period has been the change from a rural to an urban setting. The city of Madison has increased in population over five-fold, in area over two-fold. Fortunately, there has been little increase in domestic sewage input into Lake Mendota, although there have been major changes in inputs into the lower lakes (Sonzogni et al., 1975).

In earlier chapters, we have presented data on various physical, chemical, and biological parameters on Lake Mendota and have indicated some of the factors influencing these parameters. It is the purpose of the present chapter to draw comparisons, where possible, between data from the late 1970s and early 1980s and data from earlier studies. When such comparisons are made, it is very important to assess the validity of the methods used to be certain that they are reliable and comparable. Unfortunately, it has not always been possible to ascertain that methods used in various investigations have been acceptable.

Chemistry and Nutrient Loading

The first nutrient analyses on Lake Mendota were done in the 1920's and are summarized in Figure 3.1. Although nutrient analyses reveal marked changed in Lakes Monona and Waubesa since the 1920s (initially detrimental, later in the direction of abatement), it is doubtful whether any detectable changes in the chemistry of Lake Mendota can be discerned. As we discussed in Chapter 3, chemical analyses of sediment cores indicate that eutrophication in Lake Mendota occurred in the 1850s, long before any chemical analyses were made. (Unfortunately, the molybdenum blue chemical assay for phosphate did not become available until the 1920s.) Although extensive chemical analyses of the nitrogen content of plankton were carried out in the 1910–1916 period (Birge and Juday, 1922), no analyses of nitrogen in water were apparently made until the 1920s (Figure 3.1).

Oxygen in Lake Mendota

One measure of eutrophication in a lake is the rate of oxygen depletion in the hypolimnion. Data were presented in Tables 3.9 and 3.10 and Figure 3.11. Stewart (1976) concluded from his study that the oxygen deficits showed distinct variations between years, but no prominent trends over the years of measurement between 1906 and the present. Stewart believed that the differences seen might more reflect differences in the hypolimnetic temperature than cultural influences. As noted in chapter 2, there has been a marked variation from year to year in temperature of the bottom sediments of Lake Mendota (Table 2.7), depending primarily on meteorological conditions at the time of summer stratification. Since temperature influences the rate of oxygen depletion by its influence on chemical reactions and bacterial metabolism in the sediments, oxygen depletion rates could vary markedly from year to year without any long-term degradation of water quality.

Secchi Transparency

Secchi disk data for Lake Mendota were obtained by Birge and were found in his 1916 unpublished manuscript by Neess and Bunge (1956). Birge's data were monthly means and ranges from several years of measurements and thus represent conditions of clarity in Lake Mendota that were found prior to and including 1916. These data, together with the data of Stewart from the 1960s

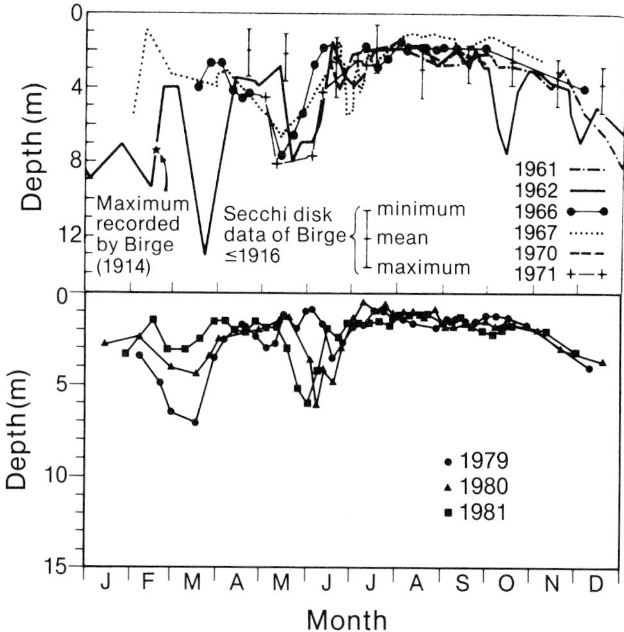

Figure 9.1. Secchi disk depths for various years. Data for 1979–81 from the present study, the other data from Stewart (1976).

and our own data from the late 1970s, are presented in Figure 9.1 and the maximum for several years are recorded in Table 9.1. As discussed by Stewart (1976), Birge's Secchi transparencies were quite similar to those found in more recent years. The maximum Secchi transparency ever recorded by Birge was 7.5 meters in 1914 and this latter maximum has been equalled or exceeded in 1949, 1962, 1966, 1971, 1976 and 1977. To quote Stewart:

Table 9.1. Maximum Secci disk reading recorded various years on Lake Mendota

Year	Day	Value meters	Source
1981	152	6.0	Present work
1980	158	6.1	"
1979	076	7.1	"
1978	159	6.5	"
1977	143	8.0	"
1976	146	8.0	"
1971	mid-May	8.0	Stewart (1976)
1968	13 June	6.3	Stewart (1976)
1949	12 May	9.5	Wohlschlag & Hasler (1951)
1962	22 March	13.2	Stewart (1976)
1914	17 February	7.5	Neess & Bunge (1956)

These data alone provide no evidence of eutrophication. On the contrary, they suggest that Lake Mendota is improving with respect to clarity and water quality. In light of known changes in nutrient inputs and sewage diversions during the past few decades, it is noteworthy that the indices (oxygen deficits, rates of depletion, clarity) examined from irregular years provide little documentary evidence of eutrophication. This was unexpected.

Phytoplankton

Although casual observations of phytoplankton in Lake Mendota were published in the 19th century (Trelease, 1889), no quantitative studies on phytoplankton apparently were done until the 1911–17 period. During those years, Birge and Juday carried out extensive studies on the plankton of Lake Mendota and the other Madison lakes (Birge and Juday, 1922). Although the emphasis in that work was on the chemical composition of the plankton, some studies were done on the abundance of phytoplankton species. The methods that were used by Birge and Juday were pioneering for their time, involving the use of large-scale continuous centrifugation to remove the plankton from the water. Birge and Juday also recognized the fact that the nannoplankton were likely to be the most abundant phytoplankton in the water and made special efforts to quantify these forms. For the quantification of nannoplankton species, continuous centrifugation (which could only be done at low speed) was not carried out. Instead, an electrical centrifuge carrying two 15 cc tubes and having a speed of about 4000 revolutions per minute was used. Although the procedure is not described in detail in Birge and Juday (1922), the method was presented by Juday in the article by Reichard (1918). According to Juday, most freshwater organisms were sedimented completely in five to eight minutes and the nannoplankton was so abundant that only a small quantity of water, not more than 15 cc, was required for the sample. Tapered glass centrifuge tubes were used and the material that was sedimented became concentrated on a small area at the bottom of the tube, from which it could be removed completely. A long tipped 1 cc pipette was used to remove the material from the centrifuge tube, which was then transferred to a Sedgwick-Rafter counting cell.

Unfortunately, the centrifuge method used for the nannoplankton is selective against the buoyant cyanobacteria (blue-green algae), so that these important members of the Lake Mendota phytoplankton were strongly underrepresented in the Birge-Juday data. Some comparative studies using a Foerst continuous-flow centrifuge have been done in the present author's laboratory. This was the type of continuous-flow centrifuge used by Birge and Juday in the 1930s. When membrane filtration is used to quantify phytoplankton there is no selectivity, so that membrane filtration can be used to compare the efficiency with which phytoplankton are sedimented by centrifugation. When a sample is passed through the Foerst continuous-flow centrifuge, both the pellet and the supernatant can be filtered through a membrane filter for quantification. We have found using this approach that the buoyant cyanobacteria float when centrifuged and hence are lost, whereas the quantification of the other phytoplankton is reasonable with the centrifuge method. Since cyanobacteria are a major part of the phytoplankton in Lake Mendota, especially during the summer months, it is impossible to make meaningful comparisons between modern data and the data of Birge and Juday.

For quantifying the net plankton, Birge and Juday pumped water from the lake, using a gasoline engine-driven pump. The water was allowed to pass through a plankton net made of silk bolting cloth, the openings of which varied from 0.001 mm^2 to 0.003 mm^2. For each sample of net plankton, the amount of water strained varied from a minimum of 2000 liters to a maximum of 38,000 liters, depending on the abundance of organisms. In the great majority of instances, the quantity strained was between 10,000 and 20,000 liters. The plankton samples collected were then quantified under the microscope using the Sedgwick-Rafter cell.

Although quantitative comparisons are not meaningful because of the problems mentioned above, it is possible to compare the list of genera in the net plankton found by Birge and Juday with the list of genera found today. These comparisons are given in Table 9.2. Also recorded in this table are the genera listed by Sager (1966), although it should be noted that Sager's data also represent only net plankton. As can be seen, most of the same species of phytoplankton are found today as were found in the period 1911–17.

Several phytoplankton observed by Birge and Juday deserve special note. The diatom species *Stephanodiscus astraea* was one of the more common members of the nannoplankton found by Birge and Juday in Lake Mendota. This diatom was found primarily during the spring period, when it reached high numbers in early April, followed by a decline in May and a complete disappearance by the first part of June. According to Birge and Juday, *S. astrea* yielded a larger number of individuals per liter of water than any other organism that was found in the nannoplankton. This diatom is still frequently the predominant member of the phytoplankton flora in the spring period, as can be seen by the data in Chapter 4 and that presented in the Appendix to this book.

One member of the phytoplankton which was found predominantly in the net plankton by Birge and Juday was *Aphanizomenon,* which was common some years but not others. Because of the size of the *Aphanizomenon* colonies, it is likely that this species would be found primarily in the net plankton. Birge and Juday found it in reasonable numbers in 1911 and 1912, it was rare during the summer of 1913, very common in the summer of 1915 and moderately common in 1916. During the late 1970s and early 1980s, *Aphanizomenon* was also very common certain years but virtually absent other years. The reasons for the marked year-to-year variability in the abundance of *Aphanizomenon* are not known.

Stewart (1976) has also discussed the historic data on phytoplankton in relation to modern data. He concluded that it was difficult to say, from the limited observations of phytoplankton in Lake Mendota, whether conditions were any better or worse than previous records indicate. When our own data (Appendix and Chapter 4) are compared with those of Birge and Juday, it is

Table 9.2. Algal genera recorded for Lake Mendota by various workers

Algal genus	Birge/Juday	Sager	Present
Net plankton			
Ceratium	+	+	+
Microcystis	+	+	+
Coelosphaerium	+	+	+
Aphanizomenon	+	+	+
Anabaena	+	+	+
Lyngbya	+		+
Staurastrum	+	+	+
Melosira	+	+	+
Tabellaria	+		
Fragilaria	+	+	+
Asterionella	+	+	+
Stephanodiscus	+	+	+
Nannoplankton			
Chlorochromonas	+		+
Cryptomonas	+		+
Euglena	+		
Aphanocapsa	+		+
Arthrospira	+		
Chroococcus	+		
Closterium	+	+	+
Cosmarium	+	+	+
Oocystis	+	+	+
Scenedesmus	+	+	+
Sphaerocystis	+	+	+
Cocconeis	+		+
Cyclotella	+		+
Stephanodiscus	+	+	+
Actinastrum		+	+
Ankistrodesmus		+	+
Aphanothece			+
Chlamydomonas			+
Chlorella			+
Chlorococcum			+
Chroococcus		+	+
Crucigenia		+	+
Dactylococcopsis			+
Dictyosphaerium		+	+
Eudorina		+	+
Glenodinium			+
Gloeobotrys			+
Gloeotrichia		+	+
Gomphosphaeria		+	+
Gymnodinium			+
Kirchneriella			+
Merismopedia		+	+
Navicula		+	+
Oscillatoria		+	+
Pandorina		+	+
Pediastrum		+	+
Peridinium			+

Table 9.2. Continued

Algal genus	Birge/Juday	Sager	Present
Planktosphaeria		+	+
Quadrigula		+	+
Schroederia		+	+
Synechocystis			+
Synura			+
Westella			+
Volvox		+	+
Gloeocystis		+	
Dimorphococcus		+	
Gonium		+	
Coelastrum		+	
Tetraedron		+	
Selenastrum		+	
Mallomonas		+	
Cyclotella		+	
Stauroneis		+	
Cymbella		+	
Surirella		+	
Synedra		+	
Gyrosigma		+	
Pinnularia		+	
Nitzschia		+	
Diatoma		+	
Gomphonema		+	

Data of Birge and Juday (1922) were for the years 1911–1916. Although both net plankton and nannoplankton were studied, the centrifuge method used for nannoplankton was selective against the buoyant cyanobacteria.
Data of Sager (1966) were obtained with a plankton net but some of the nannoplankton were obtained. Because Sager was interested in species diversity, he made special effort to detect rare forms.
Present data for the years 1976–1981 as summarized from the Appendix. See also Table 4.5.

difficult to conclude that phytoplankton blooms are any more extensive now than they were in the early part of the century. Nuisance algal blooms have been reported on Lake Mendota since 1888 (Trelease, 1889), and were found in our own work throughout the 1970s and early 1980s. Interestingly, nuisance algal blooms were virtually absent from Lake Mendota during the summer of 1983 (see Chapter 4), but returned in 1984.

Zooplankton

Detailed comparison of zooplankton for 1894–96 and 1976–81, with some species aggregated, were given in Figures 6.1 through 6.7. Note that the vertical and horizontal scales of the graphs of the two data sets were the same. In general, it can be stated that the annual cycles of the various groups were quite similar in the two data sets. The peaks tended to occur at the same time of year, and the peak heights were also similar. Although there was considerable

Table 9.3. Results of a Friedman test for significant differences in abundance of seven zooplankton groups among years studied. Groups that showed significant differences at the 1% level between each pair of years are shown

	Year				
Year	1896	1976	1978	1979	1980
1895	DP,CH CY	DP,DG	CY	DI,DG CY	DI,DG CH,CY
1896	–	DG,CH CY	DP,CH	DP,CY	DP,DG CY
1976		–	DP,DG CY	DI,DP CH,CY	DP,DG CH,CY
1978			–	nd	CY
1979				–	nd

DP *Daphnia pulicaria.*
DG *Daphnia galeata mendotae.*
CH *Chydorus sphaericus.*
DI *Diaphanosoma leuchtenbergianum.*
CY Cyclopoids.
nd no significant difference.
Diaptomids and *D. retrocurva* did not present significant differences in any pair of years.
From Pedrós-Alió and Brock (1985).

year-to-year variability in the phasing of the annual cycles, this variability was no greater in the present data set than in Birge's.

Note that certain species are much more common in some years than in others. *Daphnia pulex,* for instance, was common in 1976 and 1977 and then virtually disappeared, being replaced by *Daphnia galeata mendotae.* The latter species was quite common throughout all of Birge's study, along with *D. pulex.* If we had made our study only in 1976 and 1977, we would have concluded that *D. galeata mendotae,* so common in Birge's study, was now absent. These results point up the importance of a study over a number of years before any conclusions can be made about long-term changes in a lake ecosystem.

In order to make a statistical comparison of the historic and modern data, a Friedman test was run (Pedrós-Alió and Brock, 1985). The Friedman test was performed on the monthly averages of abundances for each zooplankton group. For the analysis, months were considered as blocks and years as treatments, thus, a check was being made for significant differences among any pair of years. Differences could result from different peak sizes, differences in the timing of peaks, or different average and total annual abundances. When significant differences were found, multiple comparison tests were performed to see which years were different. Results from these tests are summarized in Table 9.3, the significantly different groups for every pair of years being shown. It is clear from this table that 1895 was very different from 1896 and close to 1978. Also, 1976 differed more from 1979 and 1980 than from 1895 and 1896. Thus, year-to-year variations were recognized, but no clear differences were evident between Birge's study in the 1890's and that of the 1976–80 period.

It can thus be concluded that as far as species of crustacean zooplankton are concerned, Lake Mendota is quite similar now to what it was almost 90 years ago. As noted earlier in this chapter, in a comparison of historic records for oxygen and secchi disk, Stewart (1976) also suggested that Lake Mendota had not changed significantly since early in this century. These data provide further support for the hypothesis that Lake Mendota has not changed significantly in the past 80 years, but that the cultural eutrophication which has occurred was a result of agricultural activity in the Lake Mendota drainage basin in the mid-1850's. Unfortunately, no data for zooplankton exist for the period around the 1850's, although we do know from the work of Trelease (1889) that cyanobacterial blooms were occurring on Lake Mendota at least ten years before the study of Birge.

It can be suggested that detailed studies of zooplankton dynamics in lakes have the potential for providing sensitive indications of the nature and extent of long-term change in lake ecosystems. Fortunately, the historic data base provided for us by Birge has made the present analysis possible. The detailed data for 1976–80 will conceivably be useful to later workers continuing this kind of study after the year 2000.

Bacteria

Pioneering early work on the bacterioplankton of freshwater lakes was done by Professor E. B. Fred and associates at the University of Wisconsin in the 1920s and early 1930s. Unfortunately, the superior methods available for assessing bacterioplankton in aquatic systems were not available, so that it is difficult to compare these data with the data obtained by more precise methods in the 1970s and 1980s that have been presented in Chapter 7. The first work on bacteria in Lake Mendota was that of Snow and Fred (1926), which was based primarily on viable plate counting methods. Snow and Fred did recognize that plate counts furnished only an estimate of the actual number of bacteria present and that direct microscopic counting methods would provide higher numbers. They therefore attempted to develop a direct microscopic counting method to determine the number of bacteria in water from Lake Mendota. Because of the low numbers of bacteria present in the water, it was necessary to concentrate the bacteria before microscopic counting. The very efficient membrane filtration method used today was not available to Snow and Fred. Snow and Fred used aluminum hydroxide as a flocculating substance to carry down any bacteria in the lake water. After flocculation, the aluminum hydroxide precipitate which had been centrifuged was transferred to a glass slide and the bacteria stained with methylene blue/fuchsin. Snow and Fred experienced considerable difficulty with this method because of the presence of debris carried down with the colloidal suspension, which also became stained and could only be distinguished from bacterial cells with difficulty. Snow and Fred did find that the direct counts obtained were many times higher than the counts obtained with plate counting procedures. However, even with this direct microscopic counting method, the counts obtained by Snow and Fred were many times lower than those obtained with the acridine orange membrane filtration method that has been described in Chapter 7.

Although a membrane filtration method for quantitative determination of bacterioplankton had been described many years ago, this method did not find widespread use because of the commercial unavailability of membrane filters. An evaporation method for concentrating bacteria was described by Kuznetsov and Karsinkin (1931), and this method was used by Bere (1933) for determination of bacterioplankton numbers in some inland lakes of Wisconsin, including Lake Mendota. Kuznetsov and Karsinkin (1931) had obtained direct counts that were 2 to 4000 times greater than those obtained by the plate counting method and although their method was not useful for waters that were fairly high in salts or hardness (due to precipitation during the evaporation process), this method could be used in Lake Mendota. Bere found that the direct counts obtained with this method on Lake Mendota water samples were 20 to 335 times greater than plate counts. It should be noted that Bere's counts were considerably higher than those obtained by Snow and Fred, but were still almost an order of magnitude lower than those obtained by Pedrós-Alió and Brock (1982). The point to be made is that the markedly higher bacterial counts found by Pedrós-Alió and Brock (1982) were due to vastly improved methodology, not to changes in the lake. Thus, no conclusions can be drawn concerning long-term changes in bacterioplankton levels in Lake Mendota because earlier assessments were done with inadequate methods.

Another approach to obtaining an idea of bacterial activity in Lake Mendota would be from studies that were done on oxygen consumption (biochemical oxygen demand, BOD) by bacteria in lake water. ZoBell (1940) carried out a detailed study on the factors which influence oxygen consumption by bacteria in Lake Mendota. Further work was carried out by ZoBell and Stadler (1940). Earlier work on oxygen consumption by water from Wisconsin lakes had been presented by Allgeier, Peterson, and Juday (1934). The work of ZoBell (1940) on Lake Mendota is especially relevant to discussions earlier in this chapter on the hypolimnetic oxygen deficit and to the extent which Lake Mendota has undergone long-term degradation. Some of ZoBell's data obtained in 1938 and 1939 are compared in Table 9.4 with similar unpublished data obtained by David M. Ward in the present author's laboratory in 1973. Note that in both studies, the temperature of incubation was 20°C, so that variation in BOD reflects variation in the organic carbon content of the water. Although the average BOD values obtained by Ward are higher than those of ZoBell, if only values are compared which were obtained on similar sampling dates, the data of the two workers are almost identical. Thus, there is no evidence from the BOD measurements that bacterial activity is any different in the 1970s than it was in the late 1930s.

It should be noted that these data only reflect oxygen demand by the lake water and ignore the oxygen demand of the sediments. The oxygen deficit data presented in Chapter 3 (see Tables 3.9 and 3.10) indicate much higher oxygen demand than could be accounted for by the data in Table 9.4. Thus, the

Table 9.4. Comparison of oxygen uptake of epilimnetic water samples in standard 5-day biochemical oxygen demand (BOD) tests. Values are mg/l O_2 consumed during 5 day incubation at 20°C

Date	Ward (unpublished)	Date	Zobell (1940)
4/5/73	2.4	10/11/38	2.06
4/28/73	4.1	10/25/38	2.30
5/17/73	1.6	11/2/38	2.17
5/29/73	1.95	11/9/38	1.54
6/26/73	3.65	11/21/38	1.60
7/10/73	2.3	11/30/38	0.97
7/20/73	3.87	12/8/38	0.86
7/30/73	2.12	12/16/38	0.96
8/6/73	4.49	1/4/39	1.36
8/21/73	2.02	1/11/39	1.50
9/1/73	1.53	1/23/39	1.29
9/24/73	0.38	2/6/39	0.91
10/20/73	1.99	2/20/39	0.84
1/5/74	6.97	3/7/39	1.24
		3/23/39	3.11
		4/10/39	3.56
		5/6/39	3.92

Note that most of Zobell's data were obtained in winter. If only data for similar dates are compared, the results for the two studies are similar.

dominant influence of the Lake Mendota sediments on oxygen deficit is emphasized.

Fish

Data on changes in fish populations have been discussed in Chapter 8. It was noted in that chapter that both perch and cisco populations have undergone marked changes over the period of time in which Lake Mendota has been under study. However, these changes are not such as to conclude that the lake has undergone any significant degradation during the historic period. As noted, the population structure of perch has changed considerably. The perch which are present today are much larger than those that were present at earlier times. The cisco almost completely disappeared from the lake in the period 1900 to 1940 but has made a striking recovery in the 1970s and there was a large recruitment of cisco in the year 1977. Because cisco is very sensitive to reduction in oxygen concentration in the water, and because it is a coldwater fish, the habitat available to cisco can change considerably depending upon the depth of the thermocline. Also, breeding of cisco occurs during a single critical period in late November. As noted in Chapter 2, there have been considerable changes from year to year in the depth of the thermocline and the temperature of the hypolimnetic waters. Thus changes in cisco populations may arise due to changes in the physical structure of the lake rather than to any eutrophication process.

Conclusion

From the above data it is difficult to conclude that Lake Mendota has undergone any *massive* changes over the past 90 years, whether from natural causes or from human intervention. There have certainly been changes in the lake, of which the change in size and number of perch is most notable. But the causes of these changes are unclear.

In Chapter 3 we discussed the chemistry of Lake Mendota and its sediments, and noted the insights which have been provided by Bortleson's (1970) study of long sediment cores. These data had shown that massive changes in Lake Mendota occurred at the time that land in the drainage basin was cleared for agriculture, which occurred in the 1850s. By the time that the first scientific observations on Lake Mendota were underway in the 1880s the lake was probably already very eutrophic. Blue-green algal (cyanobacterial) blooms (Trelease, 1889) and fish kills (Dunning and Stevens, 1888) were occurring, and the lake had clearly begun serious deterioration.

Have there been any *further* major changes in the lake since the 1880s? Comparison of historical and current data do not support such a conclusion. Indeed, if agriculture is the main culprit in the eutrophication of Lake Mendota, it could even be possible that Lake Mendota is now *less* subject to cultural eutrophication than it had been in the early part of the century, since urbanization of the drainage basin has removed major amounts of land from agriculture. Despite well-documented effects of urbanization on eutrophication, it seems likely that residential areas with well-tended lawns and controlled storm drainage would have less impact on nutrient loading than plowed fields, where serious soil erosion can occur.

We can conclude this chapter with a quote from Stewart (1976):

> Naturally we can now wonder where this leaves us. Does this for the Madison lakes reflect an inadequate number of years for comparison (for the indices examined) or an insufficient number of indices compared? Were the Madison lakes as eutrophic several decades ago as they are now? Are the systems too complex and our data and understanding too limited to really say? Can results of coring provide a more powerful tool in differentiating between previous and present trophic conditions than most recorded year-to-year observations provide? Are we so used to thinking in terms of oligotrophic, mesotrophic, or eutrophic, that we can't appreciate the subtle changes within one category? Are oxygen deficits, once well regarded as an overall annual integrator of productivity in a lake, of limited value to use for lakes that are already eutrophic? Perhaps to all these questions we can give a qualified yes.

References

Allgeier, R.J., W.H. Peterson and C. Juday. 1934. Availability of carbon in certain aquatic materials under aerobic conditions of fermentation. *Internationale Revue der gestamten Hydrobiologie,* 30: 372–378.

Bere, Ruby. 1933. Numbers of bacteria in inland lakes of Wisconsin as shown by the direct microscopic method. *Internationale Revue der gesamten Hydrobiologie,* 29: 248–263.

Birge, E.A., and C. Juday. 1922. The Inland Lakes of Wisconsin. The Plankton. I. Its Quantity and Chemical Composition. *Wisconsin Geological and Natural History Survey,* Bulletin No. 64, Madison, Wisconsin, 222 pp.

Bortelson, G.C. 1970. *The chemical investigation of recent lake sediments from Wisconsin lakes and their interpretation.* Ph.D. Thesis, University of Wisconsin, Madison.

Dunning, Philo and B.J. Stevens. 1884. Two hundred tons of dead fish, mostly perch, at Lake Mendota, Wisconsin. *Bulletin of the U. S. Fish Commission,* 4: 439–443.

Edmondson, W.T. 1966. Changes in the oxygen deficit of Lake Washington. *Verhandlungen der Internationale Verein theoretische angewandte Limnologie,* 16: 253–158.

Edmondson, W.T. 1969. Eutrophication in North America. pp. 124–149, In: *Eutrophication: Causes, Consequences, Correctives.* National Academy of Sciences, Washington, D.C.

Edmondson, W.T. 1972. The present condition of Lake Washington. *Verhandlungen der Internationale Verein theoretische angewandte Limnologie,* 18: 284–291.

Kuznetzov, S.I. and G.S. Karsinkin. 1981. Direct method for the quantitative study of bacteria in water and some considerations on the causes which produce a zone of oxygen minimum in Lake Glubokoje. *Zentralblatt für Bakterologie,* II 83: 169–174.

Neess, J.C. and W.W. Bunge. 1956. An unpublished manuscript of E.A. Birge on the temperature of Lake Mendota: Part I. *Transactions of the Wisconsin Academy of Sciences, Arts, and Letters,* 46: 193–238.

Pedrós-Alió, C. and T.D. Brock. 1982. Assessing biomass and production of bacteria in eutrophic Lake Mendota, Wisconsin. *Applied and Environmental Microbiology,* 44: 203–218.

Pedrós-Alió, C. and T.D. Brock. 1985. Zooplankton dynamics in Lake Mendota: short-term versus long-term changes. *Freshwater Biology,* 15: 89–94.

Reichard, J. 1918. Methods of Collecting and Photographing. pp. 83–85, In: Ward, H.B. and G.C. Whipple (editors), *Fresh-Water Biology.* John Wiley, New York.

Sager, P.E. 1967. *Species diversity and community structure in lacustrine phytoplankton.* Ph.D. Thesis, University of Wisconsin-Madison. 201 pp.

Snow, L.M. and E.B. Fred. 1926. Some characteristics of the bacteria of Lake Mendota. *Transactions of the Wisconsin Academy of Sciences, Arts, and Letters,* 22: 143–154.

Sonzogni, W.C., G.P. Fitzgerald, and G. Fred Lee. 1975. Effects of wastewater diversion on the lower Madison lakes. *Journal of the Water Pollution Control Federation,* 47: 535–542.

Stewart, K.M. 1976. Oxygen deficits, clarity, and eutrophication in some Madison lakes. *Internationale Revue gesamten Hydrobiologie,* 61: 536–479.

Trelease, W. 1889. The "working" of the Madison lakes. *Transactions of the Wisconsin Academy of Sciences,* 7: 121–129.

Zobell, C.E. 1940. Some factors which influence oxygen consumption by bacteria in lake water. *Biological Bulletin,* 78: 388–402.

Zobell, C.E. and J. Stadler. 1940. The effect of oxygen tension on the oxygen uptake of lake bacteria. *Journal of Bacteriology,* 39: 309–322.

10. Energy flow in the Lake Mendota ecosystem

In previous chapters, quantitative data on various components of the Lake Mendota ecosystem have been presented. In the present chapter, these data are summarized and incorporated into a tentative model of energy relationships. In order to place all data on the same scale, values have been converted to grams of carbon per m^3 of lake surface. It is assumed that a gram of organic carbon is equivalent to 17.5 Kcal of energy (see Chapter 5).

A summary of standing crop for the main trophic groups in Lake Mendota is given in Table 10.1. This table also includes data for nonliving organic carbon dissolved in the water and for organic carbon in the sediments. There is unfortunately considerable uncertainty about some of the values given in Table 10.1, in some cases because of large year-to-year variability (for example, phytoplankton), and in other cases because of very limited quantitative data (for example, fish). From the data in Table 10.1, we can conclude that a food pyramid in the classical sense of ecological energetics does not exist in Lake Mendota.

Standing crop data, however, can be very misleading when analyzing the quantitative significance of a trophic group, because turnover is not considered. For instance, the bacterial population density in Lake Mendota is much lower than that of zooplankton, yet data on bacterial production (Chapter 7) show that a major proportion of the organic carbon produced by phytoplankton passes through the bacterial component. We thus need to know the rate of transfer (flux) from each component to the other components of the system.

Table 10.1. Standing crop of various trophic groups in Lake Mendota. Data are g C/m² of lake surface, averaged for the whole year

Group	Standing crop	Source of data
Total plankton	5.5	
Phytoplankton	1.8	This study (Chapter 4)
Bacteriopankton	0.2	" (Chapter 7)
Zooplankton	3.5	" (Chapter 6)
Dissolved organic carbon	40	" (Unpublished)
Fish (perch only)	1	Hasler and Wisby (1958) (see Chapter 8)
Bottom fauna	2	Juday (1942) (see Chapter 8)
Sediment organic carbon	10880	Fallon (1978) (see Chapter 3)

The Juday data were given as kg/hectare of organic matter on a wet-weight ash-free basis and have been converted to g C/m² using Juday's value of 90% water and the ratio of organic matter to organic carbon of 2.5:1 given by Fallon (1978). The data for dissolved organic carbon were obtained on water samples filtered through glass fiber filters of 1.2 μm nominal pore size. Sediment organic carbon was calculated from Fallon's data showing that sediment is 64 mg C/g dry weight and from Bortleson's estimate that 17 g dry sediment/m² has accumulated since the beginning of eutrophication.

Figure 10.1 presents an overall summary of the fluxes through the various components in Lake Mendota. Although some of the data in Figure 10.1 are highly tentative, the overall picture is reasonable and may not be too far from reality. This summary provides a basis for further research on the trophic-dynamic relationships in the Lake Mendota ecosystem.

Several major conclusions can be drawn from the data presented in Figure

Figure 10.1. A summary of standing crop and energy flow data for Lake Mendota. Standing crop values are in g C/m² and energy flow values are in g C/m²/year. The sizes of the boxes and the widths of the arrows reflect the quantitative differences. Standing crop data from Table 10.1. Energy flow data from earlier chapters, or by difference. One mm width of arrow represents an energy flow of 12g C/m²/year. In some cases, values have been averaged from the data of several years. Unmeasured fluxes which have been calculated by difference from measured values and from loss values are: Bacterial sedimentation (by difference from bacterial production after other losses have been deducted, as described in Chapter 7); algal sedimentation (by difference from total sedimentation, after deducting bacterial sedimentation), zooplankton consumption of phytoplankton (by difference from phytoplankton production, after deducting all other phytoplankton losses and assuming that the rest goes to zooplankton), oxygen respiration (from data in Table 3.9, assuming that 1 mole of organic carbon is oxidized per mole of oxygen consumed, and that only the lake area corresponding to the >10 meter sediments and only three spring-summer months are involved), denitrification (from Brezonik and Lee, see reference in Chapter 3, assuming 1 mole of organic carbon oxidized per mole of nitrate consumed), sulfate reduction (from Table 7.8), methanogenesis (from Tables 7.7 and 7.8).

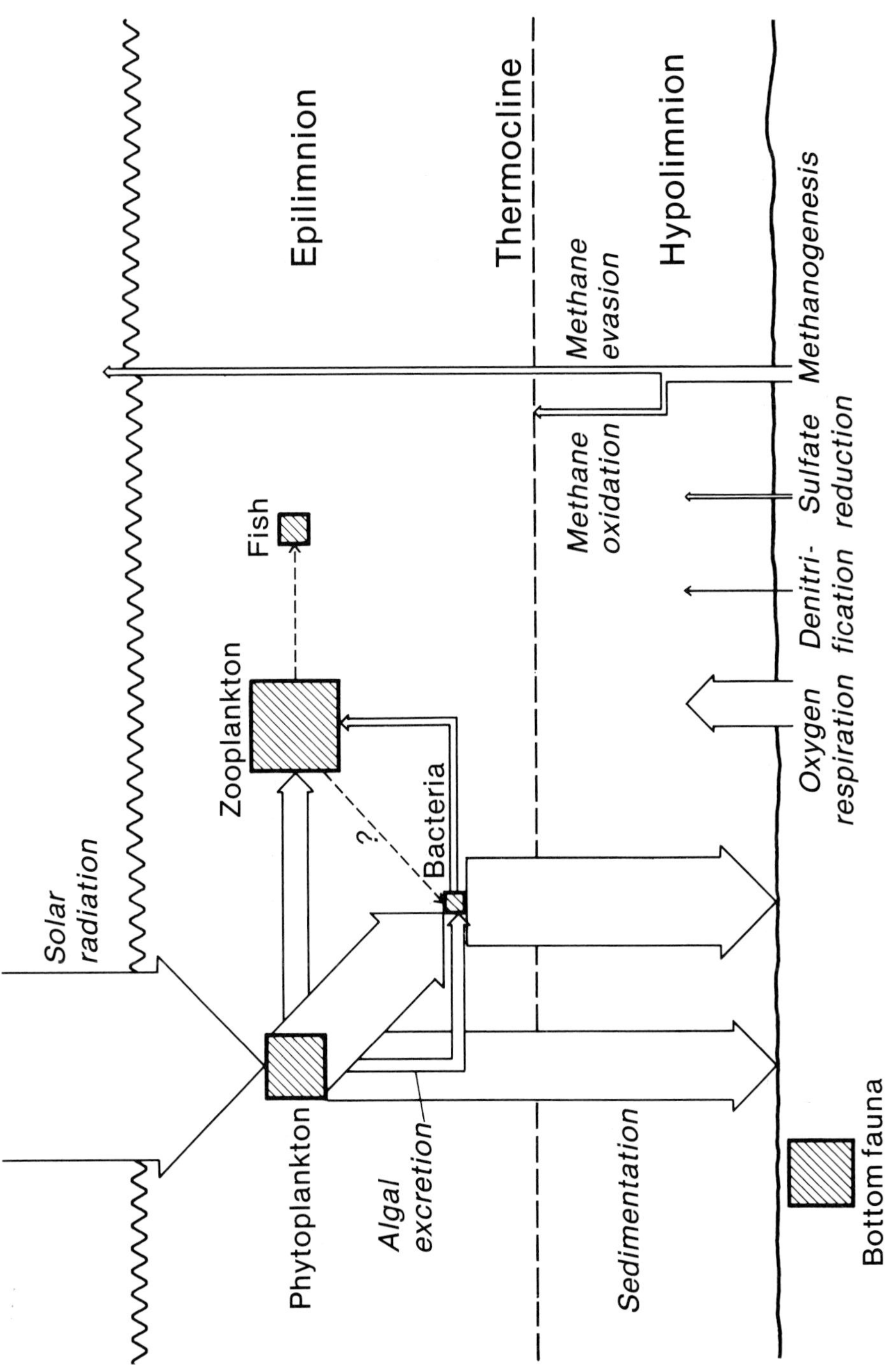
Solar radiation
Epilimnion
Thermocline
Hypolimnion
Phytoplankton
Zooplankton
Fish
Bacteria
?
Algal excretion
Sedimentation
Bottom fauna
Methane evasion
Methane oxidation
Methanogenesis
Sulfate reduction
Denitri- fication
Oxygen respiration

10.1. Because of the tentative nature of some of the data, these conclusions should be viewed primarily as providing a framework for further research.

1. Much more energy from the primary producers passes through the bacterial and detrital food chain than through the conventionally-considered food chain from phytoplankton to zooplankton and fish.
2. Bacteria consume a major amount of algal production, either indirectly via algal excretion, or directly.
3. Most of the organic carbon synthesized by the primary producers sediments to the bottom of the lake, either directly as algal material or indirectly as bacterial biomass. (Although sedimentation of bacterial biomass was not measured directly, the importance of this process was inferred from the large unaccounted losses in bacterial production described in Chapter 7.)
4. A significant amount of the organic carbon sedimenting to the bottom of the lake is recycled because of microbial processes in the lake sediments. These processes included respiration, denitrification, sulfate reduction, and methanogenesis.
5. Most of the organic carbon in Lake Mendota exists in nonliving form, either dissolved in the water or as part of the sediments. Such organic carbon may become available to the higher trophic groups via bacterial assimilation, but probably at very low rates.

The data presented in Figure 10.1 represent averages for the whole lake. It should be noted that if only the area of the lake corresponding to the deep sediments is concerned, the importance of bacterial processes is even greater. For calculating the data presented in Figure 10.1, it has been assumed that the deep sediments, where anaerobic microbial processes are expected to occur, would be only those circumscribed by the 10 meter contour, since only these sediments would be hypolimnetic and anaerobic throughout the whole summer. Such sediments account for about 39% of the total lake sediments. It is also assumed that these sediment microbial processes only occur in the summer when the hypolimnion is anaerobic, and hence only over an approximately three-month period. This assumption is probably true for denitrification, sulfate reduction, and methanogenesis, since these processes all require anaerobic conditions, but possibly may not be true for respiration, since oxygen consumption by sediments might occur throughout the year, albeit only measurable in the summer.

A major conclusion from the analysis illustrated in Figure 10.1 is that the energy flow in Lake Mendota is dominated by microbial components, both in the water column and in the sediments. The higher trophic levels, zooplankton, bottom fauna, and fish, have high standing crops, but are much less important from a dynamic viewpoint. In fact, from the relationships shown in Figure 10.1, Lake Mendota can be considered to be primarily a culture vessel for microorganisms. This is, perhaps, to be expected of such a eutrophic lake.

Appendix 1. Physical and chemical data

This Appendix provides the raw data from measurements of physical and chemical parameters in Lake Mendota for the years 1976 and 1978–1981. Graphical presentation and discussion of these data can be found primarily in Chapters 2 and 3. Details of methods have been described in those chapters or in papers cited in those chapters. The methods are summarized briefly here.

Temperature. Measured with a Yellow Springs Instrument Co. Model 518 combined temperature-oxygen probe. The thermistor was checked before use with a mercury thermometer.

Oxygen. Measured with a Yellow Springs Instrument Co. Model 518 temperature-oxygen probe.

pH. Measured in the laboratory within two hours of sampling with a Corning Model 12 Research pH Meter and glass electrode.

Light. Measured in the lake with a Li-Cor Model 185 Quantum Meter.

Soluble phosphate. Assayed by the method of Strickland and Parsons on a water sample that had been filtered successively through a Whatman GF/C glass fiber filter and a Gelman Metricel GA-6 membrane filter.

Total phosphate. Unfiltered water sample was digested with persulfate followed by soluble phosphate assay.

Ammonium. Assayed on filtered sample by the alkaline hypochlorite oxidation method of Strickland and Parsons.

Nitrite. Assayed on filtered sample by the method of Strickland and Parsons.

Nitrate. Assayed on filtered sample by the hydrazine reduction method of Strickland and Parsons.

Sulfate. Assayed by the turbidimetric method of Standard Methods for Analysis of Water and Wastewater.

Chlorophyll. The water sample was filtered through a Whatman GF/C glass fiber filter and the filter extracted overnight in the refrigerator with a mixture of equal parts of dimethyl sulfoxide and 90% acetone. The extract was analyzed spectrophotometrically by the method of Strickland and Parsons.

Depth Meter	Temp °C	O_2 mg/l	pH	Sol.P μg/l	Sulf. mg/l	NH_4^+ μg/l	NO_2^- μg/l	NO_3^- μg/l	Chlor. μg/l
Jan. 16, 1976									
0	2.7				2.5				
2	3.1								
4	3.1				2.4				
6	3.1								
8	3.1				2.6				
10	3.1								
12	3.1				2.4				
14	3.1								
16	3.2				2.3				
18	3.2								
20	3.2				2.3				
22	3.3				2.2				
Mar. 9, 1976									
0	1.5	14.8		111.6	1.6	15.0		450.9	
1	2.5	14.8		158.7	2.3			501.0	
5	2.5	14.4		164.9	2.5			534.4	0.6
10	2.5	14.4		164.9	2.5			517.7	
15	2.5	13.2		169.3	2.5			668.0	
20	2.5	6.0		322.4	2.3	15.0		634.6	
23	2.5	1.2		236.2	2.3	43.5		693.1	
Apr. 2, 1976									
0	4.7	13.0	8.0	273.4	7.8	43.8		268.0	
2	3.3	13.0							
4	3.6	13.0	8.0	274.0	7.5	45.0		301.4	
6	3.5	9.2							
8	3.5	9.2	8.0	207.7	7.5	45.0		284.7	
10	3.5	9.2							
12	3.5	9.2	8.1	205.8	11.3	42.5		225.5	
14	3.5	9.2							
16	3.5	9.2	8.0	193.4	9.0	37.5		250.5	1.4
18	2.5	9.0							
20	2.5	9.0	8.0	226.3	7.4	32.5		296.4	
Apr. 8, 1976									
0	6.5	12.2	7.9	71.3	8.6	26.3		244.5	5.2
2	5.5	12.0							
4	5.0	11.9	8.1	75.7	9.6	27.0		333.8	10.2
6	5.0	12.4							
8	4.8	12.5	8.1	92.0	7.9	26.3		293.3	7.8
10	4.8	12.5							
12	4.7	12.6	8.1	80.1	8.7	23.3		240.8	5.8
14	4.5	12.4							
16	4.5	12.6	8.1	79.2	7.6	21.8		288.8	5.8
18	4.2	12.4							
20	4.0	12.4	8.0	87.1	8.4	19.5		240.8	2.9
22	4.0	12.2							
23	3.9	12.2							
Apr. 15, 1976									
0	8.0	11.4	8.2	162.1	7.7	85.1		453.6	5.1
2	6.2	11.6							
4	6.0	11.6	8.3	152.5	7.6	98.7		502.2	7.6
6	6.0	11.6							
8	5.5	11.4	8.3	189.8	7.2	87.0		437.4	5.1
10	5.5	11.8							
12	5.5	11.8	8.3	191.5	7.6	90.7		437.4	6.5
14	5.5	11.6							
16	5.2	11.8	8.3	220.5	7.6	90.2		405.0	4.0
18	5.0	11.6							
20	5.0	11.6	8.3	178.9	7.2	78.0		437.4	2.5
22	4.5	11.3							
Apr. 27, 1976									
0	9.5	11.4	8.3	294.2	7.5	200.0		356.4	5.8
2	9.5	11.2							
4	9.5	11.2	8.3	243.8	7.2	175.4		420.4	
6	9.5	11.2							
8	9.5	11.2	8.3	248.6	7.2	180.8		405.0	5.2
10	9.5	11.2							
12	9.5	11.2	8.3	223.1	7.3	175.4		429.3	5.1
14	9.5	11.2							
16	9.3	11.1	8.3	248.6	7.8	191.8		434.3	5.5
18	9.2	11.1							
20	9.2	11.1	8.3	270.3	7.3	180.8		438.6	3.8
22	9.2	11.1							
24	9.0	5.8							
May 7, 1976									
0	9.6	11.5	7.9	259.7	7.4	168.5		419.7	3.4
2	10.0	11.5							
4	10.0	11.5	7.9	225.9	7.4	185.8		462.0	3.8
6	10.0	11.5							
8		7.9		244.5	7.4	164.2		485.1	2.7
10									
12		7.9		219.0	7.4	170.6		462.0	2.0
14									
16		7.9		223.0	7.4	177.1		492.8	2.8
18									
20	10.0	11.2	7.9	259.7	7.4	179.3		462.0	3.3
23	9.5	11.2							
May 18, 1976									
0	12.0	10.4	8.2	113.1	6.9	165.6		365.8	0.7
2									
4	12.0	10.4	8.2	84.5	6.9	175.2		434.0	0.3
6									
8	11.8	10.2	8.2	105.9	6.4	124.8		396.8	0.6
12				105.9	6.4	211.2		390.6	0.2
14									
16	11.2	9.8	8.2	83.3	6.9	168.0		421.6	
18									
20	11.0	10.0	8.1	94.0	7.1	194.4		334.8	0.0
22									
May 25, 1976									
0	14.5	10.0	8.1	64.5	8.1	207.8		351.5	1.8
2	14.0	10.0							
4	14.0	10.0	8.1	72.0	8.1	188.2		371.3	3.5
6	14.0	10.0							
8	13.5	10.0	8.1	45.0	8.4	215.6		396.0	3.6

Depth Meter	Temp °C	O_2 mg/l	pH	Sol.P µg/l	Sulf. mg/l	NH_4^+ µg/l	NO_2^- µg/l	NO_3^- µg/l	Chlor. µg/l
10	13.5	10.0							
12	13.0	10.0	8.2	72.0	8.6	227.4		326.7	1.1
14	12.0	9.4							
16	11.0	9.1	8.0	91.5	8.6	274.4		386.1	0.0
18	10.8	9.2							
20	10.0	8.5	7.9	90.0	8.1	315.6		340.6	0.0
22	10.0	8.0							
June 3, 1976									
0	18.0	9.4	8.6	87.0	8.4	172.5		338.6	3.1
2	17.5	9.3							
4	17.5	9.4	8.6	93.0	8.2	188.2		357.4	3.3
6	17.0	9.4							
8	16.0	9.4	8.5	91.5	8.4	192.1		366.3	3.3
10	14.0	9.6							
12	13.5	9.0	8.3	117.0	8.2	203.8		336.6	1.0
14	11.0	8.8							
16	10.0	8.4	8.1	133.5	8.1	256.8		362.3	0.2
18	9.5	7.6							
20	9.0	6.8	7.9	160.5	8.2	297.9		346.5	0.7
22	9.0								
June 8, 1976									
0	22.5	8.8	8.9	40.0	7.6	44.9		275.0	11.9
2	21.0	10.1							
4	20.0	9.5	8.8	54.4	7.9	58.7		379.5	2.2
6	18.0	9.3							
8	15.0	8.4	8.5	94.4	8.0	75.2		385.0	3.1
10	12.0	7.2							
12	11.5	7.4	8.2	120.0	7.7	76.6		357.5	1.2
14	10.5	6.6							
16	10.0	6.2	8.0	144.0	7.7	141.5		433.4	0.6
18	9.5	5.8							
20	9.0	4.8	7.9	168.0	7.6	179.4		411.4	0.0
June 14, 1976									
0	22.0	8.8	8.7	37.8	7.8	36.5	15.8	426.6	7.2
2	22.0	8.6							
4	22.0	8.4	8.7	46.8	7.7	9.0	15.3	360.8	10.7
6	17.0	8.2							
8	15.5	8.0	8.2	113.4	8.0	98.0	17.2	684.1	1.2
10	13.0	7.9							
12	11.5	7.4	8.0	163.8	8.0	104.2	19.0	720.5	0.4
14	10.5	6.9							
16	10.5	6.7	7.8	165.6	8.0	170.8	28.2	717.7	0.7
18	9.5	6.2							
20	9.5	5.5	7.6	181.8	7.8	162.8	27.3	533.7	0.1
22	9.5	5.0							
June 21, 1976									
0	20.5	9.1	8.8	59.8	7.5	31.8	19.4	219.9	38.3
2	20.5	8.8	8.7	30.4	7.6	67.5	19.6	232.4	7.2
4	20.0	8.4	8.7	38.0	7.5	92.8	19.6	290.8	2.5
6	20.0	8.2	8.7	38.0	7.6	72.0	19.6	263.3	2.5
8	18.0	7.8	8.7	48.6	7.6	101.3	21.8	268.3	2.2
10	16.0	7.4	8.3	91.2	7.6	204.2	32.5	311.9	1.2
12	14.0	6.7							
14	12.0	6.5							
16	10.5	5.6							
18	10.0	5.0	7.7	136.8	7.5	317.6	68.2	338.2	0.4
20	10.0	4.4							
22	9.5	3.0							
July 1, 1976									
0	21.0	8.8	8.8	59.3	7.8	46.8	20.6	152.6	46.3
2	21.0	8.9	8.9	39.0	7.7	55.0	23.6	163.9	29.2
4	21.0	8.7	8.8	31.2	7.9	38.4	17.2	130.6	34.4
6	21.0	8.5	8.8	31.2	8.0	31.0	18.7	147.2	12.7
8	21.0	8.0	8.8	39.0	9.3	52.7	19.0	133.8	12.7
10	20.0	7.8	8.8	37.4	7.7	55.0	26.1	190.4	2.6
12	13.5	4.0							
14	11.5	2.7							
16	10.0	1.8							
18	9.5	0.9	7.7	187.2	7.9	128.0	54.7	419.8	0.1
20	9.5	0.2							
July 7, 1976									
0	23.0	8.3	8.9	7.6	7.8	9.9	0.7	10.5	81.0
2	23.0	8.2	8.9	8.8	7.8	0.0	0.8	0.3	78.1
4	23.0	8.2	8.9	55.1	7.8	52.4	3.2	13.3	63.6
6	22.5	7.4	8.8	53.9	7.9	35.2	3.8	35.0	34.4
8	21.0	5.2	8.7	27.7	7.8	121.0	14.0	117.2	13.7
10	17.0	2.5	8.5	60.5	8.2	241.3	41.6	199.1	2.0
12	13.5	1.5							
14	11.0	0.8							
16	10.0	0.8							
18	9.0	0.4	7.5	178.9	7.7	228.3	31.0	613.5	0.3
20	8.8	0.3							
July 13, 1976									
0	23.0	8.4	8.8	6.7	8.0	14.9	0.3	6.2	46.6
2	23.0	8.2	8.8	6.7	8.0	0.0	0.3	6.1	44.5
4	23.0	7.9	8.7	52.0	7.7	25.2	0.2	9.0	36.1
6	23.0	7.8	8.7	9.4	8.0	3.1	0.3	22.5	27.8
8	20.5	6.4	8.7	25.4	7.8	63.7	0.6	68.0	6.7
10	17.5	3.9	8.3	81.5	8.0	248.9	0.5	237.1	0.0
12	14.8	2.0							
14	11.0	0.5							
16	9.5	0.2							
18	8.8	0.2	7.5	252.5	7.4	335.5	5.1	335.8	0.0
20	8.5	0.1							
July 20, 1976									
0	23.0		8.7	3.4	6.5	25.5	3.6	33.2	42.8
2	23.0		8.7	8.5	6.6	2.5	4.6	28.5	52.6
4	23.0		8.7	3.4	6.6	40.0	4.0	0.0	45.7
6	23.0		8.7	0.0	6.6	8.0	6.0	0.0	40.8
8	23.0		8.7	3.4	6.3	30.3	1.9	1.8	11.9
10	22.0		8.6	5.1	6.8	63.3	1.9	11.0	15.8
12	16.0								
14	14.0								
16	11.0								
18	10.0		7.4	348.5	6.1	637.3	6.6	50.4	0.0

Depth Meter	Temp °C	O_2 mg/l	pH	Sol.P μg/l	Sulf. mg/l	NH_4^+ μg/l	NO_2^- μg/l	NO_3^- μg/l	Chlor. μg/l
20	9.0								
July 26, 1976									
0	24.0	8.2	9.0	0.0	7.2	34.7	9.1	17.6	83.9
2	24.0	8.2	8.9	4.2	7.2	0.0	1.9	0.0	72.3
4	24.0	8.2	8.8	5.9	6.9	31.2	9.1	3.3	85.3
6	24.0	8.2	9.0	5.4	7.2	12.9	0.8	0.0	80.9
8	23.0	8.2	8.7	9.8	7.1	49.7	2.8	1.6	12.7
10	20.0	8.2	8.6	29.5	7.5	85.7	7.0	48.1	24.6
12	16.5	8.2							
14	14.0	8.2							
18			7.5	332.7	6.4	184.0	6.6	0.0	0.0
Aug. 2, 1976									
0	23.0		8.9	0.0	7.1	0.0	3.1	0.0	2.3
2	23.0		8.8	0.0	7.0	0.0	3.8	0.0	
4	23.0		8.8	0.0	6.8	0.0	0.7	5.4	14.2
6	23.0		8.8	0.0	7.1	0.0	0.7	0.0	15.3
8	23.0		8.8	0.0	7.0	46.7	11.5	0.0	48.3
10	22.0		8.6	0.0	7.4	105.5	10.1	44.0	6.4
12	15.5								
14	14.0								
18			7.6	248.1	6.8	337.3	109.9	51.1	20.7
Aug. 10, 1976									
0	21.9	11.6	8.8	0.0	6.9	0.0	11.5	0.0	15.6
2	21.9		8.8	0.0	6.9	0.0	12.0	0.0	15.9
4	21.9	11.1	8.6	0.0	6.7	0.0	9.6	0.0	17.3
6	21.9	11.1	8.6						
8	21.9		8.6	0.0	6.7	0.0	8.9	0.0	15.0
10	21.9	7.7	8.6	0.0	6.7	0.0	13.2	0.0	14.5
12	18.0	0.7							
14	16.0	0.5							
16		0.3							
18		0.3	7.5	301.6	6.0	183.7	15.6	0.0	7.2
20		0.3							
Aug. 16, 1976									
0	23.0		8.9	0.0	6.9		0.0	0.0	102.9
2	22.0		8.9	53.5	7.1	0.0	0.0		34.1
4	22.0		8.8	53.5	6.9	0.0	0.0		23.4
6	22.0		8.8	9.5	6.9	0.0	0.0		25.5
8	22.0		8.8	5.4	6.9	0.0	0.0		20.2
10	22.0								
12	21.0								
14	17.0								
16	15.0								
18	13.5		7.6	317.5	6.1	14.9	0.0		0.0
20	13.5								
21	13.5								
Aug. 23, 1976									
0	24.0	8.3	9.1	54.0	6.9	22.4	0.0	15.9	34.4
2	24.0	8.3	9.1	52.6	7.0	12.3	0.0	0.0	28.9
4	23.0	8.3	8.9	2.8	6.7	0.0	0.0	0.0	18.2
6	22.0	7.8	8.8	8.4	7.0	20.1	0.0	9.9	12.4
8	22.0	6.8	8.8	2.8	6.9	0.0	0.0	0.0	13.6
10	21.0	5.9	8.7	54.0	6.7	11.2	0.0	0.0	3.3
12	19.3								
14	16.5								
16	15.0								
18	12.5	1.6	7.4	449.4	5.1	131.9	0.0	0.0	3.2
20	12.0								
Aug. 30, 1976									
0	21.8		8.7		7.4	0.0	0.0		21.6
2	21.8	7.4	8.7			0.0	0.0		18.2
4	21.8		8.6			0.0	0.0		16.8
6	21.8	6.3	8.6		7.7	0.0	0.0		14.8
8	21.0		8.2			0.0	3.0		6.4
10	20.8	3.7	8.4		8.4	0.0	6.0		4.0
12	19.8	0.2			6.7				
14	16.0	0.0			6.0				
16	14.0	0.0			4.5				
18	13.0	0.0	7.7		4.0		3.0		3.3
20	12.8				3.5				
22	12.2				3.4				
Sep. 8, 1976									
0	21.0	8.4	8.8	70.5	7.6	128.2	5.0		133.0
2	20.5	8.6	8.8	28.2	7.0	147.9	1.8		5.8
4	20.5	8.6	8.8	32.4	7.3	98.6	1.2	0.0	27.5
6	20.5	7.8	8.7	35.3	7.3	88.8	0.0	0.0	24.6
8	20.5	7.4	8.7	28.2	7.4	98.6	0.0	0.0	7.5
10	20.5	7.0	8.7	28.2	7.3	108.5	0.0	0.0	16.2
12	20.5	6.4							
14	16.5	0.7							
18			7.5	40.9	5.3	290.6	13.6	52.1	2.2
Sep. 13, 1976									
0	20.5	8.3	8.7	22.2	8.2	6.6	7.2		38.2
2	19.5	8.4	8.7	34.0	8.1	0.0	0.4		36.1
4	19.5	8.2	8.7	23.7	8.1	0.0	1.2		39.9
6	19.2	8.0	8.7	38.5	8.0	0.0	0.0		54.4
8	19.2	8.0	8.7	23.7	8.3	24.6	7.2	23.0	55.8
10	19.2	7.7	8.7	22.2	7.8	0.0	0.4		44.8
12	19.2	7.6							
14	19.2	6.0							
18			7.4	478.0	5.1	223.5	9.2	29.0	2.9
Sep. 20, 1976									
0	18.5	9.0	8.7	73.5	6.9	85.1	0.4	49.3	21.7
2	18.5	9.0	8.7	0.0	6.9	85.6	0.0	45.6	12.4
4	18.3	9.1	8.7	37.5	6.6	75.8	1.4	49.3	13.3
6	18.3	8.8	8.7	30.0	6.9	48.5	0.0	33.8	13.6
8	18.3	8.9	8.7	34.5	7.0	109.4	0.0	23.6	14.2
10	18.3	8.8	8.7	34.5	7.0	47.5	0.0	11.0	17.9
12	18.3	8.7							
14	18.3	8.6							
16	15.0	0.3							
18	13.0	0.1	7.4	525.0	3.4	196.6	2.2	13.8	0.0
20	12.2	0.1							
22	12.0	0.0							

Depth Meter	Temp °C	O_2 mg/l	pH	Sol.P μg/l	Sulf. mg/l	NH_4^+ μg/l	NO_2^- μg/l	NO_3^- μg/l	Chlor. μg/l
				Sep. 27, 1976					
0	17.2	9.4	8.7	65.2	6.9	153.6	1.2	0.0	56.4
2	17.2	9.2	8.7	65.2	7.0	172.0	1.2	0.0	58.4
4	17.2	9.1	8.7	99.9	6.6	229.8	1.0	20.9	44.0
6	17.2	8.9	8.7	72.2	6.6	211.4	1.0	20.9	49.7
9	17.2	8.1	8.7	93.0	7.3	213.2	1.1	16.3	29.2
12	17.2	7.2	8.6	83.3	6.9	209.3	1.0	0.0	13.6
15	16.9	1.8							
16	16.2	0.0							
17	14.5	0.0							
18	13.5	0.0	7.5	502.4	3.6	211.0	3.2	19.2	1.9
20	12.5	0.0							
				Oct. 4, 1976					
0	16.8	9.4	8.6	42.5	6.6	188.6	1.9	0.0	22.8
2	16.5	9.2	8.6	58.7	7.0	196.0	1.2	0.0	20.2
4	16.5	9.1	8.6	76.3	6.6	174.2	1.9	0.0	16.5
6	16.5	9.1	8.6	61.6	6.8	170.7	1.9	0.0	16.5
9	16.5	9.0	8.6	58.7	7.0	152.8	1.9	0.0	22.7
12	16.5	8.9	8.6	66.0	6.8	189.0	1.6	8.3	17.6
14	16.5	8.7							
16	16.2	1.4							
17	14.0	0.3							
18	13.5	0.2							
19	13.0	0.2	7.3	645.3	2.3	161.8	1.9	0.0	1.6
20	12.5	0.2							
21	12.5	0.0							
				Oct. 12, 1976					
0	15.0	9.9	8.5	67.2	6.7	132.4	6.5		25.2
2	15.0	9.9							
4	15.0	9.6	8.5	80.0	7.0	105.2	5.4		17.9
6	15.0	9.4							
8	15.0	9.4	8.5	104.0	6.7	133.6	5.4		14.8
10	15.0	9.3							
12	15.0	9.3	8.5	72.0	6.5	168.2	5.0		15.6
14	14.8	9.3							
16	14.8	9.2	8.5	83.2	6.5	133.0	5.9	10.2	10.7
18	14.8	8.3							
20	13.8	0.2	8.2	190.4	6.4	265.4	4.6		7.5
22	12.3	0.1	7.5	456.0	4.9	229.9	22.0	23.5	3.8
				Oct. 18, 1976					
0	12.5	10.4	8.3	88.9	6.8	240.7	14.6	94.3	999.9
2	12.5	9.9							
4	12.5	9.8	8.3	113.9	6.7	427.7	9.2	19.1	32.8
6	12.5	9.8							
8	12.5	9.8	8.3	157.4	6.7	427.2	3.4		28.6
10	12.5	9.8							
12	12.5	9.0	8.3	132.3	6.7	363.7	4.0		32.4
14	12.5	9.0							
16	12.5	9.0	8.3	144.0	6.7	405.7	10.5	21.9	28.0
18	12.5	10.0							
20	12.5	10.3	8.3	127.3	6.9	332.1	7.4	25.1	10.7
22	12.5	9.6	8.3	77.0	6.6	453.0	69.3	213.0	15.0
				Oct. 25, 1976					
0	10.5	10.9	8.4	85.7	7.1	281.3	8.8	47.7	34.7
4	10.5	10.4							
5		8.4		104.4	6.9	336.3	9.3	45.4	26.0
8	10.5	10.1							
10			8.4	120.0	6.8	326.9	8.5	33.6	27.2
12	10.5	10.0							
15			8.4	109.1	6.7	292.6	8.5	30.8	29.8
16	10.5	9.8							
20	10.5	9.7	8.4	104.4	6.9	332.3	9.4	35.7	31.2
22	10.5	9.6	8.4	120.0	6.8	300.0	10.6	39.6	27.5
				Nov. 8, 1976					
0	6.0	11.9	8.4	70.9	7.0	326.0	16.0	53.7	13.9
4	6.5	11.6							
5			8.4	83.5	7.2	283.7	11.3	55.2	8.9
8	6.5	11.1							
10			8.4	94.5	6.5	325.4	26.5	55.5	9.5
12	6.5	10.7							
15			8.4	86.7	6.9	332.6	19.5	53.4	7.6
16	6.5	10.2							
20	6.5	9.8	8.4	83.5	7.2	304.1	11.7	51.0	8.1
				Nov. 23, 1976					
0	3.0	12.5	8.3	72.0	7.6	351.8	10.5	50.9	11.1
4	3.0	12.5	8.3						
5			8.3	93.6	7.9	209.9	8.7	42.1	1.9
8	3.0	12.5	8.3						
10			8.3	108.0	7.6	111.4	7.5	40.1	1.7
12	3.0	12.5	8.3						
15			8.3	100.8	7.6	218.4	6.8	46.1	1.6
16	3.0	12.0	8.3						
20	3.0	12.0	8.3	124.2	7.9	218.8	6.4	30.6	1.9
				Dec. 14, 1976					
0	0.5	12.5	8.2	89.2	6.1	164.2	8.0	87.2	0.3
5	1.0	10.9	8.2	101.0	6.1	187.7	11.2	84.0	0.5
10	1.0	10.5	8.2	114.5	5.9	425.8	7.3	87.9	0.6
15	1.5	9.5	8.2	111.1	5.9	460.0	6.8	99.0	1.0
20	2.0	8.7	8.1	111.1	6.0	469.5	5.9	84.1	0.5
23	3.0	5.5							

Depth Meter	Temp °C	O_2 mg/l	pH	Light $\mu E/m^2/sec$	Sol.P $\mu g/l$	Tot.P $\mu g/l$	NH_4^+ $\mu g/l$	NO_2^- $\mu g/l$	NO_3^- $\mu g/l$	Chlor. $\mu g/l$
Jan. 4, 1978										
0	0.0	14.3	8.8	80	108.6	150.9				31.8
2	1.0	14.9	8.8	16	213.2	132.3				14.5
4	1.0		8.8	4	332.0	129.2				11.6
6				1						
8	1.0	14.7	8.8		192.9	123.1				37.6
16	1.8	12.8	8.6		163.2	141.6				18.5
20	2.0	11.2	8.6		314.8	218.9				18.5
24	3.8	8.0								
Mar. 16, 1978										
0	0.0			850						
2	0.0			55						
4	0.5			29						
6	0.5			15						
8	1.0			8						
10	1.0			5						
12				3						
14	1.5									
18	2.0									
20	2.2									
22	3.0									
24	4.0									
Mar. 23, 1978										
0	0.0	12.6		810						
2	1.0	12.6		120						
4	1.0	12.0								
6	1.0	11.9								
8	1.0	11.8								
10	1.0	9.8								
12	1.5	8.0								
14	1.5									
16	2.0	3.8								
18	2.0									
20	2.2	1.3								
22	3.0	1.2								
24	3.5	1.1								
Mar. 31, 1978										
0	0.0	12.3		1000	105.6	160.2				28.5
2	1.0	11.7		230	109.0	70.4				27.9
4	1.0	11.2		66	95.4	163.6				29.4
6	1.0	11.5		27						
8	1.0	11.3		11	98.8	167.0				28.2
10	1.0			4						
12	1.0	11.0								
14	1.0	10.3								
16	1.0	3.6			120.9	177.2				15.9
18	1.0	2.4								
20	1.8	0.7			197.4	282.6				0.7
22	2.0	0.6								
24	3.0	0.6								
Apr. 14, 1978										
0	2.5	12.9		1150						55.5
2	2.0	12.6		140						
4				25						
5										34.7
6				5						
8	2.0	12.6		1						
10										38.7
12	2.0	12.2								
15										46.8
18	2.0	12.0								
20	2.0	8.4								26.6
May 2, 1978										
0	7.5			1700		110.9				16.2
2	6.5			180		123.4				24.9
4	6.0			50		117.2				24.3
6	5.5			25						
8	5.5			8		107.8				16.2
10	5.5			2						
12	5.5									
16	5.2					117.2				5.2
18	5.0									
May 15, 1978										
0	8.5	11.2								
2	8.5	10.6								
4	8.5	10.4								
10	8.0	10.4								
14	8.0	10.2								
18	7.8	9.8								
May 23, 1978										
0	13.2	10.3		150						
2	13.0	9.6		5						
4	12.5	9.3								
6	11.2	7.6								
8	10.5	7.0								
10	9.5	6.5								
12	9.5	6.3								
14	9.0	6.2								
16	9.0	6.0								
18	9.0	5.4								
20	8.6	4.6								
22	8.5	4.3								
23	8.3	0.6								
May 30, 1978										
0	18.0	9.3		1400						
2	18.0	9.6		450						
4	17.9	9.7		70						
6	17.0	9.4		15						
8	11.5	8.0		4						
10	10.0	7.3								
12	9.5	6.3								
14	9.0	5.5								
16	9.0	5.1								
18	9.0	3.1								

Depth Meter	Temp °C	O_2 mg/l	pH	Light $\mu E/m^2/sec$	Sol.P $\mu g/l$	Tot.P $\mu g/l$	NH_4^+ $\mu g/l$	NO_2^- $\mu g/l$	NO_3^- $\mu g/l$	Chlor. $\mu g/l$
May 31, 1978										
0	20.8	8.8		780						
2	20.2	8.6	8.5	125						
4	18.2	7.9		23						
6	11.5	7.4								
8	10.1	7.1								
10	9.5	7.3								
12	9.2	7.2								
14	9.0	6.9								
16	9.0	6.5								
18	8.8	5.6								
20	8.8	5.1								
22	8.5	3.5								
23	8.5	0.1								
June 1, 1978										
0	19.5	9.0								
2	19.5	9.0								
4	15.0	8.0								
6	12.0	7.5								
8	10.5	7.4								
10	9.5	7.9								
12	9.0	8.0								
14	9.0	7.2								
16	8.5	5.8								
18	8.2	4.5								
20	8.0	4.0								
22	7.9	3.8								
23	7.9	2.0								
June 2, 1978										
0	21.0	8.2								
2	21.0	8.3								
4	20.5	8.2								
6	19.5	8.1								
8	15.0	7.9								
10	12.0	8.0								
12	11.0	7.8								
14	11.0	7.7								
16	10.0	6.9								
18	9.5	4.8								
20	9.0	4.6								
22	9.0	3.8								
24	9.0	0.1								
June 5, 1978										
0	19.0	9.4		1700						
2	19.8	9.3		700		101.5				
4	19.2	9.1		280						
6	13.5	8.5		130						
8	10.8	8.4		50						
10	10.2	8.1		15						
12	9.8	8.0		5						
14	9.2	7.6								
16	9.2	7.3								
18	9.0	6.1								
20	8.8	5.2								
22	8.8	0.4								
June 8, 1978										
0	18.5	9.1	8.4	1250	167.9	67.1	106.3	7.7	151.7	3.0
2	18.5	8.8	8.5	420	191.6	73.4	113.9	7.1	121.1	2.3
4	18.5	8.5	8.6	160	332.2	70.2	128.0	9.1	177.4	2.3
6	15.5	8.0	8.5	64	190.0	92.4	86.1	8.3	268.8	3.2
8	12.0	6.2	8.3	27	216.9	108.2	174.3	10.8	187.2	3.0
10	10.0	5.9	8.2	12	216.9	124.0	182.1	12.0	258.4	1.3
12	9.0	5.5								
14	9.0	5.4								
16	8.5	4.9								
18	8.5	4.0	8.0		324.3	180.9			293.5	0.4
20	8.5	2.3								
22	8.5	2.1								
June 12, 1978										
0	19.5	9.1	8.6	1100	297.6	110.9	151.4	3.9		5.8
2	19.0	8.9	8.6	260	200.7	95.3	156.4	14.3		5.3
4	19.0	8.9	8.7	74	144.5	101.5	54.5	11.4	110.2	5.3
6	19.0	9.0	8.7	22	136.7	89.0	60.0	9.7	84.8	5.2
8	14.5	7.2	8.5	10	241.4	148.4	184.5	11.4	132.8	2.8
10	12.5	6.8	8.4		274.2	123.4	220.6	6.2		3.0
12	11.0	4.2								
14	10.0	3.7								
16	9.5	3.7								
18	9.5	3.5	8.1		285.1	192.1	274.0	5.6		0.3
20	9.0	2.0								
22	9.0	1.4								
23	9.0	0.4								
June 15, 1978										
0	17.2	9.2	8.5	320	170.1	85.1				7.5
2	17.2	9.1	8.5	84	223.0	91.3				7.5
4	17.2	9.0	8.6	40	146.8	88.2				8.1
6	17.0	8.7	8.6	18	140.5	82.0				7.7
8	16.8	8.5	8.6	8	153.0	103.8				7.5
10	14.2	7.0	8.5	4	176.3	106.9				6.3
12	11.8	5.6								
14	10.1	4.5								
16	9.5	4.2								
18	9.0	3.5	8.1		263.4	178.4				0.4
20	8.5	2.5								
22	8.3	0.0								
June 19, 1978										
0	18.5	9.2	8.6	1250	49.2	89.0				5.0
2	18.5	8.7	8.6	300	44.5	104.5				5.4
4	18.0	8.6	8.6	88	52.2	104.5				5.4
6	18.0	8.3	8.6	34	47.6	98.3				4.7
8	17.0	7.6	8.6	13	56.9	126.1				4.3
10	16.5	7.0	8.5		61.5	107.6				4.0
12	12.0	5.0								
14	9.5	2.0								
16	9.0	2.0								
18	9.0	1.5	7.9		165.1	218.9				0.9

Depth Meter	Temp °C	O_2 mg/l	pH	Light $\mu E/m^2/sec$	Sol.P $\mu g/l$	Tot.P $\mu g/l$	NH_4^+ $\mu g/l$	NO_2^- $\mu g/l$	NO_3^- $\mu g/l$	Chlor. $\mu g/l$
20	8.2	0.0								
22	8.2	0.0								
23	8.0	0.0								
June 22, 1978										
0	19.4	9.1	8.4	600	54.1	101.9				7.5
2	19.1	9.2	8.5	160	57.2	105.0				9.2
4	19.0	8.5	8.5	61	61.9	101.9				7.9
6	19.0	7.9	8.5	26	58.8	114.4				7.4
8	19.0	7.4	8.5	10	65.0	117.6				6.0
10	15.9	5.5	8.3		97.9	136.4				4.0
12	12.5	3.8								
14	10.6	2.6								
16	10.2	2.4								
18	9.9	1.7	8.1		190.3	242.9				0.8
20	9.7	0.4								
22	9.2	0.2								
June 26, 1978										
0	20.0	9.0	8.6	730	52.1	100.9				14.7
2	19.5	8.8	8.6	210	60.0	104.1				14.1
4	19.5	8.5	8.5	73	63.2	116.8				11.1
6	19.0	7.7	8.5	30	60.0	104.1				10.6
8	19.0	7.4	8.5	13	66.3	129.5				7.5
10	16.5	5.5	8.4		87.0	126.3				4.5
12	14.5	4.7								
14	10.5	1.9								
16	10.0	1.4								
18	9.5	0.6	8.0		210.8	269.2				0.9
20	9.0	0.3								
22	9.0	0.1								
June 29, 1978										
0	22.5	8.8	8.8	210	35.6	103.2				30.9
2	22.5	9.0	8.7	36	35.6	122.4				32.1
4	21.0	7.0	8.6	10	50.0	96.8				13.6
6	19.8	6.4	8.4		59.6	122.4				10.1
8	19.8	5.6	8.4		64.4	106.4				7.1
10	16.0	4.0	8.3		110.7	157.5				5.0
12	13.5	2.9								
14	12.0	2.2								
16	10.5	1.3								
18	9.4	0.3	7.9		214.6	272.6				0.9
20	9.0	0.2								
22	9.0	0.2								
23	8.5	0.1								
July 3, 1978										
0	20.5		8.7	200	33.7	89.8				27.2
2	20.5		8.7	44	35.3	109.0				30.3
4	20.5		8.7	11	33.7	99.4				25.7
6	20.5		8.7		35.3	105.8				25.4
8	18.6		8.7		44.9	121.8				19.1
10	16.6		8.3		96.2	144.3				8.0
12	13.2									
14	11.0									
16	10.0									

Depth Meter	Temp °C	O_2 mg/l	pH	Light $\mu E/m^2/sec$	Sol.P $\mu g/l$	Tot.P $\mu g/l$	NH_4^+ $\mu g/l$	NO_2^- $\mu g/l$	NO_3^- $\mu g/l$	Chlor. $\mu g/l$
18	9.0		7.9		229.2	275.7				1.1
20	8.5									
22	8.0									
23	8.0									
July 6, 1978										
0	21.2	8.6	8.7	500	36.6	96.0	0.0	10.2	76.6	26.0
2	21.2	9.6	8.6	140	36.6	109.0	0.0	11.3	67.4	26.3
4	21.2	8.6	8.8	30	35.0	96.0	0.0	10.6	107.5	25.7
6	21.2	8.4	8.8	10	38.2	102.5	59.1	10.2	96.0	24.9
8	19.0	6.0	8.7		48.0	99.2	40.0	15.1	90.3	19.7
10	16.0	3.3	8.2		126.2	164.4	779.8	90.5	173.6	5.6
12	12.8	1.4								
14	10.5	0.3								
16	10.0	0.1								
18	9.8	0.0	7.9		227.2	288.2				1.7
20	9.5	0.0								
22	9.0	0.0								
23	9.0	0.0								
July 10, 1978										
0	21.0	10.5	8.9	1100	22.4	96.6				43.4
2	21.0	10.1	9.0	100	24.0	103.1				48.0
4	20.7	9.4	9.0	11	22.4	106.3				45.7
6	20.8	9.1	9.0		24.0	103.1				44.5
8	19.6	7.2	9.0		33.7	116.0				28.6
10	14.8	2.8	8.1		117.0	161.2				9.5
12	11.8	0.4								
14	10.8	0.0								
16	9.8	0.0								
18	9.3	0.0	7.8		246.9	293.5				2.5
20	8.8	0.0								
22	8.8	0.0								
23	8.5	0.0								
July 13, 1978										
0	22.0	10.1	8.9	1400	16.4	80.0				39.3
2	22.0	10.1	8.9	145	18.0	83.2				38.2
4	22.0	10.0	8.9	20	16.4	83.2				35.8
6	22.0	9.8	8.9		16.4	102.1				31.2
8	21.5	7.8	8.8		18.0	73.7				21.4
10	17.0	2.5	8.9		120.0	155.0				7.5
12	15.5	1.4								
14	14.5	0.7								
16	12.0	0.5								
18	11.0	0.4	7.9		237.0	278.5				2.1
20	11.0	0.4								
22	10.2	0.4								
23	10.2	0.4								
July 17, 1978										
0	24.8	10.4	8.9	350	16.7	99.4	0.0	0.4	23.3	7.4
2	24.8	10.3	8.9	37	16.7	102.5				7.4
4	24.8	10.3	9.0	2	15.1	93.1	0.0	0.0	21.0	7.5
6	24.8	10.1	9.0		19.8	105.6				6.2
8	22.5	5.2	8.7		46.5	108.8				4.0
10	20.0	1.6	8.2		101.6	159.1				2.1

Depth Meter	Temp °C	O_2 mg/l	pH	Light $\mu E/m^2/sec$	Sol.P $\mu g/l$	Tot.P $\mu g/l$	NH_4^+ $\mu g/l$	NO_2^- $\mu g/l$	NO_3^- $\mu g/l$	Chlor. $\mu g/l$
12	15.2	0.6								
14	13.5	0.5								
16	12.5	0.5								
18	12.0	0.4	7.8		250.9	306.9	629.5	51.1	82.8	0.4
20	11.7	0.4								
22	11.2	0.4								
July 20, 1978										
0	23.3	9.7	9.1	980	39.6	76.0	0.0	0.0	2.8	42.2
2	23.3	9.2	9.0	78	44.4	98.4	0.0	0.0	0.0	37.6
4	23.1	8.8	9.0	14	41.2	79.2				27.8
6	23.1	8.5	9.0	3	36.1	79.2	0.0	0.0	0.0	23.8
8	20.0	4.5	8.7		52.4	95.2	6.3	1.0	28.7	12.4
10	17.5	2.2	8.2		98.8	133.6				7.6
12	13.7	1.0								
14	11.0	0.9								
16	9.3	0.7								
18	9.3	0.6	7.9		268.4	309.6	232.2	1.6	17.3	1.5
20	9.3	0.5								
22	9.3	0.5								
July 24, 1978										
0	24.0	11.3	9.1	600	9.7	183.1				187.9
2	22.5	6.9	9.0	20	11.3	118.9				59.0
4	22.5	6.6	9.0		16.1	86.8				20.2
6	22.0	6.5	9.0		24.2	86.8				13.3
8	20.0	4.1	8.7		57.9	112.5				7.7
10	17.0	1.1	8.2		106.0	163.8				5.1
12	14.5	0.6								
14	11.5	0.5								
16	10.0	0.6								
18	9.0	0.6	7.9		300.1	353.2				0.5
20	9.0	0.6								
22	8.5	0.5								
23	8.5	0.4								
July 27, 1978										
0	22.5	8.5	9.1	600	14.4	106.3	0.0	0.0	0.0	53.2
2	22.5	8.5	9.1	48	16.0	106.3	0.0	0.0	0.0	53.2
4	22.5	8.5	9.1	5	16.0	103.1	0.0	0.0	0.0	50.9
6	22.5	8.4	9.1		17.7	99.9	0.0	0.0	0.0	52.0
8	22.5	8.2	9.1		17.7	99.9	0.0	0.0	0.0	44.4
10	18.0	1.2	8.2		106.4	151.5	66.6	0.0	335.2	6.7
12	13.0	0.8								
14	10.5	0.8								
16	9.5	0.7								
18	9.5	0.5	7.9		317.8	383.8	200.4	5.5	0.0	2.4
20	8.5	0.4								
22	8.5	0.3								
23	8.0	0.2								
July 31, 1978										
0	21.0	8.0	9.0	485	21.6	112.7	0.0	1.0	17.3	40.5
2	20.5	7.7	9.0	26	21.6	116.0	0.0	1.6	0.0	41.0
4	20.5	7.5	9.0	3	20.0	109.4	0.0	1.0	14.4	38.7
6	20.5	7.5	9.0		20.0	112.7	0.0	0.0	5.7	41.0
8	20.5	7.5	9.0		18.3	112.7	0.0	0.0	17.2	40.5

Depth Meter	Temp °C	O_2 mg/l	pH	Light $\mu E/m^2/sec$	Sol.P $\mu g/l$	Tot.P $\mu g/l$	NH_4^+ $\mu g/l$	NO_2^- $\mu g/l$	NO_3^- $\mu g/l$	Chlor. $\mu g/l$
10	20.0	6.8	8.8		31.5	125.9	0.0	2.2	47.5	26.2
12	13.7	0.6								
14	11.4	0.6								
16	10.2	0.6								
18	9.5	0.6	7.9		304.4	377.3	213.6	0.0	22.9	2.2
20	9.5	0.5								
22	8.7	0.3								
Aug. 7, 1978										
0	23.7	10.7	9.0	1050	0.0	0.2	0.0	0.0	12.4	64.7
2	23.0	8.9	9.0	60	3.0	4.0				48.6
4	22.7	8.4	8.9	5	6.0	44.0	37.1	0.0	10.3	39.3
6	22.3	7.4	8.9		14.0	94.0	12.4	2.1	78.7	20.2
8	21.6	6.0	8.8		32.0	224.0	117.5	0.0	5.2	13.3
10	19.2	0.5	8.2		108.0	344.0				4.4
12	15.5	0.1								
14	13.0	0.1								
16	11.4	0.1								
18	10.6	0.1	7.8		374.0	484.0	234.9	0.8	9.6	1.2
20	10.1	0.1								
22	10.0	0.1								
23	10.0	0.1								
Aug. 14, 1978										
0	23.0			750	377.3	87.3	51.9	0.0	4.1	8.2
2	22.8			67		97.7	84.1	0.0	3.1	8.2
4	22.5			8		101.2	92.7	0.0	1.0	9.1
6	22.2					97.7	76.7	0.4	0.0	0.7
8	19.8					115.2	107.6	19.7	221.8	1.8
10	18.0					157.1	162.0	21.7	269.5	1.3
12	14.8									
14	12.0									
16	10.2									
18	9.8					457.6	262.1	0.0	2.1	
20	9.2									
22	9.2									
Aug. 22, 1978										
0	23.0	7.0	9.0	600	19.2	108.2	0.0	0.0	0.0	52.0
2	22.5	7.0	9.0	15	15.7	101.2	0.0	0.0	5.2	57.8
4	22.0	7.1	9.0		19.2	111.7	21.0	0.0	0.0	56.6
6	22.0	7.0	9.0		13.9	104.7	0.0	0.0	0.0	57.2
8	22.0	7.0	9.0		15.7	115.2	3.7	0.0	0.0	55.5
10	21.0	1.3	8.7		61.1	136.2	34.2	2.9	93.4	21.7
12	18.0	0.3								
14	13.5	0.2								
16	11.0	0.1								
18	10.3	0.1	7.7		351.1	436.6	265.8	0.0	0.0	2.8
20	10.0	0.0								
22	9.7	0.0								
Aug. 28, 1978										
0	23.0	8.1	8.8	320	29.1	96.0	16.1	0.0	2.1	34.7
2	22.6	7.8	8.8	35	30.9	102.9	18.6	0.0	5.2	33.9
4	22.3	7.8	8.8	6	32.6	102.9	0.0	0.0	0.0	33.4
6	22.2	7.8	8.8		32.6	109.8	44.9	0.8	7.5	33.9
8	22.0	7.6	8.8		34.3	109.8	0.0	1.7	22.4	31.0

Depth Meter	Temp °C	O_2 mg/l	pH	Light $\mu E/m^2/sec$	Sol.P $\mu g/l$	Tot.P $\mu g/l$	NH_4^+ $\mu g/l$	NO_2^- $\mu g/l$	NO_3^- $\mu g/l$	Chlor. $\mu g/l$
10	22.0	6.7	8.8		36.0	116.6	56.4	2.9	152.2	28.1
12	19.0	1.0								
14	13.3	0.6								
16	11.6	0.3								
18	10.5	0.2	7.9		260.8	329.4	263.4	1.3	19.5	5.8
20	10.0	0.2								
22	10.0	0.2								
23	10.0	0.2								
					Sep. 5, 1978					
0	24.0	9.2	9.2	1250	12.6	168.3	0.0	4.0	5.9	55.7
2	23.5	9.0	9.0	60	16.1	123.0	0.0	2.9	7.9	77.7
4	23.0	8.7	9.0	3	14.4	109.0	19.9	0.8	18.1	50.6
6	22.5	8.8	8.8		21.4	105.5	77.5	3.2	27.4	59.6
8	22.5	8.4	8.7		30.1	63.7	135.2	5.6	21.4	23.5
10	21.5	6.8	8.5		73.7	147.4	249.9	10.9	70.1	10.1
12	16.0	2.2								
14	15.0	0.8								
16	11.0	0.4								
18	11.0	0.2	7.9		429.5	611.4	558.3	2.7	0.0	7.2
20	10.5	0.1								
22	9.5	0.1								
23	10.0									
					Sep. 12, 1978					
0	24.0	8.6	8.6	80	20.3	100.3	0.0	8.4	22.9	31.8
2	23.0	7.8	8.8	7	23.8	103.5	18.6	25.6	91.9	35.1
4	23.0	7.8	8.9		22.0	94.0	3.9	9.2	26.4	31.0
6	23.0	7.8	8.8		25.6	100.3	20.8	9.4	26.2	22.7
8	23.0	7.8	8.8		23.8	109.8	4.9	9.2	60.9	22.7
10	22.0	7.0	8.9		27.3	94.0	29.3	11.9	0.0	24.6
12	20.0	2.0								
14	17.0	1.0								
16	15.5	6.0								
18	12.0	0.2	8.2		50.2	503.0	330.0	4.5	0.0	7.2
20	11.0	0.1								
22	10.0	0.1								
					Sep. 19, 1978					
0	21.0	8.0	8.7	800	39.1	116.2	128.8	18.6	62.4	24.8
2	21.0	7.6	8.6	75	39.1	132.0	126.9	20.4	62.3	23.5
4	21.0	7.4	8.7	10	48.8	141.5	129.6	17.8	51.5	28.9
6	20.0	7.0	8.7		47.0	128.8	114.7	19.4	71.5	32.5
8	20.0	6.6	8.8		48.6	132.0	126.9	20.7	48.5	21.7
10	20.0	6.2	8.8		50.2	128.8	163.0	17.8	48.8	11.6
12	18.0	2.2								
14	15.0	2.0								
16	12.0	2.0								
18	11.0	2.0	8.2		391.0	496.6	493.5	0.8	19.0	2.9
20	11.0	2.0								
22	11.0	2.0								
					Sep. 28, 1978					
0	19.0	7.2	8.5	1000	46.6	84.3	172.7	19.7	25.7	25.3
2	19.0	6.8	8.7	80	45.1	99.1	122.9	17.7	153.7	28.5
4	19.0	6.2	8.4	8	51.0	93.2	185.2	18.0	18.9	23.5
6	19.0	6.4	8.4		49.6	99.1	170.9	18.8	44.9	14.5

Depth Meter	Temp °C	O_2 mg/l	pH	Light $\mu E/m^2/sec$	Sol.P $\mu g/l$	Tot.P $\mu g/l$	NH_4^+ $\mu g/l$	NO_2^- $\mu g/l$	NO_3^- $\mu g/l$	Chlor. $\mu g/l$
8	19.0	6.2	8.5		51.0	87.3	197.6	18.2	32.8	23.5
10	19.0	6.1	8.3		51.0	102.1	206.5	19.4	48.6	20.2
12	18.0	5.0								
14	15.0	0.5								
16	13.0	0.2								
18	11.0	0.1	7.8		277.6	362.7	220.8	2.3	0.0	6.5
20	10.0	0.1								
22	10.0	0.1								
					Oct. 3, 1978					
0	17.0	6.8	8.6	700	72.5	99.9	200.0	15.0	69.0	16.3
2	17.0	6.6	8.4	60	59.0	111.9	265.2	15.0	97.1	14.5
4	17.0	6.4	8.3	6	59.0	102.9	263.6	14.6	46.8	14.5
6	17.0	6.4	8.4		57.5	111.9	260.4	14.8	64.9	14.5
8	17.0	6.2	8.4		54.5	99.9	263.8	14.4	138.6	14.5
10	17.0	6.2	8.4		56.0	105.9	194.9	14.1	120.7	13.0
12	17.0	6.2								
14	16.0	6.0								
16	12.0	1.1								
18	11.0	0.4	7.9		363.6	424.1	297.0	4.3	0.0	3.3
20	10.0	0.1								
22	10.0	0.0								
					Oct. 11, 1978					
0	14.0	7.2	7.9	70	138.6	126.5	390.5	17.1	92.2	11.6
2	14.0	7.0	8.2	9	94.3	138.3	366.4	14.5	86.6	10.1
4	14.5	6.8	8.1	2	91.3	138.3	397.8	16.4	49.8	8.7
6	14.5	6.8	8.1		95.8	135.3	371.1	16.4	54.5	11.6
8	14.5	6.8	8.1		95.8	135.3	411.1	16.4	49.8	7.2
10	14.5	6.8	8.0		95.8	138.3	431.2	16.4	53.3	4.3
12	14.5	6.8								
14	14.0	6.6								
16	14.0	5.8								
18	11.0	0.1	7.4		344.2	436.9	454.3	0.0	122.0	1.8
20	11.0	0.2								
21	10.0	0.1								
					Oct. 20, 1978					
0	10.0	10.6								
2	10.8	8.8								
4	10.6	8.0								
6	10.5	7.4								
8	10.5	7.4								
10	10.5	7.2								
12	10.5	7.2								
14	10.5	7.2								
16	10.5	7.0								
18	10.5	6.5								
19	10.5	1.4								
					Oct. 24, 1978					
0	12.0	9.1	8.0	900	105.9	148.5	314.5	12.5	246.8	21.7
2	12.0	8.4	8.1	100	115.0	154.5	349.7	13.7	142.6	20.2
4	12.0	8.1	8.1	11	113.5	160.5	351.6	14.0	109.1	23.1
6	12.0	8.1	8.1		116.5	169.6	359.5	14.9	69.9	24.6
8	12.0	8.2	8.1		115.0	160.5	332.1	13.7	115.3	23.1
10	12.0	8.2	8.1		115.0	169.6	336.1	15.1	100.2	23.1

Depth Meter	Temp °C	O_2 mg/l	pH	Light $\mu E/m^2/sec$	Sol.P $\mu g/l$	Tot.P $\mu g/l$	NH_4^+ $\mu g/l$	NO_2^- $\mu g/l$	NO_3^- $\mu g/l$	Chlor. $\mu g/l$
12	12.0	8.0								
14	12.0	8.0								
16	12.0	8.0								
18	12.0	8.0	8.1		116.5	157.5	375.8	14.0	144.8	18.8
20	12.0	8.0								
22	12.0	7.8								
					Nov. 7, 1978					
0	11.5	10.4	8.0	600	100.7	134.8	347.2	19.4	77.7	15.9
2	11.5	10.4	8.1	100	106.7	143.8	359.2	21.4	55.2	20.2
4	11.5	10.4	8.2	19	106.7	162.0	349.4	21.1	56.1	23.1
6	11.5	10.4	8.2		106.7	140.8	357.2	21.8	86.6	26.0
8	11.5	10.4	8.2		106.7	162.0	373.4	21.1	96.5	29.8
10	11.5	10.2	8.2		106.7	162.0	357.8	20.4	89.3	27.5
12	11.5	10.4								
14	11.5	10.4								
16	11.5	10.4								
18	11.5	10.4	8.3		111.3	165.0	371.3	22.4	63.4	27.5
20	11.5	10.4								
22	11.5	10.4								
					Nov. 16, 1978					
0	9.0	9.5	8.2	300	98.0	129.6	329.8	27.3	65.4	20.2
2	9.0	9.4	8.2	55	95.0	135.6	308.5	26.3	116.4	20.2
4	9.2	9.4	8.2	15	93.5	147.7	383.4	28.3	57.4	19.7
6	9.2	9.4	8.1		96.5	147.7	432.5	28.3	48.8	15.9
8	9.2	9.4	8.1		93.5	129.6	426.9	28.3	64.4	21.7
10	9.5	9.4	8.1		98.0	144.7	429.3	27.0	107.1	18.8
12	9.3	9.3								
14	9.2	9.3								
16	9.2	9.2								
18	9.5	9.1	8.1		93.5	144.7	423.6	28.3	85.5	24.6
20	9.2	9.0								
22	9.2	7.3								
					Nov. 28, 1978					
0	4.0	12.4	8.1	200	78.2	102.9	406.9	34.7	128.3	13.0
2	5.0	12.1	8.1	55	83.8	111.3	402.2	34.4	123.9	13.0
4	5.0	12.0	8.1	15	79.6	108.5	384.4	33.7	121.5	13.0
6	5.0	11.7	8.1		83.8	105.7	376.7	32.4		13.0
8	5.0	12.2	8.2		83.8	114.2	362.0	34.0	84.5	20.2
10	5.0	11.7	8.2		79.6	111.3	382.9	33.4	159.2	15.9
12	5.0	10.8								
14	5.0	11.0								
16	5.0	10.4								
18	5.0	9.8	8.3		85.2	108.5	408.4	34.4	49.8	13.0
20	5.0	9.3								
22	5.0	1.4								
					Feb. 7, 1979					
0				1340	82.9		224.9	2.4	246.6	2.4
2				9	95.1		237.5	2.2	261.3	0.2
4				2	86.0		241.5	2.2	241.9	0.6
6					85.0		234.2	2.2	243.0	0.6
8					87.0		229.2	3.1	260.3	0.0
10					81.9		229.2	3.1	233.1	0.0
12					84.0		229.6	3.0	225.4	0.0
14					81.9		218.3	3.1	235.1	0.0
16					89.0		237.2	5.3	242.8	0.0
18					81.9		230.2	4.7	302.7	0.4
					Feb. 21, 1979					
0	0.0	12.3		1050	85.3		185.1	3.8	466.9	0.5
2	0.0	12.5		29	87.5		192.2	4.8	441.8	1.0
4	0.0	12.3			90.7		186.9	4.4	522.8	
6	0.0	12.0					208.1	4.1	473.8	0.6
8	0.5	12.0					211.6	4.4	452.4	0.5
10	0.5	11.9					189.9	5.0	493.0	0.0
12	0.7	12.0					212.1	5.3	464.8	0.4
14	1.0						177.6	5.2	465.2	
16	1.0	12.1					190.9	6.2	469.7	
18	1.0	11.9					295.1	8.6	560.5	
20	1.7	5.6					584.7	7.0	49.7	
					Feb. 28, 1979					
0	0.2	11.3		870	280.7		178.8	4.5	364.2	0.8
2	0.5	10.9		46	261.8		137.6	3.9	417.5	0.6
4	0.5	10.7		14	246.1		145.9	2.6	371.0	0.2
6	0.7	10.2		6	261.8		118.4		377.1	0.2
8	0.7	10.0		3	236.7		126.7	3.6	424.2	0.2
10	0.7	9.6		1	252.4		115.7	3.6	472.6	0.2
12	1.0	9.0			252.4		132.2	3.2	480.6	0.4
14	1.0	8.4			258.7		121.2	3.6	524.0	0.2
16	1.2	7.4			249.3		140.4	3.6	345.6	0.2
18	1.5	5.4			239.8		337.9	3.2	399.4	0.3
19	2.0	0.7								
					Mar. 17, 1979					
0	1.0	11.0		180	180.0		328.1	22.8	782.7	1.6
2	1.5	10.6		30	162.1		84.0	16.1	534.2	2.9
4	1.5	10.6		14	175.3		56.9	13.4	500.0	2.7
6	1.5	10.5		9	140.8		52.6	13.9	511.1	2.4
8	1.5	10.4		5	169.7		58.6	13.7	500.0	3.3
10	1.5	10.0		2	183.7		36.4	13.9	502.8	
12	1.7	9.3			145.5		35.8	13.7	505.3	0.9
14	2.0	8.5			168.8		49.4	14.8	497.7	
16	2.0	6.0			161.3		34.8	15.3	492.9	0.4
18	2.2	2.8			172.5		217.2	54.8	458.6	
20	2.7	0.4			215.5		643.1	32.3	462.9	0.1
					Mar. 30, 1979					
0	0.5	12.6			337.4	48.7	227.8	10.1	530.8	6.2
2	0.5	11.6			193.6	110.2	68.6	7.8	412.8	25.9
4	0.5	11.5			362.2	119.4	90.3	7.4	441.0	24.4
6	0.5	11.4			189.9	110.2	60.6	7.6	482.3	25.5
8	0.5	11.4			218.0	110.2	63.2	8.1	560.5	22.6
10	0.5	10.7								23.3
12	1.0	8.6			196.4	113.2	69.0	8.9	539.8	12.1
14	1.0	9.2								
16	1.5	8.3			264.8	122.5	82.3	26.2	580.8	
18	1.5	3.3								
20	1.5	0.3			290.1	174.7	256.8	112.8	685.8	2.0
22	1.5	0.3								

Depth Meter	Temp °C	O_2 mg/l	pH	Light $\mu E/m^2/sec$	Sol.P $\mu g/l$	Tot.P $\mu g/l$	NH_4^+ $\mu g/l$	NO_2^- $\mu g/l$	NO_3^- $\mu g/l$	Chlor. $\mu g/l$
Apr. 6, 1979										
0	0.0	12.6		1600	108.6	109.0	33.7	4.5	539.6	38.3
2	0.7	12.6		130	592.5	121.3	142.8	3.4	289.5	36.3
4	1.0	12.4		31	179.7	115.2	38.7	3.4	417.2	36.7
6	1.0	12.4		9	141.0	115.2	34.7	3.1	317.3	35.8
8	1.0	12.2			213.7	112.1	57.8	3.4	122.3	
10	1.0	12.0			241.5	115.2	44.7	3.1	324.8	37.9
12	1.0	11.7			343.6	112.1	55.4	8.5	452.7	
14	1.0	8.2			278.6	112.1	0.0	6.5	732.8	14.9
16	1.0	7.6			338.9	109.0				
18	1.2	2.8			291.0	115.2	97.8	4.2	522.4	
20	1.5	0.4			462.6	186.3	273.2	84.7	688.2	2.1
22	1.7	0.4			604.8	214.5	543.5	90.1	644.1	
24	2.2	0.4			383.8	306.9	591.1	132.4	617.4	
Apr. 18, 1979										
0	3.0	13.2		1600		101.0	0.0	5.8		22.7
2	3.0	13.2		310	35.9	116.3	38.1	2.3	328.9	28.2
4	2.5	13.6		60	37.4	116.3	65.8	4.0	341.6	24.4
6	2.5	13.2		13	34.4	113.2	21.0	1.8	464.1	26.9
8	2.5	13.4		3	29.9	119.4	18.7	2.0	410.9	24.4
10	2.5	13.6								
12	2.5	14.2			32.9	110.2	18.0	1.8	411.1	29.5
14	2.5	14.0								
16	3.0	13.3			32.9	101.0	0.0	2.5	296.2	
18	3.0	13.0								
Apr. 20, 1979										
0	2.5	12.5		450	32.4	111.7	23.0	3.7	672.1	37.0
2	2.5	12.6		45	37.1	117.9	22.6	3.1	696.7	40.3
4	2.5	12.2		9	35.6	114.8	26.6	3.1	687.8	23.8
6	2.5	11.6		2	34.0	117.9	10.7	0.0	752.3	16.7
8	2.5	11.3			37.1	121.0	21.7	0.0	603.0	27.5
10	2.5	10.4								
12	2.5	10.0			37.1	117.9	20.7	0.0	754.1	29.5
14	2.5	9.8								
16	2.5				34.0	114.8	0.0	2.7	629.4	39.4
18	2.5						440.0			
20	2.5				166.6	155.3	195.8	19.1	582.1	
22	2.5				202.4		493.5	100.2	792.2	14.9
23	2.5				269.5	317.5	468.4	148.0	596.4	
Apr. 27, 1979										
0	6.0	12.6		230	10.2	80.9	43.7	2.0	539.2	32.2
2	5.7	11.6		35	11.7	77.9	73.7	2.0	648.4	25.7
4	5.7	11.6		11	22.3	84.0	44.3	3.4	496.9	27.9
6	5.7	11.4		3	13.2	93.1	32.1	2.6	504.5	33.5
8	5.7	11.0			13.2	117.3	35.5	1.2	513.8	29.5
10	5.7	10.8								
12	5.5	10.0			13.2	81.0	23.5	1.2	604.3	29.5
14	5.0	8.8								
16	5.0	8.0			13.2	78.0	0.0	8.8	629.8	36.1
18	5.0	8.5								
20	5.0	8.1			19.3	84.0	35.8	0.9	644.0	27.1
22	4.5	7.6			31.4	108.2	87.2	7.7	614.6	
24	4.5	0.6								

Depth Meter	Temp °C	O_2 mg/l	pH	Light $\mu E/m^2/sec$	Sol.P $\mu g/l$	Tot.P $\mu g/l$	NH_4^+ $\mu g/l$	NO_2^- $\mu g/l$	NO_3^- $\mu g/l$	Chlor. $\mu g/l$
May 4, 1979										
0	7.2	14.3			23.6	66.0	34.2	7.4	726.4	11.2
2	6.9	14.7			23.6	78.5	56.3	8.0	627.9	18.5
4	6.2	14.0			22.0	84.8	21.1	8.0	712.0	20.3
6	5.9	13.0			23.6	69.2	12.2	8.0	783.5	22.5
8	5.8	12.7			22.0	75.4	10.3	7.4	782.0	20.0
10	5.7	12.1								
12	5.7	11.8			23.6	81.7	47.6	11.1	476.8	15.7
14	5.7	11.6								
16	5.6	11.9			25.1	78.5	33.9	7.2	700.9	13.2
18	5.3	11.6			28.2	78.5	0.0	7.2	150.5	
May 10, 1979										
0	9.5	11.7		1800	18.3	66.8	57.6	10.8	427.1	15.9
2	9.2	10.2		250						
4	9.0	9.5		75	18.3	73.0	44.8	11.1	452.3	23.5
6	8.7	8.3		17						
8	8.7	7.2			19.8	73.0	45.1	10.8	493.8	18.4
10	8.7	6.5								
12	8.2	6.2			18.3	60.7	51.8	10.8	370.9	12.8
14	8.0	6.0								
16	7.0	5.8			26.0	60.7	74.5	10.2	432.0	
18	7.0	5.8								
20	6.7	5.7			27.5	57.6	62.2	9.1	446.2	5.7
22	6.7	5.3			27.5	60.7	76.7	9.9	430.5	
23	6.7	6.1								
May 16, 1979										
0	13.2	14.3		1450	7.6	99.2	51.5	9.2	243.1	73.7
2	11.7	10.8		72	12.2	92.6	88.6	11.1	290.7	100.6
4	11.2	9.5		14	9.1	83.6	106.9	7.8	271.2	34.9
6	11.2	9.0								
8	11.0	7.7			6.1	61.7	86.3	6.4	504.6	22.5
10	10.5	7.9								
12	9.0	6.8			9.1	86.7	85.8	7.0	281.9	41.6
14	8.0	7.0								
16	7.7	5.9			29.0	71.1	121.4	7.4	484.0	9.2
18	7.5	5.9								
20	7.2	6.0			71.9	111.7	185.8	3.7	588.7	6.3
22	7.2	5.3								
23	7.2	0.8								
May 25, 1979										
0	13.0	11.4		1450	7.6	97.8	22.9	13.7	359.6	28.5
2	13.0	11.1		190	10.7	97.8	46.0	13.7	304.3	28.9
4	13.0	11.0		24	16.8	85.6	35.4	13.7	272.8	33.9
6	13.0	10.8		4						
8	13.0	10.7			7.6	88.7	41.9	14.0	405.8	31.3
10	13.0	10.4								
12	13.0	10.3			10.7	76.4	47.0	13.7	221.0	26.0
14	13.0	10.1								
16	10.5	9.0			47.4	103.9	203.7	12.0	440.0	11.2
18	9.7	8.4								
20	8.7	7.8			99.4	165.2	279.6	12.4	441.5	7.0
22	8.5	7.2								
23	8.5	5.6								

Depth Meter	Temp °C	O_2 mg/l	pH	Light $\mu E/m^2/sec$	Sol.P $\mu g/l$	Tot.P $\mu g/l$	NH_4^+ $\mu g/l$	NO_2^- $\mu g/l$	NO_3^- $\mu g/l$	Chlor. $\mu g/l$
May 31, 1979										
0	17.5			460	1.1	38.5	45.6	10.8	186.6	28.1
2	15.8			43	2.6	74.9	75.4	11.6	201.9	95.4
4	15.2			8	5.6	50.7	116.2	11.6	211.7	37.6
6	15.0									
8	15.0				0.0	47.6	112.8	11.6	291.3	26.8
10	14.7									
12	14.0				2.6	38.5	135.5	11.6	252.9	13.6
14	13.1									
16	12.2				26.8	65.8	221.6	11.6	329.7	14.4
18	11.5									
20	10.8				66.2	105.2	336.7	11.0	433.1	9.1
21	9.5									
June 4, 1979										
0	16.5	12.2		460	6.4	72.3	29.1	11.4	36.2	66.5
2	16.5	12.0		24	4.9	75.4	29.1	12.4	17.8	90.3
4	16.5	11.7		2	4.9	81.7	20.0	10.8	45.5	75.9
6	16.5	11.4								
8	16.0	10.8			4.9	72.3	39.1	12.0	67.3	46.3
10	15.3	9.8					171.7			
12	13.0	9.0			15.8	59.8		14.3	232.7	10.3
14	12.0	7.6								
16	11.5	7.2			42.4	100.4	243.7	15.4	264.2	0.0
18	10.5	6.2								
20	9.5	5.3			95.6	153.6	237.9	15.4	337.0	12.4
22	9.2	3.7								
24	9.0	1.5								
June 11, 1979										
0	18.2		8.8	1600	6.9	62.8	58.1	11.4	43.8	31.8
2	17.8		8.8	250	6.9	87.3	106.3	11.6	15.8	30.9
4	17.0		8.8	23	6.9	72.0	73.5	11.4	15.1	31.8
6	16.8			4						
8	16.5		8.8		5.3	62.8	101.9	11.6	23.5	19.6
10	14.0		8.6		11.5	47.5	168.4	15.7	154.9	11.6
12	11.0		8.4		32.9	72.0	238.5	20.6	213.7	12.1
14	9.0									
16	8.5		8.0		101.9	154.8	273.9	24.9	360.1	6.4
18	7.0									
20			7.9		129.5	207.0	280.7	25.7	351.7	8.1
June 18, 1979										
0	19.0	8.5	8.6	170	16.8	49.1	182.4	16.3	68.1	5.8
2	19.0	8.3	8.6	40	18.4	67.6	184.1	15.7	45.5	8.7
4	19.0	7.6	8.6	15	15.3	49.1	169.6	16.6	51.1	53.5
6	19.0	7.9		6						
8	19.0	7.8	8.6		15.3	52.2	188.7	15.7	73.2	5.8
10	16.2	6.6	8.4		36.9	73.4	411.2	21.4	119.9	2.9
12	13.8	5.5	8.2		60.0	92.2	277.8	27.6	154.0	10.1
14	11.7	4.1								
16	11.0	2.4	8.0		130.8	166.1	370.1	33.0	319.4	2.9
18	10.5	2.0								
20	10.0	1.1	7.8		177.0	224.6	368.5	34.6	241.2	2.9
22	9.8	0.9								
23	9.0									
June 25, 1979										
0	19.6	9.8	8.6	1300	18.6	52.7	88.7	18.4	91.4	14.6
2	19.4	9.2	8.6	270	18.6	55.8	69.2	18.1	89.1	12.1
4	19.2	7.7	8.6	66	20.2	62.0	116.0	18.4	87.6	8.7
6	19.0	6.4		15						
8	19.0	5.7	8.6		21.7	58.9	124.1	17.9	110.0	11.6
10	18.9	5.4	8.6		21.7	58.9	116.3	19.0	90.2	14.5
12	16.3	4.9	8.3		52.7	89.9	243.7	24.7	117.8	5.8
14	13.8	4.2								
16	11.8	3.5	8.0		124.0	161.2	343.7	36.0	213.2	5.2
18	10.8	1.6			210.8					
20	10.2	1.5	7.8		210.8	244.9	400.7	97.0	128.8	0.6
22	10.0	1.3								
July 2, 1979										
0	21.0	10.9	8.9	1300	3.7	46.9	9.8	0.0	13.8	28.9
2	20.9	10.9	8.9	135	3.7	62.1	18.2	0.0	13.8	39.3
4	20.5	5.5	8.9	20	3.7	52.9	19.7	0.0	14.1	22.5
6	19.8	5.0		4						
8	19.8	4.2	8.8		5.2	43.8	32.3	15.5	16.2	19.1
10	19.1	4.2	8.7		20.4	56.0	44.8	26.8	17.4	11.6
12	16.3	3.6	8.3		70.5	116.7	131.2	35.7	23.4	15.6
14	12.9	2.7								
16	12.0	1.7	8.0		148.0	186.6	148.2	136.4	39.8	8.7
18	11.1	1.5								
20	10.4	1.3	7.9		225.5	262.6	229.2	56.1	86.3	6.9
22	10.1	1.3								
July 9, 1979										
0	23.5	12.5	9.3	1000	3.1	48.8	31.8	0.0	0.0	45.1
2	23.4	12.2	9.0	85	3.1	64.0	25.2	0.0	0.0	47.4
4	22.1	10.0	8.9	13	3.1	54.9	19.3	0.0	0.0	37.0
6	21.9	8.6								
8	21.2	6.3	8.7		7.7	42.7	86.8	4.7	18.0	17.9
10	20.5	5.5	8.6		33.5	64.0	153.4	10.0	64.9	16.2
12	18.0	4.0	8.3		71.5	103.5	303.7	16.7	108.8	9.2
14	14.4	1.9								
16	12.5	1.6	7.8		167.3	206.9	253.7	49.1	169.6	9.2
18	11.6	1.4								
20	11.0	1.4	7.7		241.8	289.0	408.2	41.5	42.2	4.0
22	10.8	1.3								
23	10.7	1.3								
July 16, 1979										
0	25.5	10.1	9.0	1250	4.2	54.1	19.3	0.0	0.0	53.2
2	25.5	7.9	9.0	50	4.2	60.2	14.1	0.0	2.5	61.8
4	25.5	7.6	9.0	3	4.2	54.1	34.7	0.0	0.8	59.5
6	24.7	5.9								
8	20.7	4.7	8.5		5.7	41.9	103.6	6.0	35.6	7.7
10	20.0	4.1	8.4		25.5	54.1	147.2	26.4	69.9	7.2
12	18.1	3.0	7.9		82.0	118.2	192.1	34.4	234.6	6.7
14	14.9	1.7								
16	12.8	1.6	7.7		173.5	215.8	435.6	36.1	129.3	4.3
18	12.0	1.5								
20	11.5	1.4	7.5		292.4	337.8	549.7	98.2	2.0	2.4
22	11.1	1.4								

Depth Meter	Temp °C	O_2 mg/l	pH	Light $\mu E/m^2/sec$	Sol.P $\mu g/l$	Tot.P $\mu g/l$	NH_4^+ $\mu g/l$	NO_2^- $\mu g/l$	NO_3^- $\mu g/l$	Chlor. $\mu g/l$
24	11.0	1.3								
July 23, 1979										
0	22.0	8.7	8.9	900	1.8	55.0	12.7	0.0	5.4	61.8
2	22.0	8.6	8.9	40	1.8	55.0	11.2	0.0	13.3	64.2
4	22.0	8.5	8.9	2	1.8	61.0	15.6	0.0	16.3	59.5
6	22.0	8.2								
8	18.1	3.6	8.6		1.8	70.1	54.7	7.2	48.5	26.8
10	17.6	2.9	8.1		21.4	55.0	75.3	0.4	166.7	7.5
12	15.0	1.8	7.9		89.4	130.5	94.7	31.9	326.3	5.8
14	12.0	1.4								
16	10.0	1.4	7.7		231.3	209.5	413.0	0.0	16.9	5.2
18	9.1	1.4								
20	9.0	1.4	7.6		306.8	375.0	765.3	0.0	9.0	6.9
22	8.9	1.4								
July 30, 1979										
0	24.5	7.5	8.8	1050	1.7	40.4	24.1	3.1	0.9	38.1
2	24.3	6.9	8.8	100	3.3	40.4	10.4	3.1	0.0	39.9
4	24.0	5.9	8.8	6	1.7	46.5	11.7	3.1	0.0	34.7
6	23.9	5.4								
8	19.9	2.0	8.0		1.7	37.3	7.3	18.9	145.8	2.3
10	18.9	1.8	7.9		21.7	58.9	8.3	18.7	172.1	2.3
12	15.9	1.6	7.8		81.8	132.7	126.6	64.9	171.5	4.6
14	13.5	1.4								
16	12.3	1.3	7.5		232.6	283.5	302.6	3.3	15.1	2.9
18	11.8	1.2								
20	11.1	1.2	7.4		309.5	375.9	425.5	0.0	8.3	2.3
22	10.9	1.1								
23	10.8	1.1								
Aug. 6, 1979										
0	24.0	7.2	9.0		1.9	43.7	3.6	3.4	3.9	28.9
2	24.0	6.5	9.0		1.9	43.7	5.2	3.4	0.0	27.2
4	24.0	5.7	9.0		0.4	46.8	3.6	3.4	0.0	27.7
6	24.0	4.9								
8	23.3	4.2	9.0		0.4	43.7	10.8	5.3	0.0	23.7
10	20.8	1.6	8.4		0.4	46.8	1.2	27.0	130.0	1.7
12	17.5	1.5	7.9		55.6	114.3	107.2	61.7	97.5	0.0
14	15.5	0.9								
16	14.3	0.2	7.9		150.8	221.8	280.2	3.4	0.0	0.6
18	12.5	0.0								
20	11.5	0.0	7.8		307.3	378.3	655.7	0.0	0.0	1.2
22	11.3	0.0								
23	11.3	0.0								
Aug. 13, 1979										
0	23.0	6.1	8.6	120	1.9	46.8	0.0	3.2	0.0	30.1
2	23.0	6.1	8.7	13	1.9	43.7	0.0	3.2	8.2	35.9
4	22.9	6.0	8.7		0.4	46.8	0.0	3.5	2.5	28.3
6	22.9	5.9								
8	22.9	5.8	8.7		0.4	40.7	0.0	2.9	2.8	28.9
10	22.8	5.5	8.7		0.4	40.7	0.0	3.2	1.9	24.9
12	17.9	0.4	8.7		37.2	89.8	213.1	42.1	46.4	5.8
14	15.8	0.3								
16	14.0	0.0	7.7		170.7	234.1	330.4	1.5	0.0	
18	12.8	0.0								

Depth Meter	Temp °C	O_2 mg/l	pH	Light $\mu E/m^2/sec$	Sol.P $\mu g/l$	Tot.P $\mu g/l$	NH_4^+ $\mu g/l$	NO_2^- $\mu g/l$	NO_3^- $\mu g/l$	Chlor. $\mu g/l$
20	12.0	0.0	7.7		295.1	350.7	798.0	0.7	0.0	2.3
22	11.8	0.0								
24	11.8	0.0								
Aug. 20, 1979										
0	21.0	7.5	8.5	380	1.7	43.5	0.0	3.0	11.1	28.9
2	21.0	7.1	8.7	35	3.3	55.8	0.0	12.0	10.0	31.2
4	20.8	6.1	8.6	6	1.7	52.7	3.5	22.7	0.0	27.8
6	20.8	5.8								
8	20.8	4.6	8.7		1.7	40.4	4.7	3.8	0.0	22.5
10	20.5	3.5	8.7		1.7	46.5	4.7	3.5	0.0	22.0
12	17.9	0.4	8.3		31.0	80.4	113.5	4.3	11.9	14.5
14	16.0	0.1								
16	14.1	0.1	7.9		203.3	234.3	338.5	0.0	5.2	4.0
18	12.8	0.0								
20	11.9	0.0	7.4		344.9	375.9	721.0	0.0	1.0	1.2
22	11.2	0.0								
24	11.0	0.0								
Aug. 28, 1979										
0	20.9	9.1		550	12.6	62.8	35.0	2.5	45.2	20.2
2	21.0	7.7		45	9.7	68.6	23.9	2.5	65.2	19.9
4	21.0	7.0		8	8.2	71.5	23.3	3.6	29.3	21.0
6	21.0			2						
8	21.0				9.7	62.8	29.5	2.5	20.2	18.4
10	20.8				11.1	65.7	34.1	3.0	54.8	13.4
12	19.7				40.1	94.7	143.5	4.4	93.6	6.5
14	14.8									
16	13.4				225.6	294.7	1144.5	0.0	17.5	2.2
18	12.8									
20	12.0				334.3	402.0	1600.7	0.0	12.5	2.5
22	11.7									
24	11.5									
Sep. 4, 1979										
0	20.5	9.0		900	12.5	74.1	17.6	0.0	17.5	30.0
2	20.0	3.9		64	9.6	77.0	13.5	0.0	47.5	29.6
4	20.0	3.8		8	9.6	79.9	11.6	0.0	17.5	30.0
6	20.0	3.6								
8	19.5	3.5			15.4	71.2	25.7	13.3	145.5	13.7
10	18.5	2.0			24.0	100.1	62.6	27.8	262.5	5.4
12	16.7	0.9			80.4	198.3	255.6	3.8	49.0	4.7
14	14.2	0.0								
16	12.2	0.1			223.4	464.1	1161.2	0.0	7.5	4.0
18	10.5	0.0								
20	10.5	0.0			379.4	417.9	1914.7	0.0	2.5	2.5
22	10.0	0.0								
24	9.0	0.0								
Sep. 10, 1979										
0	19.7	6.8		900	15.6	84.3	23.1	0.0	117.7	8.7
2	19.5	7.0		68	15.3	102.8	6.6	5.1	60.8	10.5
4	19.2	6.8		7	13.9	99.7	0.9	5.1	38.2	10.1
6	19.2	6.6								
8	19.2	6.6			15.3	105.9	0.0	6.3	140.4	7.2
10	19.2	6.5			13.9	102.8	2.2	5.4	54.2	10.8
12	17.0	0.0			79.6	142.8	147.9	10.7	29.6	5.4

Depth Meter	Temp °C	O_2 mg/l	pH	Light $\mu E/m^2/sec$	Sol.P $\mu g/l$	Tot.P $\mu g/l$	NH_4^+ $\mu g/l$	NO_2^- $\mu g/l$	NO_3^- $\mu g/l$	Chlor. $\mu g/l$
14	15.5	0.0								
16	12.5	0.0			272.3	241.3	1044.0	5.1	3.5	0.7
18	11.0	0.0								
20	10.0	0.0			243.8	512.2	1688.6	0.0	65.4	3.2
22	10.0	0.0				595.4				4.7
Sep. 17, 1979										
0	20.0	4.2	8.6	1200	22.4	149.0	4.8	5.4	12.4	35.4
2	19.5	4.2	8.6	100	21.0	84.3	3.0	5.1	18.4	31.0
4	19.5	4.2	8.7	9	18.2	90.5	1.5	5.1	18.4	33.2
6	19.2	4.1								
8	19.0	4.0	8.3		19.6	118.2	9.6	0.0	28.7	32.1
10	19.0	4.0	8.3		21.0	349.1	7.6	0.0	23.0	50.2
12	19.0	3.8	8.5		19.6	109.0	11.2	5.1	87.2	50.9
14	16.0	0.0								
16	14.0	0.0	8.3		276.7	401.4	1445.8	0.0	0.0	0.7
18	13.0	0.0								
20	12.0	0.0	7.9		428.0	512.2	2387.6	0.0	0.0	5.7
22	12.0	0.0	7.4							
23	12.0	0.0								
Sep. 24, 1979										
0	18.5	4.4	8.5	30	32.2	113.5	18.8	11.1	31.7	49.1
2	18.5	4.1	8.5	5	32.2	248.5	10.9	7.6	11.4	27.8
4	16.5	4.2	8.5		27.6	104.3	6.6	6.9	21.4	20.8
6	16.5	4.1								
8	16.5	4.3	8.5		33.7	131.9	8.5	7.2	11.7	23.7
10	16.0	4.3	8.5		29.1	101.2	19.4	6.9	2.6	16.8
12	15.5	3.8	8.5		29.1	294.6	11.9	6.5	21.7	34.1
14	15.0	0.0								
16	10.8	0.0	7.6		322.2	389.7	1069.9	7.6	2.7	1.7
18	10.8	0.0								
20	10.8	0.0	7.5		483.3	533.9	2317.6	7.2	11.7	1.2
22	10.5	0.0								
Oct. 1, 1979										
0	19.0	6.8	8.7	700	26.3	103.3	31.4	7.6	168.3	52.0
2	19.0	6.6	8.7	20	24.9	100.5	8.4	6.9	35.9	38.2
4	19.0	6.6	8.7		23.5	94.9	2.6	7.2	31.3	49.7
6	18.0	5.8								
8	18.0	5.6	8.6		22.1	111.7	8.8	6.9	25.2	33.0
10	18.0	4.8			24.9	83.6	2.0	6.9	30.9	31.2
12	17.5	3.6	8.5		33.4	86.4	12.8	12.5	16.6	28.9
14	17.5	1.1	8.4							
16	15.0	0.5	7.5		226.1	286.2	1113.8	9.7	54.5	17.9
18	13.5	0.4								
20	13.0	0.3	7.4		390.8	477.6	2330.9	6.5	5.7	1.6
22	12.0	0.3								
24	12.0	0.3								
Oct. 8, 1979										
0	16.0	5.2	8.3	420	39.4	152.1	37.2	3.9	21.0	49.1
2	15.5	5.0	8.3	27	24.0	152.1	35.1	2.1	22.6	49.7
4	15.5	3.3	8.3		47.1	132.8	37.8	2.4	19.2	48.0
6	15.5	2.7								
8	15.5	2.3	8.3		20.1	144.4	41.6	2.1	4.5	46.8
10	15.5	2.3	8.2		49.1	105.8	49.4	2.1	3.9	45.7
12	15.5	2.3	8.3		27.9	144.4	47.1	1.8	11.0	40.5
14	15.5	2.8								
16	15.5	4.0	8.3		68.3	129.0	51.6	2.1	13.2	45.7
18	14.5	1.2								
20	13.0	0.4	7.5		457.7	595.5	1976.6	0.9	0.0	9.2
22	12.0	0.2								
Oct. 16, 1979										
0	14.0		7.8	38	75.1	147.5	184.0	5.3	151.2	25.4
2	14.0		8.0	5	75.1	155.5	160.7	5.6	87.2	28.3
4	14.0		8.0		75.1	152.8	157.5	6.3	163.8	12.7
6	14.0									
8	14.0		8.0		73.7	136.8	124.2	5.3	62.0	37.0
10	14.0									
12	14.0		8.0		71.1	195.6	152.8	5.3	206.4	25.4
14	14.0									
16	14.0		8.0		71.1	139.4	145.1	4.4	195.2	6.9
18	14.0									
20	14.0		8.0		69.7	128.7	150.3	5.6	53.2	28.9
Oct. 29, 1979										
0	12.0	9.2	7.7	240	144.3	151.6	330.9	9.3	65.0	21.4
2	12.0	8.7	7.9	14	189.2	232.9	300.4	9.0	79.4	3.5
8	12.0	8.6	7.7		163.5	271.4	335.9	9.3	65.0	5.8
16	12.0	8.5	7.9		137.8	292.8	316.3	9.3	36.7	5.2
20	11.5	8.5	7.9		142.1	237.2	324.2	10.4	54.6	3.5
Nov. 16, 1979										
0	5.0				148.0	164.9	308.7	18.6	0.1	9.2
5					191.7	194.0	229.3	19.0	0.0	8.1
10					171.3	191.1	311.3	18.2	0.0	13.0
15					162.6	496.2	346.4	17.9	0.0	12.3
20					172.8	216.5	318.2	17.5	9.5	6.4
Dec. 10, 1979										
0	3.0	14.0			371.5	118.1	437.6	36.0	145.7	0.4
2	4.0	14.0								1.1
4	3.0	13.9								
5					499.9	121.5	356.6	37.3	128.5	1.1
6	2.8	13.8								
8	2.8	13.8								
10	2.8	13.9			373.2	131.7	445.4	36.7	170.8	0.4
12	2.8	13.8								
14	2.8	13.9								
15					445.3	124.9	360.1	36.7	151.5	0.4
16	1.5	13.8								
18	2.0	13.3								
Jan. 14, 1980										
0	1.0	12.9	7.9	575	134.1	175.0	493.1	34.3	141.2	30.7
2	1.3	13.7	8.1	66	120.3	165.4	443.0	35.2	161.4	16.5
4	1.3			17						
5		13.2	8.1	8	124.9	165.4	475.3	35.5	142.4	14.5
6	1.3									
8	1.5									
10	1.5	12.8	8.1		124.9	158.9	470.3	34.9	149.1	15.9
12	1.5									

Depth Meter	Temp °C	O_2 mg/l	pH	Light $\mu E/m^2/sec$	Sol.P $\mu g/l$	Tot.P $\mu g/l$	NH_4^+ $\mu g/l$	NO_2^- $\mu g/l$	NO_3^- $\mu g/l$	Chlor. $\mu g/l$
14	1.5									
15		12.7	8.1		131.0	155.7	483.7	33.7	155.1	12.7
16	1.6									
18	1.6						470.3	33.4	156.6	
20	1.6	12.8	8.1		131.0	162.1	482.5	33.7	161.2	7.2
Feb. 7, 1980										
0	1.0	14.1	7.9	300	97.5	175.3	413.4	4.7	186.6	63.3
2	1.0	13.8	8.0	62	94.2	152.4	406.5	4.7	195.1	38.0
4	1.5			14						
5		12.9	8.0	9	95.8	139.4	416.1	4.7	190.4	17.0
6	1.5									
8	1.5									
10	1.8	12.2	8.0		102.4	145.9	405.4	4.1	183.2	18.1
12	2.0									
14	2.0									
15		12.9	8.0		97.5	132.9	425.1	3.6	175.1	13.4
16	2.0									
18	2.0									
20	2.3	10.2	8.0		130.1	168.8	546.5	6.0	250.7	6.5
22	2.8	8.3					686.0	7.4	204.0	
Feb. 28, 1980										
0	1.0	10.9		250	136.3	142.5	296.9	13.6	366.0	9.4
2	1.5	11.1		65	134.6	139.3	306.0	12.5	322.0	3.6
4	1.7	11.1		19	121.6	145.8	319.5	12.5	318.0	7.2
6	1.7									
8	1.7	11.1			126.5	152.3	309.5	12.5	340.0	3.6
10	1.7									
12	1.7									
14	1.7									
16	1.7	8.9			150.9	194.5	383.5	20.2	381.5	1.8
18	1.7									
20	2.5	5.5					703.7	23.0	347.0	
22	2.5	1.0					715.1	23.0	311.0	
23	2.5	0.4					837.8	30.7	311.0	
Mar. 17, 1980										
0	0.5	11.1		500	118.1	128.2	363.0	10.1	503.0	1.8
2	1.2	11.2		145	118.1	128.2	402.0	8.3	506.0	1.8
4	1.2	11.2		55	118.1	124.9	323.0	8.1	509.0	10.8
6	1.2			24						
8	1.2	11.0			121.4	128.2	301.0	8.4	509.0	9.0
10	1.2									
12	1.2									
14	1.2									
16	1.4	8.9			145.9	151.1	390.0	7.3	548.0	1.4
18	1.4									
20	1.6	5.1			204.8		736.0	5.6	516.0	
22	2.0	0.8								
23	2.1									
Mar. 26, 1980										
0	0.5	9.3	7.3	620	99.6	104.7	189.0	7.1	455.6	0.7
2	2.0	11.1	8.0	225	135.5	137.3	183.0	10.5	605.6	10.8
4	2.0	10.9	8.0	69	130.6	153.6	199.0	9.3	541.6	18.1
6	2.0			21						
8	2.0	11.0	8.0		137.1	143.8	194.0	9.9	532.1	15.5
10	2.0									
12	2.0									
14	2.0									
16	1.8	8.3	7.7		189.3	189.5	335.0	9.1	499.9	5.4
18	2.0									
20	2.0	0.5	7.3		370.3	407.9	1277.0	12.1	455.1	0.0
22	2.0	0.1					928.0	56.4	499.2	
23	2.3	0.0					1676.0	46.7	151.0	
Apr. 1, 1980										
0	1.0	11.1		710	67.0	113.0	143.0	12.2	569.0	12.7
2	2.0	11.7		200	89.9	155.1	143.0	15.2	653.0	28.9
4	2.0	11.7		49	88.1	148.1	121.0	15.8	577.0	36.1
6	2.0			14						
8	2.0	11.8			79.3	137.6	51.0	15.5	549.0	28.2
10	2.0									
12	2.0									
14	2.0									
16	2.0	11.8			89.9	151.6	131.0	14.4	522.0	3.6
18	2.0	6.7								
20	2.0	2.2			290.2	334.4	783.0	11.7	558.0	
22	2.0	0.2					811.0	23.3	585.0	
23	2.0	0.1					398.0	33.7	563.0	
Apr. 28, 1980										
0	7.0	11.7	8.4	170	49.9	146.6	48.7	23.8	327.0	48.8
2	8.0	11.4	8.5	24	46.8	146.6	49.1	23.3	278.0	50.6
4	8.0	11.4	8.5		43.7	146.6	41.1	22.6	271.0	47.0
6	8.0	11.2								
8	8.0	11.2	8.5		45.3	131.0	35.6	22.6	269.0	47.0
10	8.0	11.2								
12	7.5	11.2								
14	7.5	11.2								
16	7.5	11.0	8.4		48.4	131.0	52.2	22.3	250.0	34.3
18	7.5	11.0								
20	7.0	11.0	8.3							
22	7.0	10.8								
24	6.5	0.8								
May 12, 1980										
0	11.5	10.5	8.5	800	39.0	133.8	49.9	14.3	160.0	43.4
2	11.0	10.5	8.5	35	47.8	186.6	54.2	13.8	134.0	68.2
4	11.0	10.5	8.5	5	41.9	177.8	42.3	15.8	142.0	38.0
6	10.2	9.9								
8	10.0	9.2	8.4		50.7	157.3	68 9	14.3	143.0	31.0
10	10.0	9.1								
12	9.0	8.9								
14	8.5	8.9								
16	8.5	8.9	8.2		62.5	119.1	119.6	17.8	197.0	26.0
18	8.5	8.4								
20	8.2	8.4	8.1		66.9	139.6	64.5	17.8	215.0	12.4
22	8.2	7.8								
24	7.0	0.4								
May 19, 1980										
0	12.0	11.1		1200	38.0	99.7	81.6	15.1	252.0	16.9

Depth Meter	Temp °C	O_2 mg/l	pH	Light μE/m²/sec	Sol.P μg/l	Tot.P μg/l	NH_4^+ μg/l	NO_2^- μg/l	NO_3^- μg/l	Chlor. μg/l
2	11.0	10.1		150	36.5	153.2	86.5	15.1	306.0	53.0
4	10.5	8.6		19						
5				10	43.9	108.7	183.4	16.8	238.0	19.3
6	10.0	8.6								
8	10.0	8.6								
10	10.0	8.0			46.9	108.7	136.9	15.8	307.0	12.1
12	9.5	8.2								
14	9.5	8.1								
15					51.4	111.6	162.2	15.8	303.0	12.1
16	9.5	8.1								
18	9.5	8.1								
20	9.5	8.0			55.8	108.7	205.3	15.6	325.0	14.5
22	9.0	6.4								
23	8.5	6.2			69.2	126.5	314.4	15.8	358.0	
June 2, 1980										
0	16.5	9.2		160	31.3	87.5	167.2	8.7	65.0	4.1
2	16.0	8.9		38	32.8	78.1	153.4	9.0	71.0	2.5
4	16.0	8.8		15	31.3	71.9	141.4	9.0	68.0	4.5
6	16.0	8.6								
8	15.0	8.5			31.3	93.8	168.5	10.4	91.0	5.4
10	14.0	8.4			36.0	78.1	175.4	11.4	93.0	7.0
12	12.0	7.5			39.1	96.9	217.6	13.6	125.0	6.2
14	11.0	7.2								
16	10.2	5.9			61.0	112.5	300.0	15.3	152.0	6.2
18	10.0	0.3			87.5	112.5	381.6	14.5	138.0	7.0
June 6, 1980										
0	18.0	7.8		800	47.9	80.5	221.6	8.4	91.4	3.3
2	17.5	7.7		155	46.3	71.4	196.9	8.1	98.3	3.3
4	17.4	7.7		65		83.5	201.7	7.1	88.6	4.1
6	17.3	7.6		22						
8	17.2	7.6			41.8	77.5	192.9	7.4	94.2	3.3
10	14.8	6.5			50.9	83.5	245.5	9.1	118.6	2.5
12	11.0	5.3			67.6	126.1	306.9	11.3	158.2	2.1
14	11.0	4.8								
16	10.5	4.2			107.2	135.2	372.3	18.7	189.6	1.7
18	10.2	3.8								
20	9.8	2.7			126.9	962.5	546.9	11.5	191.2	2.1
22	9.7	2.2					578.0	10.6	162.9	4.1
24	9.5	1.8					624.2	10.3	155.2	
June 11, 1980										
0	17.8	9.2		1000	30.1	78.1	186.0	18.3	168.5	8.3
2	17.5	8.9		300	31.6	81.1	152.0	17.6	212.6	8.7
4	17.2	8.8		100	33.1	99.1	131.8	14.1	113.2	7.8
6	16.7	8.4		34						
8	16.2	8.1			45.1	87.1	232.9	19.2	177.6	4.1
10	14.5	7.1			57.1	96.1	340.8	19.2	167.0	3.7
12	13.4	6.3			60.1	90.1	384.9	21.1	218.3	4.1
14	11.8	5.1								
16	10.0	2.3				141.1	713.2	21.1	218.3	2.1
18	9.9	2.0								
20	9.8	1.9			126.0	216.1	668.2	22.1	217.4	2.9
22	9.7	1.4								
24	9.4	0.9								2.1
June 18, 1980										
0	18.8	9.4	8.4	950	46.8	72.1	155.1	12.3	124.2	12.1
2	17.5	9.2	8.5	280	37.7	72.1	143.8	12.3	114.3	10.6
4	17.3	9.2	8.5	92	45.3	78.3	133.3	12.1	125.0	8.2
6	17.3	9.0		29						
8	17.0	8.9	8.4		43.8	65.9	134.6	12.1	131.4	3.4
10	16.3	8.7			43.8	69.0	114.9	12.1	110.8	2.9
12	13.2	6.0	8.2		80.0	96.9	301.6	15.2	134.4	2.9
14	11.5	4.7								
16	10.5	4.1	7.7		134.4	174.4	483.1	20.0	170.9	0.5
18	9.8	3.1								
20	9.5	2.9	7.6		193.4	224.0	645.3	16.6	144.0	0.0
22	9.0	2.4								
June 24, 1980										
0	21.0	8.8		980	24.1	81.4	28.1	20.5	179.7	38.6
2	20.5	8.4		180	22.6	75.2	22.5	19.5	140.8	28.0
4	20.0	7.5		35	27.2	72.1	37.8	20.2	157.3	27.0
6	19.0	6.3		8						
8	18.5	6.3			46.8	75.2	158.5	21.4	182.7	13.0
10	17.5	5.7			49.2	81.4	226.5	23.7	216.0	5.3
12	16.0	5.2			98.1	106.2	362.5	31.3	269.1	2.4
14	14.5	4.2								
16	12.5	1.9			146.5	162.0	583.7	42.8	329.5	1.5
18	11.5	1.1								
20	11.2	1.1			196.4	230.2	773.4	36.1	204.8	1.5
22	10.8	0.8								
July 1, 1980										
0	20.5	8.7	8.8	180	7.6	117.1	40.5	9.0	8.5	72.3
2	20.5	8.4	8.8	9	4.6	93.1	34.8	9.0	9.9	69.9
4	20.5	8.3	8.8		4.6	90.1	26.4	9.0	3.4	65.6
6	20.5	8.0								
8	18.0	6.1	8.6		34.6	78.1	118.5	17.3	84.1	15.4
10	16.0	4.4	8.7		24.1	96.1	103.2	15.7	71.1	45.3
12	14.0	3.0	8.0		111.0	132.1	369.8	41.8	254.3	7.2
14	13.5	2.1								
16	11.0	0.4	7.7		181.5	201.1	602.3	34.5	266.7	0.1
18	10.5	0.1								
20	10.0	0.0	7.5		232.5	264.0	825.5	59.9	50.1	0.1
22	9.5	0.0								
July 8, 1980										
0	25.0	8.1	8.7	900	9.8	92.7	20.8	0.8	136.0	101.2
2	25.0	7.7	8.8	12	11.4	89.6	33.4	0.3	134.0	102.4
4	25.0	7.4	8.7		8.3	86.6	24.4	0.3	142.0	96.6
6	23.5	7.0								
8	19.0	3.7	8.3		52.4	80.5	197.0	11.5	152.0	14.5
10	17.5	3.1	8.1		102.6	120.0	319.0	9.3	138.0	4.6
12	15.0	0.8	7.8		172.6	177.8	426.0	8.6	142.0	2.9
14	14.5	0.8								
16	14.5	0.8	7.7		177.1	217.4	546.0	9.6	123.0	1.7
18	12.0	0.2								
20	11.5	0.1	7.5		266.8	314.7	1090.0	1.2	118.0	1.7
22	11.0	0.0								

Depth Meter	Temp °C	O_2 mg/l	pH	Light $\mu E/m^2/sec$	Sol.P $\mu g/l$	Tot.P $\mu g/l$	NH_4^+ $\mu g/l$	NO_2^- $\mu g/l$	NO_3^- $\mu g/l$	Chlor. $\mu g/l$
July 16, 1980										
0	25.0	8.0	8.6	650	8.1	76.7	169.2	13.5	13.3	32.4
2	25.0	7.8	8.5	30	5.0	76.7	160.5	12.9	10.1	33.0
4	25.0	7.2	8.4	3	3.5	73.7	173.9	13.5	17.5	29.5
6	24.8	7.0	8.4		5.0	70.7	190.2	14.1	7.5	22.6
8	20.0	5.3	8.3		41.4	79.7	212.2	43.7	129.0	1.7
10	16.8	1.9	7.8		95.9	134.2	261.5	30.0	202.0	2.3
12	16.0	0.5	7.5		129.2	176.6	290.6	30.0	460.7	0.6
14	14.5	0.3								
16	13.5	0.3								
18	13.0	0.2	7.5		194.3	231.2	746.7	87.1	116.7	0.0
20	11.5	0.2								
22	11.0	0.1								
July 23, 1980										
0	25.7	8.3	8.3	850	4.9	75.8	0.0	44.7	79.8	54.4
2	25.5	7.7	8.4	38	4.9	84.8	0.0	41.8	41.9	60.7
4	25.0	6.5	8.4	3	4.9	72.8	0.0	50.7	70.0	50.9
6	25.0	6.4	8.4		6.4	84.8	37.1	38.6	57.7	42.8
8	24.0	3.9	8.3		7.9	78.8	40.4	50.4	142.5	29.5
10	18.0	0.2	7.5		99.3	150.7	99.2	51.7	560.4	2.3
12	15.5	0.2	7.5		160.8	213.6	351.3	100.7	374.7	1.2
14	14.0	0.1								
16	13.0	0.1	7.4		207.2	213.6	850.1	30.7	18.6	0.6
18	12.0	0.1								
20	11.8	0.0	7.4		255.2	297.6	1460.0	7.1	0.8	0.6
22	11.5	0.0								
24	11.2	0.0								
July 29, 1980										
0	24.5	8.2	8.8	1000	6.3	112.3	23.6	7.1	0.0	52.6
2	24.3	7.4	8.8	40	6.3	118.9	21.7	7.1	0.0	55.5
4	24.2	7.2	8.8		6.3	105.6	21.7	7.1	0.0	59.0
6	24.0	6.3	8.7		6.3	108.9	14.8	9.4	6.0	59.0
8	23.5	5.4	8.6		4.7	92.3	23.5	13.8	18.6	46.3
10	19.2	0.1	7.9		67.8	125.6	35.3	67.7	431.9	8.1
12	15.2	0.1	7.7		167.4	211.9	277.2	169.2	272.6	1.2
14	13.2	0.1								
16	12.2	0.0	7.6		203.9	215.2	1129.4	9.4	10.4	2.9
18	11.8	0.0								
20	11.3	0.0	7.4		303.5	304.9	1619.5	7.1	0.0	2.3
22	11.2	0.0								
24	11.0	0.0								
Aug. 5, 1980										
0	24.5	8.2	8.7	700	11.8	77.9	33.2	7.1	0.0	44.5
2	24.5	7.8	8.7	24	2.8	81.5	30.7	7.1	0.0	46.3
4	24.5	7.5	8.7		2.8	92.4	27.3	7.1	0.0	48.6
6	24.0	6.3	8.7		2.8	70.7	27.3	7.1	1.5	46.9
8	23.5	5.0	8.5		2.8	67.1	26.5	9.7	20.9	38.2
10	17.5	0.2	7.6		58.9	70.7	161.0	127.1	353.8	6.4
12	16.0	0.1	7.5		156.5	168.4	217.9	159.5	213.4	5.2
14	14.5	0.1								
16	12.5	0.0	7.5		266.9	287.7	1089.2	9.4	0.0	0.6
18	12.0	0.0								
20	11.5	0.0	7.3		375.4	389.0	1671.1	7.1	0.0	0.6
22	11.5	0.0								
24	11.3	0.0								
Aug. 12, 1980										
0	24.0	8.2	8.4	980	0.9	100.1	19.7	24.0	45.4	45.7
2	23.8	7.1	8.5	50	0.9	89.2	23.4	7.8	10.8	49.8
4	23.8	7.0	8.5	3	0.9	78.3	12.0	19.2	7.0	45.7
6	23.8	7.0	8.5		0.9	89.2	26.7	8.7	10.3	50.3
8	23.8	7.0	8.5		0.9	92.8	30.6	8.1	7.6	46.9
10	19.0	0.2	8.2		37.3	107.3	131.9	45.3	76.7	28.3
12	15.5	0.1	7.6		119.1	158.3	402.7	72.2	58.4	6.9
14	14.8	0.1								
16	14.2	0.0	7.5		144.6	161.9	1057.4	7.8	2.5	4.1
18	12.5	0.0								
20	12.0	0.0	7.4		379.2	423.8	1704.5	7.1	2.5	4.6
22	11.5	0.0								
24	11.5	0.0								
Aug. 19, 1980										
0	22.0	8.4		900	7.4	73.6	32.4	7.1	36.9	50.3
2	22.0	8.5		35	1.9	81.0	34.9	7.1	9.6	48.7
4	22.0	8.0		2	0.0	73.6	24.1	7.1	11.4	48.4
6	22.0	6.8			0.0	95.7	44.6	11.6	22.2	43.8
8	21.3	5.6			3.7	121.4	95.4	19.2	32.5	22.9
10	19.0	1.2			36.8	81.0	193.3	25.6	53.5	16.4
12	15.0	0.0			235.4	254.8	753.1	7.1	7.3	3.5
14	13.5	0.0								
16	13.0	0.0			314.4	334.7	1141.4	7.1	3.1	1.9
18	12.0	0.0								
20	12.0	0.0			406.4	426.6	1796.2	7.1	2.5	3.5
22	11.8	0.0								
24	11.5	0.0								
Aug. 26, 1980										
0	23.5	8.3	8.7	750	3.7	84.7	58.3	7.1	56.5	51.5
2	23.5	7.9	8.7	20	1.9	77.3	24.1	7.1	45.2	50.9
4	23.5	7.8	8.7		0.0	81.0	19.1	7.1	29.2	41.6
6	23.2	7.5	8.5		0.0	84.7	23.2	7.1	47.0	38.7
8	22.5	5.0	8.5		0.0	92.0	24.8	18.9	48.8	28.3
10	20.5	1.5	8.2		29.5	77.3	169.7	27.5	107.5	11.0
12	17.0	0.1	7.8		150.8	187.6	548.6	10.6	60.4	4.6
14	14.3	0.1								
16	13.5	0.0	7.7		308.9	331.0	1116.4	7.1	19.7	2.3
18	12.5	0.0								
20	12.0	0.0	7.6		400.9	459.7	1708.6	7.1	2.5	2.3
22	12.0	0.0								
24	12.0	0.0								
Sep. 2, 1980										
0	23.0	8.4	8.4	800	7.2	58.1	28.9	9.0	34.6	38.2
2	23.0	8.2	8.5	20	1.8	61.7	14.2	10.3	15.1	38.2
4	23.0	8.0	8.5		1.8	72.6	18.0	9.0	15.6	42.8
6	23.0	7.8	8.5		1.8	69.0	13.3	11.2	13.7	34.7
8	23.0	7.7	8.5		1.8	65.3	11.9	9.3	26.6	34.7
10	23.0	7.4	8.5		1.8	61.7	14.4	9.3	31.4	27.2
12	18.0	0.1	8.0		27.2	90.8	180.1	17.5	136.1	13.3
14	15.5	0.0								

Depth Meter	Temp °C	O_2 mg/l	pH	Light $\mu E/m^2/sec$	Sol.P $\mu g/l$	Tot.P $\mu g/l$	NH_4^+ $\mu g/l$	NO_2^- $\mu g/l$	NO_3^- $\mu g/l$	Chlor. $\mu g/l$
16	14.5	0.0	7.7		267.2	327.1	1128.0	8.1	1.7	1.2
18	13.0	0.0								
20	12.2	0.0	7.4		398.1	407.1	1866.5	7.8	2.0	1.2
22	12.0	0.0								
24	12.0	0.0								
Sep. 8, 1980										
0	23.3	8.4	8.4	1200	1.8	54.4	101.3	12.5	77.8	6.9
2	22.8	8.9	8.5	150	1.8	50.8	45.1	10.3	53.6	23.1
4	22.2	8.1	8.5	24	1.8	58.1	163.6	10.3	56.6	13.3
6	22.0	7.1	8.5		1.8	61.7	179.2	9.6	48.9	11.0
8	22.0	6.8	8.4		1.8	54.4	113.2	10.6	48.0	19.1
10	22.0	6.8	8.4		10.9	54.4	146.3	10.9	43.6	19.7
12	19.5	0.5	8.1		36.3	112.6	361.5	11.2	40.3	12.1
14	16.0	0.2								
16	13.8	0.0	7.5		254.5	385.3	1528.4	8.1	7.0	2.3
18	13.0	0.0								
20	12.7	0.0	7.4		392.6	476.2	2187.6	7.8	3.7	2.3
22	12.5	0.0								
24	12.2	0.0								
Sep. 15, 1980										
0	21.0	8.7	8.3	380	59.1	105.8	129.2	14.3	32.3	34.7
2	21.0	8.2	8.4	30	41.9	80.7	118.5	12.8	29.4	32.4
4	21.0	7.9	8.4	1	37.2	74.5	131.6	12.8	27.2	34.7
6	21.0	7.4	8.4		40.4	71.3	140.6	14.3	45.2	23.1
8	21.0	7.2	8.4		45.1	74.5	157.3	14.0	34.0	16.2
10	21.0	7.1	8.4		81.0	158.9	127.6	14.3	25.1	26.0
12	20.8	7.0	8.4		40.4	83.9	165.1	14.3	54.5	16.2
14	16.0	0.3								
16	14.8	0.0	7.5		281.2	290.3	1326.1	8.1	9.2	5.2
18	13.5	0.0								
20	12.8	0.0	7.5		372.0	374.8	1689.4	8.1	0.0	1.2
22	12.5	0.0								
24	12.5	0.0								
Sep. 22, 1980										
0	19.2	8.9	8.2	60	79.5	124.5	134.9	30.6	68.5	40.5
2	19.2	7.9	8.2	5	92.0	133.9	153.9	29.7	69.4	35.3
4	19.2	7.4	8.2		59.1	99.5	160.7	21.2	44.0	38.2
6	19.0	7.3	8.2		79.5	118.3	176.0	20.6	63.2	31.8
8	19.0	7.0	8.2		65.4	105.8	194.2	22.8	49.8	30.6
10	19.0	7.0	8.2		63.8	90.1	180.6	20.9	57.3	18.5
12	19.0	7.0	8.2		65.4	96.4	174.8	30.0	77.0	24.9
14	18.0	2.3								
16	14.0	0.0	7.4		317.2	327.9	1381.0	10.3	10.1	5.8
18	13.5	0.0								
20	13.0	0.0	7.3		381.3	396.7	1934.4	8.1	0.0	2.9
22	12.5	0.0								
24	12.2	0.0								
Sep. 29, 1980										
0	17.2	9.3	8.2	600	84.7	85.8	155.7	36.0	96.0	37.0
2	17.2	9.1	8.2	20	69.2	92.0	72.0	37.2	82.0	42.8
4	17.2	8.9	8.2	1	123.4	138.4	141.2	35.0	76.0	27.2
6	17.2	8.8	8.2		78.8	92.0	128.1	35.0	74.6	40.5
8	17.2	8.7	8.2		69.2	79.6	124.0	35.0	76.0	35.3
10	17.2	8.6	8.2		69.2	88.9	113.4	35.0	66.0	31.2
12	17.2	8.5	8.2		66.1	85.8	118.3	35.0	82.5	31.8
14	17.2	8.5								
16	15.5	0.0	7.8		124.9	126.0	460.9	32.2	76.4	19.1
18	13.8	0.0								
20	13.2	0.0	7.3		418.9	420.0	2105.9	8.1	0.0	9.8
22	12.8	0.0								
24	12.5	0.0								
Oct. 6, 1980										
0	15.5	9.7	8.1	625	87.8	126.0	244.3	49.4	190.7	20.2
2	15.5	9.3	8.1	40	103.3	126.0	223.3	50.1	71.1	25.4
4	15.5	9.3	8.1		155.9	166.3	211.5	50.4	88.0	27.2
6	15.5	9.3	8.1		124.9	178.7	181.2	49.8	67.8	26.6
8	15.5	9.3	8.1		94.0	122.9	194.4	50.4	72.3	26.0
10	15.5	9.3	8.1		123.4	116.8	228.0	49.4	58.8	20.2
12	15.5	9.2	8.1		101.7	126.0	226.9	53.8	79.9	20.2
14	15.5	8.9								
16	15.5	8.3	8.0		126.5	132.2	272.7	55.4	78.5	23.7
18	14.0	0.5								
20	13.0	0.1	7.3		494.3	491.2	2427.4	9.0	0.0	1.6
22	13.0	0.1								
24	12.5	0.0								
Oct. 13, 1980										
0	14.0	9.9	8.0	600	53.9	115.0	256.4	48.8	117.4	48.6
2	14.0	9.9	8.1	45	64.7	118.6	252.2	49.8	90.8	38.2
4	14.0	9.7	8.1	4	64.7	136.5	233.4	49.8	87.2	36.4
6	14.0	9.7	8.1		64.7	122.2	266.4	49.4	82.4	37.6
8	14.0	9.7	8.1		68.3	118.6	251.7	49.4	88.2	47.4
10	14.0	9.7	8.1		59.3	111.5	274.5	49.4	68.1	32.4
12	14.0	9.7	8.0		62.9	118.6	257.3	50.4	90.2	46.3
14	14.0	9.7								
16	14.0	9.6	8.0		62.9	118.6	203.4	50.4	76.6	43.4
18	14.0	9.6								
20	14.0	9.4	8.0		66.5	118.6	149.5	50.4	111.0	46.8
22	14.0	9.4								
24	14.0	7.8								
Nov. 4, 1980										
0	9.5	11.1	7.8	500	74.0	130.1	509.0	32.0	166.6	22.6
2	9.5	11.1	7.8		81.2	140.9	244.8	33.0	130.5	28.9
4	9.5	11.1	7.8		93.8	133.7	486.5	32.3	93.2	23.1
6	9.5	11.1								
8	9.5	10.9	7.8		92.0	126.5	414.1	32.0	107.8	32.4
10	9.5	10.8								
12	9.5	10.8	7.8		84.8	119.3	205.6	31.4	122.8	28.9
14	9.0	10.8								
16	9.0	10.8	7.8		81.2	119.3	466.2	31.4	98.4	27.8
18	9.0	10.9								
20	9.0	10.9	7.8		83.0	209.2	444.8	32.3	121.8	24.3
22	9.2	10.8								
24	9.2	7.8								
Nov. 20, 1980										
0	7.0	10.0	8.1	550	95.5	138.9	527.1	29.0		9.8
2	7.0	9.8		100	97.1	141.9	547.8	29.6		8.7

Depth Meter	Temp °C	O_2 mg/l	pH	Light $\mu E/m^2/sec$	Sol.P $\mu g/l$	Tot.P $\mu g/l$	NH_4^+ $\mu g/l$	NO_2^- $\mu g/l$	NO_3^- $\mu g/l$	Chlor. $\mu g/l$
4	7.0	9.5		22						
5				10	97.1	132.7	474.8	29.0		8.0
6	7.0	9.7								
8	7.0	9.4								
10	7.0	9.5			103.2	138.9	547.2	29.3		11.6
12	7.0	9.6								
14	7.0	9.6								
15			8.1		98.6	132.7	577.9	29.6		10.5
16	7.0	9.6								
18	7.0	9.6								
20	7.0	9.6	8.1		97.1	135.8	329.7	29.6		9.0
22	7.0	9.5								
24	7.0	9.2	8.0							
Dec. 18, 1980										
0	1.5	12.1	8.1	130	93.3	133.1	422.0	16.7	326.8	4.5
2	1.5	11.8	8.1	35	93.3	136.3	256.0	16.4	284.6	4.0
4	1.5	11.8		12						
5			8.1	7	93.3	142.6	380.0	16.7	274.2	3.5
6	1.5	11.8								
8	1.5	11.8								
10	1.5	11.8	8.1		93.3	120.5	402.0	16.7	324.4	3.9
12	1.5	11.8								
14	1.5	11.8								
15			8.1		93.3	130.0	380.0	16.0	322.8	4.1
16	1.5	11.8								
18	1.5	11.8								
20	1.5		8.1		93.3	139.4	357.0	16.4	260.6	3.8
22	1.5									
23	1.5				93.3	139.4	252.0	16.7	307.2	
Jan. 28, 1981										
0	0.4		7.7	210	84.6	731.6	207.2	6.2	532.7	4.8
2	1.3		7.7	50	86.7	145.3	177.0	6.5	557.9	7.2
4			7.7		86.6	121.4	166.8	6.5	608.7	4.8
6	1.3									
8	1.4		7.8		88.6	141.3	195.5	6.8	591.5	5.8
12			7.8		88.6	121.4	181.9	7.3	531.7	4.3
14	1.5									
18	1.8									
20	2.0									
22	2.0									
23	2.2		7.7		180.3	217.1	456.4	4.7	635.7	1.0
Feb. 16, 1981										
0	2.0		8.0	600	691.2	118.5	131.5	13.3	420.0	0.4
2	2.2		8.0	43	91.0	146.2	131.2	10.9	501.2	2.5
4	2.2			19						
5			8.0		97.0	150.2	62.9	10.2	471.5	2.5
6	2.2									
8	2.2									
10			8.0		98.9	154.2	84.1	9.9	666.3	1.7
12	2.2									
14	2.2									
15	2.2		7.9		122.8	162.1	134.0	11.8	470.0	2.1
16	2.8									
18	3.0		7.9		138.6	193.9	144.3	9.6	526.7	1.7
Feb. 26, 1981										
0	2.1	13.4	7.9	220	54.7	112.6	96.2	9.6	459.8	0.4
2	2.6	14.8	7.8	65	80.6	151.4	100.7	9.6	496.3	0.8
4	2.6	14.8		15	80.6	157.9		9.3		
5			7.8	8			76.5		478.4	0.8
6	2.6	14.8								
8	2.6	14.8								
10	2.6	14.8	7.8		82.2	151.4	30.2	9.3	691.2	0.8
12	2.6	13.2								
14	2.6	11.8								
15			7.8		96.8	154.7	83.1	11.8	543.0	0.8
16	2.6	10.2								
20	2.6	6.4	7.6		195.5	238.9	349.4	13.0	584.5	0.8
22	2.7	2.8								
Mar. 10, 1981										
0	1.5		7.9		41.7	77.0	125.2	7.3	316.0	1.7
2			7.8		72.5	138.5	33.2	10.4	647.8	19.9
4			7.8		77.3	151.4	16.5	10.4	453.0	19.5
15			7.8		95.1	154.7	18.5	9.2	557.5	13.3
Mar. 20, 1981										
0	3.0		8.4	500			7.1	9.2	350.7	14.9
2			8.3	99			6.8	9.5	484.3	17.0
4			8.3				1.8	9.2	533.2	23.2
Mar. 30, 1981										
0	7.2	11.7	8.1	1200	57.2	182.4	43.2	9.2	472.4	31.4
2	7.2	10.1	8.1	90	57.2	189.2	32.9	8.9	618.6	38.4
4	7.2	9.7	8.1	15	55.5	202.8	32.6	9.2	362.9	29.3
6	7.2	9.7								
8	7.2	9.5	8.1		52.1	209.6	38.8	9.2	362.9	32.2
10	7.2	9.8								
12	7.0	10.0	8.1		57.2	213.0	37.4	7.9	479.6	26.0
14	7.0	10.2								
16	7.0	10.4	8.0		55.5	189.2	63.2	8.5	479.1	31.0
18	6.8	10.8								
20	6.8	10.8	8.0		60.6	185.8	56.5	8.2	376.0	25.2
22	6.8	10.5								
24	6.5	9.2								
Apr. 6, 1981										
0	8.5	11.3	7.9	1200						22.3
2	8.5	9.1	8.0	86						19.0
4	8.5	9.1	8.1	12						24.4
6	8.2	9.1								
8	8.2	8.9								
10	8.2	8.9	8.1							19.8
12	8.2	8.9								
14	8.0	8.8								
16	8.0	8.7	8.0							18.2
18	8.0	8.7								
20	8.0	8.7	8.0							19.0
22	8.0	8.7								
24	7.9	2.5								

Depth Meter	Temp °C	O_2 mg/l	pH	Light $\mu E/m^2/sec$	Sol.P $\mu g/l$	Tot.P $\mu g/l$	NH_4^+ $\mu g/l$	NO_2^- $\mu g/l$	NO_3^- $\mu g/l$	Chlor. $\mu g/l$
Apr. 13, 1981										
0	10.0	9.2	7.9	240	89.0	175.3		8.8	539.6	19.4
2	9.8	9.0	8.0	33	93.1	145.5	39.3	8.2	570.6	25.6
4	9.8	9.0	8.0	7	91.7	156.3	62.6	8.5	582.5	25.2
6	9.8	8.9								
8	9.5	8.9	8.0		82.2	164.5	27.6	8.2	582.4	15.3
10	9.2	8.7								
12	9.2	8.5	8.0		99.9	161.7	72.8	8.2	473.3	12.8
14	9.1	8.4								
16	9.0	8.2	8.0		103.9	142.7	5.1	8.5	679.8	9.1
18	9.0	8.2								
20	9.0	8.2	8.0		105.3	142.7	16.4	7.9	656.0	7.4
22	8.8	5.8								
Apr. 20, 1981										
0	10.5	10.9		1300						18.2
2	10.5	10.7		120						15.7
4	10.5	10.5		20						21.9
6	10.5	10.5								
8	10.5	10.4								
10	10.5	10.3								22.7
12	10.5	10.3								
14	10.5	10.3								
16	10.5	10.2								
18	10.5	10.2								
20	10.0	10.2								35.5
22	10.0	10.2								
24	10.0	10.0								
Apr. 27, 1981										
0	11.0	8.6	8.5	850	72.7	159.0	47.3	8.1	426.4	16.5
2	11.0	8.6	8.5	55	79.5	178.0	44.4	8.4	395.2	19.0
4	10.5	8.0	8.5	7	72.7	175.3	17.2	9.0	394.6	24.0
6	10.0	7.4								
8	10.0	7.3	8.4		74.1	172.6	63.3	8.4	426.1	15.3
10	10.0	7.3								
12	10.0	7.1	8.4		83.6	85.7	54.4	8.7	348.4	13.2
14	10.0	7.1								
16	10.0	7.1	8.4		98.5	121.0	65.3	8.1	302.6	12.8
18	10.0	7.0								
20	10.0	7.0	8.4		79.5	129.2	80.7	8.4	395.2	5.8
22	10.0	6.8								
24	10.0	4.4								
May 4, 1981										
0	12.0	7.8	8.5	150						12.4
2	12.0	7.6	8.5	15						13.2
4			8.5							17.3
5	12.0	7.6								
8			8.4							5.4
10	12.0	7.4								
15	11.5	7.1								
16			8.4							7.8
20	11.0	6.7								
May 12, 1981										
0	12.5	8.1	8.5	850	61.5	72.5	102.7	9.9	238.1	27.7
2	12.3	7.7	8.6	90	46.1	95.0	100.4	10.5	129.5	98.3
4	12.1	7.8	8.6	12	47.5	89.0	22.8	10.8	124.1	90.5
6	12.1	7.8								
8	12.1	7.7	8.6		50.3	89.3	57.8	10.5	144.9	33.0
10	12.1	7.7								
12	12.1	7.7	8.6		51.7	106.2	70.9	10.5	119.2	33.9
14	12.1	7.7								
16	12.1	7.7	8.6		44.7	92.1	21.7	10.2	165.8	33.5
18	12.1	7.7								
20	12.1	7.5	8.6		46.1	86.5	31.2	11.1	144.4	16.5
22	12.1	7.6								
24	12.0	7.1								
May 18, 1981										
0	12.5	8.9	8.5	980						10.3
2	12.8	8.3	8.6	150						3.7
4	12.8	8.1	8.6	35						7.8
8	12.8	7.9	8.6							6.6
12	12.5	7.7	8.6							5.4
16	12.5	7.7	8.6							5.8
20	12.3	7.5	8.5							9.1
May 26, 1981										
0	15.4	9.3	8.6	750	39.1	55.7	40.7	10.7	227.1	4.1
2	14.5	9.5	8.7	310	51.7	72.5	49.5	7.1	204.6	4.5
4	14.2	9.3	8.7	65	57.3	72.5	79.3	7.7	204.0	9.5
6	13.9	9.3								
8	13.6	9.3	8.6		44.7	64.1	70.6	9.5	269.3	7.0
10	13.4	9.2								
12	12.5	9.0	8.6		46.1	58.5	44.5	9.5	284.7	5.4
14	11.6	8.8								
16	11.0	8.1	8.5		67.1	72.5	102.4	10.7	211.6	4.1
18	10.8	7.7								
20	10.7	7.6	8.5		81.2	86.5	73.5	9.2	279.8	4.5
22	10.7	7.5								
24	10.6	7.3								
June 1, 1981										
0	18.2	6.5	8.8	1200	55.1	101.4	101.2	13.0	256.3	0.8
2	18.5	6.4	8.8	300	56.5	80.9	79.0	12.1	238.2	1.7
4	18.0	6.4	8.8	150	69.7	171.6	104.0	12.7	275.5	1.2
6	16.0	5.8		50						
8	15.1	5.6	8.7		72.6	130.6	72.0	13.3	208.8	3.7
10	14.2	5.5								
12	13.8	5.5	8.5		91.6	59.5	66.5	13.0	227.9	0.4
14	13.5	5.3								
16	13.2	5.3	8.5		77.0	89.7	60.4	13.3	265.5	0.4
18	12.8	4.4								
20	12.5	3.9	8.4		107.7	159.9	123.8	15.1	320.6	0.4
22	12.5	3.5								
24	12.5	1.5								
June 8, 1981										
0	21.2	8.7	8.9	550						7.0

Depth Meter	Temp °C	O_2 mg/l	pH	Light $\mu E/m^2/sec$	Sol.P $\mu g/l$	Tot.P $\mu g/l$	NH_4^+ $\mu g/l$	NO_2^- $\mu g/l$	NO_3^- $\mu g/l$	Chlor. $\mu g/l$
2	21.2	8.7	8.9	135						7.0
4	21.0	8.7	8.9	46						7.0
6	21.0	8.7		12						
8	16.0	7.2	8.8							5.8
10	14.5	6.8								
12	13.5	6.1	8.5							2.1
14	13.5	5.7								
16	13.0	5.6	8.4							0.0
18	13.0	5.2								
20	13.0	3.8	8.2							3.7
22	12.5	2.1								
24	12.5	0.8								
June 15, 1981										
0	24.0	5.3	9.0	400	25.8	250.6	12.0	11.5	231.5	149.9
2	23.5	4.9	9.0	40	43.4	89.7	4.2	10.5	247.8	27.7
4	23.0	4.7	8.8	9	66.8	113.1	22.6	11.5	246.9	24.4
6	22.2	4.6								
8	20.0	4.2	8.8		59.5	189.2	78.9	11.5	282.9	7.0
10	14.5	3.0								
12	14.0	2.9	8.5		100.4	121.9	253.9	12.7	328.1	2.5
14	13.5	2.2								
16	13.2	1.9	8.5		154.6	288.6	311.3	15.2	295.0	0.4
18	13.0	1.5								
20	13.0	1.2	8.3		161.9	285.7	425.4	16.2	335.3	0.4
22	13.0	0.5								
24	12.8	0.4								
June 22, 1981										
0	21.0	6.0	9.0	950						27.1
2	21.0	5.7	9.0	150						24.9
4	21.0	5.7	9.1	30						26.0
6	21.0	5.7	9.1							26.4
8	21.0	5.7	9.1							26.7
10	21.0	5.5	9.1							26.0
12	18.0	4.0	8.9							21.0
14	14.0	1.9								
16	13.2	1.2	8.2							2.2
18	13.1	1.0								
20	13.1	0.7	8.1							1.8
22	13.0	0.5								
24	13.0	0.2								
June 29, 1981										
0	21.8	5.8	9.1	300	17.6	87.8	18.7	11.3	213.7	51.6
2	21.8	5.7	9.2	25	14.7	90.7	6.6	11.0	218.7	54.0
4	21.8	5.6	9.2		14.7	70.3	7.1	11.3	95.5	59.3
6	21.3	5.4								
8	21.3	5.3			17.6	67.4	30.0	12.4	212.7	38.2
10	19.0	4.0	8.9		49.7	111.1	168.9	17.1	161.2	25.5
12	18.0	2.5	8.7		81.8	122.8	252.8	21.5	270.7	14.9
14	15.7	2.0								
16	14.5	1.2	8.2		147.5	172.4	246.4	26.2	206.6	5.3
18	14.0	0.7								
20	14.0	0.4	8.0		181.0	207.4	247.3	25.3	273.6	4.8
22	13.4	0.3								
24	13.2	0.2								
July 6, 1981										
0	26.0	8.0	9.3	950						16.9
2	25.0	7.8	9.3	140						21.9
4	24.2	7.5	9.3	25						30.2
6	23.0	5.5								
8	21.8	4.0	9.0							42.5
10	19.3	2.6	8.9							14.0
12	16.0	1.1	8.5							9.5
14	15.0	0.9								
16	14.0	0.3	8.0							0.4
18	13.8	0.2								
20	13.5	0.1	8.0							0.4
22	13.3	0.1								
24	13.2	0.1								
July 13, 1981										
0	26.8	8.3	9.1	500	13.3	64.2	0.0	0.2	77.4	17.3
2	26.2	8.1	9.2	65	6.1	67.1	0.0	0.0	69.7	18.2
4	26.0	7.4	9.1	8	7.6	55.5	11.4	0.2	64.3	19.4
6	25.7	7.6	9.1		9.0	67.1	5.7	0.0	75.0	19.4
8	20.4	1.3	8.5		20.5	58.4	0.0	86.5	280.9	8.7
10	18.0	0.1	8.4		50.9	75.7	0.0	97.8	468.5	5.4
12	17.3	0.0	8.2		85.5		0.0	323.9		4.1
14	15.8	0.0								
16	15.1	0.0	8.1		137.4	145.0	242.9	183.9	452.7	1.7
18	14.7	0.0								
20	14.1	0.0	8.0		189.4	220.0	710.6	89.2	258.0	2.1
22	13.9	0.0								
24	13.9	0.0								
July 20, 1981										
0	25.8	8.2		700						22.7
2	25.6	7.8		75						26.0
4	25.5	6.5		10						28.1
6	24.4	3.7								25.6
8	22.4	0.5								16.9
10	20.6	0.4								7.4
12	18.8	0.4								2.9
14	15.9	0.3								
16	14.7	0.3	7.8							5.0
18	14.5	0.3								
20	14.3	0.3	7.8							5.0
22	14.1	0.3								
24	13.8	0.3								
July 27, 1981										
0	24.0	8.4	9.0	180	6.1	44.1	249.6	0.0	0.0	10.7
2	24.0	8.2	9.0	21	6.1	61.4	231.3	0.0	0.0	15.7
4	24.0	8.2	9.0	4	7.6	49.8	249.6	0.0	0.0	18.6
6	24.0	8.1	9.0		9.0	52.7	237.4	0.0	10.5	14.5
8	22.0	1.6	9.0		7.6	52.7	516.0	1.2	9.3	12.8
10	20.0	0.3	8.3		43.7	58.5	447.0	77.9	0.0	5.8
12	17.1	0.3	8.1		72.6	113.4	717.7	189.3	0.0	3.7
14	16.3	0.3								
16	15.0	0.3	8.0		165.1	205.9	763.9	178.6	442.5	1.7
18	14.7	0.3								

Depth Meter	Temp °C	O_2 mg/l	pH	Light $\mu E/m^2/sec$	Sol.P $\mu g/l$	Tot.P $\mu g/l$	NH_4^+ $\mu g/l$	NO_2^- $\mu g/l$	NO_3^- $\mu g/l$	Chlor. $\mu g/l$
20	14.3	0.3	8.0		102.9	205.9	805.3	132.5	267.2	1.7
22	14.1	0.3								
24	14.0	0.3								
Aug. 3, 1981										
0	24.2	8.4	8.8	220						29.7
2	24.1	7.9	8.8	22						28.9
4	23.8	7.2	8.6	5						21.1
6	23.1	5.5								21.1
8	22.6	4.4	8.5							9.1
10	21.0	1.4								5.8
12	18.3	0.3	8.6							2.5
14	15.7	0.3								
16	14.9	0.3	7.4							2.9
18	14.5	0.3								
20	14.1	0.3	7.5							3.3
22	14.0	0.3								
24	14.0	0.3								
Aug. 10, 1981										
0	25.7	8.3	8.8	1000	75.5	81.6	267.7	6.5	164.6	39.2
2	24.7	8.0	8.8	70	75.5	75.9	124.6	0.0	103.6	49.1
4	24.4	6.9	8.7	6	89.9	107.6	120.3	0.0	134.0	40.1
6	24.2	7.1	8.7		88.5	107.6	103.0	0.2	127.9	40.9
8	24.0	7.0	8.7		66.8	73.0	111.7	0.0	121.8	29.7
10	20.8	0.4	8.2				358.1	5.9	226.0	15.7
12	18.0	0.3	7.9		56.7	58.5	144.8	112.3	8.6	3.7
14	16.0	0.3								
16	15.3	0.3	7.6		211.3	243.4	983.9	4.3	81.5	4.5
18	14.9	0.3								
20	14.3	0.3	7.4		300.9	315.7	1227.7	0.0	85.4	4.1
22	14.1	0.3								
24	14.1	0.3								
Aug. 17, 1981										
0	24.6	8.4	8.8	880						38.8
2	23.5	8.1	8.9	92						38.4
4	23.3	7.8	8.8	10						41.3
6	23.2	7.2	8.8							38.4
8	23.1	6.7	8.8							35.5
10	22.8	6.2	8.7							40.5
12	18.2	0.4	7.8							4.1
14	16.9	0.3								
16	15.2	0.3	7.6							0.8
18	14.5	0.3								
20	14.1	0.3	7.4							2.9
22	14.0	0.3								
24	14.0	0.3								
Aug. 24, 1981										
0	24.3	8.4	8.6	230	9.8	59.5	0.0	0.0	29.3	39.2
2	24.3	7.8	8.7	18	9.8	62.4	0.0	0.0	6.0	26.4
4	24.3	7.8	8.7	3	8.3	76.7	0.0	0.0	3.4	28.5
6	23.1	3.3	8.4		11.2	59.5	0.0	0.4	68.2	19.0
8	23.0	2.5	8.3		9.8	62.4	0.0	0.0	96.8	5.8
10	22.2	1.6	8.2		11.2	53.8	0.0	7.1	76.7	2.1
12	17.9	0.4	7.9		51.2	93.8	171.2	0.0	80.3	0.8
14	16.1	0.3								
16	15.3	0.3	7.8		227.0	248.2	948.8	0.1	3.4	0.0
18	14.6	0.3								
20	14.3	0.3	7.6		301.4	391.2	1451.6	0.1	11.2	1.7
22	14.0	0.3								
24	13.9	0.3								
Sep. 1, 1981										
0	23.2	8.4	8.3	75						21.5
2	23.2	8.3	8.4	9						27.3
4	23.2	8.1	8.4							22.3
6	23.2	8.1	8.4							26.8
8	23.2	8.1	8.4							14.9
10	23.2	7.5	8.4							12.4
12	22.5	1.9	8.4							17.8
14	21.0	0.2								
16	18.0	0.1	7.6							1.7
18	15.5	0.1								
20	15.0	0.1								
22	14.5	0.1								
24	14.0	0.0								
Sep. 8, 1981										
0	22.0	8.6	8.2	550	4.3	66.5	0.0	5.5	31.9	29.7
2	21.8	8.4	8.3	45	7.2	57.8	0.0	5.5	31.9	25.2
4	21.8	8.3	8.3	5	5.7	43.3	0.0	5.2	29.2	19.4
6	21.8	8.1	8.3		14.4	69.4	0.0	9.7	19.2	26.4
8	21.8	8.1	8.3		8.6	66.5	0.0	6.1	37.4	23.1
10	21.8	8.1	8.3		8.6	63.6	0.0	4.3	33.0	28.9
12	21.8	7.3	8.3		8.6	60.7	0.0	4.0	24.3	14.5
14	16.0	0.5								
16	15.0	0.3	7.7		233.2	289.6	619.7	6.7	24.9	0.4
18	14.5	0.3								
20	14.2	0.3	7.6		339.0	396.9	1924.7	6.7	21.9	2.9
22	14.0	0.3								
24	14.0	0.3								
Sep. 14, 1981										
0	23.0	8.4	8.4	180						18.2
2	23.0	8.1	8.5	18						20.2
4	22.5	6.6	8.5							25.2
6	22.0	5.5								
8	22.0	5.3	8.5							21.5
10	21.5	5.1	8.2							11.2
12	21.3	3.7	8.2							11.2
14	19.5	0.7								
16	16.2	0.3	7.4							0.0
18	15.3	0.3								
20	14.7	0.3	7.2							0.8
22	14.3	0.3								
24	14.3	0.3								
Sep. 21, 1981										
0	20.0	9.1	8.2	500	8.6	89.7	10.9	9.1	31.7	26.4
2	19.8	8.8	8.2	35	8.6	83.9	58.8	4.9	20.5	22.7
4	19.8	8.8	8.2		8.6	83.9	0.0	5.2	29.2	19.0
6	19.8	8.6								

Depth Meter	Temp °C	O_2 mg/l	pH	Light $\mu E/m^2/sec$	Sol.P $\mu g/l$	Tot.P $\mu g/l$	NH_4^+ $\mu g/l$	NO_2^- $\mu g/l$	NO_3^- $\mu g/l$	Chlor. $\mu g/l$
8	19.8	8.5	8.2		11.5	89.7	62.4	6.1	31.4	29.3
10	19.8	8.0	8.2		13.0	81.0	107.3	4.9	14.5	16.7
12	19.1	0.5	8.2		21.6	92.6	117.9	4.0	27.3	21.1
14	17.5	0.2								
16	16.0	0.2	7.5		194.1	298.3	1293.1	2.8	16.4	0.0
18	15.0	0.2								
20	14.3	0.2	7.4		349.1	417.1	1944.1	6.7	21.9	1.7
22	14.3	0.2								
24	14.1	0.2								
Sep. 28, 1981										
0	18.0	9.3	7.9	500						7.4
2	18.0	8.9	7.9	65						16.1
4	17.8	8.8	7.9	9						9.5
6	17.8	8.6								
8	17.8	8.3	7.9							11.2
10	17.8	8.1								
12	17.8	8.1	8.0							10.7
14	17.8	7.8								
16	17.8	7.5	8.0							8.7
18	16.5	1.9								
20	15.0	0.4	7.2							3.7
22	14.8	0.3								
24	14.8	0.3								
Oct. 5, 1981										
0	16.5	9.5	7.7	130	108.1	116.9	511.8	3.6	74.8	14.5
2	16.5	9.4	7.7	21	108.1	125.7	620.6	2.0	50.3	5.8
4	16.5	9.1	7.7	5	108.1	128.6	597.0	2.3	44.4	9.9
6	16.5	8.8								
8	16.5	8.6	7.7		109.5	146.1	680.9	2.3	102.2	7.4
10	16.5	8.6								
12	16.5	8.6	7.7		112.4	134.4	648.3	2.3	96.3	5.4
14	16.5	8.4								
16	16.5	8.8	7.7		116.8	140.3	554.4	2.9	92.9	9.5
18	16.2	7.1								
20	16.0	6.0	7.7		116.8	140.3	596.0	3.3	54.9	9.9
22	16.0	4.9								
24	15.8	0.3								
Oct. 12, 1981										
0	15.2	9.6	7.7	400	110.0	134.4	645.4	5.2	79.2	9.5
2	15.2	9.5	7.7	35	109.5	125.7	541.9	6.1	81.3	12.0
4	15.2	9.5	7.7	5	100.8	102.4	499.4	6.7	124.1	7.8
6	15.2	9.5								
8	15.2	9.5	7.7		111.0	137.4	696.4	5.5	52.9	5.4
10	15.2	9.8								
12	15.2	9.6	7.7		111.0	108.2	692.0	5.2	85.0	5.0
14	15.2	9.6								
16	15.2	9.5	7.7		111.0	143.2	654.4	5.5	133.9	4.1
18	15.2	9.5								
20	15.2	9.6	7.7		111.0	108.2	748.0	5.2	108.1	6.6
22	15.2	9.5								
24	15.2	9.5								
Oct. 19, 1981										
0	14.8	10.0	7.7	550						13.0
2	14.8	9.8	7.8	36						13.0
4	14.8	9.8	7.8							11.2
6	14.8	9.5								
8	14.8	9.6	7.9							17.7
10	14.8	9.6								
12	14.8	9.7	7.9							15.2
14	14.8	9.6								
16	14.8	9.6	7.9							8.0
18	14.8	9.5								
20	14.7	9.4	7.9							15.2
22	14.7	9.5								
24	14.7	8.2								
Nov. 10, 1981										
0	11.8	10.8	7.8	425	22.1	27.2	653.9	35.0	152.7	11.2
2	11.8	10.5	7.9	50	20.9	29.6	664.3	33.2	146.6	13.2
4	11.8	10.4	7.9	5	19.7	29.6	675.0	35.3	157.5	14.4
6	11.8	10.4								
8	11.8	10.3	7.9		22.1	27.2	646.3	34.1	148.4	7.4
10	11.8	10.3								
12	11.6	10.3	7.9		22.1	24.8	595.6	33.5	131.0	4.5
14	11.6	10.3								
16	11.5	10.2	7.9		22.1	22.3	507.6	36.0	100.6	10.3
18	11.5	10.2								
20	11.5	10.2	7.9		22.1	32.1	589.8	35.0	127.1	12.0
22	11.5	10.2								
24	11.8	10.1								
Dec. 1, 1981										
0	8.0	11.6	7.7	95	30.6	37.0	610.7	5.6	86.9	12.3
2	7.5	11.3	7.8	21	30.6	41.8	571.6	6.2	250.3	15.2
4	7.5	11.3	7.8	7	30.6	44.2	564.5	4.7	182.5	11.2
6	7.5	12.2								
8	7.5	11.6	7.8		28.2	32.1	584.1	6.5	196.3	6.5
10	7.5	11.6								
12	7.5	11.6	7.9		29.4	41.8	545.2	6.9	288.1	11.9
14	7.5	11.8								
16	7.5	11.8	7.9		30.6	41.8	581.3	6.2	183.7	13.0
18	7.5	11.8								
20	7.5	11.7	7.9		30.6	32.1	530.0	5.0	187.4	5.4
22	7.5	11.7								
24	7.5	10.8								

Appendix 2. Phytoplankton data

This Appendix provides raw data for the phytoplankton analyses performed during the years 1976–1981. The methods used have been described in Chapter 4. The volumes and surface areas of all significant phytoplankton species found at each sampling depth are given. Note that when a sampling depth of 99 m is specified, an integrated tube sample of the whole water column was taken. From the raw data given, the significance of the individual counts can be assessed. The data presented in this Appendix are summarized in graphical form in Chapter 4.

M	Genus and species	Fam.	Ct.	Vol. filt. ml	Flds. ctd.	Mag.	Dim1 µm	Dim2 µm	Dim3 µm	Biovol. mm³/m³	Surf.area mm²/m³
	Apr. 2, 1976										
0	flagellate small	flag	26	40	30	40	4	16	0	9.0	11163.3
0	flagellate large	flag	6	40	30	40	15	20	0	65.1	24003.5
0	Schroederia setigera	grn	2	40	30	40	4	14	0	0.5	843.0
0	Diachros sp.	chry	61	40	30	40	8	0	0	75.3	56505.8
0	Navicula sp.	diat	1	40	30	40	3	15	0	0.6	912.2
0	Ankistrodesmus fractus	grn	2	40	30	40	2	20	0	0.2	581.8
0	Asterionella sp.	diat	10	40	30	40	1	54	0	2.5	10043.6
4	flagellate small	flag	54	40	30	40	4	16	0	18.8	23185.3
4	flagellate large	flag	21	40	30	40	15	20	0	228.0	84012.1
4	Diachros sp.	chry	74	40	30	40	8	0	0	91.4	68548.1
4	Asterionella sp.	diat	2	40	30	40	1	54	0	0.5	2008.7
8	flagellate small	flag	49	40	30	40	4	16	0	17.0	21038.5
8	flagellate large	flag	10	40	30	40	15	20	0	108.6	40005.8
8	Diachros sp.	chry	62	40	30	40	8	0	0	76.6	57432.2
8	Navicula sp.	diat	1	40	30	40	3	15	0	0.6	912.2
8	Ankistrodesmus fractus	grn	2	40	30	40	2	20	0	0.2	581.8
8	Asterionella sp.	diat	2	40	30	40	1	54	0	0.5	2008.7
12	flagellate small	flag	72	40	30	40	4	16	0	25.0	30913.7
12	flagellate large	flag	13	40	30	40	15	20	0	141.1	52007.5
12	Diachros sp.	chry	82	40	30	40	8	0	0	101.3	75958.7
12	Navicula sp.	diat	1	40	30	40	3	15	0	0.6	912.2
12	Ankistrodesmus fractus	grn	2	40	30	40	2	20	0	0.2	581.8
16	Aphanizomenon flos-aquae	bg	16	40	30	40	4	0	0	7.3	7275.3
16	flagellate small	flag	48	40	30	40	4	16	0	16.7	20609.1
16	flagellate large	flag	5	40	30	40	15	20	0	54.3	20002.9
16	Diachros sp.	chry	75	40	30	40	8	0	0	92.6	69474.4
16	Navicula sp.	diat	1	40	30	40	3	15	0	0.6	912.2
16	Ankistrodesmus fractus	grn	4	40	30	40	2	20	0	0.4	1163.7
16	Asterionella sp.	diat	13	40	30	40	1	54	0	3.2	13056.7
16	Coelosphaerium sp.	bg	168	40	30	40	4	0	0	25.9	38905.7
	Apr. 8, 1976										
0	flagellate large	flag	24	40	30	40	15	20	0	260.5	96013.8
0	flagellate small	flag	58	40	30	40	6	10	0	32.7	27702.9
0	Planktosphaeria gelatinosa	grn	75	40	30	40	8	0	0	92.6	69474.4
0	Asterionella sp.	diat	2	40	30	40	2	64	0	2.4	4791.5
4	flagellate large	flag	60	40	30	40	15	20	0	651.3	240034.6
4	flagellate small	flag	111	40	30	40	6	10	0	62.7	53017.6
4	Planktosphaeria gelatinosa	grn	62	40	30	40	8	0	0	76.6	57432.2
4	Fragillaria sp.	diat	1	40	30	40	6	30	0	5.0	1990.3
8	flagellate large	flag	25	40	30	40	15	20	0	271.4	100014.4
8	flagellate small	flag	43	40	30	40	6	10	0	24.3	20538.4
8	Planktosphaeria gelatinosa	grn	48	40	30	40	8	0	0	59.3	44463.6
12	flagellate large	flag	9	40	17	40	15	20	0	172.4	63538.6
12	flagellate small	flag	11	40	17	40	6	10	0	11.0	9271.8
12	Planktosphaeria gelatinosa	grn	12	40	17	40	8	0	0	26.2	19616.3
12	Asterionella sp.	diat	8	40	17	40	2	64	0	16.7	33822.0
16	flagellate large	flag	4	40	30	40	15	20	0	43.4	16002.3
16	flagellate small	flag	33	40	30	40	6	10	0	18.6	15762.0
16	Planktosphaeria gelatinosa	grn	111	40	30	40	8	0	0	137.1	102822.1
16	Stephanodiscus sp.	diat	1	40	30	40	66		0		
16	Schroederia setigera	grn	1	40	30	40	4	12	0	0.2	366.2
16	Asterionella sp.	diat	7	40	30	40	2	64	0	8.3	16770.1
20	flagellate large	flag	7	40	30	40	15	20	0	76.0	28004.0
20	flagellate small	flag	25	40	30	40	6	10	0	14.1	11940.9

M	Genus and species	Fam.	Ct.	Vol. filt. ml	Flds. ctd.	Mag.	Dim1 µm	Dim2 µm	Dim3 µm	Biovol. mm³/m³	Surf.area mm²/m³
20	Planktosphaeria gelatinosa	grn	48	40	30	40	8	0	0	59.3	44463.6
20	Asterionella sp.	diat	5	40	30	40	2	64	0	5.9	11978.6
	Apr. 15, 1976										
0	flagellate large	flag	13	30	30	40	15	20	0	188.2	69343.3
0	flagellate small	flag	30	30	30	40	6	10	0	22.6	19105.5
0	Diachros sp.	chry	182	30	30	40	8	0	0	299.7	224788.2
0	Asterionella sp.	diat	36	30	30	40	2	64	0	56.6	114994.9
4	flagellate large	flag	19	30	30	40	15	20	0	275.0	101347.9
4	flagellate small	flag	38	30	30	40	6	10	0	28.6	24200.2
4	Diachros sp.	chry	125	30	30	40	8	0	0	205.9	154387.5
4	Schroederia setigera	grn	1	30	30	40	4	12	0	0.3	488.2
4	Asterionella sp.	diat	42	30	30	40	2	64	0	66.0	134160.7
8	flagellate large	flag	16	30	22	40	15	20	0	315.8	116380.4
8	flagellate small	flag	53	30	22	40	6	10	0	54.4	46026.8
8	Diachros sp.	chry	182	30	22	40	8	0	0	408.7	306529.4
8	Schroederia setigera	grn	2	30	22	40	4	12	0	0.8	1331.5
8	Asterionella sp.	diat	38	30	22	40	2	64	0	81.5	165522.9
12	flagellate large	flag	21	30	30	40	15	20	0	304.0	112016.1
12	flagellate small	flag	78	30	30	40	6	10	0	58.7	49674.2
12	Diachros sp.	chry	219	30	30	40	8	0	0	360.6	270486.9
12	Stephanodiscus sp.	diat	1	30	30	40	56		0		
12	Schroederia setigera	grn	2	30	30	40	4	12	0	0.6	976.4
12	Asterionella sp.	diat	44	30	30	40	2	64	0	69.2	140549.3
16	flagellate large	flag	6	30	30	40	15	20	0	86.8	32004.6
16	flagellate small	flag	46	30	30	40	6	10	0	34.6	29295.0
16	Diachros sp.	chry	170	30	30	40	8	0	0	280.0	209967.0
16	Stephanodiscus sp.	diat	3	30	30	40	56		0		
16	Schroederia setigera	grn	4	30	30	40	4	12	0	1.2	1952.9
16	Asterionella sp.	diat	23	30	30	40	2	64	0	36.2	73469.0
20	flagellate large	flag	1	30	30	40	15	20	0	14.5	5334.1
20	flagellate small	flag	4	30	30	40	6	10	0	3.0	2547.4
20	Diachros sp.	chry	25	30	30	40	8	0	0	41.2	30877.5
20	Schroederia setigera	grn	4	30	30	40	4	12	0	1.2	1952.9
20	Asterionella sp.	diat	47	30	30	40	2	64	0	73.9	150132.2
	Apr. 27, 1976										
0	flagellate large	flag	14	40	30	40	14	22	0	145.6	55498.7
0	flagellate small	flag	18	40	30	40	6	10	0	10.2	8597.5
0	Dinobryon sp.	chry	1	40	30	40	8	14	0	1.4	1035.0
0	Schroederia setigera	grn	8	40	30	40	5	16	0	3.9	4852.5
0	Asterionella sp.	diat	95	40	30	40	2	64	0	112.0	227594.0
0	Navicula sp.	diat	1	40	30	40	8	30	40	8.8	5012.6
4	flagellate large	flag	21	40	32	40	14	22	0	204.8	78045.0
4	flagellate small	flag	22	40	32	40	6	10	0	11.6	9851.2
4	Planktosphaeria gelatinosa	grn	1	40	32	40	5	0	0	0.3	339.2
4	Dinobryon sp.	chry	1	40	32	40	8	14	0	1.3	970.3
4	Stephanodiscus sp.	diat	3	40	32	40	60		0		
4	Schroederia setigera	grn	2	40	32	40	5	16	0	0.9	1137.3
4	Asterionella sp.	diat	123	40	32	40	2	64	0	136.0	276257.2
8	flagellate large	flag	17	40	32	40	14	22	0	165.8	63179.3
8	flagellate small	flag	29	40	30	40	6	10	0	16.4	13851.5
8	Stephanodiscus sp.	diat	1	40	30	40	60		0		
8	Schroederia setigera	grn	3	40	30	40	5	16	0	1.4	1819.7
8	Asterionella sp.	diat	50	40	30	40	2	64	0	59.0	119786.3
8	Scenedesmus sp.	grn	2	40	30	40			0		
12	flagellate large	flag	13	40	30	40	14	22	0	135.2	51534.5

M	Genus and species	Fam.	Ct.	Vol. filt. ml	Flds. ctd.	Mag.	Dim1 µm	Dim2 µm	Dim3 µm	Biovol. mm³/m³	Surf.area mm²/m³
12	flagellate small	flag	12	40	30	40	6	10	0	6.8	5731.6
12	Planktosphaeria gelatinosa	grn	2	40	30	40	5	0	0	0.6	723.7
12	Stephanodiscus sp.	diat	4	40	30	40	60		0		
12	Schroederia setigera	grn	3	40	30	40	5	16	0	1.4	1819.7
12	Asterionella sp.	diat	124	40	30	40	2	64	0	146.2	297070.1
16	flagellate large	flag	17	40	30	40	14	22	0	176.8	67391.3
16	flagellate small	flag	38	40	30	40	6	10	0	21.5	18150.2
16	Planktosphaeria gelatinosa	grn	3	40	30	40	5	0	0	0.9	1085.5
16	Stephanodiscus sp.	diat	3	40	30	40	60		0		
16	Schroederia setigera	grn	1	40	30	40	5	16	0	0.5	606.6
16	Asterionella sp.	diat	87	40	30	40	2	64	0	102.6	208428.2
16	Scenedesmus sp.	grn	1	40	30	40			0		
20	flagellate large	flag	19	40	30	40	14	22	0	197.6	75319.7
20	flagellate small	flag	22	40	30	40	6	10	0	12.4	10508.0
20	Planktosphaeria gelatinosa	grn	17	40	30	40	5	0	0	5.1	6151.4
20	Dinobryon sp.	chry	2	40	30	40	8	14	0	2.8	2070.1
20	Asterionella sp.	diat	45	40	30	40	2	64	0	53.1	107807.7
	May 7, 1976										
0	flagellate large	flag	3	50	30	40	14	22	0	25.0	9514.1
0	flagellate small	flag	3	50	30	40	6	14	0	1.8	1515.7
0	Stephanodiscus sp.	diat	2	50	30	40	60		0		
0	Fragillaria sp.	diat	9	50	30	40			0		
0	Dinobryon sp.	chry	5	50	30	40	8	14	0	5.6	4140.2
0	Schroederia setigera	grn	19	50	30	40	5	14	0	6.4	8176.4
0	Asterionella sp.	diat	3	50	30	40	2	64	0	2.8	5749.7
0	Characium sp.	grn	5	50	30	40		0	0		
4	flagellate large	flag	1	50	30	40	12	22	0	6.1	2646.1
4	flagellate small	flag	2	50	30	40	6	14	0	1.2	1010.4
4	Stephanodiscus sp.	diat	1	50	30	40	60		0		
4	Fragillaria sp.	diat	9	50	30	40			0		
4	Dinobryon sp.	chry	2	50	30	40	8	14	0	2.2	1656.1
4	Schroederia setigera	grn	18	50	30	40	5	14	0	6.1	7746.1
4	Asterionella sp.	diat	4	50	30	40	2	64	0	3.8	7666.3
4	Characium sp.	grn	2	50	30	40		0	0		
8	flagellate large	flag	16	40	30	40	12	22	0	122.3	52922.2
8	flagellate small	flag	19	40	30	40	6	14	0	14.0	11998.9
8	Planktosphaeria gelatinosa	grn	14	40	30	40	5	0	0	4.2	5065.8
8	Dinobryon sp.	chry	6	40	30	40	8	14	0	8.3	6210.2
8	Schroederia setigera	grn	34	40	30	40	5	14	0	14.4	18289.4
8	Asterionella sp.	diat	171	40	30	40	2	64	0	201.7	409669.3
12	flagellate large	flag	2	40	31	40	12	22	0	14.8	6401.9
12	flagellate small	flag	9	40	31	40	6	14	0	6.4	5500.4
12	Dinobryon sp.	chry	5	40	31	40	8	14	0	6.7	5008.3
12	Stephanodiscus sp.	diat	2	40	31	40	60		0		
12	Schroederia setigera	grn	23	40	31	40	5	14	0	9.4	11973.1
12	Asterionella sp.	diat	98	40	31	40	2	64	0	111.9	227207.6
12	Microcystis aeruginosa	bg	7	40	31	40	4	0	0	1.0	1568.8
16	flagellate large	flag	10	40	30	40	12	22	0	76.4	33076.4
16	flagellate small	flag	12	40	30	40	6	14	0	8.9	7578.3
16	Dinobryon sp.	chry	2	40	30	40	8	14	0	2.8	2070.1
16	Stephanodiscus sp.	diat	5	40	30	40	60		0		
16	Schroederia setigera	grn	23	40	30	40	5	14	0	9.7	12372.2
16	Asterionella sp.	diat	103	40	30	40	2	64	0	121.5	246759.8
16	Staurastrum sp.	grn	1	40	30	40	15	32	54	119.4	45808.0
20	flagellate large	flag	7	40	30	40	12	22	0	53.5	23153.5
20	flagellate small	flag	11	40	30	40	6	14	0	8.1	6946.7
20	Planktosphaeria gelatinosa	grn	24	40	30	40	5	0	0	7.2	8684.3
20	Dinobryon sp.	chry	8	40	30	40	8	14	0	11.1	8280.3
20	Stephanodiscus sp.	diat	1	40	30	40	60		0		
20	Schroederia setigera	grn	12	40	30	40	5	14	0	5.1	6455.1
20	Asterionella sp.	diat	188	40	30	40	2	64	0	221.7	450396.6
20	Scenedesmus sp.	grn	4	40	30	40	3	6	0	1.0	1658.6
	May 18, 1976										
0	Lyngbya limnetica	bg	5	15	30	40	3	0	0	3.4	4547.1
0	Planktosphaeria gelatinosa	grn	13	15	30	40	4	0	0	5.4	8028.2
0	Schroederia setigera	grn	16	15	30	40	4	12	0	9.9	15622.9
0	flagellate large	flag	12	15	30	40	14	20	0	302.6	117610.6
0	flagellate small	flag	11	15	30	40	4	8	0	5.7	6061.8
0	Staurastrum sp.	grn	1	15	30	40	15	32	54	318.4	122154.6
0	Navicula sp.	diat	1	15	30	40	7	20	0	12.0	8084.0
4	Schroederia setigera	grn	3	15	30	40	4	12	0	1.9	2929.3
4	flagellate large	flag	4	15	30	40	14	20	0	100.9	39203.5
4	flagellate small	flag	7	15	30	40	4	8	0	3.6	3857.5
8	Planktosphaeria gelatinosa	grn	12	40	30	40	2	0	0	0.2	694.7
8	Planktosphaeria gelatinosa	grn	78	40	30	40	4	0	0	12.0	18063.3
8	Schroederia setigera	grn	8	40	30	40	4	12	0	1.9	2929.3
8	Dinobryon sp.	chry	2	40	30	40	6	12	0	1.3	1103.6
8	flagellate large	flag	14	40	30	40	14	20	0	132.4	51454.6
8	flagellate small	flag	19	40	30	40	4	8	0	3.7	3926.4
12	Aphanizomenon flos-aquae	bg	6	40	30	40	4	0	0	2.7	2728.3
12	Planktosphaeria gelatinosa	grn	37	40	30	40	4	0	0	5.7	8568.5
12	Schroederia setigera	grn	1	40	30	40	4	12	0	0.2	366.2
12	flagellate large	flag	6	40	30	40	14	20	0	56.7	22052.0
12	flagellate small	flag	26	40	30	40	4	8	0	5.0	5372.9
16	Planktosphaeria gelatinosa	grn	1	40	35	40	4	0	0	0.1	198.5
16	Schroederia setigera	grn	10	40	35	40	4	12	0	2.0	3138.5
16	Dinobryon sp.	chry	1	40	35	40	6	12	0	0.6	473.0
16	flagellate large	flag	18	40	35	40	14	20	0	145.9	56705.1
16	flagellate small	flag	33	40	35	40	4	8	0	5.5	5845.3
16	Staurastrum sp.	grn	1	40	35	40	15	32	54	102.4	39264.0
20	Aphanizomenon flos-aquae	bg	3	40	30	40	4	0	0	1.4	1364.1
20	Lyngbya limnetica	bg	4	40	30	40	3	0	0	1.0	1364.1
20	Planktosphaeria gelatinosa	grn	5	40	30	40	4	0	0	0.8	1157.9
20	Schroederia setigera	grn	12	40	30	40	4	12	0	2.8	4393.9
20	Microcystis aeruginosa	bg	56	40	30	40	4	0	0	8.6	12968.6
20	Dinobryon sp.	chry	4	40	30	40	6	12	0	2.6	2207.2
20	flagellate large	flag	3	40	30	40	14	20	0	28.4	11026.0
20	flagellate small	flag	5	40	30	40	4	8	0	1.0	1033.3
	May 25, 1976										
0	Aphanizomenon flos-aquae	bg	9	10	30	40	4	0	0	16.4	16369.5
0	flagellates	flag	20	10	30	40			0		
4	flagellates	flag	80	25	30	40			0		
8	Aphanizomenon flos-aquae	bg	31	50	30	40	4	0	0	11.3	11276.8
8	flagellates	flag	126	50	30	40			0		
12	flagellates	flag	23	50	30	40			0		
16	flagellates	flag	5	50	30	40			0		
20	flagellates	flag	1	50	30	40			0		
	June 2, 1976										
0	Aphanizomenon flos-aquae	bg	13	15	30	40	4	0	0	15.8	15763.2
0	Lyngbya limnetica	bg	10	15	30	40		0	0		

M	Genus and species	Fam.	Ct.	Vol. filt. ml	Flds. ctd.	Mag.	Dim1 μm	Dim2 μm	Dim3 μm	Biovol. mm^3/m^3	Surf.area mm^2/m^3
0	Planktosphaeria gelatinosa	grn	68	15	30	40	4	0	0	28.0	41993.4
0	flagellate large	flag	2	15	30	40	16	26	0	85.6	28375.3
0	Schroederia setigera	grn	29	15	30	40	1	8	0	0.7	4512.1
0	Dinobryon sp.	chry	1	15	30	40		0	0		
4	Planktosphaeria gelatinosa	grn	26	15	30	40	4	0	0	10.7	16056.3
4	flagellate large	flag	3	15	30	40	16	26	0	128.5	42563.0
4	Schroederia setigera	grn	6	15	30	40	1	8	0	0.2	933.5
4	Microcystis aeruginosa	bg	11	15	30	40	5	0	0	8.8	10614.1
4	Melosira sp.	diat	16	15	30	40		0	0		
4	Staurastrum sp.	grn	1	15	30	40					
8	Lyngbya limnetica	bg	8	15	30	40		0	0		
8	Planktosphaeria gelatinosa	grn	168	15	30	40	4	0	0	69.2	103748.4
8	flagellate large	flag	12	15	30	40	16	26	0	513.8	170251.9
8	Schroederia setigera	grn	14	15	30	40	1	8	0	0.4	2178.2
8	flagellate small	flag	10	15	30	40	3	10	0	3.3	5013.1
12	Aphanizomenon flos-aquae	bg	4	20	30	40	4	0	0	3.6	3637.7
12	Planktosphaeria gelatinosa	grn	2	20	30	40	8	0	0	4.9	3705.3
12	flagellate large	flag	2	20	30	40	16	26	0	64.2	21281.5
12	Microcystis aeruginosa	bg	11	20	30	40	5	0	0	6.6	7960.6
12	flagellate small	flag	17	20	30	40	3	10	0	4.2	6391.7
16	flagellate large	flag	2	20	30	40	16	26	0	64.2	21281.5
16	flagellate small	flag	12	20	30	40	3	10	0	3.0	4511.8

June 8, 1976

M	Genus and species	Fam.	Ct.	Vol. filt. ml	Flds. ctd.	Mag.	Dim1 μm	Dim2 μm	Dim3 μm	Biovol. mm^3/m^3	Surf.area mm^2/m^3
0	Aphanizomenon flos-aquae	bg	599	10	22	40	4	0	0	1486.0	1485656.2
0	Anabaena sp.	bg	95	10	22	40	5	0	0	368.2	294527.0
0	Lyngbya limnetica	bg	8	10	22	40	1	0	0	1.2	4960.5
0	Planktosphaeria gelatinosa	grn	2	10	22	40	2	0	0	0.2	631.6
0	flagellate large	flag	4	10	22	40	8	18	0	60.6	38285.2
4	Aphanizomenon flos-aquae	bg	600	10	20	40	4	0	0	1637.4	1636950.0
4	Anabaena sp.	bg	100	10	20	40	5	0	0	426.4	341031.3
4	Lyngbya limnetica	bg	57	10	20	40	1	0	0	9.7	38877.6
4	Planktosphaeria gelatinosa	grn	1	10	20	40	2	0	0	0.1	347.4
4	flagellate large	flag	4	10	20	40	8	18	0	66.7	42113.7
4	Schroederia setigera	grn	13	10	20	40	1	6	0	0.6	3433.6
4	Microcystis aeruginosa	bg	10	10	20	40	1	0	0	0.1	868.4
4	Melosira sp.	diat	20	10	20	40		0	0		
8	Lyngbya limnetica	bg	2	10	30	40	1	0	0	0.2	909.4
8	flagellate large	flag	13	10	30	40	8	18	0	144.5	91246.3
8	Schroederia setigera	grn	21	10	30	40	1	6	0	0.6	3697.7
8	Dinobryon sp.	chry	3	10	30	40		0	0		
8	Planktosphaeria gelatinosa	grn	45	10	30	40	2	0	0	3.5	10421.2
8	flagellate large	flag	6	10	30	40	8	18	0	66.7	42113.7
8	Schroederia setigera	grn	24	10	30	40	1	6	0	0.7	4226.0
16	Aphanizomenon flos-aquae	bg	50	10	30	40	4	0	0	91.0	90941.7
16	Planktosphaeria gelatinosa	grn	72	10	30	40	2	0	0	5.6	16673.9
16	Schroederia setigera	grn	24	10	30	40	1	6	0	0.7	4226.0
16	Microcystis aeruginosa	bg	24	10	30	40	1	0	0	0.2	1389.5
20	Lyngbya limnetica	bg	2	10	30	40	1	0	0	0.2	909.4
20	Schroederia setigera	grn	18	10	30	40	1	6	0	0.5	3169.5

June 14, 1976

M	Genus and species	Fam.	Ct.	Vol. filt. ml	Flds. ctd.	Mag.	Dim1 μm	Dim2 μm	Dim3 μm	Biovol. mm^3/m^3	Surf.area mm^2/m^3
8	Aphanizomenon flos-aquae	bg	94	25	33	40	4	0	0	62.2	62171.0
8	Anabaena sp.	bg	44	25	12	40	4	0	0	80.0	80028.7
8	Lyngbya limnetica	bg	63	25	12	40	2	0	0	28.7	57293.3
8	flagellates	flag	40	25	12	40		0	0		
12	Aphanizomenon flos-aquae	bg	3	25	30	40	4	0	0	2.2	2182.6
12	Anabaena sp.	bg	21	25	30	40	4	0	0	15.3	15278.2
12	Lyngbya limnetica	bg	18	25	30	40	2	0	0	3.3	6547.8
12	flagellates	flag	13	25	30	40		0	0		
12	Gloebotrys sp.	chry	1	25	30	40		0	0		
12	Staurastrum sp.	grn	3	25	30	40					
12	Schroederia setigera	grn	1	25	30	40			0		
16	Anabaena sp.	bg	11	25	30	40	4	0	0	8.0	8002.9
16	Lyngbya limnetica	bg	6	25	30	40	2	0	0	1.1	2182.6
16	flagellates	flag	5	25	30	40		0	0		
20	Aphanizomenon flos-aquae	bg	46	25	30	40	4	0	0	33.5	33466.5
20	Anabaena sp.	bg	40	25	30	40	4	0	0	29.1	29101.3
20	Lyngbya limnetica	bg	9	25	30	40	2	0	0	1.6	3273.9
20	flagellates	flag	7	25	30	40		0	0		

June 21, 1976

M	Genus and species	Fam.	Ct.	Vol. filt. ml	Flds. ctd.	Mag.	Dim1 μm	Dim2 μm	Dim3 μm	Biovol. mm^3/m^3	Surf.area mm^2/m^3
0	Aphanizomenon flos-aquae	bg	574	1	7	40	4	0	0	44754.2	44743301.0
0	Anabaena sp.	bg	36	1	7	40	4	0	0	2806.9	2806200.0
0	Lyngbya limnetica	bg	59	1	7	40	2	0	0	1150.0	2299525.0
2	Aphanizomenon flos-aquae	bg	587	10	13	40	4	0	0	2464.4	2463819.6
2	Anabaena sp.	bg	54	10	13	40	4	0	0	226.7	226654.6
2	Lyngbya limnetica	bg	56	10	13	40	2	0	0	58.8	117524.6
2	flagellates	flag	1	10	13	40	2	10	0	0.9	2134.6
4	Aphanizomenon flos-aquae	bg	409	10	30	40	4	0	0	744.1	743902.9
4	Anabaena sp.	bg	4	10	30	40	4	0	0	7.3	7275.3
4	Lyngbya limnetica	bg	56	10	30	40	2	0	0	25.5	50927.3
4	flagellates	flag	2	10	30	40	2	10	0	0.8	1850.0
6	Aphanizomenon flos-aquae	bg	545	25	23	40	4	0	0	517.3	517181.3
6	Anabaena sp.	bg	3	25	23	40	4	0	0	2.8	2846.9
6	Lyngbya limnetica	bg	52	25	23	40	2	0	0	12.3	24672.9
8	Aphanizomenon flos-aquae	bg	481	25	30	40	4	0	0	350.0	349943.5
8	Anabaena sp.	bg	17	25	30	40	4	0	0	12.4	12368.1
8	Lyngbya limnetica	bg	171	25	30	40	2	0	0	31.1	62204.1
8	flagellates	flag	2	25	30	40	3	10	0	0.7	1130.5
8	Gloebotrys sp.	chry	2	25	30	40		0	0		
10	Aphanizomenon flos-aquae	bg	151	25	30	40	4	0	0	109.9	109857.5
10	Lyngbya limnetica	bg	1	25	30	40	2	0	0	0.2	363.8
10	flagellates	flag	4	25	30	40	3	10	0	1.4	2261.0
10	Gloebotrys sp.	chry	3	25	30	40		0	0		
18	Aphanizomenon flos-aquae	bg	156	25	30	40	4	0	0	113.5	113495.2
18	flagellates	flag	6	25	30	40	3	10	0	2.1	3391.5

July 1, 1976

M	Genus and species	Fam.	Ct.	Vol. filt. ml	Flds. ctd.	Mag.	Dim1 μm	Dim2 μm	Dim3 μm	Biovol. mm^3/m^3	Surf.area mm^2/m^3
0	Aphanizomenon flos-aquae	bg	649	5	17	40	4	0	0	4167.2	4166198.3
0	Anabaena sp.	bg	6	5	17	40	5	0	0	60.2	48145.6
0	Lyngbya limnetica	bg	32	5	17	40	2	0	0	51.4	102710.6
0	Planktosphaeria gelatinosa	grn	1	5	17	40	8	0	0	17.4	13077.5
0	Microcystis aeruginosa	bg	448	5	17	40	5	0	0	1907.1	2288567.7
0	Melosira sp.	diat	26	5	17	40		0	0		
2	Aphanizomenon flos-aquae	bg	804	5	17	40	4	0	0	5162.5	5161207.2
2	Anabaena sp.	bg	115	5	17	40	5	0	0	1153.8	922790.5
2	Lyngbya limnetica	bg	25	5	17	40	2	0	0	40.1	80242.6
2	Melosira sp.	diat	11	5	17	40		0	0		
2	Ceratium hirundinella	dino	1	5	17	40	13	22	25	63.3	33940.3
4	Aphanizomenon flos-aquae	bg	498	5	13	40	4	0	0	4181.5	4180518.5
4	Anabaena sp.	bg	12	5	13	40	5	0	0	157.4	125919.2
4	Lyngbya limnetica	bg	19	5	13	40	2	0	0	39.9	79748.8
4	Schroederia setigera	grn	1	5	13	40	1	5	0	0.1	681.3

M	Genus and species	Fam.	Ct.	Vol. filt. ml	Flds. ctd.	Mag.	Dim1 µm	Dim2 µm	Dim3 µm	Biovol. mm³/m³	Surf.area mm²/m³
4	Microcystis aeruginosa	bg	8	5	13	40	5	0	0	44.5	53441.8
4	Dinobryon sp.	chry	1	5	13	40		0	0		
6	Aphanizomenon flos-aquae	bg	361	5	27	40	4	0	0	1459.5	1459108.5
6	Lyngbya limnetica	bg	25	5	27	40	2	0	0	25.3	50523.1
6	Planktosphaeria gelatinosa	grn	173	5	27	40	4	0	0	237.4	356120.5
6	Planktosphaeria gelatinosa	grn	15	5	27	40	8	0	0	164.7	123510.0
6	flagellate small	flag	4	5	27	40	2	6	0	1.2	2464.9
6	Schroederia setigera	grn	4	5	27	40	1	5	0	0.2	1312.0
6	Microcystis aeruginosa	bg	24	5	27	40	5	0	0	64.3	77193.8
8	Aphanizomenon flos-aquae	bg	181	5	31	40	4	0	0	637.3	637178.4
8	Anabaena sp.	bg	5	5	31	40	5	0	0	27.5	22002.0
8	Lyngbya limnetica	bg	24	5	31	40	2	0	0	21.1	42243.9
8	flagellate small	flag	4	5	31	40	2	6	0	1.0	2146.9
8	Schroederia setigera	grn	3	5	31	40	1	5	0	0.1	857.1
8	Microcystis aeruginosa	bg	8	5	31	40	5	0	0	18.7	22411.1
10	Aphanizomenon flos-aquae	bg	425	10	24	40	4	0	0	966.5	966255.2
10	Lyngbya limnetica	bg	9	10	24	40	2	0	0	5.1	10230.9
10	flagellate small	flag	4	10	24	40	2	6	0	0.7	1386.5
10	Schroederia setigera	grn	2	10	24	40	1	5	0	0.1	369.0
10	Microcystis aeruginosa	bg	6	10	24	40	5	0	0	9.0	10855.4
18	Aphanizomenon flos-aquae	bg	400	20	25	40	4	0	0	436.6	436520.0
18	Anabaena sp.	bg	85	20	25	40	5	0	0	145.0	115950.6
18	Lyngbya limnetica	bg	11	20	25	40	2	0	0	3.0	6002.2
18	flagellate small	flag	13	20	25	40	2	6	0	1.1	2163.0
	July 6, 1976										
0	Aphanizomenon flos-aquae	bg	1089	2	10	40	4	0	0	29717.9	29710642.8
0	Lyngbya limnetica	bg	9	2	10	40	2	0	0	61.4	122771.3
2	Aphanizomenon flos-aquae	bg	413	1	30	40	4	0	0	7513.6	7511781.8
4	Aphanizomenon flos-aquae	bg	533	2	25	40	4	0	0	5818.1	5816629.1
4	Lyngbya limnetica	bg	17	2	25	40	2	0	0	46.4	92760.5
6	Aphanizomenon flos-aquae	bg	415	2	30	40	4	0	0	3775.0	3774079.2
6	Lyngbya limnetica	bg	29	2	30	40	2	0	0	65.9	131865.4
8	Aphanizomenon flos-aquae	bg	237	3	30	40	4	0	0	1437.2	1436878.4
8	Lyngbya limnetica	bg	28	3	30	40	2	0	0	42.4	84878.9
8	Anabaena sp.	bg	9	3	30	40	4	0	0	54.6	54565.0
10	Aphanizomenon flos-aquae	bg	64	20	31	40	4	0	0	56.3	56325.2
10	Microcystis aeruginosa	bg	595	20	31	40	4	0	0	177.8	266692.0
10	Lyngbya limnetica	bg	34	20	31	40	2	0	0	7.5	14961.4
18	Aphanizomenon flos-aquae	bg	27	20	30	40	4	0	0	24.6	24554.3
18	Microcystis aeruginosa	bg	6	20	30	40	4	0	0	1.9	2779.0
18	Lyngbya limnetica	bg	55	20	30	40	2	0	0	12.5	25009.0
	July 13, 1976										
0	Aphanizomenon flos-aquae	bg	600	1	27	40	4	0	0	12128.5	12125555.6
0	Anabaena sp.	bg	33	1	27	40	4	0	0	667.1	666905.6
0	Lyngbya limnetica	bg	29	1	27	40	2	0	0	146.6	293034.3
0	flagellate small	flag	2	1	27	40			0		
0	flagellate large	flag	2	1	27	40			0		
2	Aphanizomenon flos-aquae	bg	891	5	20	40	4	0	0	4862.9	4861741.6
2	Anabaena sp.	bg	7	5	20	40	4	0	0	38.2	38195.5
2	Lyngbya limnetica	bg	52	5	20	40	2	0	0	71.0	141869.0
4	Aphanizomenon flos-aquae	bg	1164	5	20	40	4	0	0	6352.9	6351366.1
4	Anabaena sp.	bg	20	5	20	40	4	0	0	109.2	109130.0
4	Lyngbya limnetica	bg	11	5	20	40	2	0	0	15.0	30010.8
6	Aphanizomenon flos-aquae	bg	536	5	20	40	4	0	0	2925.4	2924684.0
6	Anabaena sp.	bg	59	5	20	40	4	0	0	322.0	321933.5
6	Lyngbya limnetica	bg	23	5	20	40	2	0	0	31.4	62749.8
8	Aphanizomenon flos-aquae	bg	354	5	20	40	4	0	0	1932.1	1931601.0
8	Lyngbya limnetica	bg	3	5	20	40	2	0	0	4.1	8184.8
10	Aphanizomenon flos-aquae	bg	20	5	30	40	4	0	0	72.8	72753.3
	July 20, 1976										
0	Aphanizomenon flos-aquae	bg	570	5	13	40	4	0	0	4786.1	4784930.8
0	Anabaena sp.	bg	7	5	13	40	4	0	0	58.8	58762.3
2	Aphanizomenon flos-aquae	bg	590	5	14	40	4	0	0	4600.2	4599050.1
2	Anabaena sp.	bg	48	5	14	40	4	0	0	374.3	374160.0
2	Lyngbya limnetica	bg	52	5	14	40	2	0	0	101.4	202670.0
4	Aphanizomenon flos-aquae	bg	502	5	15	40	4	0	0	3653.1	3652217.4
4	Anabaena sp.	bg	21	5	15	40	4	0	0	152.8	152782.0
4	Lyngbya limnetica	bg	8	5	15	40	2	0	0	14.6	29101.3
6	Aphanizomenon flos-aquae	bg	578	5	20	40	4	0	0	3154.6	3153857.0
6	Lyngbya limnetica	bg	10	5	20	40	2	0	0	13.6	27282.5
8	Aphanizomenon flos-aquae	bg	535	5	18	40	4	0	0	3244.4	3243586.2
8	Anabaena sp.	bg	2	5	18	40	4	0	0	12.1	12125.6
8	Lyngbya limnetica	bg	80	5	18	40	2	0	0	121.3	242511.1
8	flagellates	flag	1	5	18	40			0		
10	Aphanizomenon flos-aquae	bg	563	10	22	40	4	0	0	1396.7	1396368.0
10	Anabaena sp.	bg	29	10	22	40	4	0	0	71.9	71926.6
10	Lyngbya limnetica	bg	61	10	22	40	2	0	0	37.8	75646.9
10	flagellates	flag	4	10	22	40			0		
18	Aphanizomenon flos-aquae	bg	69	25	30	40	4	0	0	50.2	50199.8
18	Lyngbya limnetica	bg	18	25	30	40	2	0	0	3.3	6547.8
	July 26, 1976										
0	Aphanizomenon flos-aquae	bg	615	5	12	40	4	0	0	5594.3	5592912.5
0	Microcystis aeruginosa	bg	119	5	12	40	4	0	0	367.4	551163.4
0	Lyngbya limnetica	bg	8	5	12	40	2	0	0	18.2	36376.7
0	flagellates	flag	4	5	12	40			0		
2	Aphanizomenon flos-aquae	bg	701	5	5	40	4	0	0	15303.8	15300026.1
2	Microcystis aeruginosa	bg	39	5	5	40	4	0	0	289.0	433520.1
2	Lyngbya limnetica	bg	26	5	5	40	2	0	0	141.9	283738.0
2	flagellates	flag	2	5	5	40		0	0		
4	Aphanizomenon flos-aquae	bg	671	5	7	40	4	0	0	10463.4	10460890.1
4	Microcystis aeruginosa	bg	136	5	7	40	4	0	0	719.9	1079830.3
4	Anabaena sp.	bg	6	5	7	40		0	0		
6	Aphanizomenon flos-aquae	bg	525	5	9	40	4	0	0	6367.5	6365916.8
6	Microcystis aeruginosa	bg	4	5	9	40	4	0	0	16.5	24702.0
6	Lyngbya limnetica	bg	37	5	9	40	2	0	0	112.2	224322.8
6	flagellates	flag	2	5	9	40			0		
8	Aphanizomenon flos-aquae	bg	362	5	30	40	4	0	0	1317.2	1316835.3
8	Microcystis aeruginosa	bg	10	5	30	40	4	0	0	12.4	18526.5
8	Lyngbya limnetica	bg	30	5	30	40	2	0	0	27.3	54565.0
8	flagellates	flag	3	5	30	40			0		
10	Aphanizomenon flos-aquae	bg	583	20	11	40	4	0	0	1446.3	1445972.5
10	Microcystis aeruginosa	bg	63	20	11	40	4	0	0	53.1	79579.7
18	Aphanizomenon flos-aquae	bg	35	25	32	40	4	0	0	23.9	23872.2
18	Lyngbya limnetica	bg	12	25	32	40	2	0	0	2.0	4092.4
	Aug. 2, 1976										
0	Aphanizomenon flos-aquae	bg	641	5	18	40	4	0	0	3887.2	3886240.6
0	Microcystis aeruginosa	bg	132	5	18	40	5	0	0	530.7	636848.4
0	Lyngbya limnetica	bg	66	5	18	40	2	0	0	100.1	200071.7
2	Aphanizomenon flos-aquae	bg	367	3	30	40	4	0	0	2225.6	2225039.4

M	Genus and species	Fam.	Ct.	Vol. filt. ml	Flds. ctd.	Mag.	Dim1 μm	Dim2 μm	Dim3 μm	Biovol. mm³/m³	Surf.area mm²/m³
2	Microcystis aeruginosa	bg	30	3	30	40	5	0	0	120.6	144738.3
2	Lyngbya limnetica	bg	36	3	30	40	2	0	0	54.6	109130.0
4	Aphanizomenon flos-aquae	bg	806	3	22	40	4	0	0	6665.2	6663544.0
4	Microcystis aeruginosa	bg	407	3	22	40	5	0	0	2231.4	2677658.2
4	Lyngbya limnetica	bg	15	3	22	40	2	0	0	31.0	62005.7
4	Anabaena sp.	bg	7	3	22	40	4	0	0	57.9	57872.0
6	Aphanizomenon flos-aquae	bg	377	3	27	40	4	0	0	2540.3	2539630.3
6	Microcystis aeruginosa	bg	37	5	27	40	5	0	0	99.2	119007.0
8	Aphanizomenon flos-aquae	bg	1017	5	22	40	4	0	0	5046.0	5044782.3
8	Microcystis aeruginosa	bg	204	5	22	40	5	0	0	671.1	805271.2
8	Lyngbya limnetica	bg	22	5	22	40	2	0	0	27.3	54565.0
10	Aphanizomenon flos-aquae	bg	353	15	21	40	4	0	0	611.6	611474.4
10	Microcystis aeruginosa	bg	84	15	21	40	5	0	0	96.5	115790.6
10	Lyngbya limnetica	bg	38	15	21	40	2	0	0	16.5	32912.2
18	Aphanizomenon flos-aquae	bg	9	20	33	40	4	0	0	7.4	7440.7
	Aug. 10, 1976										
0	Aphanizomenon flos-aquae	bg	398	5	17	40	4	0	0	2555.6	2554925.9
0	Lyngbya limnetica	bg	4	5	17	40	2	0	0	6.4	12838.8
0	Microcystis aeruginosa	bg	381	5	17	40	5	0	0	1621.9	1946304.2
2	Aphanizomenon flos-aquae	bg	237	5	17	40	4	0	0	1521.8	1521400.6
2	Lyngbya limnetica	bg	13	5	17	40	2	0	0	20.9	41726.2
2	Anabaena sp.	bg	63	5	17	40	5	0	0	632.1	505528.7
2	Microcystis aeruginosa	bg	381	5	17	40	5	0	0	1621.9	1946304.2
4	Aphanizomenon flos-aquae	bg	586	5	22	40	4	0	0	2907.5	2906826.4
4	Lyngbya limnetica	bg	7	5	22	40	2	0	0	8.7	17361.6
4	Anabaena sp.	bg	22	5	22	40	5	0	0	170.6	136412.5
4	Microcystis aeruginosa	bg	157	5	22	40	5	0	0	516.5	619743.0
6	Aphanizomenon flos-aquae	bg	466	5	30	40	4	0	0	1695.6	1695152.7
6	Microcystis aeruginosa	bg	778	5	30	40	5	0	0	1876.8	2252127.7
8	Aphanizomenon flos-aquae	bg	615	5	16	40	4	0	0	4195.7	4194684.4
8	Lyngbya limnetica	bg	9	5	16	40	2	0	0	15.4	30692.8
8	Anabaena sp.	bg	18	5	16	40	5	0	0	191.9	153464.1
8	Microcystis aeruginosa	bg	223	5	16	40	5	0	0	1008.6	1210373.9
10	Aphanizomenon flos-aquae	bg	895	15	20	40	4	0	0	1628.3	1627855.8
10	Lyngbya limnetica	bg	42	15	20	40	2	0	0	19.1	38195.5
10	Microcystis aeruginosa	bg	854	15	20	40	5	0	0	1030.1	1236064.9
18	Aphanizomenon flos-aquae	bg	199	20	30	40	4	0	0	181.0	180973.9
18	Lyngbya limnetica	bg	4	20	30	40	2	0	0	0.9	1818.8
	Aug. 16, 1976										
0	Aphanizomenon flos-aquae	bg	21	1	14	40	4	0	0	818.7	818475.0
0	Anabaena sp.	bg	21	1	14	40	5	0	0	1279.2	1023093.8
0	Microcystis aeruginosa	bg	511	1	14	40	5	0	0	13207.4	15848842.0
2	Aphanizomenon flos-aquae	bg	59	1	16	40	4	0	0	2012.6	2012084.4
2	Lyngbya limnetica	bg	4	1	16	40		0	0		
2	Microcystis aeruginosa	bg	156	1	16	40	5	0	0	3528.0	4233594.8
4	Aphanizomenon flos-aquae	bg	6	1	30	40	4	0	0	109.2	109130.0
4	Anabaena sp.	bg	38	1	30	40	5	0	0	1080.2	863945.9
4	Microcystis aeruginosa	bg	174	1	30	40	5	0	0	2098.7	2518446.1
6	Aphanizomenon flos-aquae	bg	51	2	30	40	4	0	0	463.9	463802.5
6	Lyngbya limnetica	bg	5	2	30	40		0	0		
6	Anabaena sp.	bg	5	2	30	40	5	0	0	71.1	56838.5
6	Microcystis aeruginosa	bg	128	2	30	40	5	0	0	771.9	926325.0
8	Aphanizomenon flos-aquae	bg	64	5	17	40	4	0	0	410.9	410842.4
8	Anabaena sp.	bg	26	5	17	40	5	0	0	260.9	208630.9
8	Microcystis aeruginosa	bg	404	5	17	40	5	0	0	1719.8	2063797.6
18	Aphanizomenon flos-aquae	bg	6	20	30	40	4	0	0	5.5	5456.5
18	Microcystis aeruginosa	bg	117	20	30	40	5	0	0	70.6	84671.9
	Aug. 23, 1976										
0	Aphanizomenon flos-aquae	bg	157	3	30	40	4	0	0	952.1	951856.1
0	Anabaena sp.	bg	310	3	30	40	5	0	0	2937.4	2349326.4
0	Microcystis aeruginosa	bg	110	3	30	40	4	0	0	226.4	339652.5
0	flagellates	flag	1	3	30	40			0		
2	Aphanizomenon flos-aquae	bg	238	3	29	40	4	0	0	1493.1	1492697.7
2	Anabaena sp.	bg	276	3	29	40	5	0	0	2705.4	2163784.5
2	Microcystis aeruginosa	bg	192	3	29	40	4	0	0	408.9	613291.0
2	Characium sp.	grn	1	3	29	40		0	0		
4	Aphanizomenon flos-aquae	bg	42	3	28	40	4	0	0	272.9	272825.0
4	Lyngbya limnetica	bg	4	3	28	40	2	0	0	6.5	12991.7
4	Anabaena sp.	bg	58	3	28	40	5	0	0	588.8	470947.9
4	Microcystis aeruginosa	bg	449	3	28	40	4	0	0	990.3	1485428.3
6	Aphanizomenon flos-aquae	bg	11	3	14	40	4	0	0	142.9	142908.3
6	Anabaena sp.	bg	29	3	14	40	5	0	0	588.8	470947.9
6	Microcystis aeruginosa	bg	503	3	14	40	4	0	0	2218.8	3328153.4
8	Aphanizomenon flos-aquae	bg	32	3	30	40	4	0	0	194.1	194008.9
8	Lyngbya limnetica	bg	5	3	30	40	2	0	0	7.6	15156.9
8	Anabaena sp.	bg	50	3	30	40	5	0	0	473.8	378923.6
8	Microcystis aeruginosa	bg	386	3	30	40	4	0	0	794.6	1191871.5
10	Aphanizomenon flos-aquae	bg	35	5	28	40	4	0	0	136.4	136412.5
10	Anabaena sp.	bg	45	5	28	40	5	0	0	274.1	219234.4
10	Microcystis aeruginosa	bg	454	5	28	40	4	0	0	600.8	901181.9
10	flagellates	flag	3	5	28	40			0		
18	Aphanizomenon flos-aquae	bg	5	20	31	40	4	0	0	4.4	4400.4
18	Lyngbya limnetica	bg	5	20	31	40		0	0		
18	Anabaena sp.	bg	61	20	31	40	5	0	0	83.9	67106.2
18	Microcystis aeruginosa	bg	357	20	31	40	4	0	0	106.7	160015.2
18	flagellates	flag	3	20	31	40			0		
	Aug. 30, 1976										
0	Aphanizomenon flos-aquae	bg	89	5	20	40	4	0	0	485.7	485628.5
0	Anabaena sp.	bg	109	5	20	40	5	0	0	929.5	743448.1
0	Lyngbya limnetica	bg	51	5	20	40	2	0	0	69.6	139140.8
0	Microcystis aeruginosa	bg	452	5	20	40	4	0	0	837.4	1256096.7
0	flagellates	flag	7	5	20	40			0		
2	Aphanizomenon flos-aquae	bg	41	5	25	40	4	0	0	179.0	178973.2
2	Anabaena sp.	bg	183	5	25	40	5	0	0	1248.5	998539.5
2	Lyngbya limnetica	bg	25	5	25	40	2	0	0	27.3	54565.0
2	Microcystis aeruginosa	bg	439	5	25	40	4	0	0	650.7	975976.0
2	flagellates	flag	9	5	25	40			0		
2	Characium sp.	grn	2	5	25	40		0	0		
4	Aphanizomenon flos-aquae	bg	74	10	23	40	4	0	0	175.6	175557.0
4	Anabaena sp.	bg	188	10	23	40	5	0	0	697.1	557512.0
4	Lyngbya limnetica	bg	8	10	23	40	2	0	0	4.7	9489.6
4	Microcystis aeruginosa	bg	526	10	23	40	4	0	0	423.7	635539.5
4	flagellates	flag	2	10	23	40		0	0		
6	Aphanizomenon flos-aquae	bg	78	10	26	40	4	0	0	163.7	163695.0
6	Anabaena sp.	bg	175	10	26	40	5	0	0	574.0	459080.5
6	Lyngbya limnetica	bg	8	10	26	40	2	0	0	4.2	8394.6
6	Microcystis aeruginosa	bg	323	10	26	40	4	0	0	230.2	345234.2
6	flagellates	flag	4	10	26	40			0		

M	Genus and species	Fam.	Ct.	Vol. filt. ml	Flds. ctd.	Mag.	Dim1 µm	Dim2 µm	Dim3 µm	Biovol. mm³/m³	Surf.area mm²/m³
	Sep. 8, 1976										
0	Aphanizomenon flos-aquae	bg	40	1	30	40	4	0	0	727.7	727533.3
0	Lyngbya limnetica	bg	6	1	30	40	2	0	0	27.3	54565.0
0	Anabaena sp.	bg	14	1	30	40	6	0	0	573.1	381955.0
0	Microcystis aeruginosa	bg	158	1	30	40	4	0	0	975.7	1463593.5
2	Aphanizomenon flos-aquae	bg	131	1	30	40	4	0	0	2383.3	2382671.7
2	Lyngbya limnetica	bg	27	1	30	40	2	0	0	122.8	245542.5
2	Anabaena sp.	bg	57	1	30	40	6	0	0	2333.2	1555102.5
2	Microcystis aeruginosa	bg	321	1	30	40	4	0	0	1982.3	2973503.3
4	Aphanizomenon flos-aquae	bg	173	3	20	40	4	0	0	1573.7	1573290.9
4	Anabaena sp.	bg	135	3	20	40	6	0	0	2763.0	1841568.8
4	Microcystis aeruginosa	bg	368	3	20	40	4	0	0	1136.3	1704438.0
4	flagellates	flag	2	3	20	40			0		
4	Characium sp.	grn	1	3	20	40		0	0		
6	Aphanizomenon flos-aquae	bg	18	3	23	40	4	0	0	142.4	142343.5
6	Lyngbya limnetica	bg	10	3	23	40	2	0	0	19.8	39539.9
6	Anabaena sp.	bg	46	3	23	40	6	0	0	818.7	545650.0
6	Microcystis aeruginosa	bg	297	3	23	40	4	0	0	797.4	1196167.5
6	eucaryote colony	eu	3	3	23	40		0	0		
8	Aphanizomenon flos-aquae	bg	34	10	30	40	4	0	0	61.9	61840.3
8	Lyngbya limnetica	bg	14	10	30	40	2	0	0	6.4	12731.8
8	Anabaena sp.	bg	7	10	30	40	6	0	0	28.7	19097.8
8	Microcystis aeruginosa	bg	320	10	30	40	4	0	0	197.6	296424.0
8	flagellates	flag	2	10	30	40			0		
8	Characium sp.	grn	1	10	30	40		0	0		
8	blue green colony	bg	1	10	30	40		0	0		
10	Aphanizomenon flos-aquae	bg	21	10	30	40	4	0	0	38.2	38195.5
10	Lyngbya limnetica	bg	23	10	30	40	2	0	0	10.5	20916.6
10	Anabaena sp.	bg	31	10	30	40	6	0	0	126.9	84575.8
10	Microcystis aeruginosa	bg	339	10	30	40	4	0	0	209.3	314024.2
10	blue green colony	bg	1	10	30	40		0	0		
18	Anabaena sp.	bg	20	20	30	40	6	0	0	40.9	27282.5
18	Microcystis aeruginosa	bg	194	20	30	40	4	0	0	59.9	89853.5
18	Characium sp.	grn	3	20	30	40		0	0		
18	blue green colony	bg	2	20	30	40		0	0		
	Sep. 13, 1976										
0	Aphanizomenon flos-aquae	bg	48	1	30	40	4	0	0	873.3	873040.0
0	Anabaena sp.	bg	23	1	30	40	6	0	0	941.5	627497.5
0	Microcystis aeruginosa	bg	68	1	30	40	5	0	0	820.2	984220.3
0	flagellates	flag	2	1	30	40			0		
2	Aphanizomenon flos-aquae	bg	13	1	30	40	4	0	0	236.5	236448.3
2	Anabaena sp.	bg	20	1	30	40	6	0	0	818.7	545650.0
2	Microcystis aeruginosa	bg	52	1	30	40	5	0	0	627.2	752639.1
2	flagellates	flag	6	1	30	40			0		
4	Aphanizomenon flos-aquae	bg	157	3	30	40	4	0	0	952.1	951856.1
4	Lyngbya limnetica	bg	2	3	30	40	2	0	0	3.0	6062.8
4	Anabaena sp.	bg	71	3	30	40	6	0	0	968.8	645685.8
4	Microcystis aeruginosa	bg	319	3	30	40	5	0	0	1282.5	1539050.4
4	flagellates	flag	12	3	30	40			0		
6	Aphanizomenon flos-aquae	bg	152	3	30	40	4	0	0	921.8	921542.2
6	Lyngbya limnetica	bg	12	3	30	40	2	0	0	18.2	36376.7
6	Anabaena sp.	bg	91	3	30	40	6	0	0	1241.7	827569.2
6	Microcystis aeruginosa	bg	189	3	30	40	5	0	0	759.9	911851.2
6	flagellates	flag	3	3	30	40			0		
8	Aphanizomenon flos-aquae	bg	82	3	30	40	4	0	0	497.3	497147.8
8	Lyngbya limnetica	bg	56	3	30	40	2	0	0	84.9	169757.8
8	Anabaena sp.	bg	20	3	30	40	6	0	0	272.9	181883.3
8	Microcystis aeruginosa	bg	415	3	30	40	5	0	0	1668.5	2002212.9
8	flagellates	flag	3	3	30	40			0		
10	Aphanizomenon flos-aquae	bg	78	3	30	40	4	0	0	473.0	472896.7
10	Lyngbya limnetica	bg	24	3	30	40	2	0	0	36.4	72753.3
10	Anabaena sp.	bg	44	3	30	40	6	0	0	600.4	400143.3
10	Microcystis aeruginosa	bg	209	3	30	40	5	0	0	840.3	1008343.4
10	flagellates	flag	4	3	30	40			0		
18	Lyngbya limnetica	bg	3	20	30	40	2	0	0	0.7	1364.1
18	Microcystis aeruginosa	bg	134	20	30	40	5	0	0	80.8	96974.6
18	flagellates	flag	5	20	30	40			0		
	Sep. 20, 1976										
0	Aphanizomenon flos-aquae	bg	154	3	30	40	4	0	0	933.9	933667.8
0	Lyngbya limnetica	bg	14	3	30	40	2	0	0	21.2	42439.4
0	Anabaena sp.	bg	21	3	30	40	6	0	0	286.5	190977.5
0	Microcystis aeruginosa	bg	174	3	30	40	5	0	0	699.6	839482.0
0	flagellates	flag	9	3	30	40			0		
0	eucaryote colony	eu	1	3	30	40		0	0		
2	Aphanizomenon flos-aquae	bg	75	3	30	40	4	0	0	454.8	454708.3
2	Lyngbya limnetica	bg	4	3	30	40	2	0	0	6.1	12125.6
2	Anabaena sp.	bg	43	3	30	40	6	0	0	586.7	391049.2
2	Microcystis aeruginosa	bg	88	3	30	40	5	0	0	353.8	424565.6
2	flagellates	flag	1	3	30	40			0		
2	Characium sp.	grn	3	3	30	40		0	0		
4	Aphanizomenon flos-aquae	bg	238	5	30	40	4	0	0	866.0	865764.7
4	Lyngbya limnetica	bg	43	5	30	40	2	0	0	39.1	78209.8
4	Anabaena sp.	bg	11	5	30	40	6	0	0	90.1	60021.5
4	Microcystis aeruginosa	bg	479	5	30	40	5	0	0	1155.5	1386592.7
4	flagellates	flag	2	5	30	40			0		
6	Aphanizomenon flos-aquae	bg	236	5	30	40	4	0	0	858.7	858489.3
6	Anabaena sp.	bg	62	5	30	40	6	0	0	507.6	338303.0
6	Microcystis aeruginosa	bg	170	5	30	40	5	0	0	410.1	492110.2
6	flagellates	flag	1	5	30	40			0		
6	Characium sp.	grn	2	5	30	40		0	0		
6	blue green small	bg	13	5	30	40		0	0		
6	Pediastrum sp.	grn	1	5	30	40		0	0		
8	Aphanizomenon flos-aquae	bg	186	5	30	40	4	0	0	676.8	676606.0
8	Lyngbya limnetica	bg	5	5	30	40	2	0	0	4.5	9094.2
8	Anabaena sp.	bg	105	5	30	40	6	0	0	859.6	572932.5
8	Microcystis aeruginosa	bg	295	5	30	40	5	0	0	711.6	853955.9
8	flagellates	flag	2	5	30	40			0		
10	Aphanizomenon flos-aquae	bg	106	5	30	40	4	0	0	385.7	385592.7
10	Lyngbya limnetica	bg	12	5	30	40	2	0	0	10.9	21826.0
10	Anabaena sp.	bg	95	5	30	40	6	0	0	777.7	518367.5
10	Microcystis aeruginosa	bg	435	5	30	40	5	0	0	1049.4	1259223.0
10	flagellates	flag	2	5	30	40			0		
10	eucaryote colony	eu	2	5	30	40		0	0		
10	Characium sp.	grn	2	5	30	40		0	0		
10	blue green small	bg	25	5	30	40		0	0		
18	Aphanizomenon flos-aquae	bg	7	20	30	40	4	0	0	6.4	6365.9
18	Lyngbya limnetica	bg	12	20	30	40	2	0	0	2.7	5456.5
18	Microcystis aeruginosa	bg	546	20	30	40	5	0	0	329.3	395135.5
18	flagellates	flag	1	20	30	40			0		
18	blue green small	bg	38	20	30	40		0	0		
18	Merismopedia sp.	bg	16	20	30	40		0	0		

M	Genus and species	Fam.	Ct.	Vol. filt. ml	Flds. ctd.	Mag.	Dim1 μm	Dim2 μm	Dim3 μm	Biovol. mm³/m³	Surf.area mm²/m³
	Sep. 27, 1976										
0	Aphanizomenon flos-aquae	bg	497	3	15	40	4	0	0	6027.9	6026401.2
0	Anabaena sp.	bg	25	3	15	40	6	0	0	682.2	454708.3
0	Microcystis aeruginosa	bg	8	3	15	40	4	0	0	32.9	49404.0
2	Aphanizomenon flos-aquae	bg	412	3	20	40	4	0	0	3747.7	3746796.7
2	Lyngbya limnetica	bg	2	3	20	40	2	0	0	4.5	9094.2
2	Anabaena sp.	bg	25	3	20	40	6	0	0	511.7	341031.3
2	Microcystis aeruginosa	bg	239	3	20	40	4	0	0	738.0	1106958.4
4	Aphanizomenon flos-aquae	bg	398	3	27	40	4	0	0	2681.8	2681095.1
4	Anabaena sp.	bg	76	3	27	40	6	0	0	1152.2	767951.9
4	Microcystis aeruginosa	bg	78	3	27	40	4	0	0	178.4	267605.0
6	Aphanizomenon flos-aquae	bg	239	3	30	40	4	0	0	1449.4	1449003.9
6	Anabaena sp.	bg	74	3	30	40	6	0	0	1009.7	672968.3
6	Microcystis aeruginosa	bg	388	3	30	40	4	0	0	798.7	1198047.0
8	Aphanizomenon flos-aquae	bg	260	3	30	40	4	0	0	1576.7	1576322.2
8	Lyngbya limnetica	bg	1	3	30	40	2	0	0	1.5	3031.4
8	Anabaena sp.	bg	74	3	30	40	6	0	0	1009.7	672968.3
8	Microcystis aeruginosa	bg	116	3	30	40	4	0	0	238.8	358179.0
8	flagellates	flag	2	3	30	40	3	12	0	6.9	11194.7
10	Aphanizomenon flos-aquae	bg	53	5	30	40	4	0	0	192.8	192796.3
10	Anabaena sp.	bg	55	5	30	40	6	0	0	450.3	300107.5
10	Microcystis aeruginosa	bg	259	5	30	40	4	0	0	319.9	479836.4
10	flagellates	flag	1	5	30	40	3	12	0	2.1	3358.4
10	Characium sp.	grn	1	5	30	40		0	0		
10	blue green colony	bg	3	5	30	40		0	0		
18	Aphanizomenon flos-aquae	bg	5	20	31	40	4	0	0	4.4	4400.4
18	Anabaena sp.	bg	20	20	31	40	6	0	0	39.6	26402.4
18	Microcystis aeruginosa	bg	280	20	31	40	4	0	0	83.7	125502.1
18	flagellates	flag	7	20	31	40	3	12	0	3.5	5687.6
18	blue green colony	bg	2	20	31	40		0	0		
18	Merismopedia sp.	bg	32	20	31	40		0	0		
	Oct. 4, 1976										
0	Aphanizomenon flos-aquae	bg	338	5	30	40	4	0	0	1229.8	1229531.3
0	Lyngbya limnetica	bg	3	5	30	40	2	0	0	2.7	5456.5
0	Anabaena sp.	bg	42	5	30	40	7	0	0	468.0	267368.5
0	Microcystis aeruginosa	bg	48	5	30	40	4	0	0	59.3	88927.2
0	flagellates	flag	2	5	30	40	3	12	0	4.2	6716.8
2	Aphanizomenon flos-aquae	bg	608	7	24	40	4	0	0	1975.2	1974733.3
2	Anabaena sp.	bg	58	7	24	40	7	0	0	577.1	329663.5
2	Microcystis aeruginosa	bg	90	7	24	40	4	0	0	99.2	148873.7
2	flagellates	flag	6	7	24	40	3	12	0	11.2	17991.5
2	Characium sp.	grn	1	7	24	40		0	0		
2	blue green colony	bg	1	7	24	40		0	0		
4	Aphanizomenon flos-aquae	bg	215	7	30	40	4	0	0	558.8	558641.7
4	Lyngbya limnetica	bg	4	7	30	40	2	0	0	2.6	5196.7
4	Anabaena sp.	bg	107	7	30	40	7	0	0	851.6	486537.9
4	Microcystis aeruginosa	bg	314	7	30	40	4	0	0	277.0	415522.9
4	flagellates	flag	23	7	30	40	3	12	0	34.2	55174.0
6	Aphanizomenon flos-aquae	bg	170	7	30	40	4	0	0	441.8	441716.7
6	Lyngbya limnetica	bg	8	7	30	40	2	0	0	5.2	10393.3
6	Anabaena sp.	bg	103	7	30	40	7	0	0	819.8	468349.6
6	Microcystis aeruginosa	bg	175	7	30	40	4	0	0	154.4	231581.3
6	flagellates	flag	17	7	30	40	3	12	0	25.3	40780.7
6	Characium sp.	grn	1	7	30	40		0	0		
8	Aphanizomenon flos-aquae	bg	457	10	19	40	4	0	0	1312.8	1312431.9
8	Lyngbya limnetica	bg	7	10	19	40	2	0	0	5.0	10051.4
8	Anabaena sp.	bg	59	10	19	40	7	0	0	519.0	296517.7
8	Microcystis aeruginosa	bg	75	10	19	40	4	0	0	73.1	109696.4
8	flagellates	flag	4	10	19	40	3	12	0	6.6	10605.5
8	Characium sp.	grn	2	10	19	40		0	0		
10	Aphanizomenon flos-aquae	bg	371	10	21	40	4	0	0	964.2	963981.7
10	Lyngbya limnetica	bg	5	10	21	40	2	0	0	3.2	6495.8
10	Anabaena sp.	bg	35	10	21	40	7	0	0	278.6	159147.9
10	Microcystis aeruginosa	bg	247	10	21	40	4	0	0	217.9	326860.4
10	flagellates	flag	3	10	21	40	3	12	0	4.5	7196.6
10	Characium sp.	grn	1	10	21	40		0	0		
18	Aphanizomenon flos-aquae	bg	77	15	30	40	4	0	0	93.4	93366.8
18	Microcystis aeruginosa	bg	180	15	30	40	4	0	0	74.1	111159.0
18	flagellates	flag	4	15	30	40	3	12	0	2.8	4477.9
18	Characium sp.	grn	1	15	30	40		0	0		
	Oct. 12, 1976										
0	Aphanizomenon flos-aquae	bg	551	3	14	40	4	0	0	7160.2	7158408.4
0	Microcystis aeruginosa	bg	86	3	14	40	4	0	0	379.4	569028.2
0	Characium sp.	grn	2	3	14	40		0	0		
2	Aphanizomenon flos-aquae	bg	565	3	25	40	4	0	0	4111.6	4110563.4
2	Microcystis aeruginosa	bg	7	3	14	40	4	0	0	30.9	46316.3
2	flagellates	flag	8	3	14	40	3	12	0	59.5	95954.7
4	Aphanizomenon flos-aquae	bg	341	3	30	40	4	0	0	2067.9	2067407.2
4	Anabaena sp.	bg	60	3	30	40	6	0	0	818.7	545650.0
4	Microcystis aeruginosa	bg	52	3	30	40	4	0	0	107.0	160563.0
4	flagellates	flag	9	3	30	40	3	12	0	31.3	50376.2
4	Characium sp.	grn	1	3	30	40		0	0		
4	Stephanodiscus sp.	diat	1	3	30	40			0		
6	Aphanizomenon flos-aquae	bg	353	7	14	40	4	0	0	1965.9	1965453.6
6	Lyngbya limnetica	bg	5	7	14	40	2	0	0	7.0	13919.6
6	Microcystis aeruginosa	bg	52	7	14	40	4	0	0	98.3	147455.8
6	flagellates	flag	15	7	14	40	3	12	0	47.9	77106.5
6	Characium sp.	grn	3	7	14	40		0	0		
8	Aphanizomenon flos-aquae	bg	180	10	14	40	4	0	0	701.7	701550.0
8	Lyngbya limnetica	bg	7	10	14	40	2	0	0	6.8	13641.3
8	Anabaena sp.	bg	31	10	14	40	6	0	0	271.9	181233.8
8	Microcystis aeruginosa	bg	118	10	14	40	4	0	0	156.2	234227.9
8	flagellates	flag	10	10	14	40	3	12	0	22.3	35983.0
8	Stephanodiscus sp.	diat	1	10	14	40			0		
10	Aphanizomenon flos-aquae	bg	66	15	14	40	4	0	0	171.5	171490.0
10	Lyngbya limnetica	bg	2	15	14	40	2	0	0	1.3	2598.3
10	Microcystis aeruginosa	bg	398	15	14	40	4	0	0	351.1	526681.9
10	Merismopedia sp.	bg	32	15	14	40		0	0		
10	Stephanodiscus sp.	diat	6	15	14	40			0		
18	Aphanizomenon flos-aquae	bg	39	20	14	40	4	0	0	76.0	76001.3
18	Lyngbya limnetica	bg	2	20	14	40	2	0	0	1.0	1948.8
18	Microcystis aeruginosa	bg	439	20	14	40	4	0	0	290.5	435703.6
18	flagellates	flag	2	20	14	40	3	12	0	2.2	3598.3
18	Characium sp.	grn	3	20	14	40		0	0		
18	blue green colony	bg	4	20	14	40		0	0		
18	Merismopedia sp.	bg	16	20	14	40		0	0		
18	Stephanodiscus sp.	diat	1	20	14	40			0		
	Oct. 18, 1976										
0	Aphanizomenon flos-aquae	bg	545	0	7	40	4	0	0	0.0	0.0
2	Aphanizomenon flos-aquae	bg	436	1	26	40	4	0	0	9152.4	9150130.9

M	Genus and species	Fam.	Ct.	Vol. filt. ml	Flds. ctd.	Mag.	Dim1 µm	Dim2 µm	Dim3 µm	Biovol. mm^3/m^3	Surf.area mm^2/m^3
2	flagellates	flag	1	1	26	40		0	0		
4	Aphanizomenon flos-aquae	bg	437	2	20	40	4	0	0	5962.7	5961226.3
4	Microcystis aeruginosa	bg	30	2	20	40	4	0	0	138.9	208423.1
4	flagellates	flag	1	2	20	40		0	0		
6	Aphanizomenon flos-aquae	bg	441	3	15	40	4	0	0	5348.7	5347370.1
6	Microcystis aeruginosa	bg	4	3	15	40	4	0	0	16.5	24702.0
6	flagellates	flag	3	3	15	40		0	0		
8	Aphanizomenon flos-aquae	bg	485	3	18	40	4	0	0	4901.9	4900745.4
8	Microcystis aeruginosa	bg	168	3	18	40	4	0	0	576.4	864570.0
8	flagellates	flag	1	3	18	40		0	0		
8	Characium sp.	grn	2	3	18	40		0	0		
10	Aphanizomenon flos-aquae	bg	164	5	30	40	4	0	0	596.7	596577.3
10	akinete	bg	6	5	30	40	6	0	0	25.0	25010.8
10	Microcystis aeruginosa	bg	96	5	30	40	4	0	0	118.6	177854.4
10	flagellates	flag	7	5	30	40		0	0		
10	Stephanodiscus sp.	diat	1	5	30	40			0		
18	Aphanizomenon flos-aquae	bg	19	15	32	40	4	0	0	21.6	21598.6
18	akinete	bg	8	15	32	40	6	0	0	10.4	10421.2
18	Microcystis aeruginosa	bg	72	15	32	40	4	0	0	27.8	41684.6
18	flagellates	flag	3	15	32	40		0	0		
18	Characium sp.	grn	1	15	32	40		0	0		
18	Stephanodiscus sp.	diat	8	15	32	40			0		
18	Staurastrum sp.	grn	1	15	32	40					
	Oct. 25, 1976										
0	Aphanizomenon flos-aquae	bg	797	2	30	40	4	0	0	7249.8	7248050.9
0	akinete	bg	15	2	30	40	6	0	0	156.3	156317.3
0	Microcystis aeruginosa	bg	13	2	30	40	4	0	0	40.1	60211.1
0	flagellates	flag	3	2	30	40		0	0		
0	Stephanodiscus sp.	diat	2	2	30	40			0		
2	Aphanizomenon flos-aquae	bg	599	2	25	40	4	0	0	6538.5	6536887.1
2	akinete	bg	35	2	25	40	6	0	0	437.7	437688.6
2	flagellates	flag	4	2	25	40		0	0		
2	Stephanodiscus sp.	diat	1	2	25	40			0		
4	Aphanizomenon flos-aquae	bg	383	4	2	40	2	0	0	6532.3	13061497.3
4	akinete	bg	15	2	2	40	6	0	0	2344.8	2344760.2
4	Microcystis aeruginosa	bg	7	2	2	40	4	0	0	324.2	486320.6
6	Aphanizomenon flos-aquae	bg	248	2	30	40	4	0	0	2255.9	2255353.3
6	akinete	bg	29	2	30	40		0	0		
6	Microcystis aeruginosa	bg	144	2	30	40	4	0	0	444.6	666954.0
6	Closterium sp.	desm	1	2	30	40		0	0		
8	Aphanizomenon flos-aquae	bg	473	5	8	40	4	0	0	6453.9	6452311.3
8	Anabaena sp.	bg	6	5	8	40		0	0		
8	akinete	bg	32	5	8	40	6	0	0	500.2	500215.5
8	Microcystis aeruginosa	bg	38	5	8	40	4	0	0	176.0	264002.6
10	Aphanizomenon flos-aquae	bg	304	2	20	40	4	0	0	4148.0	4146940.0
10	Anabaena sp.	bg	7	2	20	40		0	0		
10	akinete	bg	10	2	20	40	6	0	0	156.3	156317.3
	Nov. 8, 1976										
0	Aphanizomenon flos-aquae	bg	431	5	13	40	4	0	0	3619.0	3618079.3
0	flagellate large	flag	7	5	13	40	12	20	0	897.8	394844.2
0	Microcystis aeruginosa	bg	9	5	13	40	5	0	0	50.1	60122.1
0	flagellate small	flag	1	5	13	40	2	10	0	1.0	2332.4
0	akinete	bg	11	5	13	40	6	46	0	105.8	105814.8
2	Aphanizomenon flos-aquae	bg	289	5	30	40	4	0	0	1051.5	1051285.7
2	flagellate large	flag	3	5	30	40	12	20	0	166.7	73328.2

M	Genus and species	Fam.	Ct.	Vol. filt. ml	Flds. ctd.	Mag.	Dim1 µm	Dim2 µm	Dim3 µm	Biovol. mm^3/m^3	Surf.area mm^2/m^3
2	akinete	bg	40	5	30	40	6	46	0	166.7	166738.5
4	Aphanizomenon flos-aquae	bg	418	10	13	40	4	0	0	1754.9	1754474.6
4	Microcystis aeruginosa	bg	3	10	13	40	5	0	0	8.4	10020.3
4	flagellate small	flag	1	10	13	40	2	10	0	0.5	1166.2
4	akinete	bg	4	10	13	40	6	34	0	19.2	19239.1
4	akinete	bg	30	10	13	40	6	46	0	144.3	144292.9
4	Stephanodiscus sp.	diat	1	10	13	40	50		0		
6	Aphanizomenon flos-aquae	bg	425	10	23	40	4	0	0	1008.5	1008266.3
6	flagellate large	flag	16	10	23	40	12	20	0	580.0	255054.7
6	Schroederia setigera	grn	1	10	23	40	3	6	0	0.3	759.9
6	flagellate small	flag	1	10	23	40	2	10	0	0.3	659.2
6	akinete	bg	2	10	23	40	6	34	0	5.4	5437.1
6	akinete	bg	23	10	23	40	6	46	0	62.5	62526.9
6	Stephanodiscus sp.	diat	1	10	23	40	50		0		
8	Aphanizomenon flos-aquae	bg	240	20	30	40	4	0	0	218.3	218260.0
8	flagellate large	flag	14	20	30	40	12	20	0	194.5	85549.6
8	Schroederia setigera	grn	2	20	30	40	3	6	0	0.3	582.6
8	Microcystis aeruginosa	bg	17	20	30	40	5	0	0	10.3	12302.8
8	flagellate small	flag	4	20	30	40	2	10	0	0.4	1010.7
8	akinete	bg	9	20	30	40	6	34	0	9.4	9379.0
8	akinete	bg	29	20	30	40	6	46	0	30.2	30221.4
8	Stephanodiscus sp.	diat	1	20	30	40	50		0		
	Nov. 23, 1976										
0	Aphanizomenon flos-aquae	bg	552	2	17	40	4	0	0	8861.0	8858788.4
0	akinete	bg	39	2	17	40	6	0	0	717.2	717220.8
2	Aphanizomenon flos-aquae	bg	144	10	30	40	4	0	0	262.0	261912.0
2	flagellate large	flag	4	10	30	40	14	22	0	166.4	63427.1
2	flagellate small	flag	6	10	30	40	6	10	0	13.5	11463.3
2	akinete	bg	5	10	30	40	6	0	0	10.4	10421.2
4	Aphanizomenon flos-aquae	bg	257	15	30	40	4	0	0	311.7	311626.8
4	Lyngbya limnetica	bg	6	15	30	40	2	0	0	1.8	3637.7
4	flagellate large	flag	2	15	30	40	14	22	0	55.5	21142.4
4	flagellate small	flag	15	15	30	40	6	10	0	22.6	19105.5
4	akinete	bg	30	15	30	40	6	0	0	41.7	41684.6
6	Aphanizomenon flos-aquae	bg	32	15	30	40	4	0	0	38.8	38801.8
6	flagellate large	flag	2	15	30	40	14	22	0	55.5	21142.4
6	flagellate small	flag	17	15	30	40	6	10	0	25.6	21652.8
8	Aphanizomenon flos-aquae	bg	86	15	30	40	4	0	0	104.3	104279.8
8	flagellate small	flag	13	15	30	40	6	10	0	19.6	16558.1
8	Stephanodiscus sp.	diat	1	15	30	40	50	20	0	482.5	86843.0
	Dec. 14, 1976										
0	Aphanizomenon flos-aquae	bg	6	20	30	40	5	0	0	8.5	6820.6
0	flagellate small	flag	21	20	30	40	6	10	0	23.7	20060.7
0	flagellate large	flag	2	20	30	40	15	20	0	43.4	16002.3
5	Aphanizomenon flos-aquae	bg	14	20	30	40	5	0	0	19.9	15914.8
5	flagellate small	flag	31	20	30	40	6	10	0	35.0	29613.5
5	flagellate large	flag	1	20	30	40	15	20	0	21.7	8001.2
10	Aphanizomenon flos-aquae	bg	32	20	30	40	5	0	0	45.5	36376.7
10	flagellate small	flag	1	20	30	40	6	10	0	1.1	955.3
10	Schroederia setigera	grn	1	20	30	40	7	14	0	1.7	1585.9
10	Dinobryon sp.	chry	1	20	30	40	6	10	0	1.1	955.3
10	blue green colony	bg	1	20	30	40		0	0		
10	Microcystis aeruginosa	bg	53	20	30	40		0	0		
15	Aphanizomenon flos-aquae	bg	9	20	30	40	5	0	0	12.8	10230.9
15	flagellate small	flag	12	20	30	40	6	10	0	13.5	11463.3

M	Genus and species	Fam.	Ct.	Vol. filt. ml	Flds. ctd.	Mag.	Dim1 µm	Dim2 µm	Dim3 µm	Biovol. mm³/m³	Surf.area mm²/m³
15	flagellate large	flag	1	20	30	40	15	20	0	21.7	8001.2
20	Aphanizomenon flos-aquae	bg	6	20	30	40	5	0	0	8.5	6820.6
20	flagellate small	flag	3	30	30	40	6	10	0	2.3	1910.5
20	Planktosphaeria gelatinosa	grn	5	30	30	40	6	0	0	3.5	3473.7
	Mar. 8, 1977										
0	flagellate small	flag	47	25	30	40	6	10	0	42.4	35918.3
0	Schroederia setigera	grn	4	25	30	40	8	14	0	6.9	5974.6
0	eucaryote sphere	eu	24	25	30	40	5	0	0	11.6	13894.9
0	akinete	bg	5	25	30	40	6	0	0	4.2	4168.5
0	Dactylococcopsis sp.	bg	2	25	30	40		0	0		
5	flagellate small	flag	43	25	30	40	6	10	0	38.8	32861.4
5	flagellate medium	flag	1	25	30	40	20	24	0	26.2	17933.0
5	Schroederia setigera	grn	3	25	30	40	8	14	0	5.2	4481.0
5	eucaryote sphere	eu	5	25	30	40	5	0	0	2.4	2894.8
10	flagellate small	flag	152	50	30	40	6	10	0	68.6	58080.6
10	flagellate medium	flag	13	50	30	40	20	24	0	170.6	116564.2
10	flagellate large	flag	1	50	30	40	23	38	0	38.8	8915.2
10	Schroederia setigera	grn	5	50	30	40	8	14	0	4.3	3734.1
10	eucaryote sphere	eu	14	50	30	40	5	0	0	3.4	4052.7
10	Aphanizomenon flos-aquae	bg	25	50	30	40	4	0	0	9.1	9094.2
15	flagellate small	flag	8	75	30	40	6	10	0	2.4	2037.9
15	Schroederia setigera	grn	1	75	30	40	8	14	0	0.6	497.9
15	eucaryote sphere	eu	3	75	30	40	5	0	0	0.5	579.0
	Mar. 30, 1977										
0	eucaryote sphere	eu	3	25	30	40	8	0	0	5.9	4446.4
0	Diachros sp.	chry	339	25	30	40	10	6	0	1308.4	785060.4
5	flagellate small	flag	21	20	30	40	6	10	0	23.7	20060.7
5	flagellate medium	flag	3	20	30	40	16	22	0	55.6	36589.8
5	flagellate large	flag	1	20	30	40	20	42	0	81.1	20645.6
5	eucaryote sphere	eu	23	20	30	40	8	0	0	56.8	42611.0
5	Aphanizomenon flos-aquae	bg	7	20	30	40	4	0	0	6.4	6365.9
5	Diachros sp.	chry	149	20	30	40	10	6	0	718.9	431320.1
10	flagellate small	flag	37	30	30	40	6	10	0	27.8	23563.4
10	flagellate medium	flag	3	30	30	40	16	22	0	37.1	24393.2
10	flagellate large	flag	4	30	30	40	20	42	0	216.1	55054.8
10	Schroederia setigera	grn	1	30	30	40	6	16	0	0.9	989.3
10	eucaryote sphere	eu	7	30	30	40	8	0	0	11.5	8645.7
10	Diachros sp.	chry	358	30	30	40	10	6	0	1151.5	690884.1
10	Stephanodiscus sp.	diat	1	30	30	40	40	20	0	154.4	30877.5
15	flagellate small	flag	44	25	27	40	6	10	0	44.2	37361.8
15	flagellate medium	flag	6	25	27	40	16	22	0	98.8	65048.6
15	flagellate large	flag	4	25	27	40	20	42	0	288.2	73406.4
15	Schroederia setigera	grn	1	25	27	40	6	16	0	1.2	1319.1
15	eucaryote sphere	eu	3	25	27	40	8	0	0	6.6	4940.4
15	Diachros sp.	chry	273	25	27	40	10	6	0	1170.8	702463.1
15	Stephanodiscus sp.	diat	1	25	27	40	40	20	0	205.9	41170.0
15	eucaryote small	eu	5	25	27	40	4	20	0	6.9	8222.0
20	flagellate small	flag	33	25	30	40	6	10	0	29.8	25219.2
20	flagellate medium	flag	1	25	30	40	16	22	0	14.8	9757.3
20	Microcystis aeruginosa	bg	1	25	30	40		0	0		
20	Schroederia setigera	grn	2	25	30	40	6	16	0	2.2	2374.4
20	Diachros sp.	chry	70	25	30	40	10	6	0	270.2	162106.9
20	eucaryote small	eu	2	25	30	40	4	20	0	2.5	2959.9
	Apr. 7, 1977										
0	flagellate small	flag	9	20	30	40	8	12	0	22.2	17011.0
0	flagellate medium	flag	5	20	30	40	16	22	0	92.6	60983.1
0	flagellate large	flag	1	20	30	40	22	40	0	93.4	22079.8
0	eucaryote sphere	eu	1	20	30	40	8	0	0	2.5	1852.7
0	Diachros sp.	chry	548	20	30	40	9	7	0	1927.4	1284928.6
0	Stephanodiscus sp.	diat	1	20	30	40	60	30	0	781.6	104211.6
0	eucaryote small	eu	4	20	30	40	5	20	0	9.6	9328.9
0	Diachros sp.	chry	15	20	30	40	18	10	0	422.1	140685.6
0	eucaryote filament	eu	23	20	30	40	6	0	0	47.1	31374.9
5	flagellate small	flag	7	10	23	40	8	12	0	45.1	34515.1
5	flagellate medium	flag	6	10	23	40	16	22	0	290.0	190903.5
5	flagellate large	flag	1	10	23	40	22	40	0	243.7	57599.4
5	Diachros sp.	chry	302	10	23	40	9	7	0	2770.9	1847263.2
5	Diachros sp.	chry	15	10	23	40	18	10	0	1101.0	367005.9
15	flagellate small	flag	7	10	30	40	8	12	0	34.6	26461.6
15	flagellate medium	flag	3	10	30	40	16	22	0	111.2	73179.7
15	Diachros sp.	chry	278	10	30	40	9	7	0	1955.5	1303686.6
15	Diachros sp.	chry	7	10	30	40	18	10	0	393.9	131306.6
15	flagellate small	flag	7	15	31	40	8	12	0	22.3	17072.0
15	flagellate medium	flag	2	15	31	40	16	22	0	47.8	31475.1
15	flagellate large	flag	1	15	31	40	22	40	0	120.5	28490.0
15	Diachros sp.	chry	268	15	31	40	9	7	0	1216.2	810833.2
15	Diachros sp.	chry	2	15	31	40	18	10	0	72.6	24204.0
15	Microcystis aeruginosa	bg	1	15	31	40	4	0	0	0.4	597.6
15	Merismopedia sp.	bg	20	15	31	40		0	0		
20	flagellate small	flag	2	20	16	40	8	12	0	9.3	7087.9
20	flagellate medium	flag	10	20	16	40	16	22	0	347.4	228686.5
20	Diachros sp.	chry	337	20	16	40	9	7	0	2222.4	1481595.3
20	eucaryote small	eu	1	20	16	40	5	20	0	4.5	4372.9
20	Diachros sp.	chry	9	20	16	40	18	10	0	474.8	158271.3
	Apr. 12, 1977										
99	flagellate small	flag	1	15	13	40	8	10	0	6.7	5393.6
99	flagellate medium	flag	4	15	13	40	18	30	0	375.2	221249.2
99	eucaryote sphere	eu	5	15	13	40	18	0	0	432.9	144292.9
99	Diachros sp.	chry	306	15	13	40	10	8	0	4542.6	2725533.2
99	Diachros sp.	chry	14	15	13	40	16	14	0	851.3	319225.8
99	Scenedesmus sp.	grn	6	15	13	40	4	14	0	38.1	43548.4
	Apr. 18, 1977										
99	flagellate small	flag	1	10	22	40	6	10	0	3.1	2605.3
99	flagellate medium	flag	4	10	22	40	16	24	0	215.6	136372.8
99	flagellate large	flag	1	10	22	40	20	36	0	189.5	49348.0
99	eucaryote sphere	eu	10	10	22	40	8	0	0	67.4	50526.8
99	Diachros sp.	chry	298	10	22	40	8	6	0	2007.6	1505699.2
99	Diachros sp.	chry	5	10	22	40	16	14	0	269.5	101053.6
99	Scenedesmus sp.	grn	7	10	22	40			0		
	Apr. 26, 1977										
99	flagellate small	flag	33	15	25	40	5	14	0	52.5	51613.1
99	flagellate medium	flag	9	15	25	40	16	34	0	373.5	207970.9
99	eucaryote sphere	eu	2	15	25	40	8	0	0	7.9	5928.5
99	Diachros sp.	chry	49	15	25	40	8	6	0	193.7	145247.8
99	Diachros sp.	chry	6	15	25	40	14	14	0	127.1	54467.9
99	Dinobryon sp.	chry	3	15	25	40		0	0		

M	Genus and species	Fam.	Ct.	Vol. filt. ml	Flds. ctd.	Mag.	Dim1 μm	Dim2 μm	Dim3 μm	Biovol. mm³/m³	Surf.area mm²/m³
99	Microcystis aeruginosa	bg	404	15	25	40	4	0	0	199.6	299388.2
	May 3, 1977										
99	flagellate small	flag	71	30	30	40	6	14	0	69.9	59784.1
99	flagellate medium	flag	8	30	30	40	24	34	0	340.9	206793.7
99	Diachros sp.	chry	4	30	30	40	29	14	0	313.8	64919.9
99	Diachros sp.	chry	19	30	30	40	8	6	0	31.3	23466.9
99	Coelosphaerium sp.	bg	1	30	30	40	70	0	0	1103.2	94562.3
99	Schroederia setigera	grn	1	30	30	40	6	20	0	1.2	1208.9
99	eucaryote filament	eu	66	30	30	40	2	0	0	10.0	20007.2
99	eucaryote filament	eu	23	30	30	40	5	0	0	21.8	17430.5
	May 16, 1977										
99	flagellate small	flag	12	25	30	40	5	10	0	7.2	6778.5
99	flagellate medium	flag	4	25	30	40	14	18	0	37.8	26409.8
99	flagellate large	flag	1	25	30	40	20	46	0	71.0	17892.3
99	Schroederia setigera	grn	32	25	30	40	6	28	0	62.2	63662.2
99	eucaryote filament	eu	12	25	30	40	2	0	0	2.2	4365.2
99	eucaryote filament	eu	25	25	30	40	3	0	0	10.2	13641.3
99	eucaryote colony	eu	76	25	30	40	6	0	0		
99	Dinobryon sp.	chry	19	25	30	40		0	0		
99	Characium sp.	grn	68	25	30	40	2	16	0	8.9	23152.5
	May 23, 1977										
99	flagellate small	flag	10	25	30	40	6	10	0	9.0	7642.2
99	flagellate medium	flag	2	25	30	40	12	24	0	16.7	10397.9
99	Schroederia setigera	grn	6	25	30	40	8	22	0	16.3	13010.8
99	eucaryote filament	eu	61	25	30	40	3	0	0	25.0	33284.7
99	eucaryote colony	eu	80	25	30	40	6	0	0		
99	Dinobryon sp.	chry	5	25	30	40		0	0		
99	Characium sp.	grn	46	25	30	40	3	16	0	14.0	23307.5
99	Anabaena sp.	bg	93	25	30	40	6	0	0	152.3	101490.9
	June 2, 1977										
99	flagellate small	flag	12	10	20	40	7	10	0	57.4	46600.8
99	flagellate medium	flag	2	10	20	40	12	20	0	54.2	35431.9
99	flagellate large	flag	1	10	20	40	23	48	0	367.5	81468.1
99	Aphanizomenon flos-aquae	bg	34	10	20	40	4	0	0	92.8	92760.5
99	Lyngbya limnetica	bg	10	10	20	40	1	0	0	1.7	6820.6
99	Planktosphaeria gelatinosa	grn	327	10	20	40	8	0	0	2423.3	1817449.7
99	eucaryote sphere	eu	6	10	20	40	10	0	0	86.8	52105.8
99	Microcystis aeruginosa	bg	4	10	20	40	4	0	0	3.7	5558.0
99	Characium sp.	grn	13	10	20	40	4	16	0	27.1	33489.9
99	Anabaena sp.	bg	130	10	20	40	3	0	0	199.6	266004.4
	June 7, 1977										
99	flagellate small	flag	3	15	24	40	6	10	0	5.6	4776.4
99	flagellate medium	flag	1	15	24	40	12	20	0	15.1	9842.2
99	Anabaena sp.	bg	359	15	24	40	5	0	0	850.4	680167.9
99	Aphanizomenon flos-aquae	bg	107	15	24	40	4	0	0	162.2	162179.3
99	Planktosphaeria gelatinosa	grn	5	15	24	40	7	0	0	13.8	11820.3
99	Characium sp.	grn	3	15	24	40	4	14	0	3.1	3724.0
	June 13, 1977										
99	flagellate small	flag	1	7	20	40	6	10	0	4.8	4094.0
99	Anabaena sp.	bg	223	7	20	40	5	0	0	1358.4	1086428.2
99	Aphanizomenon flos-aquae	bg	125	7	20	40	4	0	0	487.3	487187.5
99	Microcystis aeruginosa	bg	580	7	20	40	4	0	0	767.5	1151289.7
99	Characium sp.	grn	1	7	20	40	4	14	0	2.6	3192.0
99	Lyngbya limnetica	bg	6	7	20	40	2	0	0	5.8	11692.5
	June 20, 1977										
99	flagellate small	flag	5	3	13	40	5	8	0	48.7	44934.5
99	flagellate medium	flag	1	3	13	40	12	20	0	138.9	90851.1
99	Anabaena sp.	bg	182	3	13	40	5	0	0	3979.7	3182958.4
99	Aphanizomenon flos-aquae	bg	253	3	13	40	4	0	0	3540.6	3539729.5
99	Schroederia setigera	grn	1	3	13	40	4	14	0	8.3	12968.7
99	Lyngbya limnetica	bg	18	3	13	40	2	0	0	63.0	125919.2
99	eucaryote filament	eu	16	3	13	40	2	0	0	56.0	111928.2
	June 27, 1977										
99	flagellate small	flag	1	2	11	40	5	8	0	17.3	15931.3
99	Anabaena sp.	bg	203	2	11	40	5	0	0	7868.9	6293576.9
99	Aphanizomenon flos-aquae	bg	367	2	11	40	4	0	0	9104.7	9102434.1
99	Microcystis aeruginosa	bg	9	2	11	40	4	0	0	75.8	113685.3
99	Schroederia setigera	grn	2	2	11	40	6	20	0	94.7	98909.2
99	Characium sp.	grn	2	2	11	40	2	16	0	8.9	23214.4
99	Lyngbya limnetica	bg	29	2	11	40	2	0	0	179.9	359633.0
99	eucaryote sphere	eu	3	2	11	40	7	0	0	135.4	116053.8
99	Staurastrum sp.	grn	1	2	11	40	16	30	50	6031.2	2344898.7
99	eucaryote small	eu	13	2	11	40	2	6	0	41.1	100926.6
	July 5, 1977										
99	flagellate small	flag	1	1	15	40	6	10	0	45.2	38210.9
99	Anabaena sp.	bg	29	1	15	40	6	0	0	2374.2	1582385.0
99	Aphanizomenon flos-aquae	bg	316	1	15	40	4	0	0	11497.8	11495026.8
99	Planktosphaeria gelatinosa	grn	3	1	15	40	4	0	0	37.1	55579.5
99	Lyngbya limnetica	bg	37	1	15	40	2	0	0	336.6	672968.3
99	eucaryote small	eu	4	1	15	40	6	8	0	222.3	204829.5
	July 11, 1977										
99	flagellate small	flag	8	2	22	40	3	7	0	20.1	27341.3
99	Anabaena sp.	bg	9	2	22	40	6	0	0	251.2	167415.3
99	Aphanizomenon flos-aquae	bg	375	2	22	40	4	0	0	4651.6	4650426.1
99	Lyngbya limnetica	bg	44	2	22	40	2	0	0	136.4	272825.0
99	Microcystis aeruginosa	bg	41	2	22	40	4	0	0	172.6	258949.9
99	Staurastrum sp.	grn	1	2	22	40	12	28	28	1182.1	608642.7
99	Planktosphaeria gelatinosa	grn	3	2	22	40	6	0	0	42.6	42632.0
	July 18, 1977										
99	Anabaena sp.	bg	81	2	8	40	6	0	0	6216.8	4143529.8
99	Aphanizomenon flos-aquae	bg	313	2	8	40	4	0	0	10676.9	10674278.4
99	Microcystis aeruginosa	bg	1	2	8	40	4	0	0	11.6	17368.6
99	eucaryote small	eu	1	2	8	40	4	6	0	17.4	23383.5
	July 26, 1977										
99	flagellate small	flag	3	2	30	40	5	10	0	22.6	21182.8
99	flagellate medium	flag	1	2	30	40	10	16	0	50.7	35427.5
99	Anabaena sp.	bg	268	2	30	40	7	0	0	7465.9	4265164.3
99	Aphanizomenon flos-aquae	bg	85	2	30	40	4	0	0	773.2	773004.2
99	Lyngbya limnetica	bg	39	2	30	40	2	0	0	88.7	177336.3
99	Microcystis aeruginosa	bg	257	2	30	40	4	0	0	793.6	1190327.6
99	Planktosphaeria gelatinosa	grn	3	2	30	40	5	0	0	18.1	21710.7
99	Schroederia setigera	grn	1	2	30	40	2	17	0	1.6	4955.0

M	Genus and species	Fam.	Ct.	Vol. filt. ml	Flds. ctd.	Mag.	Dim1 µm	Dim2 µm	Dim3 µm	Biovol. mm³/m³	Surf.area mm²/m³
	Aug. 1, 1977										
99	flagellate small	flag	6	3	24	40	4	6	0	15.4	15907.0
99	flagellate small	flag	3	3	24	40	4	8	0	9.6	10332.6
99	Anabaena sp.	bg	126	3	24	40	7	0	0	2925.1	1671053.1
99	Aphanizomenon flos-aquae	bg	104	3	24	40	4	0	0	788.4	788161.1
99	Lyngbya limnetica	bg	58	3	24	40	2	0	0	109.9	219775.7
99	Schroederia setigera	grn	1	3	24	40	2	10	0	0.8	2460.1
99	Microcystis aeruginosa	bg	304	3	24	40	6	0	0	2640.0	2640026.3
99	Microcystis aeruginosa	bg	146	3	24	40	2	0	0	47.0	140878.6
99	eucaryote small	eu	75	3	24	40	2	6	0	72.4	177915.5
99	blue green colony	bg	7	3	24	40		0	0		
	Aug. 8, 1977										
99	flagellate small	flag	7	3	21	40	3	8	0	13.8	19453.3
99	flagellate small	flag	1	3	21	40	4	8	0	3.7	3936.2
99	Anabaena sp.	bg	83	3	21	40	6	0	0	1617.9	1078308.3
99	Aphanizomenon flos-aquae	bg	94	3	21	40	4	0	0	814.3	814144.5
99	Lyngbya limnetica	bg	37	3	21	40	2	0	0	80.1	160230.6
99	Schroederia setigera	grn	1	3	21	40	2	10	0	0.9	2811.5
99	Microcystis aeruginosa	bg	312	3	21	40	5	0	0	1792.0	2150397.3
99	Microcystis aeruginosa	bg	139	3	21	40	3	0	0	172.4	344890.6
99	eucaryote small	eu	5	3	21	40	2	4	0	3.7	9424.3
99	blue green colony	bg	1	3	21	40		0	0		
99	Gloeotrichia sp.	bg	6	3	21	40	2	0	0	2.2	6616.6
	Aug. 15, 1977										
99	flagellate small	flag	4	3	23	40	4	8	0	13.4	14375.8
99	flagellate small	flag	2	3	23	40	5	6	0	8.9	8633.0
99	Anabaena sp.	bg	117	3	23	40	7	0	0	2834.2	1619157.1
99	Aphanizomenon flos-aquae	bg	33	3	23	40	4	0	0	261.0	260963.0
99	Lyngbya limnetica	bg	9	3	23	40	2	0	0	17.8	35585.9
99	Microcystis aeruginosa	bg	514	3	23	40	6	0	0	4657.8	4657803.8
99	Microcystis aeruginosa	bg	51	3	23	40	3	0	0	57.8	115538.9
99	eucaryote small	eu	23	3	23	40	2	4	0	15.4	39581.9
99	blue green colony	bg	3	3	23	40		0	0		
99	eucaryote sphere	eu	1	3	23	40	20	0	0	335.6	100687.5
99	Diachros sp.	chry	3	3	23	40	6	0	0	27.2	27185.6
	Aug. 22, 1977										
99	Anabaena sp.	bg	26	4	7	40	6	0	0	1140.3	760012.5
99	Aphanizomenon flos-aquae	bg	26	4	7	40	4	0	0	506.8	506675.0
99	Lyngbya limnetica	bg	33	4	7	40	2	0	0	160.8	321543.8
99	Microcystis aeruginosa	bg	351	4	7	40	6	0	0	7838.2	7838198.3
99	eucaryote sphere	eu	4	4	7	40	18	0	0	2411.8	803917.8
	Aug. 29, 1977										
99	flagellate small	flag	3	12	10	40	4	10	0	6.9	7808.2
99	flagellate small	flag	3	12	10	40	8	16	0	46.3	34061.4
99	Anabaena sp.	bg	143	12	10	40		0	0		
99	Aphanizomenon flos-aquae	bg	100	12	10	40	4	0	0	454.8	454708.3
99	Lyngbya limnetica	bg	261	12	10	40	3	0	0	667.7	890091.6
99	Coelosphaerium sp.	bg	35	12	10	40		0	0		
99	Microcystis aeruginosa	bg	968	12	10	40	6	0	0	5043.8	5043839.6
99	Microcystis aeruginosa	bg	83	12	10	40	3	0	0	54.1	108119.5
99	eucaryote sphere	eu	11	12	10	40	8	0	0	135.9	101895.8
99	Actinastrum gracilimum	grn	13	12	10	40	5	0	0	0.0	29946.6
	Sep. 7, 1977										
99	flagellate small	flag	11	10	19	40	8	16	0	107.3	78879.1
99	Anabaena sp.	bg	54	10	19	40		0	0		
99	Aphanizomenon flos-aquae	bg	134	10	19	40	4	0	0	384.9	384826.8
99	Lyngbya limnetica	bg	334	10	19	40		0	0		
99	Microcystis aeruginosa	bg	2402	10	19	40	6	0	0	7904.7	7904721.3
99	Microcystis aeruginosa	bg	197	10	19	40	3	0	0	81.0	162076.4
99	eucaryote sphere	eu	8	10	19	40	20	0	0	975.1	292523.7
99	Actinastrum gracilimum	grn	6	10	19	40	5	0	0	0.0	8729.4
99	flagellate small	flag	7	10	19	40	4	0	0	1.7	9136.2
	Sep. 14, 1977										
99	Aphanizomenon flos-aquae	bg	63	10	17	40	4	0	0	202.3	202211.5
99	Anabaena sp.	bg	75	10	17	40	6	0	0	541.8	361091.9
99	Lyngbya limnetica	bg	66	10	17	40	2	0	0	53.0	105920.3
99	Microcystis aeruginosa	bg	37	10	17	40	6	0	0	136.1	136088.0
99	Microcystis aeruginosa	bg	61	10	17	40	3	0	0	28.0	56090.3
99	Coelosphaerium sp.	bg	18	10	17	40	4	5	0	19.6	29424.4
99	flagellate large	flag	1	10	17	40	5	10	0	4.3	4365.6
99	eucaryote sphere	eu	6	10	17	40	6	0	0	22.1	22068.3
99	eucaryote small	eu	312	10	17	40	3	5	0	239.1	420558.9
99	Actinastrum Hantzschii	grn	7	10	17	40		0	0		
99	Stephanodiscus sp.	diat	4	10	17	40	40	15	0	2452.0	572141.9
	Sep. 21, 1977										
99	Aphanizomenon flos-aquae	bg	24	5	15	40	4	0	0	174.7	174608.0
99	Anabaena sp.	bg	12	5	15	40	6	0	0	196.5	130956.0
99	Lyngbya limnetica	bg	56	5	15	40	2	0	0	101.9	203709.3
99	Microcystis aeruginosa	bg	23	5	15	40	5	0	0	111.0	133159.2
99	Microcystis aeruginosa	bg	92	5	15	40	3	0	0	95.9	191749.3
99	blue green colony	bg	20	5	15	40	6	0	0		
99	eucaryote sphere	eu	4	5	15	40	11	0	0	205.5	112085.3
99	eucaryote small	eu	347	5	15	40	2		0		
99	Actinastrum Hantzschii	grn	38	5	15	40	2	0	0	0.0	22409.3
99	Stephanodiscus sp.	diat	3	5	15	40	31	12	0	2002.9	592269.0
99	blue green colony	bg	67	5	15	40	3	5	0		
	Sep. 28, 1977										
99	Anabaena sp.	bg	35	5	15	40	7	0	0	780.0	445614.2
99	Aphanizomenon flos-aquae	bg	17	5	15	40	4	0	0	123.7	123680.7
99	Lyngbya limnetica	bg	45	5	15	40	2	0	0	81.9	163695.0
99	Microcystis aeruginosa	bg	2	5	15	40	5	0	0	9.6	11579.1
99	Microcystis aeruginosa	bg	72	5	15	40	3	0	0	75.0	150064.7
99	flagellate large	flag	4	5	15	40	13	18	0	469.6	197956.2
99	flagellate small	flag	1	5	15	40	6	9	0	8.3	7116.0
99	Dictyosphaerium pulchellum	grn	118	5	15	40	5	0	0	569.3	683164.7
99	eucaryote small	eu	296	5	15	40	3	5	0	514.1	904381.3
99	eucaryote filament	eu	4	5	15	40		0	0		
99	eucaryote sphere	eu	9	5	15	40	4	0	0	22.2	33347.7
	Oct. 5, 1977										
99	Anabaena sp.	bg	9	5	30	40	6	0	0	73.7	49108.5
99	Aphanizomenon flos-aquae	bg	48	5	30	40	4	0	0	174.7	174608.0
99	Lyngbya limnetica	bg	160	5	30	40	2	0	0	145.5	291013.3
99	Microcystis aeruginosa	bg	12	5	30	40	6	0	0	50.0	50021.6
99	Microcystis aeruginosa	bg	238	5	30	40	4	0	0	294.0	440930.7

M	Genus and species	Fam.	Ct.	Vol. filt. ml	Flds. ctd.	Mag.	Dim1 µm	Dim2 µm	Dim3 µm	Biovol. mm³/m³	Surf.area mm²/m³
99	flagellate large	flag	3	5	30	40	11	16	0	112.1	55228.9
99	flagellate small	flag	13	5	30	40	4	5	0	14.0	14760.0
99	Dictyosphaerium pulchellum	grn	34	5	30	40		0	0		
99	Stephanodiscus sp.	diat	2	5	30	40	30	12	0	625.3	187580.8
99	eucaryote small	eu	144	5	30	40			0		
99	Schroederia setigera	grn	1	5	30	40			0		
99	blue green colony	bg	2	5	30	40		0	0		
99	Actinastrum Hantzschii	grn	11	5	30	40	4	18	0	116.8	129737.8
99	eucaryote sphere	eu	5	5	30	40	5	0	0	12.1	14473.8
	Oct. 12, 1977										
99	Anabaena sp.	bg	20	6	30	40	6	0	0	136.4	90941.7
99	Aphanizomenon flos-aquae	bg	16	6	30	40	4	0	0	48.5	48502.2
99	Lyngbya limnetica	bg	166	6	30	40	2	0	0	125.8	251605.3
99	Microcystis aeruginosa	bg	232	6	30	40	5	0	0	466.4	559654.7
99	Microcystis aeruginosa	bg	109	6	30	40	2	0	0	14.0	42070.6
99	flagellate small	flag	8	6	30	40	12	18	0	222.3	150459.6
99	flagellate small	flag	26	6	30	40	4	6	0	26.8	27572.1
99	Dictyosphaerium pulchellum	grn	72	6	30	40	5	0	0	144.7	173685.9
99	Stephanodiscus sp.	diat	1	6	30	40	18	8	0	62.5	29526.6
99	eucaryote small	eu	203	6	30	40	3	5	0	146.9	258431.0
99	Scenedesmus sp.	grn	8	6	30	40	2	12	0	11.8	25554.4
99	Ankistrodesmus fractus	grn	1	6	30	40	2	40	0	1.3	3864.5
99	Oocystis sp.	grn	4	6	30	40	5	0	0	8.0	9649.2
99	blue green colony	bg	2	6	30	40		0	0		
99	eucaryote filament	eu	9	6	30	40	8	0	0	109.2	54565.0
99	eucaryote sphere	eu	6	6	30	40	9	0	0	70.3	46895.2
99	Schroederia setigera	grn	1	6	30	40		18	0		
	Oct. 19, 1977										
99	Aphanizomenon flos-aquae	bg	39	7	28	40	4	0	0	108.6	108573.2
99	Actinastrum Hantzschii	grn	291	7	28	40	2	0	0	0.0	65666.2
99	Microcystis aeruginosa	bg	449	7	28	40	4	0	0	424.4	636612.1
99	Microcystis aeruginosa	bg	20	7	28	40	2	0	0	2.4	7089.2
99	flagellate large	flag	2	7	28	40	10	18	0	53.2	27695.3
99	flagellate small	flag	25	7	28	40	3	5	0	10.8	13292.3
99	Dictyosphaerium pulchellum	grn	82	7	28	40	5	0	0	151.4	181661.3
99	eucaryote small	eu	113	7	28	40	2	4	0	26.7	68460.5
99	Kirchneriella sp.	grn	1	7	28	40		0	0		
99	Closterium sp.	desm	1	7	28	40	5	70	0	12.9	15547.2
99	Schroederia setigera	grn	2	7	28	40		0	0		
99			5	7	28	40	30	0	0		
99	eucaryote filament	eu	4	7	28	40	5	0	0	17.4	13919.6
99	eucaryote sphere	eu	7	7	28	40	10	0	0	103.4	62030.7
	Oct. 26, 1977										
99	Anabaena sp.	bg	16	8	30	40	6	0	0	81.9	54565.0
99	Aphanizomenon flos-aquae	bg	15	8	30	40	4	0	0	34.1	34103.1
99	Lyngbya limnetica	bg	244	8	30	40	2	0	0	138.7	277372.1
99	Microcystis aeruginosa	bg	80	8	30	40	4	0	0	61.8	92632.5
99	Microcystis aeruginosa	bg	41	8	30	40	2	0	0	4.0	11868.5
99	flagellate large	flag	4	8	30	40	13	20	0	163.1	67207.0
99	flagellate small	flag	45	8	30	40	5	6	0	57.7	55844.8
99	Dictyosphaerium pulchellum	grn	165	8	30	40		0	0		
99	Stephanodiscus sp.	diat	3	8	30	40	30	12	0	586.2	175857.0
99	Schroederia setigera	grn	2	8	30	40			0		
99	eucaryote sphere	eu	2	8	30	40	10	0	0	24.1	14473.8

M	Genus and species	Fam.	Ct.	Vol. filt. ml	Flds. ctd.	Mag.	Dim1 µm	Dim2 µm	Dim3 µm	Biovol. mm³/m³	Surf.area mm²/m³
	Nov. 2, 1977										
99	Microcystis aeruginosa	bg	506	10	28	40	4	0	0	334.8	502200.5
99	Microcystis aeruginosa	bg	46	10	28	40	2	0	0	3.8	11413.6
99	Lyngbya limnetica	bg	240	10	28	40	2	0	0	117.0	233850.0
99	flagellate large	flag	13	10	28	40	13	20	0	454.3	187219.6
99	flagellate small	flag	62	10	28	40	4	6	0	41.0	42267.2
99	Dictyosphaerium pulchellum	grn	53	10	28	40	3	5	0	14.8	29588.6
99	Schroederia setigera	grn	4	10	28	40		12	0		
99	Actinastrum Hantzschii	grn	14	10	28	40	2	17	0	18.8	39805.9
99	Stephanodiscus sp.	diat	1	10	28	40	30		0		
99	Stephanodiscus sp.	diat	1	10	28	40	60		0		
99	eucaryote sphere	eu	3	10	28	40	12	0	0	53.6	26797.3
	Nov. 14, 1977										
99	Microcystis aeruginosa	bg	18	15	25	40	4	0	0	8.9	13339.1
99	Microcystis aeruginosa	bg	32	15	25	40	2	0	0	2.0	5928.5
99	Lyngbya limnetica	bg	280	15	25	40	2	0	0	101.9	203709.3
99	Aphanizomenon flos-aquae	bg	37	15	25	40	5	0	0	84.1	67296.8
99	Anabaena sp.	bg	9	15	25	40	7	0	0	40.1	22917.3
99	flagellate large	flag	3	15	25	40	13	18	0	70.4	29693.4
99	flagellate small	flag	17	15	25	40	6	9	0	28.3	24194.4
99	Dictyosphaerium pulchellum	grn	51	15	25	40	3	5	0	10.6	21259.2
99	Schroederia setigera	grn	3	15	25	40		16	0		
99	Stephanodiscus sp.	diat	1	15	25	40	40	16	0	296.4	66695.4
99	eucaryote sphere	eu	3	15	25	40	10	0	0	23.2	13894.9
	Nov. 29, 1977										
99	Microcystis aeruginosa	bg	59	15	26	40	5	0	0	54.7	65688.9
99	Microcystis aeruginosa	bg	21	15	26	40	2	0	0	1.2	3740.9
99	Lyngbya limnetica	bg	219	15	26	40	2	0	0	76.6	153201.7
99	Aphanizomenon flos-aquae	bg	101	15	26	40	5	0	0	220.8	176636.7
99	flagellate large	flag	4	15	26	40	14	20	0	116.4	45234.8
99	flagellate small	flag	13	15	26	40	5	8	0	12.7	11683.0
99	Dictyosphaerium pulchellum	grn	118	15	26	40	5	0	0	109.5	131377.8
99	Schroederia setigera	grn	2	15	26	40	5	26	0	4.8	5895.6
99	Schroederia setigera	grn	2	15	26	40	3	16	0	1.1	2174.9
99	Stephanodiscus sp.	diat	1	15	26	40	35	14	0	190.9	49099.7
99	eucaryote sphere	eu	1	15	26	40	10	0	0	7.4	4453.5
	Mar. 16, 1978										
0	Diachros sp.	chry	23	15	20	40	8	7	0	104.6	78442.9
0	flagellate small	flag	4	15	20	40	7	10	0	11.8	9532.0
2	Diachros sp.	chry	31	15	20	40	8	7	0	141.0	105727.4
2	flagellate small	flag	4	15	20	40	7	10	0	11.8	9532.0
4	Diachros sp.	chry	24	15	20	40	8	7	0	109.1	81853.4
4	flagellate small	flag	6	15	20	40	7	10	0	17.6	14298.0
8	Diachros sp.	chry	23	15	20	40	8	7	0	104.6	78442.9
8	flagellate small	flag	8	15	20	40	7	10	0	23.5	19064.0
	Mar. 23, 1978										
0	Diachros sp.	chry	33	15	15	40			0		
2	Diachros sp.	chry	43	10	15	40			0		
2	dinoflagellate	dino	1	10	15	40	15	18	0	71.9	27235.3
2	flagellate small	flag	1	10	15	40	7	10	0	5.9	4766.0
4	Diachros sp.	chry	108	15	20	40			0		
4	flagellate small	flag	5	15	20	40	7	10	0	14.7	11915.0

M	Genus and species	Fam.	Ct.	Vol. filt. ml	Flds. ctd.	Mag.	Dim1 μm	Dim2 μm	Dim3 μm	Biovol. mm³/m³	Surf.area mm²/m³
8	Diachros sp.	chry	60	15	15	40			0		
8	flagellate small	flag	4	15	15	40	6	9	0	10.2	8733.3
	Mar. 31, 1978										
0	Diachros sp.	chry	347	10	15	40			0		
0	flagellate small	flag	5	10	0	40	5	9	0	0.0	0.0
2	Diachros sp.	chry	534	10	20	40			0		
2	flagellate small	flag	11	10	20	40	5	9	0	21.1	19528.2
4	Diachros sp.	chry	477	10	20	40			0		
4	flagellate small	flag	8	10	20	40	5	9	0	15.3	14202.3
8	Diachros sp.	chry	374	10	15	40			0		
8	flagellate small	flag	7	10	15	40	5	9	0	17.9	16569.4
16	Diachros sp.	chry	251	10	20	40			0		
16	flagellate small	flag	3	10	20	40	5	9	0	5.7	5325.9
	Apr. 14, 1978										
0	Diachros sp.	chry	828	10	15	40			0		
0	flagellate medium	flag	312	10	15	40	15	24	0	19640.0	12274177.5
0	flagellate small	flag	25	10	15	40	5	10	0	69.4	64992.7
0	Cyclotella sp.	diat	93	10	15	40	20	0	0	0.0	1982388.2
5	Diachros sp.	chry	681	10	15	40			0		
5	flagellate medium	flag	4	10	15	40	15	24	0	251.8	157361.3
5	flagellate small	flag	12	10	15	40	5	10	0	33.3	31196.5
10	Diachros sp.	chry	689	10	15	40			0		
10	flagellate medium	flag	5	10	15	40	12	24	0	191.8	119635.2
10	flagellate small	flag	20	10	15	40	5	10	0	55.5	51994.2
15	Diachros sp.	chry	632	10	12	40			0		
15	flagellate medium	flag	1	10	12	40	15	24	0	78.7	49175.4
15	flagellate small	flag	8	10	12	40	5	10	0	27.8	25997.1
20	Diachros sp.	chry	529	10	10	40			0		
20	dinoflagellate	dino	2	10	10	40	15	18	0	215.8	81705.9
20	flagellate small	flag	6	10	10	40	5	10	0	25.0	23397.4
	May 3, 1978										
0	flagellate medium	flag	22			40	16	20	0		
0	flagellate small	flag	283			40	6	10	0		
0	flagellate medium	flag	7			40	18	0	0		
0	dinoflagellate	dino	1			40	12	14	0		
0	dinoflagellate	dino	2			40	18	20	0		
0	Diachros sp.	chry	17			40			0		
0	Cyclotella sp.	diat	1			40	15	0	0		
0	Oocystis sp.	grn	4			40	8	0	0		
0	Oocystis sp.	grn	4			40	11	0	0		
0	Scenedesmus sp.	grn	2			40	5	10	0		
2	flagellate medium	flag	30			40	16	20	0		
2	flagellate small	flag	273			40	6	10	0		
2	flagellate medium	flag	8			40	18	18	0		
2	Diachros sp.	chry	39			40			0		
2	dinoflagellate	dino	1			40	14	16	0		
2	dinoflagellate	dino	1			40	15	18	0		
2	Dictyosphaerium pulchellum	grn	27			40	6	0	0		
4	flagellate medium	flag	20			40	16	20	0		
4	flagellate small	flag	241			40	6	10	0		
4	Diachros sp.	chry	26			40			0		
4	flagellate medium	flag	7			40	20	20	0		
4	dinoflagellate	dino	2			40	18	20	0		
4	Oocystis sp.	grn	4			40	7	0	0		

M	Genus and species	Fam.	Ct.	Vol. filt. ml	Flds. ctd.	Mag.	Dim1 μm	Dim2 μm	Dim3 μm	Biovol. mm³/m³	Surf.area mm²/m³
8	flagellate medium	flag	17			40	16	20	0		
8	flagellate small	flag	193			40	6	10	0		
8	flagellate medium	flag	6			40			0		
8	Diachros sp.	chry	15			40			0		
8	dinoflagellate	dino	3			40	14	16	0		
16	flagellate medium	flag	6			40	16	20	0		
16	flagellate medium	flag	4			40	20	20	0		
16	flagellate small	flag	58			40	6	10	0		
16	Diachros sp.	chry	118			40			0		
16	dinoflagellate	dino	1			40	16	18	0		
	May 23, 1978										
99	flagellate medium	flag	8	8	20	40	12	20	0	249.4	163067.4
99	flagellate medium	flag	17	8	20	40	16	16	0	869.7	688506.9
99	flagellate small	flag	10	8	20	40	6	9	0	36.0	30703.0
99	flagellate small	flag	403	8	20	40	4	5	0	375.8	394839.4
99	Cyclotella sp.	diat	11	8	20	40	14	10	0	538.6	261587.3
99	Diachros sp.	chry	36	8	20	40	6	3	0	129.5	129494.7
99	flagellate large	flag	2	5	20	40	20	28	0	596.8	163112.8
99	flagellate medium	flag	3	5	20	40	14	14	0	164.5	133171.7
99	flagellate medium	flag	4	5	20	40	16	16	0	327.4	259202.6
99	flagellate medium	flag	11	5	20	40	14	18	0	718.1	501372.7
99	flagellate medium	flag	8	5	20	40	15	24	0	755.4	472083.8
99	flagellate small	flag	125	5	20	40	5	9	0	478.8	443822.0
99	flagellate small	flag	270	5	20	40	4	5	0	402.9	423252.1
99	dinoflagellate	dino	11	5	20	40	16	20	0	1500.6	526868.1
99	Cyclotella sp.	diat	12	5	20	40	15	10	0	1079.1	503590.5
99	Diachros sp.	chry	24	5	20	40	8	7	0	327.4	245560.3
	June 5, 1978										
99	flagellate medium	flag	1	10	30	40	14	18	0	21.8	15193.1
99	flagellate small	flag	6	10	30	40	7	10	0	17.6	14298.0
99	flagellate small	flag	24	10	30	40	3	5	0	6.2	7673.8
	June 8, 1978										
99	flagellate large	flag	2	10	30	40	26	30	0	360.2	79542.1
99	flagellate medium	flag	22	10	30	40	12	16	0	309.5	219547.7
99	flagellate small	flag	39	10	30	40	4	7	0	24.9	26071.2
99	Cosmarium sp.	desm	1	10	30	40	42	50	0	1566.7	212291.2
	June 12, 1978										
99	flagellate medium	flag	5	10	30	40	16	20	0	159.2	110013.9
99	flagellate small	flag	20	10	30	40	8	12	0	90.9	69590.4
99	flagellate small	flag	60	10	30	40	3	6	0	18.0	23280.8
99	Dictyosphaerium pulchellum	grn	16	10	30	40	6	0	0	30.7	30695.0
99	Diachros sp.	chry	6	10	30	40	8	4	0	27.3	20463.4
	June 16, 1978										
99	eucaryote sphere	eu	7	15	40	40	14	0	0	85.3	36556.9
99	Schroederia setigera	grn	2	15	40	40	7	20	0	4.4	3952.2
99	Dictyosphaerium pulchellum	grn	39	15	40	40	6	0	0	37.4	37409.6
99	flagellate large	flag	1	15	40	40	22	30	0	64.5	16116.7
99	flagellate medium	flag	3	15	40	40	16	20	0	47.7	33004.2
99	flagellate small	flag	41	15	40	40	8	12	0	93.2	71330.2
99	flagellate small	flag	84	15	40	40	4	6	0	23.9	24598.0
99	Cosmarium sp.	desm	3	15	40	40	44	54	0	2785.6	357347.1
99	eucaryote filament	eu	22	15	40	40	6	0	0	41.4	27623.5

M	Genus and species	Fam.	Ct.	Vol. filt. ml	Flds. ctd.	Mag.	Dim1 µm	Dim2 µm	Dim3 µm	Biovol. mm³/m³	Surf.area mm²/m³
	June 22, 1978										
99	Cosmarium sp.	desm	1	10	40	40	40	50	0	1065.8	149678.4
99	Schroederia setigera	grn	53	10	40	40	3	10	0	15.9	33173.2
99	flagellate medium	flag	1	10	40	40	16	20	0	23.9	16502.1
99	flagellate small	flag	12	10	40	40	8	12	0	40.9	31315.7
99	flagellate small	flag	61	10	40	40	4	6	0	26.0	26794.3
99	eucaryote sphere	eu	6	10	40	40	10	0	0	40.0	23980.5
99	Dictyosphaerium pulchellum	grn	4	10	40	40	6	0	0	5.8	5755.3
	June 26, 1978										
99	Schroederia setigera	grn	130	20	40	40	4	12	0	41.6	65721.9
99	Schroederia setigera	grn	48	20	40	40	5	20	0	40.0	49437.1
99	flagellate large	flag	2	20	40	40	28	30	0	156.7	32836.9
99	flagellate large	flag	1	20	40	40	30	40	0	119.9	22094.1
99	flagellate large	flag	1	20	40	40	35	50	0	204.0	31715.4
99	flagellate large	flag	1	20	40	40	24	24	0	46.0	5755.3
99	flagellate medium	flag	3	20	40	40	16	25	0	42.2	26238.6
99	flagellate medium	flag	44	20	40	40	12	18	0	253.2	171382.9
99	Microcystis aeruginosa	bg	84	20	40	40	5	0	0	35.0	41965.9
99	eucaryote filament	eu	44	20	40	40	3	0	0	15.5	20717.6
99	Anabaena sp.	bg	7	20	40	40	6	0	0	9.9	6592.0
99	eucaryote small	eu	5	20	40	40	5	10	0	4.2	4269.5
99	eucaryote sphere	eu	2	20	40	40	8	0	0	3.4	2557.9
99	Dictyosphaerium pulchellum	grn	4	20	40	40	5	0	0	1.7	1998.4
	June 29, 1978										
99	flagellate small	flag	84	20	40	40	6	10	0	65.5	55395.0
99	Aphanizomenon flos-aquae	bg	36	20	40	40	4	0	0	22.6	22601.1
99	Lyngbya limnetica	bg	26	20	40	40	3	0	0	9.2	12242.2
99	Microcystis aeruginosa	bg	145	20	40	40	3	0	0	13.0	26078.8
99	Microcystis aeruginosa	bg	97	20	40	40	4	0	0	20.7	31014.8
99	Aphanocapsa colonies	bg	28	20	40	40	1	0	0		
99	eucaryote sphere	eu	1	20	40	40	10	0	0	3.3	1998.4
99	eucaryote sphere	eu	44	20	40	40	4	0	0	9.4	14068.6
99	eucaryote sphere	eu	10	20	40	40	5	0	0	4.2	4995.9
99	Dictyosphaerium pulchellum	grn	40	20	40	40	4	0	0	8.5	12789.6
99	Dictyosphaerium pulchellum	grn	7	20	40	40	6	0	0	5.0	5035.9
99	Cosmarium sp.	desm	1	20	40	40	30	36	0	215.8	40853.0
99	Cosmarium sp.	desm	1	20	40	40	26	30	0	135.1	29828.3
99	Pandorina sp.	grn	16	20	40	40	9	0	0	38.8	25898.9
99	Pandorina sp.	grn	16	20	40	40	6	0	0	11.5	11510.6
99	Pandorina sp.	grn	16	20	40	40	10	0	0	53.3	31974.0
99	Eudorina sp.	grn	48	20	40	40	3	0	0	4.3	8633.0
99	Oocystis sp.	grn	11	20	40	40	4	10	0	5.9	7319.5
99	Oocystis sp.	grn	1	20	40	40	20	30	0	40.0	10761.7
99	Ceratium hirundinella	dino	4	20	40	40	50	140	90	2331.4	297079.8
99	Ceratium hirundinella	dino	1	20	40	40	30	100	99	149.9	31295.5
99	Anabaena sp.	bg	12	20	40	40	5	0	0	11.8	9417.1
99	Schroederia setigera	grn	275	20	40	40	4	12	0	87.9	139027.2
	July 3, 1978										
99	Pandorina sp.	grn	140	10	40	40	8	0	0	477.5	358108.8
99	Ceratium hirundinella	dino	1	10	40	40	40	140	90	746.1	116387.1
99	Cosmarium sp.	desm	1	10	40	40	28	34	0	355.1	71792.7
99	Cosmarium sp.	desm	1	10	40	40	36	50	0	863.3	131309.1
99	flagellate large	flag	1	10	40	40	24	38	0	145.8	32366.2
99	flagellate small	flag	14	10	40	40	5	9	0	13.4	12427.0
99	Aphanocapsa colonies	bg	40	10	40	40	1	0	0		
99	Microcystis aeruginosa	bg	117	10	40	40	4	0	0	49.9	74819.2
99	Microcystis aeruginosa	bg	235	10	40	40	2	0	0	12.5	37569.5
99	Anabaena sp.	bg	120	10	40	40	4	0	0	150.7	150673.8
99	Eudorina sp.	grn	96	10	40	40	2	0	0	5.1	15347.5
99	eucaryote sphere	eu	5	10	40	40	8	0	0	17.1	12789.6
99	Schroederia setigera	grn	2	10	40	40	5	26	0	4.3	5291.0
99	Schroederia setigera	grn	148	10	40	40	4	12	0	94.6	149643.8
99	Aphanizomenon flos-aquae	bg	27	10	40	40	5	0	0	53.0	42377.0
	July 6, 1978										
99	Microcystis aeruginosa	bg	850	15	30	40	4	0	0	322.1	483162.7
99	Microcystis aeruginosa	bg	268	15	30	40	3	0	0	42.8	85690.3
99	Aphanocapsa colonies	bg	59	15	30	40	1	0	0		
99	Pandorina sp.	grn	38	15	30	40	10	0	0	225.0	135001.3
99	Pandorina sp.	grn	40	15	30	40	5	0	0	29.6	35526.7
99	eucaryote sphere	eu	11	15	30	40	5	0	0	8.1	9769.8
99	eucaryote sphere	eu	2	15	30	40	10	0	0	11.8	7105.3
99	eucaryote sphere	eu	3	15	30	40	15	0	0	60.0	23980.5
99	Coelosphaerium sp.	bg	116	15	30	40	4	0	0	44.0	65937.5
99	Staurastrum sp.	grn	2	15	30	40	15	30	50	508.9	203148.0
99	Staurastrum sp.	grn	1	15	30	40	10	15	30	50.9	35243.5
99	Ceratium hirundinella	dino	1	15	30	40	50	100	90	740.1	99300.1
99	Aphanizomenon flos-aquae	bg	28	15	30	40	5	0	0	48.8	39063.6
99	Anabaena sp.	bg	84	15	30	40	4	0	0	93.8	93752.6
99	Oocystis sp.	grn	2	15	30	40	14	26	0	60.3	22343.2
99	Stephanodiscus sp.	diat	1	15	30	40	40	16	0	227.4	51158.4
99	Dictyosphaerium pulchellum	grn	36	15	30	40	5	0	0	26.6	31974.0
	July 10, 1978										
99	Aphanizomenon flos-aquae	bg	87	15	20	40	5	0	0	227.6	182064.2
99	Anabaena sp.	bg	44	15	20	40	5	0	0	115.1	92078.4
99	Microcystis aeruginosa	bg	394	15	20	40	3	0	0	94.5	188966.4
99	Microcystis aeruginosa	bg	1120	15	20	40	4	0	0	636.6	954956.9
99	Staurastrum sp.	grn	2	15	20	40	15	25	50	636.1	275115.8
99	flagellate small	flag	42	15	20	40	6	10	0	87.3	73859.9
99	Aphanocapsa colonies	bg	36	15	20	40		0	0		
99	Ceratium hirundinella	dino	2	15	20	40	50	110	90	2442.5	321952.8
99	Pandorina sp.	grn	19	15	20	40	5	0	0	21.1	25312.8
99	eucaryote sphere	eu	8	15	20	40	10	0	0	71.1	42632.0
99	flagellate large	flag	1	15	20	40	20	24	0	85.3	24209.2
	July 13, 1978										
99	Anabaena sp.	bg	46	15	25	40	4	0	0	61.6	61608.8
99	Aphanizomenon flos-aquae	bg	436	15	25	40	5	0	0	912.6	729930.9
99	Microcystis aeruginosa	bg	1380	15	25	40	3	0	0	264.7	529489.5
99	Microcystis aeruginosa	bg	1180	15	25	40	4	0	0	536.6	804892.2
99	flagellate small	flag	74	15	25	40	5	8	0	69.0	63661.6
99	Pandorina sp.	grn	51	15	25	40	5	0	0	45.3	54355.8
99	Staurastrum sp.	grn	1	15	25	40	15	25	55	279.9	119473.4
99	Coelosphaerium sp.	bg	480	15	25	40	3	4	0	92.1	184170.3
99	eucaryote sphere	eu	3	15	25	40	15	0	0	71.9	28776.6
99	eucaryote sphere	eu	1	15	25	40	18	0	0	41.4	13812.8
99	eucaryote sphere	eu	1	15	25	40	22	0	0	75.7	20633.9
99	Oocystis sp.	grn	4	15	25	40	10	15	0	42.6	22958.3
99	Cosmarium sp.	desm	1	15	25	40	24	30	0	245.6	57476.5

M	Genus and species	Fam.	Ct.	Vol. filt. ml	Flds. ctd.	Mag.	Dim1 μm	Dim2 μm	Dim3 μm	Biovol. mm³/m³	Surf.area mm²/m³
	July 17, 1978										
99	Microcystis aeruginosa	bg	622	10	30	40	3	0	0	149.2	298317.4
99	Microcystis aeruginosa	bg	2467	10	30	40	5	0	0	2738.9	3286661.0
99	Aphanizomenon flos-aquae	bg	265	10	30	40	5	0	0	693.4	554563.3
99	Staurastrum sp.	grn	5	10	30	40	15	25	45	1431.2	628967.2
99	Ceratium hirundinella	dino	1	10	30	40	50	140	90	1554.3	198053.2
99	flagellate small	flag	99	10	30	40	5	9	0	126.4	117169.0
99	Aphanocapsa sp.	bg	5500	10	30	40	1	0	0	48.8	293095.0
99	Oocystis sp.	grn	3	10	30	40	15	20	0	119.9	44188.2
99	eucaryote sphere	eu	8	10	30	40	10	0	0	71.1	42632.0
99	eucaryote sphere	eu	1	10	30	40	25	0	0	138.8	33306.3
99	Pandorina sp.	grn	22	10	30	40	5	0	0	24.4	29309.5
99	Dinobryon sp.	chry	1	10	30	40	20	30	0	71.1	43063.0
99	Coelosphaerium sp.	bg	50	10	30	40	5	10	0	55.5	66612.5
99	Coelosphaerium sp.	bg	315	10	30	40	3	5	0	75.5	151077.2
99	Cosmarium sp.	desm	1	10	30	40	30	50	0	799.4	140615.2
	July 20, 1978										
99	Aphanizomenon flos-aquae	bg	562	10	30	40	5	0	0	1470.5	1176092.8
99	flagellate small	flag	57	10	30	40	5	9	0	72.8	67460.9
99	Microcystis aeruginosa	bg	327	10	30	40	3	0	0	78.4	156832.5
99	Microcystis aeruginosa	bg	882	10	30	40	5	0	0	979.2	1175044.6
99	Aphanocapsa sp.	bg	4600	10	30	40	1	0	0	40.9	245134.0
99	Coelosphaerium sp.	bg	237	10	30	40	3	5	0	56.8	113667.6
99	eucaryote filament	eu	8	10	30	40	9	0	0	67.8	30134.8
99	eucaryote filament	eu	26	10	30	40	4	0	0	43.5	43528.0
99	Anabaena sp.	bg	10	10	30	40	10	0	0	104.7	41853.8
99	Diachros sp.	chry	25	10	30	40	8	7	0	113.7	85264.0
99	Staurastrum sp.	grn	1	10	30	40	10	20	35	118.7	70497.6
99	Oocystis sp.	grn	1	10	30	40	10	15	0	13.3	7174.5
	July 24, 1978										
99	Aphanizomenon flos-aquae	bg	774	5	25	40	4	0	0	3110.7	3109907.4
99	Anabaena sp.	bg	41	5	25	40	6	0	0	370.7	247105.0
99	Microcystis aeruginosa	bg	17	5	25	40	3	0	0	9.8	19568.1
99	Microcystis aeruginosa	bg	140	5	25	40	4	0	0	191.0	286487.1
99	Microcystis aeruginosa	bg	880	5	25	40	5	0	0	2344.8	2813712.2
99	Coelosphaerium sp.	bg	276	5	25	40	2	5	0	47.1	141197.2
99	eucaryote filament	eu	6	5	25	40	8	0	0	96.5	48215.6
99	eucaryote sphere	eu	1	5	25	40	20	0	0	170.5	51158.4
99	eucaryote sphere	eu	1	5	25	40	15	0	0	71.9	28776.6
99	eucaryote sphere	eu	6	5	25	40	10	0	0	127.9	76737.6
99	Schroederia setigera	grn	2	5	25	40	5	20	0	10.7	13183.2
99	Dictyosphaerium pulchellum	grn	16	5	25	40	5	0	0	42.6	51158.4
99	blue green filaments	bg	18	5	25	40	3	0	0	40.7	54242.6
99	flagellate small	flag	25	5	25	40	5	7	0	63.3	58897.9
99	Aphanocapsa sp.	bg	800	5	25	40	1	0	0	17.1	102316.8
	July 27, 1978										
99	Aphanizomenon flos-aquae	bg	734	15	25	40	5	0	0	1536.4	1228828.6
99	Microcystis aeruginosa	bg	1200	15	25	40	5	0	0	1065.8	1278960.1
99	Microcystis aeruginosa	bg	65	15	25	40	3	0	0	12.5	24939.7
99	flagellate small	flag	56	15	25	40	5	9	0	57.2	53021.9
99	Coelosphaerium sp.	bg	335	15	25	40	2	5	0	19.0	57126.9
99	eucaryote filament	eu	51	15	25	40	7	0	0	209.2	119534.6
99	eucaryote filament	eu	20	15	25	40	5	0	0	41.9	33483.1
99	eucaryote sphere	eu	1	15	25	40	15	0	0	24.0	9592.2
99	blue green filaments	bg	70	15	25	40	4	0	0	93.8	93752.6
99	Ceratium hirundinella	dino	1	15	25	40	56	120	85	1336.9	158073.5
99	Dictyosphaerium pulchellum	grn	40	15	25	40	4	0	0	18.2	27284.5
99	Ceratium hirundinella	dino	1	15	25	40	60	140	90	1790.5	194805.4
99	Oocystis sp.	grn	1	15	25	40	25	35	0	155.4	33981.8
99	Anabaena sp.	bg	17	15	25	40	6	0	0	51.2	34152.7
	July 31, 1978										
99	Microcystis aeruginosa	bg	180	10	20	40	3	0	0	64.7	129494.7
99	Microcystis aeruginosa	bg	735	10	20	40	4	0	0	626.7	940035.7
99	blue green filaments	bg	150	10	20	40	2	0	0	94.2	188342.3
99	blue green filaments	bg	25	10	20	40	3	0	0	35.3	47085.6
99	Anabaena sp.	bg	30	10	20	40	8	0	0	301.4	150673.8
99	flagellate small	flag	19	10	20	40	5	9	0	36.4	33730.5
99	Aphanizomenon flos-aquae	bg	679	10	20	40	4	0	0	1705.5	1705125.3
99	eucaryote sphere	eu	24	10	20	40	5	0	0	40.0	47961.0
99	Schroederia setigera	grn	3	10	20	40	4	20	0	6.4	9782.2
99	Oocystis sp.	grn	1	10	20	40	15	20	0	60.0	22094.1
99	eucaryote sphere	eu	1	10	20	40	8	0	0	6.8	5115.8
99	blue green filaments	bg	1	10	20	40	10	0	0	15.7	6278.1
99	Ceratium hirundinella	dino	1	10	20	40	50	130	90	2164.9	278341.4
	Aug. 3, 1978										
99	Aphanizomenon flos-aquae	bg	794	10	20	40	4	0	0	1994.4	1993916.8
99	Microcystis aeruginosa	bg	395	10	20	40	4	0	0	336.8	505189.3
99	Microcystis aeruginosa	bg	80	10	20	40	3	0	0	28.8	57553.2
99	Anabaena sp.	bg	43	10	20	40	6	0	0	243.0	161974.3
99	Anabaena sp.	bg	14	10	20	40	9	0	0	178.0	79103.8
99	blue green filaments	bg	42	10	20	40	3	0	0	59.3	79103.8
99	eucaryote sphere	eu	6	10	20	40	10	0	0	79.9	47961.0
99	eucaryote filament	eu	9	10	20	40	8	0	0	90.4	45202.1
99	flagellate small	flag	11	10	20	40	5	9	0	21.1	19528.2
99	eucaryote sphere	eu	4	10	20	40	5	0	0	6.7	7993.5
99	Cosmarium sp.	desm	1	10	20	40	20	30	0	319.7	86093.8
99	Ceratium hirundinella	dino	1	10	20	40	50	130	90	2164.9	278341.4
	Aug. 7, 1978										
99	Aphanizomenon flos-aquae	bg	721	10	22	40	4	0	0	1646.4	1645997.2
99	Anabaena sp.	bg	44	10	22	40	8	0	0	401.9	200898.4
99	Anabaena sp.	bg	209	10	22	40	5	0	0	745.7	596417.2
99	Microcystis aeruginosa	bg	220	10	22	40	4	0	0	170.5	255792.0
99	Microcystis aeruginosa	bg	140	10	22	40	3	0	0	45.8	91561.9
99	Microcystis aeruginosa	bg	100	10	22	40	2	0	0	9.7	29067.3
99	eucaryote sphere	eu	10	10	22	40	10	0	0	121.1	72668.2
99	flagellate small	flag	14	10	22	40	5	9	0	24.4	22594.6
99	Staurastrum sp.	grn	3	10	22	40	20	40	50	2775.7	881856.2
99	blue green filaments	bg	30	10	22	40	2	0	0	17.1	34244.0
99	Ceratium hirundinella	dino	7	10	10	10	50	140	90	127.3	16219.4
99	Staurastrum sp.	grn	30	10	10	10	20	40	50	238.1	75657.9
	Aug. 14, 1978										
99	Anabaena sp.	bg	22	10	25	40	8	0	0	176.8	88395.3
99	Anabaena sp.	bg	110	10	25	40	6	0	0	497.3	331482.4
99	Aphanizomenon flos-aquae	bg	613	10	25	40	4	0	0	1231.8	1231507.3
99	Microcystis aeruginosa	bg	565	10	25	40	4	0	0	385.4	578090.0
99	Microcystis aeruginosa	bg	60	10	25	40	3	0	0	17.3	34531.9

M	Genus and species	Fam.	Ct.	Vol. filt. ml	Flds. ctd.	Mag.	Dim1 μm	Dim2 μm	Dim3 μm	Biovol. mm³/m³	Surf.area mm²/m³
99	eucaryote sphere	eu	9	10	25	40	10	0	0	95.9	57553.2
99	eucaryote sphere	eu	6	10	25	40	5	0	0	8.0	9592.2
99	flagellate small	flag	20	10	25	40	5	9	0	30.6	28404.6
99	blue green filaments	bg	18	10	25	40	3	0	0	20.3	27121.3
99	blue green filaments	bg	160	10	25	40	1	0	0	20.1	80359.4
99	Oocystis sp.	grn	4	10	25	40	4	8	0	5.5	6995.2
99	Oocystis sp.	grn	1	10	25	40	5	11	0	2.9	2969.0
99	Staurastrum sp.	grn	2	10	25	40	15	30	50	916.0	365666.3
99	Dictyosphaerium pulchellum	grn	42	10	25	40	5	0	0	56.0	67145.4
99	Coelosphaerium sp.	bg	50	10	25	40	2	5	0	4.3	12789.6
99	Ceratium hirundinella	dino	2	10	12	10	50	140	90	30.3	3861.8
99	Staurastrum sp.	grn	29	10	12	10	15	30	50	107.9	43077.0
	Aug. 22, 1978										
99	Aphanizomenon flos-aquae	bg	1027	10	10	40	4	0	0	5159.3	5158066.8
99	Anabaena sp.	bg	9	10	10	40	8	0	0	180.9	90404.3
99	Anabaena sp.	bg	82	10	10	40	5	0	0	643.7	514802.2
99	Staurastrum sp.	grn	1	10	10	40	10	25	35	445.3	242897.5
99	eucaryote sphere	eu	4	10	10	40	10	0	0	106.6	63948.0
99	Dictyosphaerium pulchellum	grn	120	10	10	40	4	0	0	204.6	306950.4
99	Microcystis aeruginosa	bg	40	10	10	40	3	0	0	28.8	57553.2
99	flagellate small	flag	1	10	10	40	5	9	0	3.8	3550.6
99	Ceratium hirundinella	dino	2	1	10	10	50	140	90	363.7	46341.1
99	Staurastrum sp.	grn	17	1	10	10	10	25	35	295.2	161029.5
	Aug. 28, 1978										
99	Anabaena sp.	bg	27	10	10	40	6	0	0	305.2	203409.6
99	Schroederia setigera	grn	3	10	10	40	4	20	0	12.8	19564.3
99	Aphanizomenon flos-aquae	bg	763	10	10	40	4	0	0	3833.1	3832137.2
99	eucaryote sphere	eu	8	10	10	40	10	0	0	213.2	127896.0
99	blue green filaments	bg	170	10	10	40	2	0	0	213.5	426909.1
99	Ceratium hirundinella	dino	1	10	10	40	50	140	90	4662.9	594159.6
99	Dictyosphaerium pulchellum	grn	16	10	10	40	4	0	0	27.3	40926.7
99	Actinastrum Hantzschii	grn	5	10	10	40	4	15	0	61.1	69208.0
99	Ceratium hirundinella	dino	4	10	10	10	50	140	90	72.7	9268.2
99	Staurastrum sp.	grn	6	10	10	10	10	25	35	10.4	5683.4
99	Cosmarium sp.	desm	2	10	10	10	40	50	0	33.3	4669.6
	Sep. 5, 1978										
99	Anabaena sp.	bg	538	5	20	40	5	0	0	4223.0	3377604.6
99	Aphanizomenon flos-aquae	bg	573	5	20	40	4	0	0	2878.6	2877869.8
99	Lyngbya limnetica	bg	311	5	20	40	2	0	0	390.6	780992.6
99	Microcystis aeruginosa	bg	45	5	20	40	4	0	0	76.7	115106.4
99	Microcystis aeruginosa	bg	25	5	20	40	3	0	0	18.0	35970.8
99	Coelosphaerium sp.	bg	55	5	20	40	2	5	0	11.7	35171.4
99	flagellate small	flag	13	5	20	40	5	9	0	49.8	46157.5
99	eucaryote sphere	eu	6	5	20	40	5	0	0	20.0	23980.5
	Sep. 12, 1978										
99	Aphanizomenon flos-aquae	bg	1085	10	10	40	4	0	0	5450.7	5449369.5
99	Anabaena sp.	bg	126	10	10	40	5	0	0	989.0	791037.5
99	Lyngbya limnetica	bg	440	10	10	40	2	0	0	552.6	1104941.3
99	Microcystis aeruginosa	bg		10	10	40	5	0	0		
99	Actinastrum Hantzschii	grn	46	10	10	40	2	10	0	93.6	205995.6
99	eucaryote sphere	eu	1	10	10	40	15	0	0	89.9	35970.8
99	eucaryote sphere	eu	3	10	10	40	10	0	0	79.9	47961.0
99	eucaryote sphere	eu	2	10	10	40	6	0	0	11.5	11510.6

M	Genus and species	Fam.	Ct.	Vol. filt. ml	Flds. ctd.	Mag.	Dim1 μm	Dim2 μm	Dim3 μm	Biovol. mm³/m³	Surf.area mm²/m³
99	flagellate small	flag	12	10	10	40	5	9	0	46.0	42606.9
99	eucaryote filament	eu	6	10	10	40	4	0	0	30.1	30134.8
99	Ceratium hirundinella	dino	27	10	6	10	60	210	0	1767.5	183820.3
99	Staurastrum sp.	grn	3	10	6	10	15	25	45	16.7	7358.4
99	Cosmarium sp.	desm	6	10	6	10	40	50	0	166.3	23348.2
	Sep. 20, 1978										
99	Aphanizomenon flos-aquae	bg	1553	20	10	40	4	0	0	3900.9	3899940.5
99	Anabaena sp.	bg	228	20	10	40	6	0	0	1288.6	858840.7
99	Lyngbya limnetica	bg	658	20	10	40	1	0	0	103.3	413097.4
99	Lyngbya limnetica	bg	183	20	10	40	2	0	0	114.9	229777.6
99	Coelosphaerium sp.	bg	135	20	10	40	2	5	0	14.4	43164.9
99	akinete	bg	1	20	10	40	16	20	0	68.2	23948.6
99	akinete	bg	1	20	10	40	10	16	0	21.3	11333.9
99	Microcystis aeruginosa	bg	60	20	10	40	3	0	0	21.6	43164.9
99	Staurastrum sp.	grn	1	20	10	40	8	10	14	28.5	29223.8
99	eucaryote sphere	eu	1	20	10	40	16	0	0	54.6	20463.4
99	eucaryote sphere	eu	2	20	10	40	10	0	0	26.6	15987.0
	Sep. 27, 1978										
99	Aphanizomenon flos-aquae	bg	589	10	25	40	5	0	0	1849.3	1479114.6
99	akinete	bg	1	10	25	40	8	20	0	13.6	8517.2
99	Anabaena sp.	bg	111	10	25	40	8	0	0	892.2	445994.5
99	Lyngbya limnetica	bg	303	10	25	40	2	0	0	152.2	304361.1
99	Microcystis aeruginosa	bg		10	25	40	4	0	0		
99	flagellate large	flag	1	10	25	40	25	30	0	199.8	45392.2
99	flagellate medium	flag	1	10	25	40	15	25	0	39.0	23980.5
99	flagellate medium	flag	1	10	25	40	15	20	0	33.0	22308.5
99	flagellate medium	flag	1	10	25	40	10	14	0	10.1	7376.3
99	flagellate small	flag	16	10	25	40	5	9	0	24.5	22723.7
99	eucaryote sphere	eu	21	10	25	40	5	0	0	28.0	33572.7
99	eucaryote sphere	eu	15	10	25	40	6	0	0	34.5	34531.9
99	eucaryote sphere	eu	1	10	25	40	20	0	0	85.3	25579.2
99	Actinastrum Hantzschii	grn	8	10	25	40	2	12	0	7.8	16935.6
99	eucaryote filament	eu	20	10	25	40	8	0	0	160.8	80359.4
	Oct. 3, 1978										
99	Anabaena sp.	bg	99	10	40	40	6	0	0	279.8	186458.8
99	Aphanizomenon flos-aquae	bg	134	10	40	40	5	0	0	263.0	210315.5
99	akinete	bg	3	10	40	40	8	20	0	25.6	15969.8
99	Microcystis aeruginosa	bg	181	10	40	40	4	0	0	77.2	115745.9
99	Lyngbya limnetica	bg	149	10	40	40	2	0	0	46.8	93543.3
99	Aphanocapsa sp.	bg	6000	10	40	40	1	0	0	40.0	239805.0
99	flagellate medium	flag		10	40	40	19	16	0		
99	eucaryote sphere	eu	6	10	40	40	6	0	0	8.6	8633.0
99	Actinastrum Hantzschii	grn	4	10	40	40	4	20	0	16.3	17912.7
99	flagellate small	flag	42	10	40	40	8	12	0	143.2	109604.9
99	Diachros sp.	chry	17	10	40	40	6	5	0	24.5	24460.1
	Oct. 11, 1978										
99	Microcystis aeruginosa	bg	28	15	20	40	4	0	0	15.9	23873.9
99	flagellate large	flag	1	15	20	40	20	22	0	78.2	22750.5
99	flagellate small	flag	8	15	20	40	7	10	0	23.5	19064.0
99	Chlorochromonas minuta	chry	5	15	20	40	8	12	0	34.1	22958.3
99	eucaryote small	eu	1	15	20	40	12	20	0	25.6	11249.2
99	eucaryote sphere	eu	9	15	20	40	8	0	0	40.9	30695.0
99	Schroederia setigera	grn	1	15	20	40	8	24	0	6.8	5392.6

M	Genus and species	Fam.	Ct.	Vol. filt. ml	Flds. ctd.	Mag.	Dim1 µm	Dim2 µm	Dim3 µm	Biovol. mm³/m³	Surf.area mm²/m³
99	Ceratium hirundinella	dino	14	15	6	10	96	300	0	2234.4	146628.4
99	Stephanodiscus sp.	diat	4	15	6	10	64	38	0	107.8	12413.6
99	Staurastrum sp.	grn	3	15	6	10	24	40	48	30.5	9055.4
99	Cosmarium sp.	desm	1	15	6	10	60	80	0	66.5	6127.0
	Oct. 24, 1978										
99	Coelosphaerium sp.	bg	31	20	20	40	2	3	0	1.7	4956.0
99	flagellate small	flag	58	20	20	40	5	9	0	55.5	51483.4
99	flagellate medium	flag	1	20	20	40	15	20	0	20.6	13942.8
99	Lyngbya limnetica	bg	32	20	20	40	2	0	0	10.0	20089.8
99	Microcystis aeruginosa	bg	37	20	20	40	2	0	0	2.0	5915.2
99	eucaryote sphere	eu	16	20	20	40	5	0	0	13.3	15987.0
99	Stephanodiscus sp.	diat	2	20	20	40	45	20	0	809.3	152875.7
99	Stephanodiscus sp.	diat	1	20	20	40	50	20	0	499.6	89926.9
99	eucaryote filament	eu	22	20	20	40	5	0	0	43.2	34529.4
	Nov. 7, 1978										
99	Aphanizomenon flos-aquae	bg	12	15	20	40	4	0	0	20.1	20089.8
99	flagellate medium	flag	1	15	20	40	15	15	0	22.5	17985.4
99	flagellate small	flag	37	15	20	40	5	9	0	47.2	43790.4
99	Microcystis aeruginosa	bg	104	15	20	40	4	0	0	59.1	88674.6
99	Stephanodiscus sp.	diat	1	15	20	40	55	22	0	886.6	145082.0
99	Stephanodiscus sp.	diat	2	15	20	40	40	16	0	682.1	153475.2
	Nov. 14, 1978										
99	Aphanizomenon flos-aquae	bg	150	20	20	40	4	0	0	188.4	188342.3
99	Dictyosphaerium pulchellum	grn	30	20	20	40	4	0	0	12.8	19184.4
99	flagellate small	flag	37	20	20	40	5	9	0	35.4	32842.8
99	flagellate large	flag	1	20	20	40	22	26	0	83.8	21727.7
99	Lyngbya limnetica	bg	13	20	20	40	2	0	0	4.1	8161.5
99	Microcystis aeruginosa	bg	17	20	20	40	2	0	0	0.9	2717.8
99	Stephanodiscus sp.	diat	1	20	20	40	45	20	0	404.7	76437.9
99	Stephanodiscus sp.	diat	2	20	20	40	40	16	0	511.6	115106.4
99	Schroederia setigera	grn	1	20	20	40	4	20	0	1.1	1630.4
99	Oocystis sp.	grn	2	20	20	40	22	35	0	225.7	54603.6
99	Pediastrum sp.	grn	30	20	20	40	5	10	0	74.9	74939.1
	Nov. 28, 1978										
99	Coelosphaerium sp.	bg	50	20	25	40	2	3	0	2.1	6394.8
99	Dictyosphaerium pulchellum	grn	20	20	25	40	5	0	0	13.3	15987.0
99	flagellate small	flag	61	20	25	40	5	9	0	46.7	43317.0
99	Microcystis aeruginosa	bg	53	20	25	40	2	0	0	2.3	6778.5
99	eucaryote sphere	eu	1	20	25	40	10	0	0	5.3	3197.4
99	eucaryote sphere	eu	1	20	25	40	7	0	0	1.8	1566.7
99	Staurastrum sp.	grn	1	20	25	40	9	14	20	25.6	20397.6
99	Schroederia setigera	grn	1	20	25	40	4	20	0	0.9	1304.3
99	Stephanodiscus sp.	diat	1	20	25	40	50	25	0	499.6	79935.0
99	Stephanodiscus sp.	diat	2	20	25	40	30	20	0	287.8	67145.4
99	Stephanodiscus sp.	diat	1	20	25	40	45	20	0	323.7	61150.3
	Feb. 7, 1979										
0	flagellate large	flag	1	30	20	40	22	30	0	64.5	16116.7
0	flagellate small	flag	29	30	20	40	5	10	0	20.1	18847.9
0	eucaryote sphere	eu	2	30	20	40	10	0	0	8.9	5329.0
0	dinoflagellate	dino	1	30	20	40	12	15	0	9.6	4490.4
2	flagellate small	flag	23	30	15	40	5	10	0	21.3	19931.1
2	eucaryote sphere	eu	1	30	15	40	10	0	0	5.9	3552.7

M	Genus and species	Fam.	Ct.	Vol. filt. ml	Flds. ctd.	Mag.	Dim1 µm	Dim2 µm	Dim3 µm	Biovol. mm³/m³	Surf.area mm²/m³
4	flagellate small	flag	36	30	20	40	5	10	0	25.0	23397.4
4	eucaryote sphere	eu	1	30	20	40	10	0	0	4.4	2664.5
6	flagellate small	flag	22	30	20	40	5	10	0	15.3	14298.4
10	flagellate small	flag	12	20	15	40	5	10	0	16.7	15598.3
16	flagellate small	flag	10	20	20	40	5	10	0	10.4	9748.9
16	eucaryote sphere	eu	1	20	20	40	10	0	0	6.7	3996.8
	Feb. 21, 1979										
0	Aphanizomenon flos-aquae	bg	11	30	20	40	4	0	0	9.2	9207.8
0	flagellate small	flag	35	30	20	40	5	10	0	24.3	22747.5
0	eucaryote sphere	eu	3	30	20	40	10	0	0	13.3	7993.5
2	flagellate small	flag	22	30	15	40	5	10	0	20.4	19064.5
2	eucaryote sphere	eu	1	30	15	40	15	0	0	20.0	7993.5
4	flagellate small	flag	22	25	15	40	5	10	0	24.4	22877.4
8	Aphanizomenon flos-aquae	bg	17	30	15	40	4	0	0	19.0	18973.7
8	flagellate small	flag	13	30	15	40	5	10	0	12.0	11265.4
16	flagellate small	flag	26	30	20	40	5	10	0	18.0	16898.1
16	eucaryote sphere	eu	1	30	20	40	10	0	0	4.4	2664.5
	Feb. 28, 1979										
0	flagellate small	flag	23	15	15	40	5	10	0	42.6	39862.2
4	flagellate small	flag	31	20	25	40	5	10	0	25.8	24177.3
12	flagellate small	flag	38	30	25	40	5	10	0	21.1	19757.8
12	Aphanizomenon flos-aquae	bg	11	30	25	40	4	0	0	7.4	7366.3
20	flagellate small	flag	17	20	15	40	5	10	0	23.6	22097.5
	Mar. 17, 1979										
0	Diachros sp.	chry	81	20	20	40	6	6	0	116.5	116545.2
0	flagellate small	flag	15	20	20	40	5	9	0	14.4	13314.7
0	eucaryote sphere	eu	1	20	20	40	12	0	0	11.5	5755.3
4	Diachros sp.	chry	67	30	15	40	6	6	0	85.7	85690.3
4	flagellate medium	flag	1	30	15	40	15	20	0	18.3	12393.6
4	flagellate small	flag	11	30	15	40	5	9	0	9.4	8679.2
8	Diachros sp.	chry	62	20	20	40	6	6	0	89.2	89207.5
8	flagellate large	flag	1	20	20	40	20	25	0	66.6	18709.8
8	flagellate small	flag	12	20	20	40	5	9	0	11.5	10651.7
8	flagellate small	flag	11	20	20	40	6	10	0	17.1	14508.2
12	Diachros sp.	chry	8	20	15	40	6	6	0	15.3	15347.5
12	flagellate small	flag	5	20	15	40	5	9	0	6.4	5917.6
16	Diachros sp.	chry	2	20	15	40	6	6	0	3.8	3836.9
16	flagellate small	flag	8	20	15	40	5	9	0	10.2	9468.2
16	Staurastrum sp.	grn	1	20	15	40	10	15	22	56.0	40752.0
20	Diachros sp.	chry	3	20	15	40	6	6	0	5.8	5755.3
20	flagellate small	flag	4	20	15	40	5	9	0	5.1	4734.1
	Mar. 30, 1979										
0	Diachros sp.	chry	14	20	20	40	7	7	0	32.0	27417.7
0	flagellate small	flag		20	20	40	6	10	0		
4	Diachros sp.	chry	448	15	15	40	7	7	0	1819.7	1559762.9
4	flagellate large	flag		15	15	40	20	25	0		
4	flagellate medium	flag		15	15	40	16	20	0		
8	Diachros sp.	chry	267	10	15	40	7	7	0	1626.8	1394386.3
8	flagellate small	flag		10	15	40	12	12	0		
8	flagellate large	flag		10	15	40	18	30	0		
12	Diachros sp.	chry	84	10	20	40	7	7	0	383.8	329012.5
12	flagellate medium	flag		10	20	40	14	14	0		
16	Diachros sp.	chry	37	10	15	40	7	7	0	225.4	193229.6

M	Genus and species	Fam.	Ct.	Vol. filt. ml	Flds. ctd.	Mag.	Dim1 µm	Dim2 µm	Dim3 µm	Biovol. mm³/m³	Surf.area mm²/m³
	Apr. 6, 1979										
0	Diachros sp.	chry	260	5	20	40	8	7	0	3547.0	2660237.0
0	flagellate small	flag	5	5	20	40	6	9	0	28.8	24562.4
0	flagellate medium	flag	1	5	20	40	14	18	0	65.3	45579.3
2	Diachros sp.	chry	227	5	20	40	8	7	0	3096.8	2322591.5
2	flagellate small	flag	3	5	20	40	6	9	0	17.3	14737.4
2	flagellate large	flag	4	5	20	40			0		
4	Diachros sp.	chry	327	5	20	40	8	7	0	4461.0	3345759.7
4	flagellate small	flag	8	5	20	40	6	9	0	46.0	39299.8
4	flagellate large	flag	3	5	20	40			0		
4	dinoflagellate	dino	1	5	20	40	18	20	0	172.7	55674.9
8	Diachros sp.	chry	120	5	20	40	8	7	0	1637.1	1227801.7
8	flagellate small	flag	5	5	20	40	6	9	0	28.8	24562.4
8	flagellate medium	flag	1	5	20	40	16	16	0	81.9	64800.6
12	Diachros sp.	chry	245	5	20	40	8	7	0	3342.3	2506761.8
12	flagellate small	flag	6	5	20	40	6	9	0	34.5	29474.8
12	flagellate medium	flag	1	5	20	40	16	22	0	102.3	67358.6
12	eucaryote sphere	eu	1	5	20	40	12	0	0	46.0	23021.3
16	Diachros sp.	chry	115	5	20	40	8	7	0	1568.9	1176643.3
16	flagellate small	flag	2	5	20	40	6	9	0	11.5	9824.9
16	flagellate medium	flag	1	5	20	40	15	18	0	76.4	54648.9
20	Diachros sp.	chry	5	5	20	40	8	7	0	68.2	51158.4
	Apr. 17, 1979										
0	Diachros sp.	chry	194	5	20	40	8	7	0	2646.6	1984946.1
0	flagellate small	flag	15	5	20	40	6	9	0	86.3	73687.1
0	flagellate large	flag	2	5	20	40			0		
0	dinoflagellate	dino	1	5	20	40	18	20	0	172.7	55674.9
2	Diachros sp.	chry	378	5	25	40	8	7	0	4125.4	3094060.3
2	flagellate small	flag	31	5	25	40	6	9	0	142.7	121829.3
2	flagellate large	flag	8	5	25	40			0		
2	dinoflagellate	dino	2	5	25	40			0		
2	eucaryote sphere	eu	1	5	25	40	13	0	0	46.8	21614.4
2	eucaryote sphere	eu	1	5	25	40	18	0	0	124.3	41438.3
4	Diachros sp.	chry	242	5	20	40	8	7	0	3301.4	2476066.8
4	flagellate small	flag	21	5	20	40	6	9	0	120.9	103161.9
4	flagellate large	flag	4	5	20	40			0		
6	Diachros sp.	chry	253	5	20	40	8	7	0	3451.5	2588615.3
6	flagellate small	flag	16	5	20	40	6	9	0	92.1	78599.6
6	flagellate medium	flag	2	5	20	40	18	20	0	250.4	181924.4
6	flagellate large	flag	1	5	20	40	20	22	0	234.5	68251.5
8	Diachros sp.	chry	312	5	20	40	8	7	0	4256.4	3192284.5
8	flagellate small	flag	18	5	20	40	6	9	0	103.6	88424.5
8	flagellate large	flag	5	5	20	40			0		
12	Diachros sp.	chry	341	5	20	40	8	7	0	4652.0	3489003.2
12	flagellate small	flag	20	5	20	40	6	9	0	115.1	98249.5
12	flagellate large	flag	2	5	20	40			0		
16	Diachros sp.	chry	205	5	20	40	8	7	0	2796.7	2097494.6
16	flagellate small	flag	13	5	20	40	6	9	0	74.8	63862.2
16	flagellate large	flag	2	5	20	40	16	18	0	245.6	88754.2
	Apr. 26, 1979										
0	Diachros sp.	chry	92	5	20	40	8	7	0	1255.1	941314.7
0	flagellate small	flag	48	5	20	40	6	9	0	276.3	235798.7
0	flagellate large	flag	9	5	20	40			0		
0	eucaryote sphere	eu	4	5	20	40	12	0	0	184.2	92085.1
0	Oocystis sp.	grn	3	5	20	40	6	8	0	23.0	21210.3
4	Diachros sp.	chry	94	5	20	40	8	7	0	1282.4	961778.0
4	flagellate small	flag	49	5	20	40	6	9	0	282.0	240711.2
4	flagellate large	flag	9	5	20	40			0		
4	eucaryote sphere	eu	3	5	20	40	12	0	0	138.1	69063.8
6	Diachros sp.	chry	79	5	20	40	8	7	0	1077.7	808302.8
6	flagellate small	flag	31	5	20	40	6	9	0	178.4	152286.7
6	flagellate large	flag	7	5	20	40			0		
8	Diachros sp.	chry	78	5	20	40	8	7	0	1064.1	798071.1
8	flagellate small	flag	28	5	20	40	6	9	0	161.1	137549.3
8	flagellate large	flag	2	5	20	40			0		
8	dinoflagellate	dino	1	5	20	40			0		
8	eucaryote sphere	eu	2	5	20	40	12	0	0	92.1	46042.6
12	Diachros sp.	chry	117	5	20	40	8	7	0	1596.1	1197106.7
12	flagellate small	flag	18	5	20	40	6	9	0	103.6	88424.5
12	flagellate large	flag	2	5	20	40			0		
12	dinoflagellate	dino	1	5	20	40			0		
16	Diachros sp.	chry	95	5	20	40	8	7	0	1296.0	972009.7
16	flagellate small	flag	36	5	20	40	6	9	0	207.2	176849.0
16	flagellate large	flag	5	5	20	40			0		
16	dinoflagellate	dino	2	5	20	40			0		
20	Diachros sp.	chry	67	5	20	40	8	7	0	914.0	685522.6
20	flagellate small	flag	18	5	20	40	6	9	0	103.6	88424.5
20	flagellate large	flag	4	5	20	40			0		
20	dinoflagellate	dino	1	5	20	40			0		
20	Oocystis sp.	grn	4	5	20	40	6	8	0	30.7	28280.4
20	eucaryote sphere	eu	4	5	20	40	12	0	0	184.2	92085.1
	May 4, 1979										
0	Diachros sp.	chry	51	5	30	40	8	7	0	463.8	347877.2
0	Dictyosphaerium pulchellum	grn	24	5	30	40	5	6	0	53.3	63948.0
0	flagellate small	flag	44	5	30	40	6	9	0	168.8	144099.2
0	eucaryote sphere	eu	4	5	30	40	10	0	0	71.1	42632.0
0	Stephanodiscus sp.	diat	1	5	30	40	20	16	0	170.5	55421.6
0	eucaryote filament	eu	1	5	30	40	8	200	0	13.4	6696.6
2	Diachros sp.	chry	61	5	25	40	8	7	0	665.7	499306.0
2	Dictyosphaerium pulchellum	grn	3	5	25	40	5	6	0	8.0	9592.2
2	flagellate small	flag	42	5	25	40	6	9	0	193.4	165059.1
2	flagellate large	flag	2	5	25	40			0		
2	dinoflagellate	dino	2	5	25	40	18	20	0	276.3	89079.8
2	eucaryote sphere	eu	2	5	25	40	10	0	0	42.6	25579.2
2	eucaryote sphere	eu	1	5	25	40	12	0	0	36.8	18417.0
2	eucaryote sphere	eu	3	5	25	40	14	0	0	175.5	75202.9
4	Diachros sp.	chry	44	5	20	40	8	7	0	600.3	450194.0
4	Dictyosphaerium pulchellum	grn	8	5	20	40	5	6	0	26.6	31974.0
4	flagellate small	flag	35	5	20	40	6	9	0	201.4	171936.6
4	flagellate large	flag	3	5	20	40	20	24	0	767.4	217882.4
4	flagellate medium	flag	1	5	20	40	15	18	0	76.4	54648.9
4	Microcystis aeruginosa	bg	35	5	20	40	2	0	0	7.5	22381.8
4	eucaryote sphere	eu	2	5	20	40	12	0	0	92.1	46042.6
8	Diachros sp.	chry	56	5	20	40	8	7	0	764.0	572974.1
8	flagellate small	flag	36	5	20	40	6	9	0	207.2	176849.0
8	flagellate large	flag	3	5	20	40			0		
8	dinoflagellate	dino	1	5	20	40	16	18	0	122.8	44377.1
8	Microcystis aeruginosa	bg	200	5	20	40	2	0	0	42.6	127896.0
8	eucaryote sphere	eu	1	5	20	40	8	0	0	13.6	10231.7
8	eucaryote sphere	eu	1	5	20	40	10	0	0	26.6	15987.0

M	Genus and species	Fam.	Ct.	Vol. filt. ml	Flds. ctd.	Mag.	Dim1 µm	Dim2 µm	Dim3 µm	Biovol. mm³/m³	Surf.area mm²/m³
12	Diachros sp.	chry	51	5	20	40	8	7	0	695.8	521815.7
12	flagellate small	flag	30	5	20	40	6	9	0	172.7	147374.2
12	flagellate large	flag	1	5	20	40	18	30	0	259.0	75932.2
16	Diachros sp.	chry	63	5	20	40	8	7	0	859.5	644595.9
16	flagellate small	flag	20	5	20	40	6	9	0	115.1	98249.5
16	flagellate medium	flag	1	5	20	40	12	14	0	38.4	29087.9
16	flagellate medium	flag	1	5	20	40	16	22	0	102.3	67358.6
16	eucaryote sphere	eu	1	5	20	40	10	0	0	26.6	15987.0
18	Diachros sp.	chry	44	5	20	40	8	7	0	600.3	450194.0
18	flagellate small	flag	12	5	20	40	6	9	0	69.1	58949.7
18	flagellate small	flag	1	5	20	40	10	12	0	22.6	17627.1
18	flagellate medium	flag	1	5	20	40	16	20	0	95.5	66008.3
18	dinoflagellate	dino	1	5	20	40	17	22	0	169.4	55489.9
	May 10, 1979										
0	Diachros sp.	chry	32	5	20	40	8	7	0	436.6	327413.8
0	flagellate small	flag	45	5	20	40	6	9	0	259.0	221061.3
0	flagellate large	flag	7	5	20	40			0		
0	dinoflagellate	dino	2	5	20	40			0		
0	Stephanodiscus sp.	diat	2	5	20	40	20	16	0	511.6	166264.8
2	Diachros sp.	chry	15	5	25	40	8	7	0	163.7	122780.2
2	flagellate small	flag	85	5	25	40	6	9	0	391.4	334048.2
2	flagellate large	flag	4	5	25	40			0		
2	dinoflagellate	dino	6	5	25	40			0		
2	Oocystis sp.	grn	11	5	25	40	6	8	0	67.5	62217.0
4	Diachros sp.	chry	19	5	25	40	8	7	0	207.4	155521.5
4	flagellate small	flag	74	5	25	40	6	9	0	340.7	290818.4
4	flagellate large	flag	4	5	25	40			0		
4	dinoflagellate	dino	1	5	25	40	20	22	0	187.6	54601.2
4	Stephanodiscus sp.	diat	1	5	25	40	20	16	0	204.6	66505.9
4	Oocystis sp.	grn	5	5	25	40	6	8	0	30.7	28280.4
8	Diachros sp.	chry	22	5	25	40	8	7	0	240.1	180077.6
8	flagellate small	flag	58	5	25	40	6	9	0	267.0	227938.8
8	flagellate large	flag	6	5	25	40			0		
8	dinoflagellate	dino	1	5	25	40	16	18	0	98.2	35501.7
8	eucaryote sphere	eu	2	5	25	40	12	0	0	73.7	36834.1
8	eucaryote sphere	eu	1	5	25	40	15	0	0	71.9	28776.6
12	Diachros sp.	chry	23	5	20	40	8	7	0	313.8	235328.7
12	flagellate small	flag	25	5	20	40	6	9	0	143.9	122811.8
12	flagellate large	flag	4	5	20	40			0		
12	dinoflagellate	dino	2	5	20	40	22	25	0	644.8	169000.7
12	eucaryote sphere	eu	1	5	20	40	14	0	0	73.1	31334.5
16	Diachros sp.	chry	32	10	20	40	8	7	0	218.3	163706.9
16	flagellate small	flag	18	10	20	40	6	9	0	51.8	44212.3
16	dinoflagellate	dino	1	10	20	40	22	26	0	167.7	43455.3
16	Stephanodiscus sp.	diat	1	10	20	40	18	14	0	90.6	33093.1
16	Oocystis sp.	grn	4	10	20	40	8	10	0	34.1	23948.6
20	Diachros sp.	chry	30	10	25	40	8	7	0	163.7	122780.2
20	flagellate small	flag	25	10	25	40	6	9	0	57.6	49124.7
20	flagellate medium	flag	1	10	25	40	17	20	0	43.9	31080.9
20	Stephanodiscus sp.	diat	1	10	25	40	16	14	0	57.3	22509.7
	May 16, 1979										
0	Diachros sp.	chry	6	5	20	40	8	7	0	81.9	61390.1
0	flagellate small	flag	374	5	20	40	6	9	0	2152.5	1837265.0
0	flagellate large	flag	11	5	20	40			0		
0	Lyngbya limnetica	bg	17	5	20	40	1	0	0	5.3	21345.5

M	Genus and species	Fam.	Ct.	Vol. filt. ml	Flds. ctd.	Mag.	Dim1 µm	Dim2 µm	Dim3 µm	Biovol. mm³/m³	Surf.area mm²/m³
0	eucaryote sphere	eu	6	5	20	40	14	0	0	438.7	188007.1
0	Stephanodiscus sp.	diat	4	5	20	40	20	16	0	1023.2	332529.6
0	dinoflagellate	dino	4	5	20	40			0		
2	flagellate small	flag	379	3	20	40	6	9	0	3635.4	3103045.7
2	flagellate large	flag	6	3	20	40			0		
2	eucaryote sphere	eu	2	3	20	40	14	0	0	243.7	104448.4
2	Stephanodiscus sp.	diat	2	3	20	40	20	16	0	852.6	277108.0
2	dinoflagellate	dino	1	3	20	40	18	20	0	287.8	92791.4
2	Oocystis sp.	grn	3	3	20	40	8	10	0	85.3	59871.4
4	Dictyosphaerium pulchellum	grn	13	3	20	40	6	8	0	124.7	124698.6
4	flagellate small	flag	91	3	20	40	6	9	0	872.9	745058.4
4	eucaryote sphere	eu	2	3	20	40	12	0	0	153.5	76737.6
4	Stephanodiscus sp.	diat	2	3	20	40	20	16	0	852.6	277108.0
4	dinoflagellate	dino	1	3	20	40	18	20	0	287.8	92791.4
8	Diachros sp.	chry	3	3	20	40	8	7	0	68.2	51158.4
8	flagellate small	flag	44	3	20	40	6	9	0	422.1	360248.0
8	flagellate large	flag	5	3	20	40			0		
8	eucaryote sphere	eu	2	3	20	40	14	0	0	243.7	104448.4
12	Diachros sp.	chry	3	5	16	40	8	7	0	51.2	38368.8
12	Dictyosphaerium pulchellum	grn	4	5	16	40	6	0	0	28.8	28776.6
12	flagellate small	flag	224	5	16	40	6	9	0	1611.5	1375492.5
12	flagellate large	flag	3	5	16	40			0		
12	eucaryote sphere	eu	5	5	16	40	14	0	0	457.0	195840.8
12	Stephanodiscus sp.	diat	4	5	16	40	20	16	0	1279.0	415662.0
12	dinoflagellate	dino	2	5	16	40	20	22	0	586.2	170628.8
16	Aphanizomenon flos-aquae	bg	16	5	20	40	5	0	0	125.6	100449.2
16	Diachros sp.	chry	1	5	20	40	8	7	0	13.6	10231.7
16	flagellate small	flag	19	5	20	40	6	9	0	109.4	93337.0
16	eucaryote sphere	eu	2	5	20	40	14	0	0	146.2	62669.0
20	Diachros sp.	chry	3	10	20	40	8	7	0	20.5	15347.5
20	flagellate small	flag	22	10	20	40	6	9	0	63.3	54037.2
20	Stephanodiscus sp.	diat	1	10	20	40	35	28	0	685.4	127296.5
	May 25, 1979										
0	Diachros sp.	chry	20	10	20	40	7	7	0	91.4	78336.3
0	flagellate small	flag	658	10	20	40	6	9	0	1893.5	1616203.7
0	flagellate large	flag	7	10	20	40			0		
0	Stephanodiscus sp.	diat	4	10	20	40	20	16	0	511.6	166264.8
0	Oocystis sp.	grn	4	10	20	40	6	10	0	19.2	16873.8
0	Oocystis sp.	grn	1	10	20	40	10	15	0	20.0	10761.7
0	Scenedesmus sp.	grn	4	10	20	40	3	10	0	9.2	14045.2
0	eucaryote sphere	eu	100	10	20	40	3	0	0	36.0	71941.5
2	Diachros sp.	chry	12	5	20	40	7	7	0	109.7	94003.6
2	flagellate small	flag	378	5	20	40	6	9	0	2175.5	1856914.9
2	flagellate large	flag	5	5	20	40			0		
2	eucaryote sphere	eu	19	5	20	40	3	0	0	13.7	27337.8
2	Ankistrodesmus fractus	grn	4	5	20	40	1	35	0	1.9	11195.5
4	Diachros sp.	chry	16	5	20	40	7	7	0	146.2	125338.1
4	Dictyosphaerium pulchellum	grn	18	5	20	40	3	3	0	12.9	25898.9
4	Dictyosphaerium pulchellum	grn	4	5	20	40	3	7	0	2.9	5755.3
4	flagellate small	flag	357	5	20	40	6	9	0	2054.6	1753753.0
4	flagellate large	flag	2	5	20	40			0		
4	eucaryote sphere	eu	2	5	20	40	10	0	0	53.3	31974.0
4	eucaryote sphere	eu	38	5	20	40	3	0	0	27.3	54675.5
8	Diachros sp.	chry	21	5	20	40	7	7	0	191.9	164506.2
8	Dictyosphaerium pulchellum	grn	12	5	20	40	3	3	0	8.6	17266.0
8	flagellate small	flag	332	5	20	40	6	9	0	1910.8	1630941.1

M	Genus and species	Fam.	Ct.	Vol. filt. ml	Flds. ctd.	Mag.	Dim1 μm	Dim2 μm	Dim3 μm	Biovol. mm³/m³	Surf.area mm²/m³
8	flagellate large	flag	4	5	20	40			0		
16	Diachros sp.	chry	3	5	20	40	7	7	0	27.4	23500.9
16	Dictyosphaerium pulchellum	grn	2	5	20	40	3	0	0	1.4	2877.7
16	flagellate small	flag	32	5	20	40	6	9	0	184.2	157199.1
16	eucaryote sphere	eu	2	5	20	40	10	0	0	53.3	31974.0
16	Stephanodiscus sp.	diat	1	5	20	40	20	16	0	255.8	83132.4
	May 31, 1979										
0	Aphanizomenon flos-aquae	bg	21	5	20	40	5	0	0	164.8	131839.6
0	Diachros sp.	chry	11	5	20	40	8	7	0	150.1	112548.5
0	Dictyosphaerium pulchellum	grn	4	5	20	40	5	0	0	13.3	15987.0
0	flagellate small	flag	209	5	20	40	5	8	0	730.9	674253.5
0	flagellate large	flag	1	5	20	40	20	25	0	266.5	74839.2
0	eucaryote sphere	eu	2	5	20	40	10	0	0	53.3	31974.0
0	eucaryote sphere	eu	2	5	20	40	8	0	0	27.3	20463.4
0	blue green sphere	bg	48	5	20	40	5	0	0	159.9	191844.0
0	Stephanodiscus sp.	diat	1	5	20	40	20	16	0	255.8	83132.4
2	Aphanizomenon flos-aquae	bg	6	5	10	40	5	0	0	94.2	75336.9
2	Diachros sp.	chry	11	5	10	40	8	7	0	300.1	225097.0
2	Dictyosphaerium pulchellum	grn	16	5	10	40	5	0	0	106.6	127896.0
2	flagellate small	flag	339	5	10	40	6	9	0	3902.1	3330656.9
2	flagellate medium	flag	1	5	10	40	20	24	0	362.4	247597.1
2	flagellate medium	flag	1	5	10	40	15	15	0	134.9	107912.3
2	eucaryote sphere	eu	15	5	10	40	12	0	0	1381.3	690638.5
2	eucaryote sphere	eu	32	5	10	40	5	0	0	213.2	255792.0
2	dinoflagellate	dino	1	5	10	40	18	20	0	345.3	111349.7
4	Aphanizomenon flos-aquae	bg	4	3	20	40	5	0	0	52.3	41853.8
4	Diachros sp.	chry	10	3	20	40	8	7	0	227.4	170528.0
4	Dictyosphaerium pulchellum	grn	16	3	20	40	4	6	0	45.5	68211.2
4	flagellate small	flag	73	3	20	40	5	8	0	425.5	392508.0
4	flagellate small	flag	1	3	20	40	10	10	0	33.3	28865.4
4	flagellate medium	flag	1	3	20	40	15	20	0	137.4	92952.2
4	eucaryote sphere	eu	3	3	20	40	10	0	0	133.2	79935.0
4	eucaryote sphere	eu	23	3	20	40	5	0	0	127.7	153208.8
8	Diachros sp.	chry	6	5	20	40	8	7	0	81.9	61390.1
8	flagellate small	flag	74	5	20	40	5	8	0	258.8	238730.9
8	eucaryote sphere	eu	34	5	20	40	5	0	0	113.2	135889.5
8	eucaryote sphere	eu	2	5	20	40	12	0	0	92.1	46042.6
16	Diachros sp.	chry	7	5	20	40	8	7	0	95.5	71621.8
16	Dictyosphaerium pulchellum	grn	12	5	20	40	3	5	0	8.6	17266.0
16	flagellate small	flag	21	5	20	40	5	8	0	73.4	67748.0
16	eucaryote sphere	eu	16	5	20	40	5	0	0	53.3	63948.0
16	Stephanodiscus sp.	diat	1	5	20	40	45	36	0	2913.6	420857.8
20	Diachros sp.	chry	7	10	20	40	8	7	0	47.7	35810.9
20	Dictyosphaerium pulchellum	grn	6	10	20	40	4	6	0	5.1	7673.8
20	flagellate small	flag	26	10	20	40	5	8	0	45.5	41939.2
20	eucaryote sphere	eu	10	10	20	40	5	0	0	16.7	19983.8
20	Stephanodiscus sp.	diat	3	10	20	40	20	16	0	383.7	124698.6
20	Stephanodiscus sp.	diat	1	10	20	40	60	45	0	3237.4	359707.5
	June 4, 1979										
0	Aphanizomenon flos-aquae	bg	16	3	20	40	5	0	0	209.3	167415.3
0	Diachros sp.	chry	8	3	20	40	8	7	0	181.9	136422.4
0	Dictyosphaerium pulchellum	grn	8	3	20	40	5	0	0	44.4	53290.0
0	flagellate small	flag	488	3	20	40	5	9	0	3115.2	2887801.8
0	flagellate large	flag	1	3	20	40	28	30	0	1044.5	218912.6
0	flagellate medium	flag	1	3	20	40	20	24	0	302.0	206330.9
0	eucaryote sphere	eu	4	3	20	40	12	0	0	307.0	153475.2
0	eucaryote sphere	eu	13	3	20	40	5	0	0	72.2	86596.3
0	Oocystis sp.	grn	2	3	20	40	6	13	0	41.6	35153.3
2	Diachros sp.	chry	6	3	20	40	8	7	0	136.4	102316.8
2	flagellate small	flag	464	3	20	40	6	9	0	4450.8	3798979.3
2	flagellate medium	flag	1	3	20	40	16	18	0	147.8	108525.9
2	flagellate large	flag	1	3	20	40	26	40	0	1200.8	247444.1
2	eucaryote sphere	eu	27	3	20	40	5	0	0	149.9	179853.8
2	eucaryote sphere	eu	2	3	20	40	10	0	0	88.8	53290.0
4	Aphanizomenon flos-aquae	bg	19	3	20	40	5	0	0	248.6	198805.7
4	Diachros sp.	chry	6	3	20	40	8	7	0	136.4	102316.8
4	flagellate small	flag	435	3	20	40	6	9	0	4172.6	3561543.2
4	eucaryote sphere	eu	36	3	20	40	5	0	0	199.8	239805.0
4	eucaryote sphere	eu	1	3	20	40	12	0	0	76.7	38368.8
8	Diachros sp.	chry	4	3	20	40	8	7	0	90.9	68211.2
8	Dictyosphaerium pulchellum	grn	12	3	20	40	4	6	0	34.1	51158.4
8	flagellate small	flag	285	3	20	40	6	9	0	2733.8	2333424.8
8	flagellate medium	flag	1	3	20	40	15	15	0	112.4	89926.9
8	flagellate medium	flag	1	3	20	40	16	16	0	136.4	108001.1
8	eucaryote sphere	eu	17	3	20	40	5	0	0	94.4	113241.3
8	eucaryote sphere	eu	1	3	20	40	12	0	0	76.7	38368.8
12	Diachros sp.	chry	4	3	20	40	8	7	0	90.9	68211.2
12	flagellate small	flag	14	3	20	40	6	9	0	134.3	114624.4
16	Diachros sp.	chry	4	10	20	40	8	7	0	27.3	20463.4
16	flagellate small	flag	28	10	20	40	6	9	0	80.6	68774.6
16	eucaryote sphere	eu	18	10	20	40	5	0	0	30.0	35970.8
16	Stephanodiscus sp.	diat	1	10	20	40	20	16	0	127.9	41566.2
16	Stephanodiscus sp.	diat	1	10	20	40	20	12	0	95.9	35171.4
20	Diachros sp.	chry	6	10	20	40	8	7	0	40.9	30695.0
20	flagellate small	flag	120	10	20	40	6	9	0	345.3	294748.4
20	eucaryote sphere	eu	12	10	20	40	5	0	0	20.0	23980.5
20	eucaryote sphere	eu	3	10	20	40	10	0	0	40.0	23980.5
20	Stephanodiscus sp.	diat	1	10	20	40	64	50	0	4092.7	419498.9
	June 11, 1979										
0	Aphanizomenon flos-aquae	bg	4	10	8	40	5	0	0	39.2	31390.4
0	Dictyosphaerium pulchellum	grn	4	10	8	40	5	0	0	16.7	19983.8
0	flagellate small	flag	244	10	8	40	5	8	0	1066.6	983958.5
0	flagellate large	flag	6	10	8	40			0		
0	eucaryote sphere	eu	1	10	8	40	10	0	0	33.3	19983.8
0	Oocystis sp.	grn	1	10	8	40	10	15	0	50.0	26904.3
2	flagellate small	flag	347	5	25	40	5	8	0	970.8	895563.6
2	flagellate large	flag	8	5	25	40			0		
4	Dictyosphaerium pulchellum	grn	26	5	20	40	5	0	0	86.6	103915.5
4	flagellate small	flag	298	5	20	40	5	8	0	1042.2	961375.9
4	flagellate medium	flag	3	5	20	40	14	16	0	180.2	134112.1
4	eucaryote sphere	eu	4	5	20	40	10	0	0	106.6	63948.0
4	Oocystis sp.	grn	7	5	20	40	6	8	0	53.7	49490.8
8	Dictyosphaerium pulchellum	grn	35	5	20	40	4	0	0	59.7	89527.2
8	flagellate small	flag	170	5	20	40	5	8	0	594.5	548435.9
8	flagellate large	flag	3	5	20	40			0		
8	eucaryote sphere	eu	2	5	20	40	10	0	0	53.3	31974.0
12	Aphanizomenon flos-aquae	bg	7	5	20	40	5	0	0	54.9	43946.5
12	Dictyosphaerium pulchellum	grn	9	5	20	40	4	0	0	15.3	23021.3
12	flagellate small	flag	31	5	20	40	5	8	0	108.4	100008.9
16	flagellate small	flag	30	10	20	40	5	8	0	52.5	48391.4

M	Genus and species	Fam.	Ct.	Vol. filt. ml	Flds. ctd.	Mag.	Dim1 μm	Dim2 μm	Dim3 μm	Biovol. mm³/m³	Surf.area mm²/m³
	June 18, 1979										
0	flagellate small	flag	7	15	30	40	5	9	0	6.0	5523.1
0	Pandorina sp.	grn	12	15	30	40	8	0	0	36.4	27284.5
0	Pandorina sp.	grn	8	15	30	40	5	0	0	5.9	7105.3
0	Oocystis sp.	grn	2	15	30	40	9	12	0	11.5	7070.1
0	Staurastrum sp.	grn	1	15	8	10	6	12	20	0.2	237.7
0	Stephanodiscus sp.	diat	2	15	8	10	20	16	0	1.7	540.3
0	Cosmarium sp.	desm	2	15	8	10	25	40	0	8.7	1841.6
2	Aphanizomenon flos-aquae	bg	35	20	25	40	5	0	0	54.9	43946.5
2	flagellate small	flag	3	20	25	40	5	9	0	2.3	2130.3
2	Pandorina sp.	grn	16	20	25	40	6	0	0	18.4	18417.0
2	Cosmarium sp.	desm	1	20	25	40	15	25	0	60.0	21092.3
4	Aphanizomenon flos-aquae	bg	31	20	35	40	5	0	0	34.8	27802.9
4	flagellate small	flag	10	20	35	40	5	9	0	5.5	5072.3
4	flagellate medium	flag	1	20	35	40	16	20	0	13.6	9429.8
4	Pandorina sp.	grn	72	20	35	40	5	0	0	34.3	41109.4
4	Pandorina sp.	grn	24	20	35	40	6	0	0	19.7	19732.5
4	Pandorina sp.	grn	16	20	35	40	8	0	0	31.2	23386.7
4	Oocystis sp.	grn	3	20	35	40	9	14	0	12.9	7692.0
4	eucaryote filament	eu	36	20	35	40	5	0	0	40.4	32287.2
8	Aphanizomenon flos-aquae	bg	16	20	35	40	5	0	0	17.9	14349.9
8	flagellate small	flag	11	20	35	40	5	9	0	6.0	5579.5
8	Pandorina sp.	grn	32	20	35	40	5	0	0	15.2	18270.9
8	Pandorina sp.	grn	6	20	35	40	8	0	0	11.7	8770.0
8	Pandorina sp.	grn	2	20	35	40	10	0	0	7.6	4567.7
8	Pandorina sp.	grn	8	20	35	40	4	0	0	1.9	2923.3
8	Oocystis sp.	grn	2	20	35	40	18	30	0	74.0	21694.9
	June 25, 1979										
0	Aphanizomenon flos-aquae	bg	184	10	25	40	4	0	0	369.7	369653.1
0	flagellate small	flag	1	10	25	40	5	9	0	1.5	1420.2
0	flagellate large	flag	1	10	25	40	25	28	0	186.5	43200.8
0	flagellate large	flag	1	10	25	40	25	40	0	266.5	56669.7
0	Microcystis aeruginosa	bg	76	10	25	40	3	0	0	21.9	43740.4
0	Microcystis aeruginosa	bg	6640	10	25	40	1	0	0	70.8	424614.8
0	eucaryote sphere	eu	1	10	25	40	10	0	0	10.7	6394.8
0	Oocystis sp.	grn	1	10	25	40	15	20	0	48.0	17675.3
0	Oocystis sp.	grn	5	10	25	40	10	12	0	63.9	36313.7
0	Schroederia setigera	grn	55	10	25	40	2	10	0	11.7	35867.9
0	Cosmarium sp.	desm	1	10	25	40	18	32	0	221.0	64083.9
2	Aphanizomenon flos-aquae	bg	302	10	20	40	4	0	0	758.6	758391.5
2	flagellate small	flag	2	10	20	40	5	9	0	3.8	3550.6
2	Microcystis aeruginosa	bg	14	10	20	40	3	0	0	5.0	10071.8
2	Microcystis aeruginosa	bg	4880	10	20	40	1	0	0	65.0	390082.8
2	eucaryote sphere	eu	4	10	20	40	5	0	0	6.7	7993.5
2	Staurastrum sp.	grn	7	10	20	40	10	15	32	854.9	586963.3
2	Oocystis sp.	grn	3	10	20	40	6	10	0	14.4	12655.4
2	Schroederia setigera	grn	48	10	20	40	2	10	0	12.8	39128.7
2	Cosmarium sp.	desm	1	10	20	40	15	26	0	155.9	54466.8
2	Pandorina sp.	grn	12	10	20	40	10	0	0	159.9	95922.0
2	blue green small	bg	2	10	20	40	14	18	0	73.1	31334.5
2	blue green small	bg	1	10	20	40	10	14	0	13.3	7993.5
2	blue green small	bg	2	10	20	40	12	15	0	46.0	23021.3
4	Aphanizomenon flos-aquae	bg	197	10	20	40	4	0	0	494.8	494712.3
4	flagellate large	flag	1	10	20	40	18	30	0	129.5	37966.1
4	Microcystis aeruginosa	bg	8700	10	20	40	1	0	0	115.9	695434.6

M	Genus and species	Fam.	Ct.	Vol. filt. ml	Flds. ctd.	Mag.	Dim1 μm	Dim2 μm	Dim3 μm	Biovol. mm³/m³	Surf.area mm²/m³
4	Oocystis sp.	grn	17	10	20	40	10	16	0	362.4	192677.1
4	Schroederia setigera	grn	15	10	20	40	2	10	0	4.0	12227.7
8	Aphanizomenon flos-aquae	bg	242	20	20	40	4	0	0	303.9	303858.8
8	flagellate medium	flag	1	20	20	40	12	15	0	10.1	7364.0
8	Microcystis aeruginosa	bg	7400	20	20	40	1	0	0	49.3	295759.5
8	eucaryote sphere	eu	12	20	20	40	5	0	0	10.0	11990.3
8	Oocystis sp.	grn	16	20	20	40	10	16	0	170.5	90671.6
8	Oocystis sp.	grn	12	20	20	40	6	10	0	28.8	25310.7
8	Schroederia setigera	grn	9	20	20	40	2	10	0	1.2	3668.3
8	Cosmarium sp.	desm	3	20	20	40	15	26	0	233.8	81700.2
8	Pandorina sp.	grn	64	20	20	40	10	0	0	426.3	255792.0
12	eucaryote sphere	eu	1	20	32	40	10	0	0	4.2	2498.0
12	Oocystis sp.	grn	4	20	32	40	16	25	0	106.6	35580.2
12	Schroederia setigera	grn	8	20	32	40	2	10	0	0.7	2038.0
12	Pandorina sp.	grn	16	20	32	40	8	0	0	34.1	25579.2
	July 2, 1979										
0	Aphanizomenon flos-aquae	bg	333	3	20	40	4	0	0	2788.1	2787465.4
0	flagellate small	flag	6	3	20	40	6	10	0	62.3	52757.1
0	flagellate medium	flag	1	3	20	40	18	22	0	223.0	153112.8
0	Microcystis aeruginosa	bg	10	3	20	40	3	0	0	12.0	23980.5
0	Microcystis aeruginosa	bg	2120	3	20	40	1	0	0	94.1	564874.1
0	Oocystis sp.	grn	1	3	20	40	16	20	0	227.4	79828.5
0	Schroederia setigera	grn	21	3	20	40	2	10	0	18.7	57062.6
0	eucaryote filament	eu	18	3	20	40	5	0	0	235.5	188342.3
0	Ceratium hirundinella	dino	10	5	12	10	40	60	0	83.1	14985.8
0	Staurastrum sp.	grn	9	5	12	10					
0	Stephanodiscus sp.	diat	5	5	12	10			0		
0	Cosmarium sp.	desm	28	5	12	10			0		
2	Aphanizomenon flos-aquae	bg	559	5	20	40	4	0	0	2808.2	2807555.3
2	flagellate small	flag	14	5	20	40	6	10	0	87.3	73859.9
2	flagellate medium	flag	3	5	20	40	15	18	0	229.3	163946.6
2	Microcystis aeruginosa	bg	135	5	20	40	3	0	0	97.1	194242.1
2	Microcystis aeruginosa	bg	4820	5	20	40	1	0	0	128.4	770573.5
2	Schroederia setigera	grn	56	5	20	40	2	10	0	29.8	91300.2
2	Ceratium hirundinella	dino	5	5	8	10			0		
2	Staurastrum sp.	grn	8	5	8	10					
2	Stephanodiscus sp.	diat	1	5	8	10	70	40	0	76.4	8182.8
2	Cosmarium sp.	desm	20	5	8	10			0		
4	Aphanizomenon flos-aquae	bg	380	5	20	40	4	0	0	1909.0	1908534.9
4	flagellate small	flag	16	5	20	40	6	10	0	99.8	84411.4
4	Microcystis aeruginosa	bg	93	5	20	40	3	0	0	66.9	133811.2
4	Microcystis aeruginosa	bg	3800	5	20	40	1	0	0	101.3	607506.1
4	eucaryote sphere	eu	1	5	20	40	10	0	0	26.6	15987.0
4	Schroederia setigera	grn	35	5	20	40	2	10	0	18.7	57062.6
4	Ceratium hirundinella	dino	5	5	11	10			0		
4	Cosmarium sp.	desm	6	5	11	10	25	41	0	58.1	12299.1
8	Aphanizomenon flos-aquae	bg	139	5	20	40	4	0	0	698.3	698122.0
8	Microcystis aeruginosa	bg	3000	5	20	40	1	0	0	79.9	479610.0
8	eucaryote sphere	eu	26	5	20	40	10	0	0	692.8	415662.0
8	Oocystis sp.	grn	3	5	20	40	8	14	0	71.6	46849.7
8	Schroederia setigera	grn	36	5	20	40	2	10	0	19.2	58693.0
8	Cosmarium sp.	desm	1	5	20	40	20	35	0	746.1	195207.2
8	Staurastrum sp.	grn	3	5	7	10					
8	Stephanodiscus sp.	diat	4	5	7	10	80	50	0	570.0	51300.9
8	Cosmarium sp.	desm	18	5	7	10			0		
12	Aphanizomenon flos-aquae	bg	11	10	20	40	4	0	0	27.6	27623.5

M	Genus and species	Fam.	Ct.	Vol. filt. ml	Flds. ctd.	Mag.	Dim1 µm	Dim2 µm	Dim3 µm	Biovol. mm³/m³	Surf.area mm²/m³
12	Microcystis aeruginosa	bg	3900	10	20	40	1	0	0	52.0	311746.5
12	eucaryote sphere	eu	2	10	20	40	10	0	0	26.6	15987.0
12	Stephanodiscus sp.	diat	1	10	20	40	50	36	0	1798.5	243801.8
12	Oocystis sp.	grn	6	10	20	40	6	10	0	28.8	25310.7
12	Pandorina sp.	grn	6	10	20	40	10	0	0	79.9	47961.0
12	Schroederia setigera	grn	13	10	20	40	2	10	0	3.5	10597.3
16	Microcystis aeruginosa	bg	3000	10	25	40	1	0	0	32.0	191844.0
16	Microcystis aeruginosa	bg	18	10	25	40	2	0	0	1.5	4604.3
16	eucaryote sphere	eu	1	10	25	40	5	0	0	1.3	1598.7
16	Oocystis sp.	grn	9	10	25	40	6	10	0	34.5	30372.9
16	Schroederia setigera	grn	5	10	25	40	2	10	0	1.1	3260.7
	July 9, 1979										
0	Aphanizomenon flos-aquae	bg	1156	3	20	40	4	0	0	9679.0	9676606.9
0	Aphanizomenon flos-aquae	bg	27	3	20	40	5	9	0	353.2	282513.4
0	flagellate small	flag	43	3	20	40	5	9	0	274.5	254457.9
0	Ceratium hirundinella	dino	13	10	7	10	40	45	0	69.5	13942.2
2			11	2	20	40	8	15	0		
2	Aphanizomenon flos-aquae	bg	458	2	20	40	4	0	0	5752.1	5750717.1
2	Aphanizomenon heterocyst	bg	6	2	20	40	5	9	0	89.9	93684.1
2	flagellate small	flag	48	2	20	40	5	9	0	459.6	426069.1
2	Microcystis aeruginosa	bg	30	2	20	40	2	0	0	16.0	47961.0
4	Anabaena sp.	bg	32	2	20	40	4	0	0	401.9	401796.8
4	Aphanizomenon flos-aquae	bg	280	2	20	40	4	0	0	3516.6	3515722.3
4	Aphanizomenon heterocyst	bg	4	2	20	40	5	9	0	60.0	62456.0
4	flagellate small	flag	32	2	20	40	5	9	0	306.4	284046.1
4	blue green filaments	bg	4	2	20	40	6	0	0	113.0	75336.9
4	Oocystis sp.	grn	4	2	20	40	8	12	0	204.6	137750.0
8	Aphanizomenon flos-aquae	bg	141	5	20	40	4	0	0	708.3	708166.9
8	Aphanizomenon heterocyst	bg	211	5	20	40	5	9	0	1265.0	1317822.6
8	Coelosphaerium sp.	bg	40	5	20	40	1	2	0	1.1	6394.8
8	flagellate small	flag	16	5	20	40	5	9	0	61.3	56809.2
8	Microcystis aeruginosa	bg	110	5	20	40	2	0	0	23.4	70342.8
8	eucaryote sphere	eu	17	5	20	40	5	0	0	56.6	67944.8
	July 16, 1979										
0	Anabaena sp.	bg	13	2	20	40	5	0	0	255.1	204037.5
0	Aphanizomenon flos-aquae	bg	743	2	20	40	4	0	0	9331.5	9329220.0
0	Aphanizomenon heterocyst	bg	4	2	20	40	5	9	0	60.0	62456.0
0	Coelosphaerium sp.	bg	29	2	20	40	2	3	0	15.5	46362.3
0	flagellate small	flag	3	2	20	40	5	9	0	28.7	26629.3
0	Microcystis aeruginosa	bg	8	2	20	40	2	0	0	4.3	12789.6
0	Cosmarium sp.	desm	2	2	9	10			0		
2	Anabaena sp.	bg	8	2	20	40	5	0	0	157.0	125561.5
2	Aphanizomenon flos-aquae	bg	663	2	20	40	4	0	0	8326.8	8324728.1
2	Aphanizomenon heterocyst	bg	9	2	20	40	5	9	0	134.9	140526.1
2	flagellate small	flag	16	2	20	40	5	9	0	153.2	142023.0
2	Microcystis aeruginosa	bg	14	2	20	40	2	0	0	7.5	22381.8
4	Anabaena sp.	bg	9	2	20	40	5	0	0	176.6	141256.7
4	Aphanizomenon flos-aquae	bg	962	2	20	40	4	0	0	12082.0	12079017.3
4	Aphanizomenon heterocyst	bg	17	2	20	40	5	9	0	254.8	265438.2
4	Coelosphaerium sp.	bg	38	2	20	40	2	3	0	20.3	60750.6
4	Microcystis aeruginosa	bg	6	2	20	40	2	0	0	3.2	9592.2
8	Aphanizomenon flos-aquae	bg	60	5	20	40	4	0	0	301.4	301347.6
8	flagellate small	flag	12	5	20	40	5	9	0	46.0	42606.9
8	blue green sphere	bg	1	5	20	40	8	0	0	13.6	10231.7
8	Oocystis sp.	grn	10	5	20	40	4	6	0	25.6	34437.5

M	Genus and species	Fam.	Ct.	Vol. filt. ml	Flds. ctd.	Mag.	Dim1 µm	Dim2 µm	Dim3 µm	Biovol. mm³/m³	Surf.area mm²/m³
20	Aphanizomenon flos-aquae	bg	21	10	20	40	4	0	0	52.7	52735.8
20	flagellate small	flag	3	10	20	40	5	9	0	5.7	5325.9
20	Microcystis aeruginosa	bg	66	10	20	40	2	0	0	7.0	21102.8
20	Oocystis sp.	grn	1	10	20	40	11	15	0	24.2	12087.5
	July 23, 1979										
0	Aphanizomenon flos-aquae	bg	512	2	20	40	4	0	0	6430.3	6428749.3
0	Aphanizomenon heterocyst	bg	5	2	20	40	5	9	0	74.9	78070.1
0	dinoflagellate	dino	1	2	20	40	10	13	0	86.6	48165.8
0	flagellate medium	flag	1	2	20	40	11	11	0	66.5	56420.8
0	flagellate small	flag	48	2	20	40	5	9	0	459.6	426069.1
0	Microcystis aeruginosa	bg	12	2	20	40	4	0	0	51.2	76737.6
0	Microcystis aeruginosa	bg	40	2	20	40	2	0	0	21.3	63948.0
0	Ceratium hirundinella	dino	4	2	11	10	50	60	0	141.7	22133.1
0	Staurastrum sp.	grn	2	2	11	10	10	15	25	6.8	4816.6
0	Cosmarium sp.	desm	1	2	11	10			0		
2	Anabaena sp.	bg	12	2	11	40	5	0	0	428.2	342440.5
2	Aphanizomenon flos-aquae	bg	300	2	11	40	4	0	0	6850.5	6848809.5
2	Aphanizomenon heterocyst	bg	3	2	11	40	5	9	0	81.8	85167.3
2	flagellate medium	flag	1	2	11	40	15	18	0	347.4	248403.9
2	flagellate small	flag	34	2	11	40	5	9	0	591.9	548725.4
2	Microcystis aeruginosa	bg	41	2	11	40	2	0	0	39.7	119175.8
4	Anabaena sp.	bg	29	2	11	40	5	0	0	1034.7	827564.5
4	Aphanizomenon flos-aquae	bg	260	2	11	40	4	0	0	5937.1	5935635.0
4	flagellate small	flag	16	2	11	40	5	9	0	278.6	258223.7
4	Microcystis aeruginosa	bg	10	2	11	40	2	0	0	9.7	29067.3
4	Stephanodiscus sp.	diat	1	2	11	40	14	8	0	284.9	152603.2
8	Anabaena sp.	bg	4	10	11	40	5	0	0	28.5	22829.4
8			1	10	11	40	10	20	0		
8	Aphanizomenon flos-aquae	bg	416	10	11	40	4	0	0	1899.9	1899403.2
8	Aphanizomenon heterocyst	bg	8	10	11	40	5	9	0	43.6	45422.6
8	Coelosphaerium sp.	bg	20	10	11	40	2	3	0	3.9	11626.9
8	Microcystis aeruginosa	bg	288	10	11	40	4	0	0	446.5	669710.0
8	Microcystis aeruginosa	bg	128	10	11	40	2	0	0	24.8	74412.2
8	Microcystis aeruginosa	bg	40	10	11	40	5	0	0	121.1	145336.4
8	Oocystis sp.	grn	8	10	11	40	4	10	0	31.0	38714.8
8	Oocystis sp.	grn	1	10	11	40	8	12	0	18.6	12522.7
8	Oocystis sp.	grn	5	10	11	40	5	9	0	27.3	28389.1
12	Anabaena sp.	bg	7	10	11	40	5	0	0	50.0	39951.4
12	Aphanizomenon flos-aquae	bg	7	10	11	40	4	0	0	32.0	31961.1
12	Coelosphaerium sp.	bg	200	10	11	40	2	3	0	38.8	116269.1
12	flagellate small	flag	9	10	11	40	5	9	0	31.3	29050.2
12	Microcystis aeruginosa	bg	17	10	11	40	2	0	0	3.3	9882.9
	July 30, 1979										
0	Aphanizomenon flos-aquae	bg	446	3	25	40	4	0	0	2987.4	2986689.8
0	flagellate medium	flag	1	3	25	40	16	18	0	118.2	86820.7
0	flagellate small	flag	113	3	25	40	6	10	0	939.4	794873.7
0	Lyngbya limnetica	bg	239	3	25	40	2	0	0	400.2	800245.4
0	Microcystis aeruginosa	bg	189	3	25	40	3	0	0	181.3	362585.2
2	Anabaena sp.	bg	11	3	20	40	5	0	0	143.9	115098.1
2	Aphanizomenon flos-aquae	bg	274	3	20	40	4	0	0	2294.2	2293590.2
2	Aphanizomenon heterocyst	bg	2	3	20	40	5	9	0	20.0	20818.7
2	Coelosphaerium sp.	bg	60	3	20	40	2	3	0	21.3	63948.0
2	flagellate small	flag	84	3	20	40	5	9	0	536.2	497080.6
2	flagellate medium	flag	2	3	20	40	18	22	0	446.0	306225.5
2	Lyngbya limnetica	bg	140	3	20	40	2	0	0	293.0	585953.7

M	Genus and species	Fam.	Ct.	Vol. filt. ml	Flds. ctd.	Mag.	Dim1 μm	Dim2 μm	Dim3 μm	Biovol. mm³/m³	Surf.area mm²/m³
2	Microcystis aeruginosa	bg	70	3	20	40	2	0	0	24.9	74606.0
2	eucaryote sphere	eu	7	3	20	40	5	0	0	38.9	46628.8
2	Stephanodiscus sp.	diat	2	3	20	40	17	11	0	423.5	176656.4
2	Stephanodiscus sp.	diat	1	3	20	40	35	20	0	1632.0	349715.6
4	Anabaena sp.	bg	22	3	20	40	5	0	0	287.8	230196.1
4	Aphanizomenon flos-aquae	bg	350	3	20	40	4	0	0	2930.5	2929768.5
4	Aphanizomenon heterocyst	bg	1	3	20	40	5	9	0	10.0	10409.3
4	Dictyosphaerium pulchellum	grn	12	3	20	40	4	0	0	34.1	51158.4
4	flagellate small	flag	68	3	20	40	5	9	0	434.1	402398.6
4	flagellate medium	flag	7	3	20	40	16	20	0	1114.1	770097.2
4	Lyngbya limnetica	bg	100	3	20	40	2	0	0	209.3	418538.4
4	Microcystis aeruginosa	bg	15	3	20	40	2	0	0	5.3	15987.0
4	Microcystis aeruginosa	bg	116	3	20	40	4	0	0	329.7	494531.3
8	Aphanizomenon flos-aquae	bg	45	10	20	40	4	0	0	113.0	113005.4
8	flagellate small	flag	13	10	20	40	5	9	0	24.9	23078.7
8	Lyngbya limnetica	bg	6	10	20	40	2	0	0	3.8	7533.7
8	eucaryote filament	eu	14	10	20	40	5	0	0	54.9	43946.5
	Aug. 13, 1979										
0	Anabaena sp.	bg	19	5	20	40	5	0	0	149.1	119283.4
0	Aphanizomenon flos-aquae	bg	373	5	20	40	4	0	0	1873.8	1873377.7
0	Aphanizomenon heterocyst	bg	5	5	20	40	5	9	0	30.0	31228.0
0	flagellate medium	flag	1	5	20	40	12	15	0	40.3	29455.9
0	flagellate small	flag	66	5	20	40	4	8	0	140.7	150649.0
0	Lyngbya limnetica	bg	362	5	20	40	1	0	0	113.7	454532.7
0	Lyngbya limnetica	bg	52	5	20	40	3	0	0	146.9	195876.0
0	eucaryote sphere	eu	11	5	20	40	4	0	0	18.8	28137.1
0	Schroederia setigera	grn	1	5	20	40	4	15	0	3.2	4963.7
2	Anabaena sp.	bg	47	3	30	40	5	0	0	409.9	327855.1
2	Aphanizomenon flos-aquae	bg	155	3	30	40	4	0	0	865.2	864979.3
2	flagellate small	flag	56	3	30	40	5	9	0	238.3	220924.7
2	Lyngbya limnetica	bg	379	3	30	40	1	0	0	132.2	528753.5
2	eucaryote sphere	eu	8	3	30	40	4	0	0	15.2	22737.1
2	eucaryote sphere	eu	1	3	30	40	16	0	0	121.3	45474.1
2	Actinastrum Hantzschii	grn	40	3	30	40	2	14	0	126.7	271404.0
4	Anabaena sp.	bg	4	5	15	40	5	0	0	41.9	33483.1
4	Aphanizomenon flos-aquae	bg	309	5	15	40	4	0	0	2069.8	2069253.7
4	Dictyosphaerium pulchellum	grn	8	5	15	40	5	0	0	35.5	42632.0
4	flagellate small	flag	55	5	15	40	5	9	0	280.9	260375.6
4	Lyngbya limnetica	bg	97	5	15	40	3	0	0	365.5	487178.6
4	Lyngbya limnetica	bg	363	5	15	40	1	0	0	152.0	607717.7
4	Microcystis aeruginosa	bg	30	5	15	40	3	0	0	28.8	57553.2
4	Actinastrum Hantzschii	grn	16	5	15	40	2	14	0	60.8	130273.9
8	Anabaena sp.	bg	31	5	20	40	5	0	0	243.3	194620.3
8	Aphanizomenon flos-aquae	bg	199	5	20	40	4	0	0	999.7	999469.6
8	Dictyosphaerium pulchellum	grn	48	5	20	40	5	0	0	159.9	191844.0
8	flagellate small	flag	74	5	20	40	5	9	0	283.4	262742.6
8	Lyngbya limnetica	bg	148	5	20	40	3	0	0	418.2	557493.1
8	Lyngbya limnetica	bg	523	5	20	40	1	0	0	164.2	656686.7
8	Actinastrum Hantzschii	grn	40	5	20	40	2	14	0	114.0	244263.6
12	Aphanizomenon flos-aquae	bg	21	5	20	40	4	0	0	105.5	105471.7
12	flagellate small	flag	8	5	20	40	5	9	0	30.6	28404.6
12	Lyngbya limnetica	bg	7	5	20	40	3	0	0	19.8	26367.9
12	Lyngbya limnetica	bg	145	5	20	40	1	0	0	45.5	182064.2
12	Microcystis aeruginosa	bg	21	5	20	40	5	0	0	69.9	83931.8
12	eucaryote sphere	eu	1	5	20	40	8	0	0	13.6	10231.7
16	Anabaena sp.	bg	2	5	20	40	5	0	0	15.7	12556.2

M	Genus and species	Fam.	Ct.	Vol. filt. ml	Flds. ctd.	Mag.	Dim1 μm	Dim2 μm	Dim3 μm	Biovol. mm³/m³	Surf.area mm²/m³
16	Aphanizomenon flos-aquae	bg	31	5	20	40	4	0	0	155.7	155696.3
16	flagellate small	flag	7	5	20	40	5	9	0	26.8	24854.0
16	Lyngbya limnetica	bg	18	5	20	40	3	0	0	50.9	67803.2
16	Lyngbya limnetica	bg	29	5	20	40	1	0	0	9.1	36412.8
16	Microcystis aeruginosa	bg	180	5	20	40	4	0	0	307.0	460425.6
16	Actinastrum Hantzschii	grn	8	5	20	40	2	14	0	22.8	48852.7
20	Aphanizomenon flos-aquae	bg	46	10	20	40	4	0	0	115.5	115516.6
20	Coelosphaerium sp.	bg	40	10	20	40	3	5	0	14.4	28776.6
20	flagellate small	flag	6	10	20	40	5	9	0	11.5	10651.7
20	Lyngbya limnetica	bg	123	10	20	40	1	0	0	19.3	77220.3
20	Oocystis sp.	grn	1	10	20	40	8	14	0	11.9	7808.3
	Aug. 20, 1979										
0	Anabaena sp.	bg	82	5	25	40	5	0	0	514.9	411841.8
0	Aphanizomenon flos-aquae	bg	230	5	25	40	4	0	0	924.4	924132.7
0	Aphanizomenon heterocyst	bg	1	5	25	40	5	9	0	4.8	4996.5
0	Coelosphaerium sp.	bg	100	5	25	40	2	3	0	17.1	51158.4
0	Dictyosphaerium pulchellum	grn	12	5	25	40	4	0	0	16.4	24556.0
0	flagellate small	flag	52	5	25	40	5	9	0	159.3	147704.0
0	flagellate medium	flag	1	5	25	40	16	20	0	76.4	52806.7
0	Lyngbya limnetica	bg	143	5	25	40	3	0	0	323.3	430927.1
0	Lyngbya limnetica	bg	38	5	25	40	1	0	0	9.5	38170.7
0	eucaryote sphere	eu	17	5	25	40	5	0	0	45.3	54355.8
0	Actinastrum Hantzschii	grn	20	5	25	40	3	22	0	161.2	229607.8
0	Actinastrum Hantzschii	grn	8	5	25	40	2	14	0	18.2	39082.2
2	Anabaena sp.	bg	35	5	25	40	5	0	0	219.8	175786.1
2	Aphanizomenon flos-aquae	bg	153	5	25	40	4	0	0	614.9	614749.1
2	Aphanizomenon heterocyst	bg	1	5	25	40	5	9	0	4.8	4996.5
2	flagellate small	flag	23	5	25	40	5	9	0	70.5	65330.6
2	Lyngbya limnetica	bg	32	5	25	40	1	0	0	8.0	32143.7
2	Lyngbya limnetica	bg	146	5	25	40	3	0	0	330.1	439967.5
2	Microcystis aeruginosa	bg	260	5	25	40	3	0	0	149.6	299276.7
2	eucaryote sphere	eu	2	5	25	40	12	0	0	73.7	36834.1
2	eucaryote sphere	eu	13	5	25	40	4	0	0	17.7	26602.4
2	eucaryote sphere	eu	1	5	25	40	6	0	0	4.6	4604.3
2	eucaryote sphere	eu	1	5	25	40	10	0	0	21.3	12789.6
2	Actinastrum Hantzschii	grn	48	5	25	40	2	14	0	109.4	234493.1
2	Actinastrum Hantzschii	grn	24	5	25	40	3	22	0	193.5	275529.3
4	Anabaena sp.	bg	32	5	25	40	5	0	0	200.9	160718.7
4	Aphanizomenon flos-aquae	bg	158	5	25	40	4	0	0	635.0	634839.0
4	Coelosphaerium sp.	bg	80	5	25	40	2	3	0	13.6	40926.7
4	flagellate small	flag	37	5	25	40	5	9	0	113.4	105097.0
4	Lyngbya limnetica	bg	38	5	25	40	1	0	0	9.5	38170.7
4	Lyngbya limnetica	bg	147	5	25	40	3	0	0	332.3	442981.0
4	Microcystis aeruginosa	bg	30	5	25	40	4	0	0	40.9	61390.1
4	eucaryote sphere	eu	1	5	25	40	10	0	0	21.3	12789.6
4	eucaryote sphere	eu	2	5	25	40	4	0	0	2.7	4092.7
4	Actinastrum Hantzschii	grn	24	5	25	40	3	22	0	193.5	275529.3
4	Actinastrum Hantzschii	grn	8	5	25	40	2	14	0	18.2	39082.2
8	Anabaena sp.	bg	34	10	15	40	5	0	0	177.9	142303.0
8	Aphanizomenon flos-aquae	bg	422	10	15	40	4	0	0	1413.3	1412985.5
8	Dictyosphaerium pulchellum	grn	28	10	15	40	4	0	0	31.8	47747.8
8	flagellate small	flag	40	10	15	40	5	9	0	102.1	94682.0
8	Lyngbya limnetica	bg	136	10	15	40	3	0	0	256.2	341527.3
8	Lyngbya limnetica	bg	22	10	15	40	1	0	0	4.6	18415.7
8	Microcystis aeruginosa	bg	195	10	15	40	2	0	0	27.7	83132.4
8	Microcystis aeruginosa	bg	230	10	15	40	5	0	0	510.7	612835.0

M	Genus and species	Fam.	Ct.	Vol. filt. ml	Flds. ctd.	Mag.	Dim1 μm	Dim2 μm	Dim3 μm	Biovol. mm³/m³	Surf.area mm²/m³
8	eucaryote sphere	eu	1	10	15	40	10	0	0	17.8	10658.0
8	eucaryote sphere	eu	10	10	15	40	4	0	0	11.4	17052.8
8	Actinastrum Hantzschii	grn	32	10	15	40	3	22	0	215.0	306143.7
8	Oocystis sp.	grn	6	10	15	40	5	8	0	21.3	22667.9
12	Aphanizomenon flos-aquae	bg	197	10	20	40	4	0	0	494.8	494712.3
12	Dictyosphaerium pulchellum	grn	20	10	20	40	3	0	0	7.2	14388.3
12	flagellate small	flag	17	10	20	40	5	9	0	32.6	30179.9
12	Lyngbya limnetica	bg	104	10	20	40	3	0	0	146.9	195876.0
12	Lyngbya limnetica	bg	22	10	20	40	1	0	0	3.5	13811.8
12	Stephanodiscus sp.	diat	1	10	20	40	20	16	0	127.9	41566.2
12	Actinastrum Hantzschii	grn	8	10	20	40	2	14	0	11.4	24426.4
16	flagellate small	flag	4	10	25	40	5	9	0	6.1	5680.9
16	Microcystis aeruginosa	bg	160	10	25	40	4	0	0	109.1	163706.9
	Aug. 28, 1979										
0	Anabaena sp.	bg	64	10	20	40	5	0	0	251.2	200898.4
0	Aphanizomenon flos-aquae	bg	159	10	20	40	4	0	0	399.4	399285.6
0	Dictyosphaerium pulchellum	grn	60	10	20	40	2	0	0	6.4	19184.4
0	Dictyosphaerium pulchellum	grn	6	10	20	40	4	0	0	5.1	7673.8
0	flagellate small	flag	63	10	20	40	5	9	0	120.7	111843.1
0	Lyngbya limnetica	bg	118	10	20	40	3	0	0	166.7	222243.9
0	Lyngbya limnetica	bg	27	10	20	40	1	0	0	4.2	16950.8
0	Microcystis aeruginosa	bg	40	10	20	40	3	0	0	14.4	28776.6
0	Microcystis aeruginosa	bg	43	10	20	40	4	0	0	36.7	54995.3
0	eucaryote sphere	eu	4	10	20	40	10	0	0	53.3	31974.0
0	Actinastrum Hantzschii	grn	40	10	20	40	2	14	0	57.0	122131.8
0	Actinastrum Hantzschii	grn	20	10	20	40	3	22	0	100.8	143504.9
2	Anabaena sp.	bg	20	10	20	40	5	0	0	78.5	62780.8
2	Aphanizomenon flos-aquae	bg	305	10	20	40	4	0	0	766.1	765925.2
2	Coelosphaerium sp.	bg	200	10	20	40	3	5	0	71.9	143883.0
2	Coelosphaerium sp.	bg	60	10	20	40	2	3	0	6.4	19184.4
2	Dictyosphaerium pulchellum	grn	44	10	20	40	3	4	0	15.8	31654.3
2	flagellate small	flag	38	10	20	40	5	9	0	72.8	67460.9
2	Lyngbya limnetica	bg	132	10	20	40	1	0	0	20.7	82870.6
2	Lyngbya limnetica	bg	198	10	20	40	3	0	0	279.8	372917.7
2	Microcystis aeruginosa	bg	500	10	20	40	3	0	0	179.9	359707.5
2	Actinastrum Hantzschii	grn	8	10	20	40	2	14	0	11.4	24426.4
2	Actinastrum Hantzschii	grn	8	10	20	40	3	22	0	40.3	57401.9
4	Anabaena sp.	bg	20	10	20	40	5	0	0	78.5	62780.8
4	Aphanizomenon flos-aquae	bg	149	10	20	40	4	0	0	374.3	374173.3
4	flagellate medium	flag	1	10	20	40	16	20	0	47.7	33004.2
4	flagellate medium	flag	1	10	20	40	12	15	0	20.1	14728.0
4	flagellate small	flag	38	10	20	40	5	9	0	72.8	67460.9
4	Lyngbya limnetica	bg	248	10	20	40	3	0	0	350.4	467088.8
4	Lyngbya limnetica	bg	158	10	20	40	1	0	0	24.8	99193.6
4	Microcystis aeruginosa	bg	80	10	20	40	3	0	0	28.8	57553.2
4	eucaryote sphere	eu	9	10	20	40	5	0	0	15.0	17985.4
4	Actinastrum Hantzschii	grn	56	10	20	40	3	22	0	282.1	401813.6
4	Actinastrum Hantzschii	grn	8	10	20	40	2	14	0	11.4	24426.4
4	Fragillaria sp.	diat	57	10	20	40	4	35	0	812.2	452498.3
8	Anabaena sp.	bg	94	10	35	40	5	0	0	210.8	168611.2
8	Aphanizomenon flos-aquae	bg	391	10	35	40	4	0	0	561.2	561080.6
8	Coelosphaerium sp.	bg	65	10	35	40	3	5	0	13.4	26721.1
8	Coelosphaerium sp.	bg	70	10	35	40	2	3	0	4.3	12789.6
8	flagellate small	flag	56	10	35	40	5	9	0	61.3	56809.2
8	flagellate medium	flag	1	10	35	40	10	12	0	6.5	5036.3
8	Lyngbya limnetica	bg	203	10	35	40	1	0	0	18.2	72825.7
8	Lyngbya limnetica	bg	346	10	35	40	3	0	0	279.4	372379.6
8	Microcystis aeruginosa	bg	30	10	35	40	5	0	0	28.5	34257.9
8	Microcystis aeruginosa	bg	420	10	35	40	4	0	0	204.6	306950.4
8	Microcystis aeruginosa	bg	25	10	35	40	3	0	0	5.1	10277.4
8	Microcystis aeruginosa	bg	1270	10	35	40	2	0	0	77.3	232039.9
8	eucaryote sphere	eu	2	10	35	40	8	0	0	7.8	5846.7
8	eucaryote sphere	eu	4	10	35	40	10	0	0	30.5	18270.9
8	eucaryote sphere	eu	1	10	35	40	15	0	0	25.7	10277.4
8	Actinastrum Hantzschii	grn	48	10	35	40	2	14	0	39.1	83747.5
8	Actinastrum Hantzschii	grn	28	10	35	40	3	22	0	80.6	114803.9
8	Melosira sp.	diat	25	10	35	40	5	0	0	56.1	44843.4
12	Anabaena sp.	bg	15	10	30	40	5	0	0	39.2	31390.4
12	Aphanizomenon flos-aquae	bg	161	10	30	40	4	0	0	269.6	269538.7
12	flagellate small	flag	21	10	30	40	5	9	0	26.8	24854.0
12	Lyngbya limnetica	bg	34	10	30	40	3	0	0	32.0	42690.9
12	Lyngbya limnetica	bg	37	10	30	40	1	0	0	3.9	15485.9
12	eucaryote sphere	eu	6	10	30	40	5	0	0	6.7	7993.5
12	eucaryote sphere	eu	20	10	30	40	4	0	0	11.4	17052.8
12	Actinastrum Hantzschii	grn	20	10	30	40	3	22	0	67.2	95669.9
12	Actinastrum Hantzschii	grn	104	10	30	40	2	14	0	98.8	211695.1
16	Aphanizomenon flos-aquae	bg	14	10	30	40	4	0	0	23.4	23438.1
16	Coelosphaerium sp.	bg	20	10	30	40	2	3	0	1.4	4263.2
16	flagellate small	flag	7	10	30	40	5	9	0	8.9	8284.7
16	Lyngbya limnetica	bg	22	10	30	40	1	0	0	2.3	9207.8
16	Lyngbya limnetica	bg	22	10	30	40	3	0	0	20.7	27623.5
16	Microcystis aeruginosa	bg	105	10	30	40	2	0	0	7.5	22381.8
16	Actinastrum Hantzschii	grn	4	10	30	40	3	22	0	13.4	19134.0
	Sep. 4, 1979										
0	Anabaena sp.	bg	35	5	20	40	5	0	0	274.7	219732.6
0	Aphanizomenon flos-aquae	bg	306	5	20	40	4	0	0	1537.2	1536872.9
0	Dictyosphaerium pulchellum	grn	40	5	20	40	4	0	0	68.2	102316.8
0	flagellate medium	flag	1	5	20	40	15	20	0	82.4	55771.3
0	flagellate small	flag	40	5	20	40	5	9	0	153.2	142023.0
0	Lyngbya limnetica	bg	99	5	20	40	3	0	0	279.8	372917.7
0	Lyngbya limnetica	bg	77	5	20	40	1	0	0	24.2	96682.4
0	Microcystis aeruginosa	bg	220	5	20	40	2	0	0	46.9	140685.6
0	eucaryote sphere	eu	1	5	20	40	15	0	0	89.9	35970.8
0	eucaryote sphere	eu	1	5	20	40	10	0	0	26.6	15987.0
0	eucaryote sphere	eu	16	5	20	40	5	0	0	53.3	63948.0
0	Actinastrum Hantzschii	grn	20	5	20	40	3	15	0	137.4	201517.5
0	Actinastrum Hantzschii	grn	24	5	20	40	2	10	0	48.9	107476.0
0	Ceratium hirundinella	dino	20	5	10	10			0		
2	Aphanizomenon flos-aquae	bg	362	5	20	40	4	0	0	1818.6	1818130.6
2	flagellate medium	flag	1	5	20	40	12	12	0	34.5	28776.6
2	flagellate small	flag	29	5	20	40	5	9	0	111.1	102966.7
2	Lyngbya limnetica	bg	39	5	20	40	1	0	0	12.2	48969.0
2	Lyngbya limnetica	bg	141	5	20	40	3	0	0	398.4	531125.2
2	Microcystis aeruginosa	bg	60	5	20	40	2	0	0	12.8	38368.8
2	Microcystis aeruginosa	bg	90	5	20	40	4	0	0	153.5	230212.8
2	eucaryote sphere	eu	2	5	20	40	14	0	0	146.2	62669.0
2	Stephanodiscus sp.	diat	1	5	20	40	20	16	0	255.8	83132.4
2	Actinastrum Hantzschii	grn	24	5	20	40	2	10	0	48.9	107476.0
2	Ceratium hirundinella	dino	18	5	9	10	50	60	0	311.7	48692.9
4	Anabaena sp.	bg	3	10	25	40	5	0	0	9.4	7533.7
4	Aphanizomenon flos-aquae	bg	260	10	25	40	4	0	0	522.5	522335.9
4	Coelosphaerium sp.	bg	30	10	25	40	3	5	0	8.6	17266.0

M	Genus and species	Fam.	Ct.	Vol. filt. ml	Flds. ctd.	Mag.	Dim1 μm	Dim2 μm	Dim3 μm	Biovol. mm³/m³	Surf.area mm²/m³
4	Dictyosphaerium pulchellum	grn	16	10	25	40	3	0	0	4.6	9208.5
4	flagellate small	flag	48	10	25	40	5	9	0	73.5	68171.1
4	Lyngbya limnetica	bg	133	10	25	40	3	0	0	150.3	200396.2
4	Lyngbya limnetica	bg	80	10	25	40	1	0	0	10.0	40179.7
4	Microcystis aeruginosa	bg	970	10	25	40	2	0	0	82.7	248118.3
4	eucaryote sphere	eu	12	10	25	40	5	0	0	16.0	19184.4
4	Actinastrum Hantzschii	grn	16	10	25	40	2	10	0	13.0	28660.3
4	Ceratium hirundinella	dino	24	5	12	10			0		
8	Aphanizomenon flos-aquae	bg	222	10	20	40	4	0	0	557.6	557493.1
8	Coelosphaerium sp.	bg	70	10	20	40	3	4	0	25.2	50359.1
8	Dictyosphaerium pulchellum	grn	16	10	20	40	4	0	0	13.6	20463.4
8	flagellate small	flag	31	10	20	40	5	9	0	59.4	55033.9
8	flagellate medium	flag	1	10	20	40	16	16	0	40.9	32400.3
8	Lyngbya limnetica	bg	121	10	20	40	3	0	0	171.0	227894.1
8	Lyngbya limnetica	bg	76	10	20	40	1	0	0	11.9	47713.4
8	Microcystis aeruginosa	bg	192	10	20	40	4	0	0	163.7	245560.3
8	eucaryote sphere	eu	1	10	20	40	14	16	0	36.6	15667.3
8	eucaryote sphere	eu	2	10	20	40	15	0	0	89.9	35970.8
8	eucaryote sphere	eu	4	10	20	40	8	0	0	27.3	20463.4
8	Actinastrum Hantzschii	grn	56	10	20	40	2	12	0	68.4	148186.6
8	Actinastrum Hantzschii	grn	8	10	20	40	4	16	0	52.1	58623.3
12	Aphanizomenon flos-aquae	bg	10	10	20	40	4	0	0	25.1	25112.3
12	flagellate small	flag	20	10	20	40	5	9	0	38.3	35505.8
12	Lyngbya limnetica	bg	80	10	20	40	1	0	0	12.6	50224.6
12	Microcystis aeruginosa	bg	163	10	20	40	4	0	0	139.0	208470.5
12	eucaryote small	eu	1	10	20	40	15	20	0	60.0	22094.1
12	eucaryote small	eu	4	10	20	40	12	25	0	191.8	81526.8
12	eucaryote small	eu	2	10	20	40	10	20	0	53.3	27325.0
12	eucaryote small	eu	1	10	20	40	12	20	0	38.4	16873.8
12	Actinastrum Hantzschii	grn	11	10	20	40	4	18	0	80.6	89563.3
12	Actinastrum Hantzschii	grn	16	10	20	40	2	10	0	16.3	35825.3
16	Aphanizomenon flos-aquae	bg	35	20	20	40	4	0	0	44.0	43946.5
16	flagellate small	flag	9	20	20	40	5	9	0	8.6	7988.8
16	Lyngbya limnetica	bg	24	20	20	40	3	0	0	17.0	22601.1
16	Lyngbya limnetica	bg	22	20	20	40	1	0	0	1.7	6905.9
16	eucaryote sphere	eu	9	20	20	40	4	0	0	3.8	5755.3
16	eucaryote sphere	eu	1	20	20	40	15	0	0	22.5	8992.7
16	Actinastrum Hantzschii	grn	8	20	20	40	2	10	0	4.1	8956.3
16	Actinastrum Hantzschii	grn	5	20	20	40	4	16	0	16.3	18319.8

Sep. 10, 1979

M	Genus and species	Fam.	Ct.	Vol. filt. ml	Flds. ctd.	Mag.	Dim1 μm	Dim2 μm	Dim3 μm	Biovol. mm³/m³	Surf.area mm²/m³
0	Dictyosphaerium pulchellum	grn	6	5	20	40	3	4	0	4.3	8633.0
0	eucaryote sphere	eu	5	5	20	40	5	0	0	16.7	19983.8
0	flagellate small	flag	22	5	20	40	4	8	0	46.9	50216.3
0	Anabaena sp.	bg	8	5	20	40	4	0	0	40.2	40179.7
0	Aphanizomenon flos-aquae	bg	358	5	20	40	4	0	0	1798.5	1798040.8
0	Lyngbya limnetica	bg	25	5	20	40	2	0	0	31.4	62780.8
0	Microcystis aeruginosa	bg	40	5	20	40	3	0	0	28.8	57553.2
0	blue green filaments	bg	96	5	20	40	1	0	0	30.1	120539.0
0	Ceratium hirundinella	dino	56	5	8	10	35	50	0	445.5	93224.4
0	dinoflagellate	dino	41	5	8	10	30	40	0	383.4	70651.9
0	Stephanodiscus sp.	diat	2	5	8	10			0		
0	Cosmarium sp.	desm	1	5	8	10			0		
2	Dictyosphaerium pulchellum	grn	6	5	15	40	3	0	0	5.8	11510.6
2	eucaryote sphere	eu	4	5	15	40	5	0	0	17.8	21316.0
2	flagellate small	flag	18	5	15	40	4	8	0	51.2	54781.5
2	Fragillaria sp.	diat	33	5	15	40	3	40	0	806.1	577683.4
2	Melosira sp.	diat	22	5	15	40	5	0	0	230.3	184156.9
2	Anabaena sp.	bg	4	5	15	40	4	0	0	26.8	26786.5
2	Aphanizomenon flos-aquae	bg	424	5	15	40	4	0	0	2840.1	2839364.2
2	Lyngbya limnetica	bg	15	5	15	40	2	0	0	25.1	50224.6
2	Microcystis aeruginosa	bg	40	5	15	40	2	0	0	11.4	34105.6
2	blue green filaments	bg	126	5	15	40	1	0	0	52.7	210943.3
2	Ceratium hirundinella	dino	36	5	9	10	35	50	0	254.6	53271.1
2	dinoflagellate	dino	37	5	9	10	30	40	0	307.6	56674.7
2	Stephanodiscus sp.	diat	1	5	9	10			0		
4	Dictyosphaerium pulchellum	grn	12	5	15	40	3	0	0	11.5	23021.3
4	eucaryote sphere	eu	5	5	15	40	5	0	0	22.2	26645.0
4	flagellate small	flag	33	5	15	40	4	8	0	93.8	100432.7
4	flagellate medium	flag	2	5	15	40	12	15	0	107.4	78549.1
4	Fragillaria sp.	diat	46	5	15	40		0	0		
4	Anabaena sp.	bg	7	5	15	40	4	0	0	46.9	46876.3
4	Aphanizomenon flos-aquae	bg	368	5	15	40	4	0	0	2465.0	2464353.8
4	Lyngbya limnetica	bg	34	5	15	40	2	0	0	56.9	113842.4
4	blue green filaments	bg	145	5	15	40	1	0	0	60.7	242752.3
4	Ceratium hirundinella	dino	99	5	10	10	35	50	0	630.1	131846.0
4	dinoflagellate	dino	56	5	10	10	30	40	0	419.0	77200.1
8	Dictyosphaerium pulchellum	grn	36	5	15	40	3	0	0	34.5	69063.8
8	eucaryote sphere	eu	14	5	15	40	5	0	0	62.2	74606.0
8	flagellate small	flag	31	5	15	40	4	8	0	88.1	94345.9
8	Aphanizomenon flos-aquae	bg	481	5	15	40	4	0	0	3221.9	3221071.3
8	Lyngbya limnetica	bg	16	5	15	40	2	0	0	26.8	53572.9
8	Microcystis aeruginosa	bg	30	5	15	40	4	0	0	68.2	102316.8
8	blue green filaments	bg	128	5	15	40	1	0	0	53.6	214291.6
8	Ceratium hirundinella	dino	317	5	10	10	35	50	0	2017.5	422173.6
8	dinoflagellate	dino	169	5	10	10	30	40	0	1264.4	232978.8
12	eucaryote small	eu	5	5	25	40	10	20	0	213.2	109299.9
12	flagellate small	flag	6	5	25	40	4	8	0	10.2	10956.3
12	Aphanizomenon flos-aquae	bg	17	5	25	40	4	0	0	68.3	68305.5
12	Microcystis aeruginosa	bg	50	5	25	40	2	0	0	8.5	25579.2
12	blue green filaments	bg	105	5	25	40	1	0	0	26.4	105471.7
12	Ceratium hirundinella	dino	40	5	11	10			0		
12	dinoflagellate	dino	9	5	11	10			0		

Sep. 17, 1979

M	Genus and species	Fam.	Ct.	Vol. filt. ml	Flds. ctd.	Mag.	Dim1 μm	Dim2 μm	Dim3 μm	Biovol. mm³/m³	Surf.area mm²/m³
0	eucaryote filament	eu	35	10	20	40	8	0	0	351.7	175786.1
0	flagellate small	flag	4	10	20	40	5	9	0	7.7	7101.2
0	Fragillaria sp.	diat	33	10	20	40	3	0	0	0.0	15113.8
0	blue green filaments	bg	6	10	20	40	2	0	0	3.8	7533.7
2	Dictyosphaerium pulchellum	grn	40	5	20	40	5	0	0	133.2	159870.0
2	eucaryote sphere	eu	3	5	20	40	5	0	0	10.0	11990.3
2	eucaryote sphere	eu	4	5	20	40	10	0	0	106.6	63948.0
2	eucaryote filament	eu	28	5	20	40	7	0	0	430.8	246100.6
2	flagellate small	flag	20	5	20	40	5	9	0	76.6	71011.5
2	flagellate medium	flag	5	5	20	40	12	16	0	211.0	149691.6
2	Fragillaria sp.	diat	81	5	20	40	4	40	0	2638.0	1450925.8
2	Aphanizomenon flos-aquae	bg	332	5	20	40	4	0	0	1667.9	1667456.8
2	Lyngbya limnetica	bg	61	5	20	40	2	0	0	76.6	153185.0
2	Microcystis aeruginosa	bg	50	5	20	40	4	0	0	85.3	127896.0
2	Microcystis aeruginosa	bg	40	5	20	40	3	0	0	28.8	57553.2
2	Microcystis aeruginosa	bg	20	5	20	40	2	0	0	4.3	12789.6
2	blue green filaments	bg	267	5	20	40	1	0	0	83.8	335249.2
2	Ceratium hirundinella	dino	149	5	9	10	35	50	0	1053.7	220483.2
2	dinoflagellate	dino	96	5	9	10	30	40	0	798.0	147047.8

M	Genus and species	Fam.	Ct.	Vol. filt. ml	Flds. ctd.	Mag.	Dim1 µm	Dim2 µm	Dim3 µm	Biovol. mm³/m³	Surf.area mm²/m³
4	Dictyosphaerium pulchellum	grn	12	3	25	40	5	0	0	53.3	63948.0
4	eucaryote sphere	eu	1	3	25	40	5	0	0	4.4	5329.0
4	eucaryote filament	eu	13	3	25	40	6	0	0	195.9	130584.0
4	flagellate small	flag	14	3	25	40	5	9	0	71.5	66277.4
4	flagellate small	flag	2	3	25	40	8	10	0	31.8	25815.9
4	Anabaena sp.	bg	26	3	25	40	5	0	0	272.1	217640.0
4	Microcystis aeruginosa	bg	40	3	25	40	4	0	0	90.9	136422.4
4	Microcystis aeruginosa	bg	35	3	25	40	3	0	0	33.6	67145.4
4	blue green filaments	bg	160	3	25	40	1	0	0	67.0	267864.6
4	Ceratium hirundinella	dino	8	3	7	10			0		
4	dinoflagellate	dino	10	3	7	10			0		
8	eucaryote sphere	eu	4	5	20	40	10	0	0	106.6	63948.0
8	eucaryote sphere	eu	2	5	20	40	5	0	0	6.7	7993.5
8	flagellate small	flag	28	5	20	40	5	9	0	107.2	99416.1
8	Fragillaria sp.	diat	167	5	20	40	3	40	0	3059.4	2192571.1
8	Aphanizomenon flos-aquae	bg	303	5	20	40	4	0	0	1522.2	1521805.5
8	akinete	bg	1	5	20	40	6	20	0	19.2	15608.8
8	Microcystis aeruginosa	bg	60	5	20	40	4	0	0	102.3	153475.2
8	blue green filaments	bg	242	5	20	40	1	0	0	76.0	303858.8
8	Ceratium hirundinella	dino	274	5	10	10	35	50	0	1743.8	364907.1
8	dinoflagellate	dino	142	5	10	10	30	40	0	1062.4	195757.4
12	Dictyosphaerium pulchellum	grn	16	5	15	40	5	0	0	71.1	85264.0
12	Dictyosphaerium pulchellum	grn	30	5	15	40	3	0	0	28.8	57553.2
12	flagellate small	flag	28	5	15	40	5	9	0	143.0	132554.8
12	flagellate medium	flag	1	5	15	40	12	16	0	56.3	39917.8
12	Aphanizomenon flos-aquae	bg	267	5	15	40	4	0	0	1788.4	1787995.9
12	Coelosphaerium sp.	bg	30	5	15	40	3	5	0	28.8	57553.2
12	Microcystis aeruginosa	bg	40	5	15	40	3	0	0	38.4	76737.6
12	blue green filaments	bg	159	5	15	40	1	0	0	66.6	266190.4
12	Ceratium hirundinella	dino	119	5	10	10			0		
12	dinoflagellate	dino	66	5	10	10			0		
12	Stephanodiscus sp.	diat	6	5	10	10			0		
20	flagellate small	flag	2	5	20	40	5	9	0	7.7	7101.2
20	Fragillaria sp.	diat	3	5	20	40	4	40	0	97.7	53738.0
20	Aphanizomenon flos-aquae	bg	18	5	20	40	4	0	0	90.4	90404.3
20	blue green filaments	bg	17	5	20	40	1	0	0	5.3	21345.5
	Sep. 24, 1979										
0	Actinastrum Hantzschii	grn	8	5	20	40	4	15	0	97.7	110732.8
0	eucaryote filament	eu	30	5	20	40	5	0	0	235.5	188342.3
0	flagellate small	flag	33	5	20	40	5	9	0	126.4	117169.0
0	Fragillaria sp.	diat	95	5	20	40	4	45	0	3480.8	1895078.4
0	akinete	bg	1	5	20	40	7	16	0	20.9	15047.5
0	Aphanizomenon flos-aquae	bg	375	5	20	40	4	0	0	1883.9	1883422.6
0	akinete	bg	2	5	20	40	5	16	0	21.3	20866.7
0	Coelosphaerium sp.	bg	180	5	20	40	2	3	0	38.4	115106.4
0	Lyngbya limnetica	bg	52	5	20	40	2	0	0	65.3	130584.0
0	Microcystis aeruginosa	bg	15	5	20	40	4	0	0	25.6	38368.8
0	blue green filaments	bg	222	5	20	40	1	0	0	69.7	278746.5
0	Ceratium hirundinella	dino	85	5	8	10	35	50	0	676.2	141501.4
0	dinoflagellate	dino	61	5	8	10	30	40	0	570.5	105116.2
0	Stephanodiscus sp.	diat	5	5	8	10	55	55	0	324.1	35361.2
2	flagellate small	flag	30	5	20	40	5	9	0	114.9	106517.3
2	Anabaena sp.	bg	77	5	20	40	5	0	0	604.4	483411.8
2	Aphanizomenon flos-aquae	bg	275	5	20	40	4	0	0	1381.5	1381176.6
2	Lyngbya limnetica	bg	49	5	20	40	2	0	0	61.5	123050.3
2	Microcystis aeruginosa	bg	310	5	20	40	2	0	0	66.1	198238.8

M	Genus and species	Fam.	Ct.	Vol. filt. ml	Flds. ctd.	Mag.	Dim1 µm	Dim2 µm	Dim3 µm	Biovol. mm³/m³	Surf.area mm²/m³
2	blue green filaments	bg	92	5	20	40	1	0	0	28.9	115516.6
2	Ceratium hirundinella	dino	17	5	10	10			0		
2	dinoflagellate	dino	20	5	10	10			0		
2	Stephanodiscus sp.	diat	1	5	10	10			0		
4	eucaryote sphere	eu	5	5	20	40	5	0	0	16.7	19983.8
4	flagellate small	flag	30	5	20	40	5	9	0	114.9	106517.3
4	Fragillaria sp.	diat	13	5	20	40	3	40	0	238.2	170679.2
4	Anabaena sp.	bg	4	5	20	40	5	0	0	31.4	25112.3
4	Aphanizomenon flos-aquae	bg	291	5	20	40	4	0	0	1461.9	1461535.9
4	Lyngbya limnetica	bg	77	5	20	40	2	0	0	96.7	193364.7
4	Microcystis aeruginosa	bg	450	5	20	40	4	0	0	767.4	1151064.1
4	Microcystis aeruginosa	bg	110	5	20	40	2	0	0	23.4	70342.8
4	blue green filaments	bg	355	5	20	40	1	0	0	111.5	445743.4
4	Ceratium hirundinella	dino	71	5	11	10			0		
4	dinoflagellate	dino	44	5	11	10			0		
4	Stephanodiscus sp.	diat	4	5	11	10			0		
8	Dictyosphaerium pulchellum	grn	40	5	25	40	5	0	0	106.6	127896.0
8	eucaryote sphere	eu	2	5	25	40	15	0	0	143.9	57553.2
8	eucaryote filament	eu	6	5	25	40	6	0	0	54.3	36161.7
8	flagellate small	flag	66	5	25	40	5	9	0	202.2	187470.4
8	Fragillaria sp.	diat	83	5	25	40	4	40	0	2162.5	1189400.9
8	Aphanizomenon flos-aquae	bg	130	5	25	40	4	0	0	522.5	522335.9
8	Coelosphaerium sp.	bg	110	5	25	40	3	5	0	63.3	126617.1
8	Lyngbya limnetica	bg	28	5	25	40	2	0	0	28.1	56251.6
8	Microcystis aeruginosa	bg	220	5	25	40	2	0	0	37.5	112548.5
8	blue green filaments	bg	286	5	25	40	1	0	0	71.8	287284.7
8	Ceratium hirundinella	dino	55	5	12	10			0		
8	dinoflagellate	dino	25	5	12	10			0		
8	Stephanodiscus sp.	diat	4	5	12	10			0		
12	eucaryote filament	eu	13	5	20	40	6	0	0	146.9	97938.0
12	flagellate small	flag	25	5	20	40	5	9	0	95.8	88764.4
12	Anabaena sp.	bg	11	5	20	40	5	0	0	86.3	69058.8
12	Aphanizomenon flos-aquae	bg	422	5	20	40	4	0	0	2120.0	2119478.3
12	Lyngbya limnetica	bg	15	5	20	40	2	0	0	18.8	37668.5
12	Microcystis aeruginosa	bg	15	5	20	40	5	0	0	50.0	59951.3
12	Microcystis aeruginosa	bg	90	5	20	40	3	0	0	64.7	129494.7
12	blue green filaments	bg	331	5	20	40	1	0	0	103.9	415608.6
12	Ceratium hirundinella	dino	82	5	10	10			0		
12	dinoflagellate	dino	33	5	10	10			0		
12	Stephanodiscus sp.	diat	3	5	10	10			0		
	Jan. 14, 1980										
0	Chlorella ellipsoidea	grn	10	15	30	40	4	0	0	3.8	5684.3
0	Aphanizomenon flos-aquae	bg	20	15	30	40	4	0	0	22.3	22322.0
0	Cosmarium sp.	desm	7	15	30	40	10	12	0	99.5	56488.0
0	Chlamydomonas sp. 1	grn	5	15	30	40	12	16	0	68.2	31422.7
0	Dictyosphaerium pulchellum	grn	88	15	30	40	4	0	0	33.3	50021.6
0	Stephanodiscus sp.	diat	1	15	30	40	50	12	0	266.5	65724.3
2	Chlorella ellipsoidea	grn	193	15	30	40	4	0	0	73.1	109706.4
2	Cosmarium sp.	desm	49	15	30	40	10	12	0	696.3	395416.3
2	Dictyosphaerium pulchellum	grn	114	15	30	40	4	0	0	43.2	64800.6
2	Chlamydomonas sp. 1	grn	17	15	30	40	15	20	0	453.0	166933.1
2	Cosmarium sp.	desm	7	15	30	40	8	10	0	53.1	37253.3
2	Stephanodiscus sp.	diat	1	15	30	40	40	20	0	284.2	56842.7
2	Stephanodiscus sp.	diat	3	15	30	40	20	20	0	213.2	63948.0
2	Actinastrum Hantzschii	grn	16	15	30	40	4	26	0	75.3	81059.3
5	Chlorella ellipsoidea	grn	245	15	30	40	4	0	0	92.8	139264.5

M	Genus and species	Fam.	Ct.	Vol. filt. ml	Flds. ctd.	Mag.	Dim1 µm	Dim2 µm	Dim3 µm	Biovol. mm³/m³	Surf.area mm²/m³
5	Chlamydomonas sp. 1	grn	5	15	30	40	15	20	0	133.2	49098.0
5	Stephanodiscus sp.	diat	1	15	30	40	20	20	0	71.1	21316.0
5	Staurastrum sp.	grn	1	15	30	40	10	30	60	203.6	94197.8
5	Actinastrum Hantzschii	grn	14	15	30	40	4	20	0	50.7	55728.3
5	Cryptomonas sp.	cryp	1	15	30	40	20	30	0	71.1	19132.0
5	Chlamydomonas sp. 1	grn	7	15	30	40	10	12	0	49.7	28244.0
10	Chlorella ellipsoidea	grn	38	15	30	40	4	0	0	14.4	21600.2
10	Cosmarium sp.	desm	1	15	30	40	10	12	0	14.2	8069.7
10	Dictyosphaerium pulchellum	grn	72	15	30	40	4	0	0	27.3	40926.7
10	Stephanodiscus sp.	diat	3	15	30	40	50	12	0	799.4	197173.0
10	Chlamydomonas sp. 1	grn	6	15	30	40	15	20	0	159.9	58917.6
10	Cosmarium sp.	desm	6	15	30	40	8	10	0	45.5	31931.4
10	Stephanodiscus sp.	diat	6	15	30	40	20	20	0	426.3	127896.0
10	Staurastrum sp.	grn	3	15	30	40	10	30	60	610.7	282593.4
10	Actinastrum Hantzschii	grn	15	15	30	40	4	20	0	54.3	59708.9
10	Chlamydomonas sp. 1	grn	1	15	30	40	10	12	0	7.1	4034.9
15	Chlorella ellipsoidea	grn	20	15	30	40	4	0	0	7.6	11368.5
15	Dictyosphaerium pulchellum	grn	109	15	30	40	4	0	0	41.3	61958.5
15	Stephanodiscus sp.	diat	1	15	30	40	50	12	0	266.5	65724.3
15	Cosmarium sp.	desm	2	15	30	40	8	10	0	15.2	10643.8
15	Stephanodiscus sp.	diat	1	15	30	40	40	20	0	284.2	56842.7
15	Stephanodiscus sp.	diat	1	15	30	40	20	20	0	71.1	21316.0
15	Actinastrum Hantzschii	grn	2	15	30	40	4	20	0	7.2	7961.2
15	Chlamydomonas sp. 1	grn	3	15	30	40	10	12	0	21.3	12104.6
20	Chlorella ellipsoidea	grn	54	20	30	40	4	0	0	15.3	23021.3
20	Aphanizomenon flos-aquae	bg	28	20	30	40	4	0	0	23.4	23438.1
20	Cosmarium sp.	desm	6	20	30	40	10	12	0	63.9	36313.7
20	Dictyosphaerium pulchellum	grn	134	20	30	40	4	0	0	38.1	57126.9
20	Stephanodiscus sp.	diat	2	20	30	40	50	12	0	399.7	98586.5
20	Stephanodiscus sp.	diat	2	20	30	40	40	20	0	426.3	85264.0
20	Stephanodiscus sp.	diat	3	20	30	40	20	20	0	159.9	47961.0
20	Staurastrum sp.	grn	1	20	30	40	10	30	60	152.7	70648.3
20	Actinastrum Hantzschii	grn	10	20	30	40	4	20	0	27.1	29854.4
20	Chlamydomonas sp. 1	grn	4	20	30	40	10	12	0	21.3	12104.6
	Feb. 7, 1980										
0	Cryptomonas sp.	cryp	3	10	30	40	24	30	0	460.4	107768.5
0	Cryptomonas sp.	cryp	3	10	30	40	20	24	0	255.8	72627.5
0	Chlamydomonas sp. 1	grn	49	10	30	40	18	20	0	2820.1	909356.0
0	Chlamydomonas sp. 1	grn	40	10	30	40	8	10	0	227.4	159657.0
0	Chlorochromonas minuta	chry	9	10	30	40	4	8	0	10.2	13116.0
0	Stephanodiscus astrea	diat	41	10	30	40	4	6	0	52.4	69916.5
0	Dictyosphaerium pulchellum	grn	38	10	30	40	4	0	0	21.6	32400.3
0	Cosmarium sp.	desm	13	10	30	40	12	20	0	665.1	292479.6
0	Actinastrum gracilimum	grn	4	10	30	40	2	20	0	5.4	11399.0
0	Stephanodiscus sp.	diat	2	10	30	40	22	10	0	129.0	49240.0
2	Cryptomonas sp.	cryp	3	10	30	40	24	30	0	460.4	107768.5
2	Cryptomonas sp.	cryp	4	10	30	40	20	24	0	341.1	96836.6
2	Chlamydomonas sp. 1	grn	37	10	30	40	18	20	0	2129.5	686656.5
2	Chlamydomonas sp. 1	grn	16	10	30	40	8	10	0	90.9	63862.8
2	Chlorochromonas minuta	chry	26	10	30	40	4	8	0	29.6	37890.6
2	Stephanodiscus astrea	diat	7	10	30	40	4	6	0	9.0	11937.0
2	Dictyosphaerium pulchellum	grn	53	10	30	40	4	0	0	30.1	45189.9
2	Cosmarium sp.	desm	7	10	30	40	12	20	0	358.1	157489.0
2	Actinastrum gracilimum	grn	7	10	30	40	2	20	0	9.5	19948.2
2	Staurastrum sp.	grn	1	10	30	40	10	20	50	169.6	95663.3
2	Chlamydomonas sp. 1	grn	12	10	30	40	6	10	0	38.4	33747.6
5	Chlamydomonas sp. 1	grn	2	15	30	40	18	20	0	76.7	24744.4
5	Chlorochromonas minuta	chry	6	15	30	40	4	8	0	4.5	5829.3
5	Dictyosphaerium pulchellum	grn	102	15	30	40	4	0	0	38.7	57979.5
5	Cosmarium sp.	desm	22	15	30	40	12	20	0	750.3	329977.0
5	Actinastrum gracilimum	grn	11	15	30	40	2	20	0	10.0	20898.1
5	Stephanodiscus sp.	diat	2	15	30	40	22	10	0	86.0	32826.6
5	Chlamydomonas sp. 1	grn	3	15	30	40	6	10	0	6.4	5624.6
5	Stephanodiscus sp.	diat	1	15	30	40	50	24	0	532.9	87040.3
5	Stephanodiscus sp.	diat	2	15	30	40	20	8	0	56.8	25579.2
10	Chlamydomonas sp. 1	grn	4	20	30	40	18	20	0	115.1	37116.6
10	Chlorochromonas minuta	chry	12	20	30	40	4	8	0	6.8	8744.0
10	Dictyosphaerium pulchellum	grn	24	20	30	40	4	0	0	6.8	10231.7
10	Cosmarium sp.	desm	20	20	30	40	12	20	0	511.6	224984.3
15	Chlorochromonas minuta	chry	29	25	30	40	4	8	0	13.2	16905.1
15	Dictyosphaerium pulchellum	grn	107	25	30	40	4	0	0	24.3	36493.0
15	Cosmarium sp.	desm	10	25	30	40	12	20	0	204.6	89993.7
15	Actinastrum gracilimum	grn	9	25	30	40	2	20	0	4.9	10259.1
20	Dictyosphaerium pulchellum	grn	59	30	30	40	4	0	0	11.2	16768.6
20	Actinastrum gracilimum	grn	8	30	30	40	2	20	0	3.6	7599.3
20	Stephanodiscus sp.	diat	1	30	30	40	22	10	0	21.5	8206.7
20	Stephanodiscus sp.	diat	2	30	30	40	40	12	0	170.5	45474.1
	Feb. 28, 1980										
0	Chlamydomonas sp. 1	grn	8	15	30	40	10	12	0	56.8	32278.9
0	Chlamydomonas sp. 1	grn	5	15	30	40	12	16	0	68.2	31422.7
0	Chlamydomonas sp. 1	grn	5	15	30	40	18	20	0	191.8	61860.9
0	Cosmarium sp.	desm	2	15	30	40	12	20	0	68.2	29997.9
0	Oocystis sp.	grn	7	15	30	40	8	12	0	31.8	21427.8
0	Dictyosphaerium pulchellum	grn	20	15	30	40	4	0	0	7.6	11368.5
2	Chlamydomonas sp. 1	grn	5	25	30	40	10	12	0	21.3	12104.6
2	Chlamydomonas sp. 1	grn	3	25	30	40	12	16	0	24.6	11312.2
2	Chlamydomonas sp. 1	grn	3	25	30	40	18	20	0	69.1	22269.9
2	Cryptomonas sp.	cryp	2	25	30	40	28	30	0	167.1	35026.0
2	Cosmarium sp.	desm	10	25	30	40	12	20	0	204.6	89993.7
2	Oocystis sp.	grn	4	25	30	40	8	12	0	10.9	7346.7
2	Dictyosphaerium pulchellum	grn	146	25	30	40	4	0	0	33.2	49794.2
2	Actinastrum Hantzschii	grn	8	25	30	40	3	18	0	8.8	12701.7
4	Chlamydomonas sp. 1	grn	13	25	30	40	10	12	0	55.4	31471.9
4	Chlamydomonas sp. 1	grn	10	25	30	40	12	16	0	81.9	37707.3
4	Cryptomonas sp.	cryp	2	25	30	40	28	30	0	167.1	35026.0
4	Dictyosphaerium pulchellum	grn	37	25	30	40	4	0	0	8.4	12619.1
4	Actinastrum Hantzschii	grn	8	25	30	40	3	18	0	8.8	12701.7
8	Chlamydomonas sp. 1	grn	11	30	30	40	10	12	0	39.1	22191.7
8	Chlamydomonas sp. 1	grn	9	30	30	40	12	16	0	61.4	28280.4
8	Chlamydomonas sp. 1	grn	2	30	30	40	18	20	0	38.4	12372.2
8	Cryptomonas sp.	cryp	3	30	30	40	28	30	0	208.9	43782.5
8	Dictyosphaerium pulchellum	grn	20	30	30	40	4	0	0	3.8	5684.3
16	Chlamydomonas sp. 1	grn	11	30	30	40	10	12	0	39.1	22191.7
	Mar. 17, 1980										
0	Chlamydomonas sp. 1	grn	3	25	30	40	16	20	0	54.6	19158.8
0	Chlamydomonas sp. 1	grn	7	25	30	40	4	5	0	2.0	2794.0
0	Staurastrum sp.	grn	1	25	30	40	10	14	28	26.6	19238.3
0	Stephanodiscus astrea	diat	6	25	30	40	4	6	0	3.1	4092.7
0	Oocystis sp.	grn	37	25	30	40	4	5	0	10.5	14768.3
0	Dictyosphaerium pulchellum	grn	94	25	30	40	4	0	0	21.4	32059.3
2	Chlamydomonas sp. 1	grn	9	25	30	40	16	20	0	163.7	57476.5

M	Genus and species	Fam.	Ct.	Vol. filt. ml	Flds. ctd.	Mag.	Dim1 μm	Dim2 μm	Dim3 μm	Biovol. mm³/m³	Surf.area mm²/m³
2	Chlamydomonas sp. 1	grn	13	25	30	40	4	5	0	3.7	5188.9
2	Cryptomonas sp.	cryp	1	25	30	40	20	30	0	42.6	11479.2
2	Stephanodiscus astrea	diat	11	25	30	40	4	6	0	5.6	7503.2
2	Oocystis sp.	grn	27	25	30	40	4	5	0	7.7	10776.8
2	Dictyosphaerium pulchellum	grn	115	25	30	40	4	0	0	26.1	39221.4
2	Cosmarium sp.	desm	7	25	30	40	12	20	0	143.2	62995.6
4	Chlamydomonas sp. 1	grn	11	10	30	40	16	20	0	500.2	175622.7
4	Chlamydomonas sp. 1	grn	8	10	30	40	4	5	0	5.7	7982.9
4	Cryptomonas sp.	cryp	3	10	30	40	20	30	0	319.7	86093.8
4	Oocystis sp.	grn	8	10	30	40	4	5	0	5.7	7982.9
4	Dictyosphaerium pulchellum	grn	16	10	30	40	4	0	0	9.1	13642.2
4	Cosmarium sp.	desm	3	10	30	40	12	20	0	153.5	67495.3
4	Fragillaria sp.	diat	6	10	30	40	3	40	0	36.6	26258.3
8	Chlamydomonas sp. 1	grn	7	15	30	40	16	20	0	212.2	74506.6
8	Chlamydomonas sp. 1	grn	6	15	30	40	4	5	0	2.8	3991.4
8	Cryptomonas sp.	cryp	2	15	30	40	20	30	0	142.1	38263.9
8	Staurastrum sp.	grn	1	15	30	40	10	14	28	44.3	32063.8
8	Dictyosphaerium pulchellum	grn	69	15	30	40	4	0	0	26.1	39221.4
8	Cosmarium sp.	desm	4	15	30	40	12	20	0	136.4	59995.8
8	Aphanizomenon flos-aquae	bg	21	15	30	40	3	0	0	13.2	17578.6
16	Chlamydomonas sp. 1	grn	1	25	30	40	16	20	0	18.2	6386.3
16	Oocystis sp.	grn	12	25	30	40	4	5	0	3.4	4789.7
16	Dictyosphaerium pulchellum	grn	20	25	30	40	4	0	0	4.5	6821.1
	Mar. 26, 1980										
0	Chlamydomonas sp. 1	grn	3	15	30	40	16	20	0	90.9	31931.4
0	Chlamydomonas sp. 1	grn	14	15	30	40	4	5	0	6.6	9313.3
0	Cosmarium sp.	desm	1	15	30	40	12	20	0	34.1	14999.0
0	Anabaena sp.	bg	4	15	30	40	4	0	0	1.5	2273.7
2	Chlamydomonas sp. 1	grn	40	15	30	40	16	20	0	1212.6	425752.0
2	Chlamydomonas sp. 1	grn	5	15	30	40	7	10	0	14.5	11276.6
2	Cryptomonas sp.	cryp	5	15	30	40	20	30	0	355.3	95659.8
2	Cosmarium sp.	desm	3	15	30	40	12	20	0	102.3	44996.9
2	Stephanodiscus astrea	diat	181	15	30	40	4	5	0	128.6	180049.2
2	Dictyosphaerium pulchellum	grn	47	15	30	40	4	0	0	17.8	26716.1
4	Chlamydomonas sp. 1	grn	32	15	30	40	16	20	0	970.1	340601.6
4	Chlamydomonas sp. 1	grn	39	15	30	40	4	5	0	18.5	25944.3
4	Chlamydomonas sp. 1	grn	20	15	30	40	7	10	0	58.0	45106.3
4	Cryptomonas sp.	cryp	8	15	30	40	20	30	0	568.4	153055.6
4	Cosmarium sp.	desm	3	15	30	40	12	20	0	102.3	44996.9
4	Stephanodiscus astrea	diat	232	15	30	40	4	5	0	164.8	230781.2
4	Dictyosphaerium pulchellum	grn	27	15	30	40	4	0	0	10.2	15347.5
8	Chlamydomonas sp. 1	grn	36	15	30	40	16	20	0	1091.4	383176.8
8	Chlamydomonas sp. 1	grn	12	15	30	40	4	5	0	5.7	7982.9
8	Chlamydomonas sp. 1	grn	19	15	30	40	7	10	0	55.1	42851.0
8	Cryptomonas sp.	cryp	4	15	30	40	20	30	0	284.2	76527.8
8	Cosmarium sp.	desm	2	15	30	40	12	20	0	68.2	29997.9
8	Stephanodiscus astrea	diat	234	15	30	40	4	5	0	166.3	232770.7
8	Dictyosphaerium pulchellum	grn	80	15	30	40	4	0	0	30.3	45474.1
16	Chlamydomonas sp. 1	grn	4	15	30	40	16	20	0	121.3	42575.2
16	Chlamydomonas sp. 1	grn	3	15	30	40	4	5	0	1.4	1995.7
16	Chlamydomonas sp. 1	grn	4	15	30	40	7	10	0	11.6	9021.3
16	Cryptomonas sp.	cryp	1	15	30	40	20	30	0	71.1	19132.0
16	Stephanodiscus astrea	diat	47	15	30	40	4	5	0	33.4	46753.1
16	Dictyosphaerium pulchellum	grn	20	15	30	40	4	0	0	7.6	11368.5

M	Genus and species	Fam.	Ct.	Vol. filt. ml	Flds. ctd.	Mag.	Dim1 μm	Dim2 μm	Dim3 μm	Biovol. mm³/m³	Surf.area mm²/m³
	Apr. 1, 1980										
0	Chlamydomonas sp. 1	grn	11	15	30	40	16	20	0	333.5	117081.8
0	Cosmarium sp.	desm	2	15	30	40	12	20	0	68.2	29997.9
0	Stephanodiscus astrea	diat	153	15	30	40	4	6	0	130.5	173938.6
0	Dictyosphaerium pulchellum	grn	37	15	30	40	4	0	0	14.0	21031.8
2	Cryptomonas sp.	cryp	7	10	30	40	20	30	0	746.1	200885.5
2	Chlamydomonas sp. 1	grn	36	10	30	40	16	20	0	1637.1	574765.2
2	Chlamydomonas sp. 1	grn	21	10	30	40	7	10	0	91.4	71042.5
2	Chlamydomonas sp. 1	grn	12	10	30	40	4	5	0	8.5	11974.3
2	Cosmarium sp.	desm	2	10	30	40	12	20	0	102.3	44996.9
2	Stephanodiscus astrea	diat	193	10	30	40	4	6	0	246.8	329119.1
2	Dictyosphaerium pulchellum	grn	75	10	30	40	4	0	0	42.6	63948.0
4	Cryptomonas sp.	cryp	6	10	30	40	20	30	0	639.5	172187.6
4	Chlamydomonas sp. 1	grn	19	10	30	40	16	20	0	864.0	303348.3
4	Chlamydomonas sp. 1	grn	11	10	30	40	7	10	0	47.9	37212.7
4	Cosmarium sp.	desm	2	10	30	40	12	20	0	102.3	44996.9
4	Stephanodiscus astrea	diat	111	10	30	40	4	6	0	142.0	189286.1
4	Dictyosphaerium pulchellum	grn	16	10	30	40	4	0	0	9.1	13642.2
8	Chlamydomonas sp. 1	grn	25	10	30	40	16	20	0	1136.9	399142.5
8	Chlamydomonas sp. 1	grn	13	10	30	40	7	10	0	56.6	43978.7
8	Chlamydomonas sp. 1	grn	2	10	30	40	4	5	0	1.4	1995.7
8	Cosmarium sp.	desm	2	10	30	40	12	20	0	102.3	44996.9
8	Stephanodiscus astrea	diat	76	10	30	40	4	6	0	97.2	129601.3
8	Dictyosphaerium pulchellum	grn	16	10	30	40	4	0	0	9.1	13642.2
16	Cryptomonas sp.	cryp	9	15	30	40	20	30	0	639.5	172187.6
16	Chlamydomonas sp. 1	grn	22	15	30	40	16	20	0	667.0	234163.6
16	Chlamydomonas sp. 1	grn	11	15	30	40	7	10	0	31.9	24808.5
16	Cosmarium sp.	desm	5	15	30	40	12	20	0	170.5	74994.8
16	Stephanodiscus astrea	diat	212	15	30	40	4	6	0	180.8	241012.9
16	Dictyosphaerium pulchellum	grn	9	15	30	40	4	0	0	3.4	5115.8
	Apr. 28, 1980										
0	Cryptomonas sp.	cryp	3	10	30	40	20	30	0	319.7	86093.8
0	Chlamydomonas sp. 1	grn	44	10	30	40	16	20	0	2000.9	702490.9
0	Chlamydomonas sp. 1	grn	34	10	30	40	7	10	0	148.0	115021.1
0	Chlamydomonas sp. 1	grn	3	10	30	40	4	5	0	2.1	2993.6
0	Cosmarium sp.	desm	15	10	30	40	12	18	0	690.6	309937.6
0	Stephanodiscus astrea	diat	38	10	30	40	4	5	0	40.5	56700.6
0	Dictyosphaerium pulchellum	grn	28	10	30	40	4	0	0	15.9	23873.9
0	Actinastrum Hantzschii	grn	15	10	30	40	3	20	0	45.8	65645.8
0	Stephanodiscus sp.	diat	8	10	30	40	18	10	0	345.3	145801.5
0	Stephanodiscus sp.	diat	7	10	30	40	10	8	0	74.6	48493.9
2	Cryptomonas sp.	cryp	7	10	30	40	20	30	0	746.1	200885.5
2	Chlamydomonas sp. 1	grn	31	10	30	40	16	20	0	1409.7	494936.7
2	Chlamydomonas sp. 1	grn	39	10	30	40	7	10	0	169.7	131936.0
2	Cosmarium sp.	desm	7	10	30	40	12	18	0	322.3	144637.5
2	Stephanodiscus astrea	diat	5	10	30	40	4	5	0	5.3	7460.6
2	Dictyosphaerium pulchellum	grn	52	10	30	40	4	0	0	29.6	44337.3
2	Stephanodiscus sp.	diat	7	10	30	40	18	10	0	302.2	127576.3
2	Stephanodiscus sp.	diat	6	10	30	40	10	8	0	63.9	41566.2
4	Cryptomonas sp.	cryp	15	10	30	40	20	30	0	1598.7	430468.9
4	Chlamydomonas sp. 1	grn	63	10	30	40	16	20	0	2864.9	1005839.2
4	Chlamydomonas sp. 1	grn	37	10	30	40	7	10	0	161.0	125170.0
4	Chlamydomonas sp. 1	grn	21	10	30	40	4	5	0	14.9	20955.0
4	Cosmarium sp.	desm	5	10	30	40	12	18	0	230.2	103312.5
4	Stephanodiscus astrea	diat	10	10	30	40	4	5	0	10.7	14921.2

M	Genus and species	Fam.	Ct.	Vol. filt. ml	Flds. ctd.	Mag.	Dim1 μm	Dim2 μm	Dim3 μm	Biovol. mm³/m³	Surf.area mm²/m³
4	Dictyosphaerium pulchellum	grn	24	10	30	40	4	0	0	13.6	20463.4
4	Stephanodiscus sp.	diat	11	10	30	40	10	8	0	117.2	76204.7
4	Stephanodiscus sp.	diat	6	10	30	40	18	16	0	414.4	143883.0
8	Cryptomonas sp.	cryp	3	10	30	40	20	30	0	319.7	86093.8
8	Chlamydomonas sp. 1	grn	31	10	30	40	16	20	0	1409.7	494936.7
8	Chlamydomonas sp. 1	grn	23	10	30	40	7	10	0	100.1	77808.4
8	Cosmarium sp.	desm	6	10	30	40	12	18	0	276.3	123975.0
8	Dictyosphaerium pulchellum	grn	71	10	30	40	4	0	0	40.4	60537.4
8	Stephanodiscus sp.	diat	10	10	30	40	10	8	0	106.6	69277.0
8	Stephanodiscus sp.	diat	3	10	30	40	18	16	0	207.2	71941.5
	May 12, 1980										
0	Cryptomonas sp.	cryp	16	10	30	40	20	30	0	1705.3	459166.8
0	Chlamydomonas sp. 1	grn	57	10	30	40	16	20	0	2592.0	910045.0
0	Chlamydomonas sp. 1	grn	28	10	30	40	7	10	0	121.9	94723.3
0	Chlamydomonas sp. 1	grn	183	10	30	40	5	7	0	284.4	310933.8
0	Stephanodiscus sp.	diat	6	10	30	40	18	12	0	310.8	120861.7
0	Dictyosphaerium pulchellum	grn	27	10	30	40	4	0	0	15.3	23021.3
0	Aphanizomenon flos-aquae	bg	15	10	30	40	3	0	0	14.1	18834.2
2	Cryptomonas sp.	cryp	6	10	30	40	20	30	0	639.5	172187.6
2	Chlamydomonas sp. 1	grn	95	10	30	40	16	20	0	4320.0	1516741.6
2	Chlamydomonas sp. 1	grn	75	10	30	40	7	10	0	326.4	253723.1
2	Chlamydomonas sp. 1	grn	132	10	30	40	5	7	0	205.2	224280.1
2	Stephanodiscus sp.	diat	5	10	30	40	18	12	0	259.0	100718.1
2	Dictyosphaerium pulchellum	grn	56	10	30	40	4	0	0	31.8	47747.8
2	Aphanizomenon flos-aquae	bg	11	10	30	40	3	0	0	10.4	13811.8
4	Cryptomonas sp.	cryp	2	10	30	40	20	30	0	213.2	57395.9
4	Chlamydomonas sp. 1	grn	30	10	30	40	16	20	0	1364.2	478971.0
4	Chlamydomonas sp. 1	grn	55	10	30	40	7	10	0	239.4	186063.6
4	Chlamydomonas sp. 1	grn	105	10	30	40	5	7	0	163.2	178404.6
4	Stephanodiscus sp.	diat	7	10	30	40	18	12	0	362.6	141005.4
4	Dictyosphaerium pulchellum	grn	24	10	30	40	4	0	0	13.6	20463.4
8	Chlamydomonas sp. 1	grn	7	10	30	40	16	20	0	318.3	111759.9
8	Chlamydomonas sp. 1	grn	19	10	30	40	7	10	0	82.7	64276.5
8	Stephanodiscus sp.	diat	4	10	30	40	18	12	0	207.2	80574.5
8	Dictyosphaerium pulchellum	grn	16	10	30	40	4	0	0	9.1	13642.2
8	Ankistrodesmus fractus	grn	1	10	30	40	2	70	0	1.2	3731.8
8	dinoflagellate	dino	1	10	30	40	18	20	0	57.6	18558.3
16	Chlamydomonas sp. 1	grn	4	10	30	40	16	20	0	181.9	63862.8
16	Chlamydomonas sp. 1	grn	6	10	30	40	7	10	0	26.1	20297.8
16	Stephanodiscus sp.	diat	5	10	30	40	18	12	0	259.0	100718.1
16	Stephanodiscus sp.	diat	1	10	30	40	45	12	0	323.7	82732.7
16	Ankistrodesmus fractus	grn	1	10	30	40	2	70	0	1.2	3731.8
16	dinoflagellate	dino	1	10	30	40	18	20	0	57.6	18558.3
	May 19, 1980										
0	Cryptomonas sp.	cryp	2	10	30	40	20	30	0	213.2	57395.9
0	Chlamydomonas sp. 1	grn	3	10	30	40	16	20	0	136.4	47897.1
0	Dictyosphaerium pulchellum	grn	50	10	30	40	4	0	0	28.4	42632.0
0	blue green sphere	bg	487	10	30	40	3	0	0	116.8	233570.1
0	Ankistrodesmus fractus	grn	3	10	30	40	2	60	0	3.2	9597.5
2	Chlamydomonas sp. 1	grn	2	5	30	40	16	20	0	181.9	63862.8
2	Chlamydomonas sp. 1	grn	1	5	30	40	7	10	0	8.7	6765.9
2	Chlamydomonas sp. 1	grn	5	5	30	40	5	7	0	15.5	16990.9
2	Dictyosphaerium pulchellum	grn	97	5	30	40	4	0	0	110.3	165412.2
2	Ankistrodesmus fractus	grn	3	5	30	40	2	60	0	6.4	19195.1
2	Oocystis sp.	grn	4	5	30	40	4	6	0	6.8	9183.3

M	Genus and species	Fam.	Ct.	Vol. filt. ml	Flds. ctd.	Mag.	Dim1 μm	Dim2 μm	Dim3 μm	Biovol. mm³/m³	Surf.area mm²/m³
2	Microcystis aeruginosa	bg	24	5	30	40	3	0	0	11.5	23021.3
2	Chlorella ellipsoidea	grn	21	5	30	40	4	8	0	23.9	35810.9
2	Staurastrum sp.	grn	1	5	30	40	10	14	28	133.0	96191.4
2	Stephanodiscus sp.	diat	7	5	30	40	20	14	0	1044.5	358108.8
2	Stephanodiscus sp.	diat	2	5	30	40	40	16	0	1364.2	306950.4
5	Chlamydomonas sp. 1	grn	8	10	30	40	16	20	0	363.8	127725.6
5	Chlamydomonas sp. 1	grn	2	10	30	40	7	10	0	8.7	6765.9
5	Chlamydomonas sp. 1	grn	7	10	30	40	5	7	0	10.9	11893.6
5	Dictyosphaerium pulchellum	grn	19	10	30	40	4	0	0	10.8	16200.2
5	Ankistrodesmus fractus	grn	3	10	30	40	2	60	0	3.2	9597.5
5	Oocystis sp.	grn	5	10	30	40	4	6	0	4.3	5739.6
5	Chlorella ellipsoidea	grn	36	10	30	40	4	8	0	20.5	30695.0
5	Stephanodiscus sp.	diat	3	10	30	40	18	12	0	155.4	60430.9
5	Actinastrum Hantzschii	grn	7	10	30	40	3	20	0	21.4	30634.7
10	Cryptomonas sp.	cryp	1	10	30	40	20	30	0	106.6	28697.9
10	Chlamydomonas sp. 1	grn	4	10	30	40	16	20	0	181.9	63862.8
10	Chlamydomonas sp. 1	grn	4	10	30	40	7	10	0	17.4	13531.9
10	Dictyosphaerium pulchellum	grn	8	10	30	40	4	0	0	4.5	6821.1
10	Ankistrodesmus fractus	grn	3	10	30	40	2	60	0	3.2	9597.5
10	Oocystis sp.	grn	3	10	30	40	4	6	0	2.6	3443.8
10	Chlorella ellipsoidea	grn	10	10	30	40	4	8	0	5.7	8526.4
10	Stephanodiscus sp.	diat	1	10	30	40	40	16	0	341.1	76737.6
15	blue green sphere	bg	30	10	30	40	3	0	0	7.2	14388.3
	June 6, 1980										
0	Chlamydomonas sp. 1	grn	6	20	30	40	15	18	0	107.9	40853.0
0	Dictyosphaerium pulchellum	grn	4	20	30	40	4	0	0	1.1	1705.3
0	Aphanizomenon flos-aquae	bg	9	20	30	40	3	0	0	4.2	5650.3
0	Staurastrum sp.	grn	1	20	10	10		20	50		
2	Chlamydomonas sp. 1	grn	3	20	30	40	15	18	0	54.0	20426.5
2	Aphanizomenon flos-aquae	bg	28	20	30	40	3	0	0	13.2	17578.6
2	Oocystis sp.	grn	9	20	30	40	8	10	0	25.6	17961.4
2	Chlamydomonas sp. 1	grn	3	20	30	40	5	7	0	2.3	2548.6
2	Pediastrum sp.	grn	1	20	30	40	50	4	0	66.6	38635.3
2	Ceratium hirundinella	dino	3	20	10	10	50	50	0	9.7	1653.2
4	Chlamydomonas sp. 1	grn	4	20	30	40	15	18	0	71.9	27235.3
4	Cryptomonas sp.	cryp	1	20	30	40	22	30	0	64.5	16116.7
4	Ceratium hirundinella	dino	2	20	10	10	50	50	0	6.5	1102.1
8	Chlamydomonas sp. 1	grn	3	20	30	40	15	18	0	54.0	20426.5
8	Chlamydomonas sp. 1	grn	3	20	30	40	5	7	0	2.3	2548.6
8	Ceratium hirundinella	dino	1	20	10	10	50	50	0	3.2	551.1
	June 10, 1980										
0	Aphanizomenon flos-aquae	bg	14	10	30	40	4	0	0	23.4	23438.1
0	Chlamydomonas sp. 1	grn	13	10	30	40	18	20	0	748.2	241257.7
0	Chlamydomonas sp. 1	grn	7	10	30	40	12	16	0	143.2	65987.7
0	Staurastrum sp.	grn	1	10	30	40	10	20	50	169.6	95663.3
0	Pediastrum sp.	grn	1	10	5	10	99	4	0	12.2	6604.2
0	Ceratium hirundinella	dino	1	10	5	10	50	50	0	13.0	2204.2
2	Aphanizomenon flos-aquae	bg	29	10	30	40	4	0	0	48.6	48550.5
2	Chlamydomonas sp. 1	grn	8	10	30	40	18	20	0	460.4	148466.3
2	Chlamydomonas sp. 1	grn	4	10	30	40	12	16	0	81.9	37707.3
2	Planktosphaeria gelatinosa	grn	15	10	30	40	5	0	0	16.7	19983.8
2	Ceratium hirundinella	dino	3	10	5	10	50	50	0	39.0	6612.7
4	Chlamydomonas sp. 1	grn	11	20	30	40	18	20	0	316.5	102070.6
4	Chlamydomonas sp. 1	grn	12	20	30	40	12	16	0	122.8	56560.9
4	Planktosphaeria gelatinosa	grn	69	20	30	40	5	0	0	38.3	45962.6

M	Genus and species	Fam.	Ct.	Vol. filt. ml	Flds. ctd.	Mag.	Dim1 μm	Dim2 μm	Dim3 μm	Biovol. mm³/m³	Surf.area mm²/m³
4	Pediastrum sp.	grn	1	20	30	40	55	4	0	80.6	46162.5
4	Ceratium hirundinella	dino	3	20	5	10	50	50	0	19.5	3306.3
8	Planktosphaeria gelatinosa	grn	8	20	30	40	5	0	0	4.4	5329.0
	June 18, 1980										
0	Pandorina sp.	grn	6	10	30	40	8	0	0	27.3	20463.4
0	Aphanizomenon flos-aquae	bg	97	10	30	40	4	0	0	162.4	162392.9
0	Planktosphaeria gelatinosa	grn	34	10	30	40	5	0	0	37.7	45296.5
0	Microcystis aeruginosa	bg	78	10	30	40	3	0	0	18.7	37409.6
0	Chlamydomonas sp. 1	grn	3	10	30	40	18	20	0	172.7	55674.9
0	Cryptomonas sp.	cryp	1	10	30	40	22	40	0	171.9	40646.8
0	Chlamydomonas sp. 1	grn	1	10	30	40	7	10	0	4.4	3383.0
0	Schroederia setigera	grn	2	10	30	40	3	50	0	4.0	8007.9
0	Ceratium hirundinella	dino	1	10	3	10	50	50	0	21.6	3673.7
0	Staurastrum sp.	grn	3	10	3	10	10	16	32	10.2	6759.8
2	Aphanizomenon flos-aquae	bg	119	10	30	40	4	0	0	199.3	199224.3
2	Chlamydomonas sp. 1	grn	5	10	30	40	18	20	0	287.8	92791.4
2	Cryptomonas sp.	cryp	4	10	30	40	22	40	0	687.8	162587.3
2	Cosmarium sp.	desm	1	10	30	40	18	30	0	172.7	50621.5
2	Planktosphaeria gelatinosa	grn	100	10	4	10	5	0	0	3.2	3896.6
2	Ceratium hirundinella	dino	6	10	4	10	50	50	0	97.4	16531.7
2	Pediastrum sp.	grn	1	10	4	10	99	4	0	15.3	8255.2
4	Aphanizomenon flos-aquae	bg	125	10	30	40	4	0	0	209.3	209269.2
4	Chlamydomonas sp. 1	grn	5	10	30	40	18	20	0	287.8	92791.4
4	Chlamydomonas sp. 1	grn	2	10	30	40	12	28	0	71.6	30025.8
4	Oocystis sp.	grn	6	10	30	40	10	20	0	106.6	54650.0
4	Planktosphaeria gelatinosa	grn	76	10	4	10	5	0	0	2.5	2961.4
4	Ceratium hirundinella	dino	3	10	4	10	50	50	0	48.7	8265.8
4	Staurastrum sp.	grn	1	10	4	10	10	16	32	2.5	1690.0
8	Planktosphaeria gelatinosa	grn	48	20	30	40	5	0	0	26.6	31974.0
8	Chlamydomonas sp. 1	grn	1	20	30	40	7	10	0	2.2	1691.5
8	Stephanodiscus sp.	diat	1	20	30	40	40	20	0	213.2	42632.0
8	Ceratium hirundinella	dino	1	20	4	10	50	50	0	8.1	1377.6
8	Staurastrum sp.	grn	5	20	4	10	10	16	32	6.4	4224.9
10	Chlamydomonas sp. 1	grn	1	20	30	40	18	20	0	28.8	9279.1
10	Oocystis sp.	grn	8	20	30	40	10	20	0	71.1	36433.3
	June 24, 1980										
0	Aphanizomenon flos-aquae	bg	875	10	10	40	4	0	0	4395.7	4394652.8
0	Microcystis aeruginosa	bg	529	10	10	40	3	0	0	380.6	761141.1
0	Schroederia setigera	grn	1	10	10	40	3	50	0	6.0	12011.8
0	Staurastrum sp.	grn	1	10	5	10	10	16	32	2.0	1352.0
0	Stephanodiscus sp.	diat	1	10	5	10	40	20	0	10.0	1995.0
2	Aphanizomenon flos-aquae	bg	533	10	15	40	4	0	0	1785.1	1784647.6
2	Microcystis aeruginosa	bg	48	10	15	40	3	0	0	23.0	46042.6
2	Planktosphaeria gelatinosa	grn	18	10	15	40	5	0	0	40.0	47961.0
2	Ceratium hirundinella	dino	12	10	5	10	30	40	0	44.9	11222.1
2	Ceratium hirundinella	dino	8	10	5	10	50	60	0	124.7	19477.2
4	Aphanizomenon flos-aquae	bg	219	10	30	40	4	0	0	366.7	366639.6
4	Staurastrum sp.	grn	1	10	30	40	10	16	32	86.8	57780.3
4	Planktosphaeria gelatinosa	grn	38	10	30	40	5	0	0	42.2	50625.5
4	Cryptomonas sp.	cryp	1	10	30	40	22	40	0	171.9	40646.8
4	Chlamydomonas sp. 1	grn	1	10	30	40	18	20	0	57.6	18558.3
4	Stephanodiscus sp.	diat	1	10	5	10	40	20	0	10.0	1995.0
4	Ceratium hirundinella	dino	10	10	5	10	30	40	0	37.4	9351.7
4	Ceratium hirundinella	dino	8	10	5	10	50	60	0	124.7	19477.2
8	Aphanizomenon flos-aquae	bg	6	10	30	40	4	0	0	10.0	10044.9

M	Genus and species	Fam.	Ct.	Vol. filt. ml	Flds. ctd.	Mag.	Dim1 μm	Dim2 μm	Dim3 μm	Biovol. mm³/m³	Surf.area mm²/m³
8	Planktosphaeria gelatinosa	grn	24	10	30	40	5	0	0	26.6	31974.0
8	Staurastrum sp.	grn	1	10	5	10	10	16	32	2.0	1352.0
8	Ceratium hirundinella	dino	3	10	5	10	30	40	0	11.2	2805.5
	July 1, 1980										
0	Anabaena sp.	bg	10	5	15	40	4	0	0	22.7	34105.6
0	Aphanizomenon flos-aquae	bg	1614	5	15	40	4	0	0	10811.0	10808334.5
0	Microcystis aeruginosa	bg	164	5	15	40	3	0	0	157.3	314624.2
0	Oocystis sp.	grn	7	5	15	40	7	10	0	121.9	94723.3
0	Planktosphaeria gelatinosa	grn	8	5	15	40	7	0	0	97.5	83558.7
0	Chlamydomonas sp. 1	grn	1	5	40	40	16	30	0	102.3	33098.2
0	Cosmarium sp.	desm	1	5	5	10	18	30	0	8.1	2368.9
0	Ceratium hirundinella	dino	3	5	5	10	50	50	0	77.9	13225.3
2	Aphanizomenon flos-aquae	bg	911	5	15	40	4	0	0	6102.1	6100615.1
2	Microcystis aeruginosa	bg	39	5	15	40	3	0	0	37.4	74819.2
2	Planktosphaeria gelatinosa	grn	16	5	15	40	7	0	0	195.0	167117.5
2	Dictyosphaerium pulchellum	grn	12	5	15	40	4	0	0	27.3	40926.7
2	Ceratium hirundinella	dino	16	5	10	10	50	50	0	207.8	35267.6
4	Aphanizomenon flos-aquae	bg	1127	5	15	40	4	0	0	7548.9	7547083.7
4	Microcystis aeruginosa	bg	170	5	15	40	3	0	0	163.1	326134.8
4	Stephanodiscus sp.	diat	1	5	15	40	60	20	0	3836.9	639480.1
4	Dictyosphaerium pulchellum	grn	56	5	45	40	4	0	0	42.4	63663.8
4	Ceratium hirundinella	dino	9	5	10	10	50	50	0	116.9	19838.0
4	Closterium sp.	desm	2	5	10	10	50	250	0	129.9	15894.9
8	Aphanizomenon flos-aquae	bg	239	10	30	40	4	0	0	400.2	400122.7
8	Microcystis aeruginosa	bg	28	10	30	40	3	0	0	6.7	13429.1
8	Oocystis sp.	grn	2	10	30	40	7	10	0	8.7	6765.9
8	Planktosphaeria gelatinosa	grn	8	10	30	40	7	0	0	24.4	20889.7
8	Dictyosphaerium pulchellum	grn	20	10	30	40	4	0	0	11.4	17052.8
8	Ceratium hirundinella	dino	4	10	10	10	50	50	0	26.0	4408.4
10	Aphanizomenon flos-aquae	bg	1299	10	20	40	4	0	0	3262.9	3262088.0
10	Microcystis aeruginosa	bg	20	10	20	40	3	0	0	7.2	14388.3
10	Ceratium hirundinella	dino	6	10	10	10	50	50	0	39.0	6612.7
10	Closterium sp.	desm	1	10	10	10	50	250	0	32.5	3973.7
12	Aphanizomenon flos-aquae	bg	114	20	30	40	4	0	0	95.5	95426.7
12	Cosmarium sp.	desm	1	20	30	40	18	30	0	86.3	25310.7
12	Fragillaria sp.	diat	79	20	30	40	3	20	0	120.6	92463.9
12	Closterium sp.	desm	1	20	10	10	50	250	0	16.2	1986.9
	July 8, 1980										
0	Aphanizomenon flos-aquae	bg	1182	5	10	40	4	0	0	11876.0	11873096.3
0	Microcystis aeruginosa	bg	132	5	10	40	3	0	0	189.9	379851.2
0	Planktosphaeria gelatinosa	grn	24	5	10	40	7	0	0	438.7	376014.3
0	Ceratium hirundinella	dino	12	5	10	10	50	50	0	155.9	26450.7
2	Aphanizomenon flos-aquae	bg	1044	5	10	40	4	0	0	10489.5	10486897.1
2	Microcystis aeruginosa	bg	40	5	10	40	3	0	0	57.6	115106.4
2	Pediastrum sp.	grn	1	5	10	40	25	4	0	199.8	131892.8
2	Ceratium hirundinella	dino	19	5	10	10	50	50	0	246.8	41880.2
2	Cosmarium sp.	desm	3	5	10	10	18	30	0	12.1	3553.4
4	Aphanizomenon flos-aquae	bg	1098	5	10	40	4	0	0	11032.0	11029322.9
4	Chlamydomonas sp. 1	grn	1	5	10	40	18	20	0	345.3	111349.7
4	Ceratium hirundinella	dino	12	5	10	10	50	50	0	155.9	26450.7
4	Cosmarium sp.	desm	3	5	10	10	18	30	0	12.1	3553.4
8	Aphanizomenon flos-aquae	bg	727	10	30	40	4	0	0	1217.4	1217109.5
8	Microcystis aeruginosa	bg	36	10	30	40	3	0	0	8.6	17266.0
8	Ceratium hirundinella	dino	7	10	10	10	50	50	0	45.5	7714.8
8	Cosmarium sp.	desm	4	10	10	10	18	30	0	8.1	2368.9

M	Genus and species	Fam.	Ct.	Vol. filt. ml	Flds. ctd.	Mag.	Dim1 µm	Dim2 µm	Dim3 µm	Biovol. mm³/m³	Surf.area mm²/m³
8	Staurastrum sp.	grn	2	10	10	10	10	20	50	4.0	2238.4
10	Aphanizomenon flos-aquae	bg	500	20	30	40	4	0	0	418.6	418538.4
10	Planktosphaeria gelatinosa	grn	10	20	30	40	7	0	0	15.2	13056.1
10	Cosmarium sp.	desm	1	20	30	40	10	20	0	17.8	9108.3
10	Ceratium hirundinella	dino	3	20	10	10	50	50	0	9.7	1653.2
10	Cosmarium sp.	desm	10	20	10	10	18	30	0	10.1	2961.1
10	Staurastrum sp.	grn	4	20	10	10	10	20	50	4.0	2238.4
10	Stephanodiscus sp.	diat	1	20	10	10	50	20	0	3.9	701.4
12	Aphanizomenon flos-aquae	bg	114	20	30	40	4	0	0	95.5	95426.7
12	Ceratium hirundinella	dino	1	20	10	10	50	50	0	3.2	551.1
12	Cosmarium sp.	desm	1	20	10	10	18	30	0	1.0	296.1
	July 16, 1980										
0	Chroococcus sp.	bg	12	15	15	40	5	0	0	17.8	21316.0
0	Aphanizomenon flos-aquae	bg	710	15	15	40	4	0	0	1585.3	1584865.3
0	Chlamydomonas sp. 1	grn	1	15	15	40	20	22	0	104.2	30334.0
0	Anabaena sp.	bg	22	15	15	40	4	0	0	16.7	25010.8
0	Ceratium hirundinella	dino	22	15	10	10	50	50	0	95.2	16164.3
0	Cosmarium sp.	desm	10	15	10	10	18	30	0	13.5	3948.2
2	Aphanizomenon flos-aquae	bg	842	15	15	40	4	0	0	1880.0	1879516.2
2	Anabaena sp.	bg	105	15	15	40	4	0	0	79.6	119369.6
2	Microcystis aeruginosa	bg	115	15	15	40	3	0	0	36.8	73540.2
2	Cosmarium sp.	desm	1	15	15	40	12	20	0	68.2	29997.9
2	Chlamydomonas sp. 1	grn	1	15	15	40	10	15	0	17.8	9566.0
2	Ceratium hirundinella	dino	21	15	10	10	50	50	0	90.9	15429.6
4	Aphanizomenon flos-aquae	bg	926	15	15	40	4	0	0	2067.5	2067021.5
4	Anabaena sp.	bg	21	15	15	40	4	0	0	15.9	23873.9
4	Ceratium hirundinella	dino	18	15	10	10	50	50	0	77.9	13225.3
4	Cosmarium sp.	desm	4	15	10	10	18	30	0	5.4	1579.3
6	Aphanizomenon flos-aquae	bg	470	15	15	40	4	0	0	1049.4	1049136.2
6	Microcystis aeruginosa	bg	58	15	15	40	3	0	0	18.5	37089.8
6	Chlamydomonas sp. 1	grn	1	15	15	40	10	15	0	17.8	9566.0
6	Ceratium hirundinella	dino	24	15	10	10	50	50	0	103.9	17633.8
6	Cosmarium sp.	desm	4	15	10	10	18	30	0	5.4	1579.3
8	Aphanizomenon flos-aquae	bg	10	15	30	40	4	0	0	11.2	11161.0
8	Ceratium hirundinella	dino	1	15	10	10	50	50	0	4.3	734.7
8	Cosmarium sp.	desm	1	15	10	10	18	30	0	1.3	394.8
	July 23, 1980										
0	Anabaena sp.	bg	112	10	15	40	4	0	0	127.3	190991.4
0	Aphanizomenon flos-aquae	bg	913	10	15	40	4	0	0	3057.8	3057004.2
0	Microcystis aeruginosa	bg	196	10	15	40	3	0	0	94.0	188007.1
0	Chlamydomonas sp. 1	grn	5	10	15	40	5	6	0	13.3	15130.7
0	Chlamydomonas sp. 1	grn	3	10	15	40	10	12	0	63.9	36313.7
0	Chlamydomonas sp. 1	grn	4	10	15	40	12	18	0	184.2	82650.0
0	Chlamydomonas sp. 1	grn	2	10	15	40	18	20	0	230.2	74233.1
0	Dictyosphaerium pulchellum	grn	46	10	15	40	4	0	0	52.3	78442.9
0	Planktosphaeria gelatinosa	grn	48	10	15	40	5	0	0	106.6	127896.0
0	Ceratium hirundinella	dino	12	10	10	10	50	50	0	77.9	13225.3
0	Cosmarium sp.	desm	1	10	10	10	18	30	0	2.0	592.2
2	Anabaena sp.	bg	139	10	15	40	4	0	0	158.0	237033.9
2	Aphanizomenon flos-aquae	bg	678	10	15	40	4	0	0	2270.7	2270152.1
2	Microcystis aeruginosa	bg	159	10	15	40	3	0	0	76.3	152516.0
2	Chlamydomonas sp. 1	grn	1	10	15	40	10	12	0	21.3	12104.6
2	Chlamydomonas sp. 1	grn	4	10	15	40	12	18	0	184.2	82650.0
2	Dictyosphaerium pulchellum	grn	51	10	15	40	4	0	0	58.0	86969.3
2	Planktosphaeria gelatinosa	grn	4	10	15	40	5	0	0	8.9	10658.0

M	Genus and species	Fam.	Ct.	Vol. filt. ml	Flds. ctd.	Mag.	Dim1 µm	Dim2 µm	Dim3 µm	Biovol. mm³/m³	Surf.area mm²/m³
2	Ceratium hirundinella	dino	44	10	10	10	50	50	0	285.7	48492.9
2	Cosmarium sp.	desm	2	10	10	10	18	30	0	4.0	1184.5
4	Anabaena sp.	bg	196	10	15	40	4	0	0	222.8	334234.9
4	Aphanizomenon flos-aquae	bg	784	10	15	40	4	0	0	2625.7	2625072.6
4	Microcystis aeruginosa	bg	242	10	15	40	3	0	0	116.1	232131.3
4	Dictyosphaerium pulchellum	grn	16	10	15	40	4	0	0	18.2	27284.5
4	Planktosphaeria gelatinosa	grn	61	10	15	40	5	0	0	135.4	162534.5
4	Ceratium hirundinella	dino	8	10	10	10	50	50	0	52.0	8816.9
4	Cosmarium sp.	desm	1	10	10	10	18	30	0	2.0	592.2
6	Anabaena sp.	bg	216	10	15	40	4	0	0	245.6	368340.5
6	Aphanizomenon flos-aquae	bg	583	10	15	40	4	0	0	1952.5	1952062.9
6	Microcystis aeruginosa	bg	81	10	15	40	3	0	0	38.8	77696.8
6	Dictyosphaerium pulchellum	grn	106	10	15	40	4	0	0	120.5	180759.7
6	Schroederia setigera	grn	1	10	15	40	4	50	0	7.1	10692.1
6	Actinastrum Hantzschii	grn	20	10	15	40	2	10	0	27.1	59708.9
6	Ceratium hirundinella	dino	4	10	10	10	50	50	0	26.0	4408.4
6	Cosmarium sp.	desm	2	10	10	10	18	30	0	4.0	1184.5
8	Anabaena sp.	bg	66	10	30	40	4	0	0	37.5	56274.2
8	Aphanizomenon flos-aquae	bg	711	10	30	40	4	0	0	1190.6	1190323.1
8	Microcystis aeruginosa	bg	6	10	30	40	3	0	0	1.4	2877.7
8	Dictyosphaerium pulchellum	grn	106	10	30	40	4	0	0	60.3	90379.8
8	Planktosphaeria gelatinosa	grn	92	10	30	40	5	0	0	102.1	122567.0
8	Schroederia setigera	grn	1	10	30	40	4	50	0	3.6	5346.0
8	Actinastrum Hantzschii	grn	12	10	30	40	2	10	0	8.1	17912.7
8	Ceratium hirundinella	dino	1	10	10	10	50	50	0	6.5	1102.1
	July 29, 1980										
0	Anabaena sp.	bg	188	10	15	40	4	0	0	213.7	320592.7
0	Aphanizomenon flos-aquae	bg	948	10	15	40	4	0	0	3175.0	3174195.0
0	Microcystis aeruginosa	bg	121	10	15	40	3	0	0	58.0	116065.6
0	Chlamydomonas sp. 1	grn	42	10	15	40	5	7	0	130.6	142723.7
0	Chlamydomonas sp. 1	grn	1	10	15	40	10	12	0	21.3	12104.6
0	Chlamydomonas sp. 1	grn	8	10	15	40	18	20	0	920.9	296932.6
0	Dictyosphaerium pulchellum	grn	28	10	15	40	4	0	0	31.8	47747.8
0	Planktosphaeria gelatinosa	grn	9	10	15	40	5	0	0	20.0	23980.5
0	Actinastrum Hantzschii	grn	16	10	15	40	2	10	0	21.7	47767.1
0	Ceratium hirundinella	dino	9	10	10	10	50	50	0	58.4	9919.0
0	Cosmarium sp.	desm	1	10	10	10	18	30	0	2.0	592.2
2	Anabaena sp.	bg	266	10	15	40	4	0	0	302.4	453604.5
2	Aphanizomenon flos-aquae	bg	692	10	15	40	4	0	0	2317.6	2317028.3
2	Microcystis aeruginosa	bg	219	10	15	40	3	0	0	105.0	210069.2
2	Chlamydomonas sp. 1	grn	1	10	15	40	10	12	0	21.3	12104.6
2	Chlamydomonas sp. 1	grn	8	10	15	40	12	16	0	327.4	150829.0
2	Chlamydomonas sp. 1	grn	1	10	15	40	18	20	0	115.1	37116.6
2	Planktosphaeria gelatinosa	grn	49	10	15	40	5	0	0	108.8	130560.5
2	Ceratium hirundinella	dino	8	10	10	10	50	50	0	52.0	8816.9
4	Anabaena sp.	bg	375	10	15	40	4	0	0	426.3	639480.1
4	Aphanizomenon flos-aquae	bg	1180	10	15	40	4	0	0	3952.0	3951002.2
4	Microcystis aeruginosa	bg	87	10	15	40	3	0	0	41.7	83452.1
4	Chlamydomonas sp. 1	grn	1	10	15	40	5	7	0	3.1	3398.2
4	Chlamydomonas sp. 1	grn	7	10	15	40	12	16	0	286.5	131975.4
4	Dictyosphaerium pulchellum	grn	21	10	15	40	4	0	0	23.9	35810.9
4	Planktosphaeria gelatinosa	grn	57	10	15	40	5	0	0	126.6	151876.5
4	Actinastrum Hantzschii	grn	24	10	15	40	2	10	0	32.6	71650.7
4	Ceratium hirundinella	dino	15	10	10	10	50	50	0	97.4	16531.7
4	Cosmarium sp.	desm	1	10	10	10	18	30	0	2.0	592.2
6	Anabaena sp.	bg	182	10	15	40	4	0	0	206.9	310361.0

M	Genus and species	Fam.	Ct.	Vol. filt. ml	Flds. ctd.	Mag.	Dim1 μm	Dim2 μm	Dim3 μm	Biovol. mm³/m³	Surf.area mm²/m³
6	Aphanizomenon flos-aquae	bg	1164	10	15	40	4	0	0	3898.4	3897429.2
6	Microcystis aeruginosa	bg	354	10	15	40	3	0	0	169.8	339563.9
6	Chlamydomonas sp. 1	grn	2	10	15	40	5	7	0	6.2	6796.4
6	Chlamydomonas sp. 1	grn	1	10	15	40	10	12	0	21.3	12104.6
6	Chlamydomonas sp. 1	grn	9	10	15	40	12	16	0	368.3	169682.6
6	Dictyosphaerium pulchellum	grn	28	10	15	40	4	0	0	31.8	47747.8
6	Fragillaria sp.	diat	5	10	15	40	3	20	0	30.5	23408.6
6	Ceratium hirundinella	dino	6	10	10	10	50	50	0	39.0	6612.7
8	Anabaena sp.	bg	71	10	15	40	4	0	0	80.7	121074.9
8	Aphanizomenon flos-aquae	bg	830	10	15	40	4	0	0	2779.8	2779094.7
8	Microcystis aeruginosa	bg	53	10	15	40	3	0	0	25.4	50838.7
8	Chlamydomonas sp. 1	grn	1	10	15	40	12	16	0	40.9	18853.6
8	Chlamydomonas sp. 1	grn	4	10	15	40	18	20	0	460.4	148466.3
8	Dictyosphaerium pulchellum	grn	12	10	15	40	4	0	0	13.6	20463.4
8	Planktosphaeria gelatinosa	grn	8	10	15	40	5	0	0	17.8	21316.0
8	Ceratium hirundinella	dino	1	10	10	10	50	50	0	6.5	1102.1
10	Anabaena sp.	bg	24	10	30	40	4	0	0	13.6	20463.4
10	Aphanizomenon flos-aquae	bg	87	10	30	40	4	0	0	145.7	145651.3
10	Microcystis aeruginosa	bg	160	10	30	40	3	0	0	38.4	76737.6
10	Planktosphaeria gelatinosa	grn	14	10	30	40	5	0	0	15.5	18651.5
10	Oocystis sp.	grn	6	10	30	40	5	10	0	13.3	13662.5
10	Ceratium hirundinella	dino	1	10	10	10	50	50	0	6.5	1102.1

Aug. 5, 1980

M	Genus and species	Fam.	Ct.	Vol. filt. ml	Flds. ctd.	Mag.	Dim1 μm	Dim2 μm	Dim3 μm	Biovol. mm³/m³	Surf.area mm²/m³
0	Anabaena sp.	bg	46	10	15	40	4	0	0	52.3	78442.9
0	Aphanizomenon flos-aquae	bg	711	10	15	40	3	0	0	1339.4	1785484.6
0	Microcystis aeruginosa	bg	161	10	15	40	3	0	0	77.2	154434.4
0	Chlamydomonas sp. 1	grn	1	10	15	40	12	18	0	46.0	20662.5
0	Staurastrum sp.	grn	1	10	15	40	10	10	22	74.6	66494.0
0	Dictyosphaerium pulchellum	grn	20	10	15	40	4	0	0	22.7	34105.6
0	Planktosphaeria gelatinosa	grn	16	10	15	40	4	0	0	18.2	27284.5
0	Cosmarium sp.	desm	1	10	10	10	18	30	0	2.0	592.2
0	Ceratium hirundinella	dino	8	10	10	10	50	50	0	52.0	8816.9
2	Anabaena sp.	bg	350	10	15	40	4	0	0	397.9	596848.0
2	Aphanizomenon flos-aquae	bg	973	10	15	40	3	0	0	1833.0	2443426.9
2	Microcystis aeruginosa	bg	332	10	15	40	3	0	0	159.2	318461.1
2	Chlamydomonas sp. 1	grn	2	10	15	40	12	18	0	92.1	41325.0
2	Dictyosphaerium pulchellum	grn	4	10	15	40	4	0	0	4.5	6821.1
2	Crucigenia Lauterbornii	grn	16	10	15	40	5	8	0	56.8	60447.7
2	Chlamydomonas sp. 1	grn	1	10	15	40	10	12	0	21.3	12104.6
2	Ceratium hirundinella	dino	16	10	10	10	50	50	0	103.9	17633.8
4	Anabaena sp.	bg	138	10	15	40	4	0	0	156.9	235328.7
4	Aphanizomenon flos-aquae	bg	678	10	15	40	3	0	0	1277.3	1702614.0
4	Microcystis aeruginosa	bg	455	10	15	40	3	0	0	218.2	436445.1
4	Chlamydomonas sp. 1	grn	2	10	15	40	12	18	0	92.1	41325.0
4	Planktosphaeria gelatinosa	grn	8	10	15	40	4	0	0	9.1	13642.2
4	Chlamydomonas sp. 1	grn	1	10	15	40	10	12	0	21.3	12104.6
4	Ceratium hirundinella	dino	12	10	10	10	50	50	0	77.9	13225.3
6	Anabaena sp.	bg	127	10	15	40	4	0	0	144.4	216570.6
6	Aphanizomenon flos-aquae	bg	730	10	15	40	3	0	0	1375.2	1833198.0
6	Microcystis aeruginosa	bg	169	10	15	40	3	0	0	81.1	162108.2
6	Planktosphaeria gelatinosa	grn	8	10	15	40	4	0	0	9.1	13642.2
6	Chlamydomonas sp. 1	grn	1	10	15	40	10	12	0	21.3	12104.6
6	Actinastrum Hantzschii	grn	8	10	15	40	2	10	0	10.9	23883.6
6	Ceratium hirundinella	dino	2	10	10	10	50	50	0	13.0	2204.2
8	Anabaena sp.	bg	55	10	15	40	4	0	0	62.5	93790.4
8	Aphanizomenon flos-aquae	bg	851	10	15	40	3	0	0	1603.2	2137056.9
8	Microcystis aeruginosa	bg	441	10	15	40	3	0	0	211.5	423016.1
8	Dictyosphaerium pulchellum	grn	16	10	15	40	4	0	0	18.2	27284.5
10	Aphanizomenon flos-aquae	bg	55	10	30	40	3	0	0	51.8	69058.8
10	Dictyosphaerium pulchellum	grn	16	10	30	40	4	0	0	9.1	13642.2

Aug. 12, 1980

M	Genus and species	Fam.	Ct.	Vol. filt. ml	Flds. ctd.	Mag.	Dim1 μm	Dim2 μm	Dim3 μm	Biovol. mm³/m³	Surf.area mm²/m³
0	Aphanizomenon flos-aquae	bg	1046	10	15	40	3	0	0	1970.5	2626746.7
0	Microcystis aeruginosa	bg	461	10	15	40	3	0	0	221.1	442200.5
0	Anabaena sp.	bg	34	10	15	40	4	0	0	38.7	57979.5
0	Actinastrum Hantzschii	grn	10	10	15	40	2	10	0	13.6	29854.4
0	Chlamydomonas sp. 1	grn	7	10	15	40	4	5	0	9.9	13970.0
0	Chlamydomonas sp. 1	grn	2	10	15	40	10	12	0	42.6	24209.2
0	Chlamydomonas sp. 1	grn	2	10	15	40	14	20	0	139.3	54127.6
0	Staurastrum sp.	grn	1	10	15	40	16	16	25	217.1	125931.5
0	Dictyosphaerium pulchellum	grn	16	10	15	40	4	0	0	18.2	27284.5
0	Ceratium hirundinella	dino	13	10	10	10	50	50	0	84.4	14327.5
0	Cosmarium sp.	desm	1	10	10	10	18	30	0	2.0	592.2
2	Aphanizomenon flos-aquae	bg	686	10	15	40	3	0	0	1292.3	1722703.9
2	Microcystis aeruginosa	bg	277	10	15	40	3	0	0	132.9	265704.0
2	Anabaena sp.	bg	32	10	15	40	4	0	0	36.4	54569.0
2	Chlamydomonas sp. 1	grn	4	10	15	40	4	5	0	5.7	7982.9
2	Chlamydomonas sp. 1	grn	1	10	15	40	10	12	0	21.3	12104.6
2	Ceratium hirundinella	dino	18	10	10	10	50	50	0	116.9	19838.0
4	Aphanizomenon flos-aquae	bg	605	10	15	40	3	0	0	1139.7	1519294.3
4	Microcystis aeruginosa	bg	224	10	15	40	3	0	0	107.4	214865.3
4	Anabaena sp.	bg	77	10	15	40	4	0	0	87.5	131306.6
4	Chlamydomonas sp. 1	grn	1	10	15	40	10	12	0	21.3	12104.6
4	Ceratium hirundinella	dino	8	10	10	10	50	50	0	52.0	8816.9
4	Cosmarium sp.	desm	1	10	10	10	18	30	0	2.0	592.2
6	Aphanizomenon flos-aquae	bg	601	10	15	40	3	0	0	1132.2	1509249.3
6	Microcystis aeruginosa	bg	275	10	15	40	3	0	0	131.9	263785.5
6	Anabaena sp.	bg	85	10	15	40	4	0	0	96.6	144948.8
6	Actinastrum Hantzschii	grn	12	10	15	40	2	10	0	16.3	35825.3
6	Chlamydomonas sp. 1	grn	4	10	15	40	4	5	0	5.7	7982.9
6	Chlamydomonas sp. 1	grn	2	10	15	40	10	12	0	42.6	24209.2
6	Ceratium hirundinella	dino	7	10	10	10	50	50	0	45.5	7714.8
6	Cosmarium sp.	desm	1	10	10	10	18	30	0	2.0	592.2
8	Aphanizomenon flos-aquae	bg	361	10	15	40	3	0	0	680.1	906554.1
8	Microcystis aeruginosa	bg	224	10	15	40	3	0	0	107.4	214865.3
8	Anabaena sp.	bg	52	10	15	40	4	0	0	59.1	88674.6
8	Actinastrum Hantzschii	grn	8	10	15	40	2	10	0	10.9	23883.6
8	Planktosphaeria gelatinosa	grn	8	10	15	40	5	0	0	17.8	21316.0
8	Cosmarium sp.	desm	2	10	10	10	18	30	0	4.0	1184.5
10	Aphanizomenon flos-aquae	bg	694	10	30	40	3	0	0	653.7	871396.9
10	Microcystis aeruginosa	bg	159	10	30	40	3	0	0	38.1	76258.0
10	Anabaena sp.	bg	25	10	30	40	4	0	0	14.2	21316.0
10	Dictyosphaerium pulchellum	grn	8	10	30	40	4	0	0	4.5	6821.1
10	Planktosphaeria gelatinosa	grn	8	10	30	40	5	0	0	8.9	10658.0

Aug. 19, 1980

M	Genus and species	Fam.	Ct.	Vol. filt. ml	Flds. ctd.	Mag.	Dim1 μm	Dim2 μm	Dim3 μm	Biovol. mm³/m³	Surf.area mm²/m³
0	Aphanizomenon flos-aquae	bg	259	10	15	40	3	0	0	487.9	650408.6
0	Lyngbya limnetica	bg	618	10	15	40	2	0	0	517.4	1034626.8
0	Microcystis aeruginosa	bg	232	10	15	40	3	0	0	111.3	222539.1
0	Anabaena sp.	bg	59	10	15	40	4	0	0	67.1	100611.5
0	Actinastrum Hantzschii	grn	16	10	15	40	2	10	0	21.7	47767.1
0	Ceratium hirundinella	dino	5	10	10	10	40	40	0	16.6	3526.8
2	Aphanizomenon flos-aquae	bg	165	10	15	40	3	0	0	310.8	414353.0

M	Genus and species	Fam.	Ct.	Vol. filt. ml	Flds. ctd.	Mag.	Dim1 μm	Dim2 μm	Dim3 μm	Biovol. mm³/m³	Surf.area mm²/m³
2	Lyngbya limnetica	bg	521	10	15	40	2	0	0	436.2	872234.0
2	Microcystis aeruginosa	bg	643	10	15	40	3	0	0	308.4	616778.5
2	Anabaena sp.	bg	52	10	15	40	4	0	0	59.1	88674.6
2	Ceratium hirundinella	dino	8	10	10	10	40	40	0	26.6	5642.8
2	Cosmarium sp.	desm	1	10	10	10	18	30	0	2.0	592.2
4	Aphanizomenon flos-aquae	bg	344	10	10	40	3	0	0	972.1	1295794.8
4	Lyngbya limnetica	bg	464	10	10	40	2	0	0	582.7	1165210.8
4	Microcystis aeruginosa	bg	154	10	10	40	3	0	0	110.8	221579.8
4	Anabaena sp.	bg	99	10	10	40	4	0	0	168.8	253234.1
4	Actinastrum Hantzschii	grn	24	10	10	40	2	10	0	48.9	107476.0
4	Planktosphaeria gelatinosa	grn	13	10	10	40	4	0	0	22.2	33253.0
4	Dictyosphaerium pulchellum	grn	28	10	10	40	4	0	0	47.7	71621.8
4	Ceratium hirundinella	dino	7	10	10	10	40	40	0	23.3	4937.5
6	Aphanizomenon flos-aquae	bg	472	10	15	40	3	0	0	889.2	1185300.6
6	Lyngbya limnetica	bg	488	10	15	40	2	0	0	408.6	816986.9
6	Microcystis aeruginosa	bg	87	10	15	40	3	0	0	41.7	83452.1
6	Anabaena sp.	bg	82	10	15	40	4	0	0	93.2	139833.0
6	Actinastrum Hantzschii	grn	8	10	15	40	2	10	0	10.9	23883.6
6	Ceratium hirundinella	dino	10	10	10	10	40	40	0	33.3	7053.5
8	Aphanizomenon flos-aquae	bg	76	10	20	40	3	0	0	107.4	143140.1
8	Lyngbya limnetica	bg	475	10	20	40	2	0	0	298.3	596417.2
10	Aphanizomenon flos-aquae	bg	82	10	30	40	3	0	0	77.2	102960.4
10	Lyngbya limnetica	bg	58	10	30	40	2	0	0	24.3	48550.5
10	Microcystis aeruginosa	bg	28	10	30	40	3	0	0	6.7	13429.1
10	Anabaena sp.	bg	24	10	30	40	4	0	0	13.6	20463.4
10	Actinastrum Hantzschii	grn	8	10	30	40	2	10	0	5.4	11941.8
	Aug. 26, 1980										
0	Aphanizomenon flos-aquae	bg	93	10	15	40	3	0	0	175.2	233544.4
0	Microcystis aeruginosa	bg	834	10	15	40	3	0	0	400.0	799989.6
0	Anabaena sp.	bg	43	10	15	40	4	0	0	48.9	73327.0
0	Actinastrum Hantzschii	grn	62	10	15	40	2	10	0	84.1	185097.5
0	Lyngbya limnetica	bg	800	10	15	40	2	0	0	669.8	1339322.7
0	Ceratium hirundinella	dino	75	10	10	10	40	40	0	249.4	52901.4
0	Peridinium sp.	dino	2	10	10	10	30	40	0	7.5	1378.6
2	Aphanizomenon flos-aquae	bg	237	10	15	40	3	0	0	446.5	595161.5
2	Microcystis aeruginosa	bg	369	10	15	40	3	0	0	177.0	353952.2
2	Anabaena sp.	bg	62	10	15	40	4	0	0	70.5	105727.4
2	Actinastrum Hantzschii	grn	32	10	15	40	2	10	0	43.4	95534.2
2	Lyngbya limnetica	bg	941	10	15	40	2	0	0	787.9	1575378.4
2	Ceratium hirundinella	dino	25	10	10	10	40	40	0	83.1	17633.8
2	Peridinium sp.	dino	1	10	10	10	30	40	0	3.7	689.3
4	Aphanizomenon flos-aquae	bg	124	10	10	40	3	0	0	350.4	467088.8
4	Microcystis aeruginosa	bg	16	10	10	40	3	0	0	11.5	23021.3
4	Actinastrum Hantzschii	grn	20	10	10	40	2	10	0	40.7	89563.3
4	Lyngbya limnetica	bg	550	10	10	40	2	0	0	690.8	1381176.6
4	Dictyosphaerium pulchellum	grn	32	10	10	40	4	0	0	54.6	81853.4
4	Ceratium hirundinella	dino	11	10	10	10	40	40	0	36.6	7758.9
4	Peridinium sp.	dino	2	10	10	10	30	40	0	7.5	1378.6
6	Aphanizomenon flos-aquae	bg	198	10	15	40	3	0	0	373.0	497223.6
6	Microcystis aeruginosa	bg	309	10	15	40	3	0	0	148.2	296399.0
6	Anabaena sp.	bg	68	10	15	40	4	0	0	77.3	115959.1
6	Actinastrum Hantzschii	grn	40	10	15	40	2	10	0	54.3	119417.8
6	Lyngbya limnetica	bg	618	10	15	40	2	0	0	517.4	1034626.8
6	Dictyosphaerium pulchellum	grn	32	10	15	40	4	0	0	36.4	54569.0
6	Chlamydomonas sp. 1	grn	1	10	15	40	10	12	0	21.3	12104.6
6	Ceratium hirundinella	dino	1	10	10	10	40	40	0	3.3	705.4
8	Aphanizomenon flos-aquae	bg	98	10	15	40	3	0	0	184.6	246100.6
8	Microcystis aeruginosa	bg	24	10	15	40	3	0	0	11.5	23021.3
8	Anabaena sp.	bg	8	10	15	40	4	0	0	9.1	13642.2
8	Actinastrum Hantzschii	grn	24	10	15	40	2	10	0	32.6	71650.7
8	Lyngbya limnetica	bg	812	10	15	40	2	0	0	679.9	1359412.6
8	Fragillaria sp.	diat	214	10	15	40	3	20	0	1306.8	1001887.9
8	Ceratium hirundinella	dino	1	10	10	10	40	40	0	3.3	705.4
8	Peridinium sp.	dino	1	10	10	10	30	40	0	3.7	689.3
10	Aphanizomenon flos-aquae	bg	72	10	30	40	3	0	0	67.8	90404.3
10	Microcystis aeruginosa	bg	24	10	30	40	3	0	0	5.8	11510.6
10	Actinastrum Hantzschii	grn	24	10	30	40	2	10	0	16.3	35825.3
10	Lyngbya limnetica	bg	96	10	30	40	2	0	0	40.2	80359.4
10	Closterium sp.	desm	1	10	30	40	4	60	0	4.3	6409.0
	Sep. 2, 1980										
0	Lyngbya limnetica	bg	776	10	10	40	2	0	0	974.6	1948714.6
0	Aphanizomenon flos-aquae	bg	147	10	10	40	3	0	0	415.4	553726.3
0	Microcystis aeruginosa	bg	144	10	10	40	3	0	0	103.6	207191.5
0	Staurastrum sp.	grn	1	10	10	40	10	15	20	152.7	113207.1
0	Anabaena sp.	bg	44	10	10	40	4	0	0	75.0	112548.5
0	Planktosphaeria gelatinosa	grn	29	10	10	40	5	0	0	96.6	115905.8
0	Actinastrum Hantzschii	grn	6	10	10	40	2	10	0	12.2	26869.0
0	Chlamydomonas sp. 1	grn	1	10	10	40	8	10	0	17.1	11974.3
0	Ceratium hirundinella	dino	5	10	10	10	40	40	0	16.6	3526.8
2	Lyngbya limnetica	bg	897	10	10	40	2	0	0	1126.6	2252573.4
2	Aphanizomenon flos-aquae	bg	80	10	10	40	3	0	0	226.1	301347.6
2	Microcystis aeruginosa	bg	48	10	10	40	3	0	0	34.5	69063.8
2	Anabaena sp.	bg	19	10	10	40	4	0	0	32.4	48600.5
2	Planktosphaeria gelatinosa	grn	28	10	10	40	5	0	0	93.3	111909.0
2	Chlamydomonas sp. 1	grn	1	10	10	40	8	10	0	17.1	11974.3
2	Dictyosphaerium pulchellum	grn	12	10	10	40	4	0	0	20.5	30695.0
2	Ceratium hirundinella	dino	15	10	10	10	40	40	0	49.9	10580.3
4	Lyngbya limnetica	bg	657	10	15	40	2	0	0	550.1	1099918.8
4	Aphanizomenon flos-aquae	bg	39	10	15	40	3	0	0	73.5	97938.0
4	Microcystis aeruginosa	bg	42	10	15	40	3	0	0	20.1	40287.2
4	Anabaena sp.	bg	20	10	15	40	4	0	0	22.7	34105.6
4	Actinastrum Hantzschii	grn	6	10	15	40	2	10	0	8.1	17912.7
6	Lyngbya limnetica	bg	686	10	15	40	2	0	0	574.4	1148469.3
6	Aphanizomenon flos-aquae	bg	84	10	15	40	3	0	0	158.2	210943.3
6	Microcystis aeruginosa	bg	102	10	15	40	3	0	0	48.9	97840.4
6	Anabaena sp.	bg	54	10	15	40	4	0	0	61.4	92085.1
6	Actinastrum Hantzschii	grn	4	10	15	40	2	10	0	5.4	11941.8
6	Ceratium hirundinella	dino	6	10	10	10	40	40	0	20.0	4232.1
8	Lyngbya limnetica	bg	908	10	15	40	2	0	0	760.3	1520131.3
8	Aphanizomenon flos-aquae	bg	67	10	15	40	3	0	0	126.2	168252.4
8	Microcystis aeruginosa	bg	71	10	15	40	3	0	0	34.1	68104.6
8	Anabaena sp.	bg	20	10	15	40	4	0	0	22.7	34105.6
8	Fragillaria sp.	diat	34	10	15	40	3	18	0	186.9	145336.8
8	Ceratium hirundinella	dino	9	10	10	10	40	40	0	29.9	6348.2
10	Lyngbya limnetica	bg	793	10	15	40	2	0	0	664.0	1327603.7
10	Aphanizomenon flos-aquae	bg	54	10	15	40	3	0	0	101.7	135606.4
10	Microcystis aeruginosa	bg	82	10	15	40	3	0	0	39.3	78656.0
10	Anabaena sp.	bg	35	10	15	40	4	0	0	39.8	59684.8
10	Ceratium hirundinella	dino	10	10	10	10	40	40	0	33.3	7053.5
12	Lyngbya limnetica	bg	365	10	15	40	2	0	0	305.6	611066.0
12	Microcystis aeruginosa	bg	15	10	15	40	3	0	0	7.2	14388.3
12	Anabaena sp.	bg	12	10	15	40	4	0	0	13.6	20463.4

M	Genus and species	Fam.	Ct.	Vol. filt. ml	Flds. ctd.	Mag.	Dim1 μm	Dim2 μm	Dim3 μm	Biovol. mm³/m³	Surf.area mm²/m³
	Sep. 8, 1980										
0	Lyngbya limnetica	bg	291	15	20	40	2	0	0	121.8	243589.3
0	Aphanizomenon flos-aquae	bg	365	15	20	40	3	0	0	343.8	458299.5
0	Microcystis aeruginosa	bg	64	15	20	40	3	0	0	15.3	30695.0
0	Anabaena sp.	bg	33	15	20	40	4	0	0	18.8	28137.1
0	Actinastrum Hantzschii	grn	4	15	20	40	2	10	0	2.7	5970.9
0	Dictyosphaerium pulchellum	grn	8	15	20	40	4	0	0	4.5	6821.1
0	Chlamydomonas sp. 1	grn	3	15	20	40	18	20	0	172.7	55674.9
0	Chlamydomonas sp. 1	grn	1	15	20	40	10	12	0	10.7	6052.3
0	Peridinium sp.	dino	9	15	10	10	40	45	0	44.9	6489.7
0	Ceratium hirundinella	dino	8	15	10	10	40	40	0	17.7	3761.9
2	Lyngbya limnetica	bg	553	15	15	40	2	0	0	308.7	617204.6
2	Aphanizomenon flos-aquae	bg	241	15	15	40	3	0	0	302.7	403471.0
2	Anabaena sp.	bg	6	15	15	40	4	0	0	4.5	6821.1
2	Chlamydomonas sp. 1	grn	2	15	15	40	18	20	0	153.5	49488.8
2	Chlamydomonas sp. 1	grn	5	15	15	40	10	12	0	71.1	40348.6
2	Chlamydomonas sp. 1	grn	7	15	15	40	5	7	0	14.5	15858.2
2	Planktosphaeria gelatinosa	grn	40	15	15	40	4	0	0	30.3	45474.1
2	Staurastrum sp.	grn	1	15	15	40	10	15	20	67.9	50314.3
2	Peridinium sp.	dino	3	15	10	10	40	45	0	15.0	2163.2
2	Ceratium hirundinella	dino	79	15	10	10	40	40	0	175.1	37148.5
4	Lyngbya limnetica	bg	441	15	20	40	2	0	0	184.6	369150.8
4	Aphanizomenon flos-aquae	bg	647	15	20	40	3	0	0	609.4	812383.0
4	Anabaena sp.	bg	28	15	20	40	4	0	0	15.9	23873.9
4	Chlamydomonas sp. 1	grn	1	15	20	40	18	20	0	57.6	18558.3
4	Chlamydomonas sp. 1	grn	1	15	20	40	10	12	0	10.7	6052.3
4	Peridinium sp.	dino	2	15	10	10	40	45	0	10.0	1442.2
4	Ceratium hirundinella	dino	31	15	10	10	40	40	0	68.7	14577.3
4	Cosmarium sp.	desm	1	15	10	10	18	30	0	1.3	394.8
6	Lyngbya limnetica	bg	118	15	30	40	2	0	0	32.9	65850.0
6	Aphanizomenon flos-aquae	bg	95	15	30	40	3	0	0	59.7	79522.3
6	Actinastrum Hantzschii	grn	8	15	30	40	2	10	0	3.6	7961.2
6	Chlamydomonas sp. 1	grn	1	15	30	40	18	20	0	38.4	12372.2
6	Chlamydomonas sp. 1	grn	1	15	30	40	10	12	0	7.1	4034.9
6	Peridinium sp.	dino	3	15	10	10	40	45	0	15.0	2163.2
6	Ceratium hirundinella	dino	7	15	10	10	40	40	0	15.5	3291.6
8	Lyngbya limnetica	bg	350	15	30	40	2	0	0	97.7	195317.9
8	Aphanizomenon flos-aquae	bg	257	15	30	40	3	0	0	161.4	215128.7
8	Microcystis aeruginosa	bg	93	15	30	40	3	0	0	14.9	29735.8
8	Anabaena sp.	bg	34	15	30	40	4	0	0	12.9	19326.5
8	Actinastrum Hantzschii	grn	14	15	30	40	2	10	0	6.3	13932.1
8	Chlamydomonas sp. 1	grn	2	15	30	40	18	20	0	76.7	24744.4
8	Chlamydomonas sp. 1	grn	2	15	30	40	10	12	0	14.2	8069.7
8	Ceratium hirundinella	dino	6	15	10	10	40	40	0	13.3	2821.4
10	Lyngbya limnetica	bg	162	15	30	40	2	0	0	45.2	90404.3
10	Aphanizomenon flos-aquae	bg	150	15	30	40	3	0	0	94.2	125561.5
10	Anabaena sp.	bg	20	15	30	40	4	0	0	7.6	11368.5
10	Actinastrum Hantzschii	grn	8	15	30	40	2	10	0	3.6	7961.2
10	Ceratium hirundinella	dino	10	15	10	10	40	40	0	22.2	4702.3
12	Lyngbya limnetica	bg	15	15	30	40	2	0	0	4.2	8370.8
12	Aphanizomenon flos-aquae	bg	28	15	30	40	3	0	0	17.6	23438.1
12	Actinastrum Hantzschii	grn	8	15	30	40	2	10	0	3.6	7961.2
12	Oocystis sp.	grn	4	15	30	40	6	10	0	8.5	7499.5
12	Fragillaria sp.	diat	50	15	30	40	3	18	0	91.6	71243.5

M	Genus and species	Fam.	Ct.	Vol. filt. ml	Flds. ctd.	Mag.	Dim1 μm	Dim2 μm	Dim3 μm	Biovol. mm³/m³	Surf.area mm²/m³
	Sep. 14, 1980										
0	Lyngbya limnetica	bg	67	10	15	40	2	0	0	56.1	112168.3
0	Aphanizomenon flos-aquae	bg	482	10	15	40	3	0	0	908.0	1210412.9
0	Microcystis aeruginosa	bg	363	10	15	40	3	0	0	174.1	348196.9
0	Actinastrum Hantzschii	grn	18	10	15	40	2	10	0	24.4	53738.0
0	Chlamydomonas sp. 1	grn	1	10	15	40	18	20	0	115.1	37116.6
0	Peridinium sp.	dino	6	10	10	10	40	45	0	44.9	6489.7
0	Ceratium hirundinella	dino	44	10	10	10	40	40	0	146.3	31035.5
2	Lyngbya limnetica	bg	58	10	15	40	2	0	0	48.6	97100.9
2	Aphanizomenon flos-aquae	bg	303	10	15	40	3	0	0	570.8	760902.7
2	Microcystis aeruginosa	bg	411	10	15	40	3	0	0	197.1	394239.5
2	Actinastrum Hantzschii	grn	8	10	15	40	2	10	0	10.9	23883.6
2	Chlamydomonas sp. 1	grn	3	10	15	40	10	12	0	63.9	36313.7
2	Chlamydomonas sp. 1	grn	3	10	15	40	5	7	0	9.3	10194.5
2	Crucigenia Lauterbornii	grn	57	10	15	40	5	8	0	202.5	215345.0
2	Planktosphaeria gelatinosa	grn	24	10	15	40	4	0	0	27.3	40926.7
2	Dictyosphaerium pulchellum	grn	50	10	15	40	4	0	0	56.8	85264.0
2	Fragillaria sp.	diat	25	10	15	40	3	24	0	183.2	137398.3
2	Peridinium sp.	dino	5	10	10	10	40	45	0	37.4	5408.1
2	Ceratium hirundinella	dino	34	10	10	10	40	40	0	113.1	23982.0
4	Lyngbya limnetica	bg	37	10	15	40	2	0	0	31.0	61943.7
4	Aphanizomenon flos-aquae	bg	364	10	15	40	3	0	0	685.7	914087.8
4	Microcystis aeruginosa	bg	157	10	15	40	3	0	0	75.3	150597.6
4	Chlamydomonas sp. 1	grn	1	10	15	40	18	20	0	115.1	37116.6
4	Chlamydomonas sp. 1	grn	1	10	15	40	10	12	0	21.3	12104.6
4	Planktosphaeria gelatinosa	grn	12	10	15	40	4	0	0	13.6	20463.4
4	Peridinium sp.	dino	8	10	10	10	40	45	0	59.9	8652.9
4	Ceratium hirundinella	dino	56	10	10	10	40	40	0	186.2	39499.7
6	Lyngbya limnetica	bg	170	10	30	40	2	0	0	71.2	142303.0
6	Aphanizomenon flos-aquae	bg	126	10	30	40	3	0	0	118.7	158207.5
6	Microcystis aeruginosa	bg	73	10	30	40	3	0	0	17.5	35011.5
6	Chlamydomonas sp. 1	grn	1	10	30	40	10	12	0	10.7	6052.3
6	Planktosphaeria gelatinosa	grn	7	10	30	40	4	0	0	4.0	5968.5
6	Peridinium sp.	dino	3	10	10	10	40	45	0	22.4	3244.8
6	Ceratium hirundinella	dino	16	10	10	10	40	40	0	53.2	11285.6
8	Lyngbya limnetica	bg	11	10	30	40	2	0	0	4.6	9207.8
8	Aphanizomenon flos-aquae	bg	336	10	30	40	3	0	0	316.5	421886.7
8	Microcystis aeruginosa	bg	133	10	30	40	3	0	0	31.9	63788.1
8	Actinastrum Hantzschii	grn	6	10	30	40	2	10	0	4.1	8956.3
8	Fragillaria sp.	diat	53	10	30	40	3	24	0	194.2	145642.2
8	Oocystis sp.	grn	6	10	30	40	4	6	0	5.1	6887.5
8	Peridinium sp.	dino	5	10	10	10	40	45	0	37.4	5408.1
8	Ceratium hirundinella	dino	6	10	10	10	40	40	0	20.0	4232.1
10	Lyngbya limnetica	bg	124	10	15	40	2	0	0	103.8	207595.0
10	Aphanizomenon flos-aquae	bg	333	10	15	40	3	0	0	627.3	836239.6
10	Microcystis aeruginosa	bg	286	10	15	40	3	0	0	137.2	274336.9
10	Crucigenia Lauterbornii	grn	8	10	15	40	5	8	0	28.4	30223.9
10	Anabaena sp.	bg	47	10	15	40	4	0	0	53.4	80148.2
10	Peridinium sp.	dino	1	10	10	10	40	45	0	7.5	1081.6
10	Ceratium hirundinella	dino	8	10	10	10	40	40	0	26.6	5642.8
12	Aphanizomenon flos-aquae	bg	47	10	30	40	3	0	0	44.3	59013.9
12	Microcystis aeruginosa	bg	24	10	30	40	3	0	0	5.8	11510.6
	Sep. 22, 1980										
0	Lyngbya limnetica	bg	181	10	15	40	2	0	0	151.5	303021.8
0	Anabaena sp.	bg	37	10	15	40	4	0	0	42.1	63095.4

M	Genus and species	Fam.	Ct.	Vol. filt. ml	Flds. ctd.	Mag.	Dim1 μm	Dim2 μm	Dim3 μm	Biovol. mm³/m³	Surf.area mm²/m³
0	Aphanizomenon flos-aquae	bg	198	10	15	40	3	0	0	373.0	497223.6
0	Microcystis aeruginosa	bg	332	10	15	40	3	0	0	159.2	318461.1
0	Chlamydomonas sp. 1	grn	52	10	15	40	4	6	0	88.7	119383.4
0	Chlamydomonas sp. 1	grn	4	10	15	40	10	12	0	85.3	48418.3
0	Actinastrum Hantzschii	grn	8	10	15	40	2	10	0	10.9	23883.6
0	Fragillaria sp.	diat	141	10	15	40	3	24	0	1033.2	774926.3
0	Ceratium hirundinella	dino	47	10	10	10	40	40	0	156.3	33151.5
0	Peridinium sp.	dino	6	10	10	10	40	45	0	44.9	6489.7
2	Lyngbya limnetica	bg	300	10	20	40	2	0	0	188.4	376684.5
2	Anabaena sp.	bg	69	10	20	40	4	0	0	58.8	88248.2
2	Aphanizomenon flos-aquae	bg	126	10	20	40	3	0	0	178.0	237311.3
2	Microcystis aeruginosa	bg	174	10	20	40	3	0	0	62.6	125178.2
2	Chlamydomonas sp. 1	grn	2	10	20	40	10	12	0	32.0	18156.9
2	Actinastrum Hantzschii	grn	8	10	20	40	2	10	0	8.1	17912.7
2	Dictyosphaerium pulchellum	grn	52	10	20	40	4	0	0	44.3	66505.9
2	Planktosphaeria gelatinosa	grn	16	10	20	40	4	0	0	13.6	20463.4
2	Ceratium hirundinella	dino	48	10	10	10	40	40	0	159.6	33856.9
2	Peridinium sp.	dino	13	10	10	10	40	45	0	97.3	14061.0
4	Lyngbya limnetica	bg	246	10	30	40	2	0	0	103.0	205920.9
4	Anabaena sp.	bg	52	10	30	40	4	0	0	29.6	44337.3
4	Aphanizomenon flos-aquae	bg	116	10	30	40	3	0	0	109.3	145651.3
4	Microcystis aeruginosa	bg	255	10	30	40	3	0	0	61.2	122300.6
4	Chlamydomonas sp. 1	grn	60	10	30	40	4	6	0	51.2	68875.0
4	Chlamydomonas sp. 1	grn	1	10	30	40	10	12	0	10.7	6052.3
4	Chlamydomonas sp. 1	grn	3	10	30	40	18	20	0	172.7	55674.9
4	Fragillaria sp.	diat	18	10	30	40	3	24	0	66.0	49463.4
4	Planktosphaeria gelatinosa	grn	6	10	30	40	4	0	0	3.4	5115.8
4	Ankistrodesmus fractus	grn	5	10	30	40	3	60	0	12.0	24010.5
4	Ankistrodesmus fractus	grn	2	10	30	40	6	140	0	44.8	44804.7
4	Ceratium hirundinella	dino	5	10	10	10	40	40	0	16.6	3526.8
4	Peridinium sp.	dino	4	10	10	10	40	45	0	29.9	4326.5
6	Lyngbya limnetica	bg	520	10	30	40	2	0	0	217.7	435279.9
6	Aphanizomenon flos-aquae	bg	137	10	30	40	3	0	0	129.0	172019.3
6	Microcystis aeruginosa	bg	149	10	30	40	3	0	0	35.7	71461.9
6	Planktosphaeria gelatinosa	grn	32	10	30	40	4	0	0	18.2	27284.5
6	Ankistrodesmus fractus	grn	1	10	30	40	3	60	0	2.4	4802.1
6	Ankistrodesmus fractus	grn	2	10	30	40	6	140	0	44.8	44804.7
6	Ceratium hirundinella	dino	37	10	10	10	40	40	0	123.0	26098.0
6	Peridinium sp.	dino	4	10	10	10	40	45	0	29.9	4326.5
8	Lyngbya limnetica	bg	290	10	20	40	2	0	0	182.1	364128.4
8	Anabaena sp.	bg	71	10	20	40	4	0	0	60.5	90806.2
8	Aphanizomenon flos-aquae	bg	327	10	20	40	3	0	0	462.0	615879.2
8	Microcystis aeruginosa	bg	391	10	20	40	3	0	0	140.6	281291.3
8	Chlamydomonas sp. 1	grn	13	10	20	40	4	6	0	16.6	22384.4
8	Chlamydomonas sp. 1	grn	3	10	20	40	10	12	0	48.0	27235.3
8	Dictyosphaerium pulchellum	grn	16	10	20	40	4	0	0	13.6	20463.4
8	Ankistrodesmus fractus	grn	5	10	20	40	3	60	0	18.0	36015.7
8	Ceratium hirundinella	dino	4	10	10	10	40	40	0	13.3	2821.4
8	Peridinium sp.	dino	1	10	10	10	40	45	0	7.5	1081.6
10	Lyngbya limnetica	bg	315	10	30	40	2	0	0	131.9	263679.2
10	Anabaena sp.	bg	38	10	30	40	4	0	0	21.6	32400.3
10	Aphanizomenon flos-aquae	bg	138	10	30	40	3	0	0	130.0	173274.9
10	Microcystis aeruginosa	bg	32	10	30	40	3	0	0	7.7	15347.5
10	Chlamydomonas sp. 1	grn	3	10	30	40	10	12	0	32.0	18156.9
10	Actinastrum Hantzschii	grn	12	10	30	40	2	10	0	8.1	17912.7
10	Fragillaria sp.	diat	64	10	30	40	3	24	0	234.5	175869.8
10	Dictyosphaerium pulchellum	grn	8	10	30	40	4	0	0	4.5	6821.1
10	Ankistrodesmus fractus	grn	1	10	30	40	3	60	0	2.4	4802.1
10	Ceratium hirundinella	dino	10	10	10	10	40	40	0	33.3	7053.5
10	Peridinium sp.	dino	2	10	10	10	40	45	0	15.0	2163.2
12	Lyngbya limnetica	bg	225	10	30	40	2	0	0	94.2	188342.3
12	Aphanizomenon flos-aquae	bg	111	10	30	40	3	0	0	104.6	139373.3
12	Microcystis aeruginosa	bg	169	10	30	40	3	0	0	40.5	81054.1
12	Ankistrodesmus fractus	grn	1	10	30	40	3	60	0	2.4	4802.1
12	Actinastrum Hantzschii	grn	8	10	30	40	2	10	0	5.4	11941.8
12	Dictyosphaerium pulchellum	grn	16	10	30	40	4	0	0	9.1	13642.2
12	Crucigenia Lauterbornii	grn	16	10	30	40	5	8	0	28.4	30223.9
12	Ceratium hirundinella	dino	7	10	10	10	40	40	0	23.3	4937.5
	Sep. 29, 1980										
0	Lyngbya limnetica	bg	559	10	15	40	2	0	0	468.0	935851.8
0	Aphanizomenon flos-aquae	bg	196	10	15	40	3	0	0	369.2	492201.1
0	Anabaena sp.	bg	32	10	15	40	4	0	0	36.4	54569.0
0	Oocystis sp.	grn	8	10	15	40	6	10	0	51.2	44996.9
0	Microcystis aeruginosa	bg	20	10	15	40	3	0	0	9.6	19184.4
0	Actinastrum Hantzschii	grn	8	10	15	40	2	10	0	10.9	23883.6
0	Cryptomonas sp.	cryp	1	10	15	40	20	28	0	198.9	54370.9
0	Chlamydomonas sp. 1	grn	2	10	15	40	18	20	0	230.2	74233.1
0	Chlamydomonas sp. 1	grn	2	10	15	40	12	16	0	81.9	37707.3
0	Ceratium hirundinella	dino	69	10	10	10	40	40	0	229.4	48669.3
0	Peridinium sp.	dino	11	10	10	10	40	45	0	82.3	11897.8
2	Lyngbya limnetica	bg	615	10	20	40	2	0	0	386.2	772203.3
2	Aphanizomenon flos-aquae	bg	243	10	20	40	3	0	0	343.3	457671.7
2	Anabaena sp.	bg	32	10	20	40	4	0	0	27.3	40926.7
2	Microcystis aeruginosa	bg	95	10	20	40	3	0	0	34.2	68344.4
2	Actinastrum Hantzschii	grn	16	10	20	40	2	10	0	16.3	35825.3
2	Chlamydomonas sp. 1	grn	1	10	20	40	18	20	0	86.3	27837.4
2	Chlamydomonas sp. 1	grn	3	10	20	40	12	16	0	92.1	42420.7
2	Fragillaria sp.	diat	32	10	20	40	4	20	0	260.5	156328.7
2	Ceratium hirundinella	dino	27	10	10	10	40	40	0	89.8	19044.5
2	Peridinium sp.	dino	5	10	10	10	40	45	0	37.4	5408.1
4	Lyngbya limnetica	bg	277	10	20	40	2	0	0	173.9	347805.4
4	Aphanizomenon flos-aquae	bg	78	10	20	40	3	0	0	110.2	146907.0
4	Microcystis aeruginosa	bg	396	10	20	40	3	0	0	142.4	284888.4
4	Fragillaria sp.	diat	34	10	20	40	4	20	0	276.8	166099.2
4	Ceratium hirundinella	dino	14	10	10	10	40	40	0	46.6	9874.9
6	Lyngbya limnetica	bg	483	10	20	40	2	0	0	303.3	606462.1
6	Aphanizomenon flos-aquae	bg	166	10	20	40	3	0	0	234.5	312648.2
6	Anabaena sp.	bg	6	10	20	40	4	0	0	5.1	7673.8
6	Microcystis aeruginosa	bg	289	10	20	40	3	0	0	104.0	207911.0
6	Actinastrum Hantzschii	grn	24	10	20	40	2	10	0	24.4	53738.0
6	Chlamydomonas sp. 1	grn	2	10	20	40	12	16	0	61.4	28280.4
6	Chlamydomonas sp. 1	grn	2	10	20	40	10	12	0	32.0	18156.9
6	Fragillaria sp.	diat	55	10	20	40	3	22	0	277.1	209914.0
6	Ceratium hirundinella	dino	17	10	10	10	40	40	0	56.5	11991.0
8	Lyngbya limnetica	bg	412	10	15	40	2	0	0	345.0	689751.2
8	Aphanizomenon flos-aquae	bg	307	10	15	40	3	0	0	578.4	770947.7
8	Anabaena sp.	bg	14	10	15	40	4	0	0	15.9	23873.9
8	Microcystis aeruginosa	bg	328	10	15	40	3	0	0	157.3	314624.2
8	Chlamydomonas sp. 1	grn	1	10	15	40	12	16	0	40.9	18853.6
8	Planktosphaeria gelatinosa	grn	20	10	15	40	4	0	0	22.7	34105.6
8	Ceratium hirundinella	dino	46	10	10	10	40	40	0	153.0	32446.2
8	Peridinium sp.	dino	4	10	10	10	40	45	0	29.9	4326.5
10	Lyngbya limnetica	bg	288	10	30	40	2	0	0	120.6	241078.1

M	Genus and species	Fam.	Ct.	Vol. filt. ml	Flds. ctd.	Mag.	Dim1 μm	Dim2 μm	Dim3 μm	Biovol. mm^3/m^3	Surf.area mm^2/m^3
10	Aphanizomenon flos-aquae	bg	137	10	30	40	3	0	0	129.0	172019.3
10	Anabaena sp.	bg	34	10	30	40	4	0	0	19.3	28989.8
10	Microcystis aeruginosa	bg	105	10	30	40	3	0	0	25.2	50359.1
10	Actinastrum Hantzschii	grn	8	10	30	40	2	10	0	5.4	11941.8
10	Fragillaria sp.	diat	36	10	30	40	3	22	0	120.9	91598.8
10	Ceratium hirundinella	dino	12	10	10	10	40	40	0	39.9	8464.2
10	Peridinium sp.	dino	2	10	10	10	40	45	0	15.0	2163.2
12	Lyngbya limnetica	bg	331	10	30	40	2	0	0	138.6	277072.4
12	Aphanizomenon flos-aquae	bg	147	10	30	40	3	0	0	138.5	184575.4
12	Anabaena sp.	bg	28	10	30	40	4	0	0	15.9	23873.9
12	Microcystis aeruginosa	bg	174	10	30	40	3	0	0	41.7	83452.1
12	Fragillaria sp.	diat	21	10	30	40	3	22	0	70.5	53432.7
12	Oocystis sp.	grn	8	10	30	40	6	10	0	25.6	22498.4
12	Dictyosphaerium pulchellum	grn	24	10	30	40	4	0	0	13.6	20463.4
12	Ceratium hirundinella	dino	10	10	10	10	40	40	0	33.3	7053.5
16	Lyngbya limnetica	bg	421	10	30	40	2	0	0	176.2	352409.3
16	Aphanizomenon flos-aquae	bg	22	10	30	40	3	0	0	20.7	27623.5
16	Anabaena sp.	bg	17	10	30	40	4	0	0	9.7	14494.9
16	Microcystis aeruginosa	bg	22	10	30	40	3	0	0	5.3	10551.4
16	Fragillaria sp.	diat	87	10	30	40	3	22	0	292.2	221363.9
16	Actinastrum Hantzschii	grn	32	10	30	40	2	10	0	21.7	47767.1
16	Planktosphaeria gelatinosa	grn	8	10	30	40	4	0	0	4.5	6821.1
	Oct. 6, 1980										
0	Dictyosphaerium pulchellum	grn	12	10	15	40	4	0	0	13.6	20463.4
0	Aphanizomenon flos-aquae	bg	140	10	15	40	3	0	0	263.7	351572.2
0	Lyngbya limnetica	bg	273	10	15	40	2	0	0	228.6	457043.9
0	Fragillaria sp.	diat	151	10	15	40	3	24	0	1106.5	829885.6
0	Planktosphaeria gelatinosa	grn	8	10	15	40	8	0	0	72.8	54569.0
0	Microcystis aeruginosa	bg	392	10	15	40	3	0	0	188.0	376014.3
0	Oocystis sp.	grn	4	10	15	40	10	16	0	113.7	60447.7
0	Cryptomonas sp.	cryp	1	10	15	40	20	24	0	170.5	48418.3
0	Chlamydomonas sp. 1	grn	6	10	15	40	10	12	0	127.9	72627.5
0	Cryptomonas sp.	cryp	1	10	15	40	20	30	0	213.2	57395.9
0	Anabaena sp.	bg	27	10	15	40	4	0	0	30.7	46042.6
0	Actinastrum Hantzschii	grn	8	10	15	40	2	10	0	10.9	23883.6
0	Ankistrodesmus fractus	grn	1	10	15	40	4	140	0	19.9	29854.6
0	Ceratium hirundinella	dino	47	10	10	10	40	40	0	156.3	33151.5
0	Peridinium sp.	dino	5	10	10	10	40	45	0	37.4	5408.1
2	Aphanizomenon flos-aquae	bg	254	10	15	40	3	0	0	478.5	637852.5
2	Lyngbya limnetica	bg	382	10	15	40	2	0	0	319.8	639526.6
2	Fragillaria sp.	diat	60	10	15	40	3	24	0	439.7	329755.9
2	Microcystis aeruginosa	bg	36	10	15	40	3	0	0	17.3	34531.9
2	Chlamydomonas sp. 1	grn	1	10	15	40	10	12	0	21.3	12104.6
2	Anabaena sp.	bg	12	10	15	40	4	0	0	13.6	20463.4
2	Actinastrum Hantzschii	grn	5	10	15	40	2	10	0	6.8	14927.2
2	Ceratium hirundinella	dino	4	10	10	10	40	40	0	13.3	2821.4
2	Peridinium sp.	dino	1	10	10	10	40	45	0	7.5	1081.6
4	Dictyosphaerium pulchellum	grn	56	10	25	40	4	0	0	38.2	57297.4
4	Aphanizomenon flos-aquae	bg	239	10	25	40	3	0	0	270.1	360110.4
4	Lyngbya limnetica	bg	454	10	25	40	2	0	0	228.1	456039.4
4	Fragillaria sp.	diat	76	10	25	40	3	24	0	334.2	250614.5
4	Microcystis aeruginosa	bg	79	10	25	40	3	0	0	22.7	45467.0
4	Cryptomonas sp.	cryp	1	10	25	40	20	24	0	102.3	29051.0
4	Chlamydomonas sp. 1	grn	2	10	25	40	10	12	0	25.6	14525.5
4	Anabaena sp.	bg	19	10	25	40	4	0	0	13.0	19440.2
4	Ceratium hirundinella	dino	3	10	10	10	40	40	0	10.0	2116.1
6	Dictyosphaerium pulchellum	grn	19	10	30	40	4	0	0	10.8	16200.2
6	Aphanizomenon flos-aquae	bg	94	10	30	40	3	0	0	88.5	118027.8
6	Lyngbya limnetica	bg	547	10	30	40	2	0	0	229.0	457881.0
6	Fragillaria sp.	diat	124	10	30	40	3	24	0	454.3	340747.7
6	Microcystis aeruginosa	bg	140	10	30	40	3	0	0	33.6	67145.4
6	Cryptomonas sp.	cryp	1	10	30	40	20	30	0	106.6	28697.9
6	Anabaena sp.	bg	52	10	30	40	4	0	0	29.6	44337.3
6	Actinastrum Hantzschii	grn	8	10	30	40	2	10	0	5.4	11941.8
6	Planktosphaeria gelatinosa	grn	12	10	30	40	4	0	0	6.8	10231.7
6	Oocystis sp.	grn	12	10	30	40	6	10	0	38.4	33747.6
6	Ceratium hirundinella	dino	3	10	10	10	40	40	0	10.0	2116.1
8	Dictyosphaerium pulchellum	grn	96	10	30	40	4	0	0	54.6	81853.4
8	Aphanizomenon flos-aquae	bg	175	10	30	40	3	0	0	164.8	219732.6
8	Lyngbya limnetica	bg	473	10	30	40	2	0	0	198.0	395937.3
8	Fragillaria sp.	diat	27	10	30	40	3	24	0	98.9	74195.1
8	Microcystis aeruginosa	bg	104	10	30	40	3	0	0	24.9	49879.4
8	Chlamydomonas sp. 1	grn	6	10	30	40	10	12	0	63.9	36313.7
8	Anabaena sp.	bg	22	10	30	40	4	0	0	12.5	18758.1
8	Actinastrum Hantzschii	grn	6	10	30	40	2	10	0	4.1	8956.3
8	Planktosphaeria gelatinosa	grn	27	10	30	40	4	0	0	15.3	23021.3
8	Ceratium hirundinella	dino	9	10	10	10	40	40	0	29.9	6348.2
8	Peridinium sp.	dino	1	10	10	10	40	45	0	7.5	1081.6
10	Aphanizomenon flos-aquae	bg	68	10	30	40	3	0	0	64.1	85381.8
10	Lyngbya limnetica	bg	400	10	30	40	2	0	0	167.5	334830.7
10	Fragillaria sp.	diat	252	10	30	40	3	24	0	923.3	692487.3
10	Microcystis aeruginosa	bg	29	10	30	40	3	0	0	7.0	13908.7
10	Chlamydomonas sp. 1	grn	2	10	30	40	10	12	0	21.3	12104.6
10	Chlamydomonas sp. 1	grn	3	10	30	40	12	16	0	61.4	28280.4
10	Anabaena sp.	bg	42	10	30	40	4	0	0	23.9	35810.9
10	Actinastrum Hantzschii	grn	12	10	30	40	2	10	0	8.1	17912.7
10	Oocystis sp.	grn	2	10	30	40	6	10	0	6.4	5624.6
10	Ceratium hirundinella	dino	13	10	10	10	40	40	0	43.2	9169.6
12	Dictyosphaerium pulchellum	grn	79	10	30	40	4	0	0	44.9	67358.6
12	Aphanizomenon flos-aquae	bg	100	10	30	40	3	0	0	94.2	125561.5
12	Lyngbya limnetica	bg	659	10	30	40	2	0	0	275.9	551633.6
12	Fragillaria sp.	diat	130	10	30	40	3	24	0	476.3	357235.5
12	Planktosphaeria gelatinosa	grn	5	10	30	40	4	0	0	2.8	4263.2
12	Microcystis aeruginosa	bg	14	10	30	40	3	0	0	3.4	6714.5
12	Chlamydomonas sp. 1	grn	2	10	30	40	12	16	0	40.9	18853.6
12	Anabaena sp.	bg	57	10	30	40	4	0	0	32.4	48600.5
12	Actinastrum Hantzschii	grn	20	10	30	40	2	10	0	13.6	29854.4
12	Ceratium hirundinella	dino	9	10	10	10	40	40	0	29.9	6348.2
16	Dictyosphaerium pulchellum	grn	28	10	30	40	4	0	0	15.9	23873.9
16	Aphanizomenon flos-aquae	bg	247	10	30	40	3	0	0	232.7	310136.9
16	Lyngbya limnetica	bg	523	10	30	40	2	0	0	218.9	437791.1
16	Fragillaria sp.	diat	224	10	30	40	3	24	0	820.7	615544.3
16	Microcystis aeruginosa	bg	91	10	30	40	3	0	0	21.8	43644.5
16	Chlamydomonas sp. 1	grn	1	10	30	40	12	16	0	20.5	9426.8
16	Cryptomonas sp.	cryp	2	10	30	40	20	24	0	170.5	48418.3
16	Anabaena sp.	bg	5	10	30	40	4	0	0	2.8	4263.2
16	Stephanodiscus sp.	diat	2	10	30	40	25	10	0	166.5	59951.3
16	Ankistrodesmus fractus	grn	1	10	30	40	4	70	0	5.0	7472.8
16	Ceratium hirundinella	dino	5	10	10	10	40	40	0	16.6	3526.8
	Oct. 13, 1980										
0	Microcystis aeruginosa	bg	110	10	30	40	3	0	0	26.4	52757.1
0	Aphanizomenon flos-aquae	bg	133	10	30	40	3	0	0	125.3	166996.8

M	Genus and species	Fam.	Ct.	Vol. filt. ml	Flds. ctd.	Mag.	Dim1 μm	Dim2 μm	Dim3 μm	Biovol. mm³/m³	Surf.area mm²/m³
0	Fragillaria sp.	diat	125	10	30	40	4	20	0	678.5	407106.0
0	Fragillaria sp.	diat	20	10	30	40	3	24	0	73.3	54959.3
0	Dictyosphaerium pulchellum	grn	88	10	30	40	4	0	0	50.0	75032.3
0	Lyngbya limnetica	bg	108	10	30	40	2	0	0	45.2	90404.3
0	Actinastrum Hantzschii	grn	16	10	30	40	2	10	0	10.9	23883.6
0	Chlamydomonas sp. 1	grn	4	10	30	40	14	16	0	111.4	45810.9
0	Chlamydomonas sp. 1	grn	2	10	30	40	8	10	0	11.4	7982.9
0	Cryptomonas sp.	cryp	3	10	30	40	20	24	0	255.8	72627.5
0	Chlamydomonas sp. 1	grn	7	10	30	40	3	5	0	2.8	4921.5
0	Anabaena sp.	bg	24	10	30	40	4	0	0	13.6	20463.4
0	Staurastrum sp.	grn	1	10	30	40	10	14	14	33.2	27384.8
0	Ceratium hirundinella	dino	8	10	10	10	40	40	0	26.6	5642.8
2	Microcystis aeruginosa	bg	125	10	30	40	3	0	0	30.0	59951.3
2	Aphanizomenon flos-aquae	bg	159	10	30	40	3	0	0	149.8	199642.8
2	Fragillaria sp.	diat	133	10	30	40	4	20	0	721.9	433160.8
2	Fragillaria sp.	diat	114	10	30	40	3	24	0	417.7	313268.1
2	Dictyosphaerium pulchellum	grn	34	10	30	40	4	0	0	19.3	28989.8
2	Lyngbya limnetica	bg	77	10	30	40	2	0	0	32.2	64454.9
2	Actinastrum Hantzschii	grn	8	10	30	40	2	10	0	5.4	11941.8
2	Chlamydomonas sp. 1	grn	12	10	30	40	14	16	0	334.2	137432.7
2	Chlamydomonas sp. 1	grn	4	10	30	40	8	10	0	22.7	15965.7
2	Cryptomonas sp.	cryp	2	10	30	40	20	24	0	170.5	48418.3
2	Chlamydomonas sp. 1	grn	3	10	30	40	3	5	0	1.2	2109.2
2	Anabaena sp.	bg	48	10	30	40	4	0	0	27.3	40926.7
2	Crucigenia Lauterbornii	grn	8	10	30	40	6	10	0	25.6	22498.4
2	Ceratium hirundinella	dino	8	10	10	10	40	40	0	26.6	5642.8
4	Microcystis aeruginosa	bg	271	10	30	40	3	0	0	65.0	129974.3
4	Aphanizomenon flos-aquae	bg	88	10	30	40	3	0	0	82.9	110494.1
4	Fragillaria sp.	diat	82	10	30	40	4	20	0	445.1	267061.5
4	Fragillaria sp.	diat	37	10	30	40	3	24	0	135.6	101674.7
4	Dictyosphaerium pulchellum	grn	12	10	30	40	4	0	0	6.8	10231.7
4	Lyngbya limnetica	bg	90	10	30	40	2	0	0	37.7	75336.9
4	Chlamydomonas sp. 1	grn	3	10	30	40	14	16	0	83.6	34358.2
4	Chlamydomonas sp. 1	grn	2	10	30	40	8	10	0	11.4	7982.9
4	Cryptomonas sp.	cryp	3	10	30	40	20	24	0	255.8	72627.5
4	Chlamydomonas sp. 1	grn	2	10	30	40	3	5	0	0.8	1406.2
4	Staurastrum sp.	grn	1	10	30	40	10	14	14	33.2	27384.8
4	Merismopedia sp.	bg	64	10	30	40	3	0	0	15.3	30695.0
4	Ceratium hirundinella	dino	9	10	10	10	40	40	0	29.9	6348.2
6	Aphanizomenon flos-aquae	bg	83	10	30	40	3	0	0	78.2	104216.1
6	Fragillaria sp.	diat	125	10	30	40	3	24	0	458.0	343495.7
6	Dictyosphaerium pulchellum	grn	24	10	30	40	4	0	0	13.6	20463.4
6	Lyngbya limnetica	bg	37	10	30	40	2	0	0	15.5	30971.8
6	Actinastrum Hantzschii	grn	8	10	30	40	2	10	0	5.4	11941.8
6	Chlamydomonas sp. 1	grn	2	10	30	40	14	16	0	55.7	22905.5
6	Chlamydomonas sp. 1	grn	5	10	30	40	8	10	0	28.4	19957.1
6	Cryptomonas sp.	cryp	2	10	30	40	20	24	0	170.5	48418.3
6	Chlamydomonas sp. 1	grn	3	10	30	40	3	5	0	1.2	2109.2
6	Anabaena sp.	bg	38	10	30	40	4	0	0	21.6	32400.3
6	Ceratium hirundinella	dino	5	10	10	10	40	40	0	16.6	3526.8
6	Cosmarium sp.	desm	1	10	10	10	12	26	0	0.8	329.0
8	Microcystis aeruginosa	bg	296	10	30	40	3	0	0	71.0	141964.6
8	Aphanizomenon flos-aquae	bg	112	10	30	40	3	0	0	105.5	140628.9
8	Fragillaria sp.	diat	76	10	30	40	4	20	0	412.5	247520.4
8	Fragillaria sp.	diat	65	10	30	40	3	24	0	238.2	178617.8
8	Dictyosphaerium pulchellum	grn	91	10	30	40	4	0	0	51.7	77590.2
8	Lyngbya limnetica	bg	55	10	30	40	2	0	0	23.0	46039.2
8	Actinastrum Hantzschii	grn	8	10	30	40	2	10	0	5.4	11941.8
8	Chlamydomonas sp. 1	grn	3	10	30	40	14	16	0	83.6	34358.2
8	Chlamydomonas sp. 1	grn	2	10	30	40	8	10	0	11.4	7982.9
8	Cryptomonas sp.	cryp	1	10	30	40	20	24	0	85.3	24209.2
8	Chlamydomonas sp. 1	grn	2	10	30	40	3	5	0	0.8	1406.2
8	Anabaena sp.	bg	21	10	30	40	4	0	0	11.9	17905.4
8	Ceratium hirundinella	dino	12	10	10	10	40	40	0	39.9	8464.2
8	Cosmarium sp.	desm	1	10	10	10	12	26	0	0.8	329.0
10	Microcystis aeruginosa	bg	123	10	30	40	3	0	0	29.5	58992.0
10	Aphanizomenon flos-aquae	bg	51	10	30	40	3	0	0	48.0	64036.4
10	Fragillaria sp.	diat	380	10	30	40	4	20	0	2062.7	1237602.2
10	Fragillaria sp.	diat	129	10	30	40	3	24	0	472.7	354487.5
10	Dictyosphaerium pulchellum	grn	40	10	30	40	4	0	0	22.7	34105.6
10	Lyngbya limnetica	bg	172	10	30	40	2	0	0	72.0	143977.2
10	Actinastrum Hantzschii	grn	16	10	30	40	2	10	0	10.9	23883.6
10	Chlamydomonas sp. 1	grn	7	10	30	40	14	16	0	195.0	80169.1
10	Chlamydomonas sp. 1	grn	4	10	30	40	8	10	0	22.7	15965.7
10	Cryptomonas sp.	cryp	6	10	30	40	20	24	0	511.6	145255.0
10	Chlamydomonas sp. 1	grn	6	10	30	40	3	5	0	2.4	4218.5
10	Oocystis sp.	grn	1	10	30	40	25	30	0	166.5	37826.8
10	Anabaena sp.	bg	63	10	30	40	4	0	0	35.8	53716.3
10	Sphaerocystis sp.	grn	64	10	30	40	4	0	0	36.4	54569.0
10	Chroococcus sp.	bg	52	10	30	40	4	0	0	29.6	44337.3
10	Stephanodiscus sp.	diat	4	10	10	10	40	15	0	15.0	3491.3
10	Ceratium hirundinella	dino	9	10	10	10	40	40	0	29.9	6348.2
12	Microcystis aeruginosa	bg	101	10	30	40	3	0	0	24.2	48440.6
12	Aphanizomenon flos-aquae	bg	13	10	30	40	3	0	0	12.2	16323.0
12	Fragillaria sp.	diat	189	10	30	40	3	24	0	692.5	519365.5
12	Dictyosphaerium pulchellum	grn	16	10	30	40	4	0	0	9.1	13642.2
12	Lyngbya limnetica	bg	128	10	30	40	2	0	0	53.6	107145.8
12	Chlamydomonas sp. 1	grn	5	10	30	40	14	16	0	139.3	57263.6
12	Chlamydomonas sp. 1	grn	3	10	30	40	8	10	0	17.1	11974.3
12	Cryptomonas sp.	cryp	1	10	30	40	20	24	0	85.3	24209.2
12	Anabaena sp.	bg	19	10	30	40	4	0	0	10.8	16200.2
12	Ceratium hirundinella	dino	3	10	10	10	40	40	0	10.0	2116.1
12	Stephanodiscus sp.	diat	12	10	10	10	40	15	0	44.9	10473.9
16	Microcystis aeruginosa	bg	233	10	30	40	3	0	0	55.9	111749.1
16	Aphanizomenon flos-aquae	bg	101	10	30	40	3	0	0	95.1	126817.1
16	Fragillaria sp.	diat	130	10	30	40	4	20	0	705.7	423390.2
16	Fragillaria sp.	diat	61	10	30	40	3	24	0	223.5	167625.9
16	Dictyosphaerium pulchellum	grn	64	10	30	40	4	0	0	36.4	54569.0
16	Lyngbya limnetica	bg	66	10	30	40	2	0	0	27.6	55247.1
16	Chlamydomonas sp. 1	grn	6	10	30	40	14	16	0	167.1	68716.4
16	Chlamydomonas sp. 1	grn	4	10	30	40	8	10	0	22.7	15965.7
16	Cryptomonas sp.	cryp	1	10	30	40	20	24	0	85.3	24209.2
16	Anabaena sp.	bg	19	10	30	40	4	0	0	10.8	16200.2
16	Ceratium hirundinella	dino	7	10	10	10	40	40	0	23.3	4937.5
16	Stephanodiscus sp.	diat	7	10	10	10	40	15	0	26.2	6109.8
16	Staurastrum sp.	grn	2	10	10	10	10	14	14	0.8	640.8
16	Cosmarium sp.	desm	1	10	10	10	12	26	0	0.8	329.0
16	Peridinium sp.	dino	1	10	10	10	40	45	0	7.5	1081.6
20	Microcystis aeruginosa	bg	37	10	30	40	3	0	0	8.9	17745.6
20	Aphanizomenon flos-aquae	bg	144	10	30	40	3	0	0	135.6	180808.6
20	Fragillaria sp.	diat	115	10	30	40	4	20	0	624.2	374537.5
20	Fragillaria sp.	diat	87	10	30	40	3	24	0	318.8	239073.0
20	Dictyosphaerium pulchellum	grn	52	10	30	40	4	0	0	29.6	44337.3
20	Lyngbya limnetica	bg	24	10	30	40	2	0	0	10.0	20089.8

M	Genus and species	Fam.	Ct.	Vol. filt. ml	Flds. ctd.	Mag.	Dim1 µm	Dim2 µm	Dim3 µm	Biovol. mm³/m³	Surf.area mm²/m³
20	Chlamydomonas sp. 1	grn	2	10	30	40	14	16	0	55.7	22905.5
20	Cryptomonas sp.	cryp	1	10	30	40	20	24	0	85.3	24209.2
20	Anabaena sp.	bg	19	10	30	40	4	0	0	10.8	16200.2
20	Ceratium hirundinella	dino	4	10	10	10	40	40	0	13.3	2821.4
20	Stephanodiscus sp.	diat	9	10	10	10	40	15	0	33.7	7855.5
	Nov. 3, 1980										
0	Dictyosphaerium pulchellum	grn	92	10	30	40	4	0	0	52.3	78442.9
0	Fragillaria sp.	diat	128	10	30	40	4	20	0	694.8	416876.5
0	Staurastrum sp.	grn	1	10	30	40	10	15	20	50.9	37735.7
0	Stephanodiscus sp.	diat	5	10	10	10	40	20	0	24.9	4987.6
0	Ceratium hirundinella	dino	2	10	10	10	40	40	0	6.7	1410.7
2	Dictyosphaerium pulchellum	grn	28	10	30	40	4	20	0	15.9	23873.9
2	Fragillaria sp.	diat	69	10	30	40	4	20	0	374.5	224722.5
2	Microcystis aeruginosa	bg	50	10	30	40	3	0	0	12.0	23980.5
2	Lyngbya limnetica	bg	33	10	30	40	2	0	0	13.8	27623.5
2	Stephanodiscus sp.	diat	1	10	10	10	40	20	0	5.0	997.5
2	Ceratium hirundinella	dino	1	10	10	10	40	40	0	3.3	705.4
4	Dictyosphaerium pulchellum	grn	8	10	30	40	4	0	0	4.5	6821.1
4	Fragillaria sp.	diat	203	10	30	40	4	20	0	1101.9	661140.1
4	Lyngbya limnetica	bg	57	10	30	40	2	0	0	23.9	47713.4
4	Actinastrum Hantzschii	grn	4	10	30	40	2	10	0	2.7	5970.9
4	Stephanodiscus sp.	diat	2	10	10	10	40	20	0	10.0	1995.0
8	Dictyosphaerium pulchellum	grn	20	10	30	40	4	0	0	11.4	17052.8
8	Fragillaria sp.	diat	30	10	30	40	4	20	0	162.8	97705.4
8	Stephanodiscus sp.	diat	10	10	10	10	40	20	0	49.9	9975.2
8	Ceratium hirundinella	dino	2	10	10	10	40	40	0	6.7	1410.7
12	Dictyosphaerium pulchellum	grn	48	10	30	40	4	0	0	27.3	40926.7
12	Fragillaria sp.	diat	112	10	30	40	4	20	0	607.9	364767.0
12	Lyngbya limnetica	bg	55	10	30	40	2	0	0	23.0	46039.2
12	Actinastrum Hantzschii	grn	4	10	30	40	2	10	0	2.7	5970.9
12	Stephanodiscus sp.	diat	10	10	10	10	40	20	0	49.9	9975.2
12	Ceratium hirundinella	dino	2	10	10	10	40	40	0	6.7	1410.7
16	Dictyosphaerium pulchellum	grn	32	10	30	40	4	0	0	18.2	27284.5
16	Fragillaria sp.	diat	111	10	30	40	4	20	0	602.5	361510.1
16	Staurastrum sp.	grn	2	10	30	40	10	15	20	101.8	75471.4
16	Lyngbya limnetica	bg	26	10	30	40	2	0	0	10.9	21764.0
16	Pandorina sp.	grn	16	10	30	40	4	5	0	9.1	13642.2
16	Sphaerocystis sp.	grn	126	10	30	40	2	0	0	9.0	26858.2
16	Stephanodiscus sp.	diat	37	10	10	25	40	20	0	184.5	36908.2
16	Ceratium hirundinella	dino	1	10	10	25	40	40	0	3.3	705.4
20	Sphaerocystis sp.	grn	20	10	30	40	2	0	0	1.4	4263.2
20	Dictyosphaerium pulchellum	grn	36	10	30	40	4	0	0	20.5	30695.0
20	Fragillaria sp.	diat	84	10	30	40	4	20	0	456.0	273575.2
20	Lyngbya limnetica	bg	44	10	30	40	2	0	0	18.4	36831.4
20	Stephanodiscus sp.	diat	3	10	10	10	40	20	0	15.0	2992.6
20	Ceratium hirundinella	dino	1	10	10	10	40	40	0	3.3	705.4
	Nov. 20, 1980										
0	Dictyosphaerium pulchellum	grn	24	20	30	40	5	0	0	13.3	15987.0
0	Aphanizomenon flos-aquae	bg	152	20	30	40	3	0	0	71.6	95426.7
0	Chroococcus sp.	bg	8	20	30	40	8	0	0	18.2	13642.2
0	Chlorella ellipsoidea	grn	7	20	30	40	4	0	0	2.0	2984.2
0	Anabaena sp.	bg	8	20	30	40	4	0	0	2.3	3410.6
0	Microcystis aeruginosa	bg	64	20	30	40	3	0	0	7.7	15347.5
0	Stephanodiscus sp.	diat	13	20	10	10	40	20	0	32.4	6483.9
0	Cosmarium sp.	desm	2	20	10	10	18	24	0	1.6	496.3

M	Genus and species	Fam.	Ct.	Vol. filt. ml	Flds. ctd.	Mag.	Dim1 µm	Dim2 µm	Dim3 µm	Biovol. mm³/m³	Surf.area mm²/m³
2	Aphanizomenon flos-aquae	bg	189	20	30	40	3	0	0	89.0	118655.6
2	Chlorella ellipsoidea	grn	15	20	30	40	4	0	0	4.3	6394.8
2	Stephanodiscus sp.	diat	48	20	10	10	40	20	0	119.7	23940.4
2	Cosmarium sp.	desm	1	20	10	10	18	24	0	0.8	248.1
5	Dictyosphaerium pulchellum	grn	28	20	30	40	5	0	0	15.5	18651.5
5	Aphanizomenon flos-aquae	bg	103	20	30	40	3	0	0	48.5	64664.2
5	Chlorella ellipsoidea	grn	13	20	30	40	4	0	0	3.7	5542.2
5	Microcystis aeruginosa	bg	112	20	30	40	3	0	0	13.4	26858.2
5	Cryptomonas sp.	cryp	4	20	30	40	20	30	0	213.2	57395.9
5	Chlamydomonas sp. 1	grn	1	20	30	40	18	20	0	28.8	9279.1
5	Actinastrum Hantzschii	grn	17	20	30	40	2	10	0	5.8	12688.1
5	Stephanodiscus sp.	diat	30	20	10	10	40	20	0	74.8	14962.8
10	Dictyosphaerium pulchellum	grn	8	20	30	40	5	0	0	4.4	5329.0
10	Aphanizomenon flos-aquae	bg	54	20	30	40	3	0	0	25.4	33901.6
10	Actinastrum Hantzschii	grn	8	20	30	40	2	10	0	2.7	5970.9
10	Sphaerocystis sp.	grn	40	20	30	40	2	0	0	1.4	4263.2
10	Ankistrodesmus fractus	grn	2	20	30	40	3	20	0	0.8	1616.6
10	Oocystis sp.	grn	1	20	30	40	20	25	0	44.4	12473.2
10	Stephanodiscus sp.	diat	40	20	10	10	40	20	0	99.8	19950.4
10	Cosmarium sp.	desm	4	20	10	10	18	24	0	3.2	992.6
15	Dictyosphaerium pulchellum	grn	8	20	30	40	5	0	0	4.4	5329.0
15	Aphanizomenon flos-aquae	bg	31	20	30	40	3	0	0	14.6	19462.0
15	Stephanodiscus sp.	diat	20	20	10	10	40	20	0	49.9	9975.2
20	Dictyosphaerium pulchellum	grn	4	20	30	40	5	0	0	2.2	2664.5
20	Aphanizomenon flos-aquae	bg	33	20	30	40	3	0	0	15.5	20717.6
20	Stephanodiscus sp.	diat	18	20	10	10	40	20	0	44.9	8977.7
20	Cosmarium sp.	desm	1	20	10	10	18	24	0	0.8	248.1
	Dec. 18, 1980										
0	Merismopedia sp.	bg	21	50	30	40	2	3	0	0.3	895.3
0	Actinastrum Hantzschii	grn	7	50	30	40	2	10	0	0.9	2089.8
0	flagellate medium	flag	2	50	30	40	10	12	0	3.0	2350.3
0	teardrop-shaped	flag	34	50	30	40	3	5	0	1.8	2174.2
0	Coelosphaerium sp.	bg	66	50	30	40	2	0	0	0.9	2813.7
0	Synechocystis sp.	bg	5	50	30	40	10	0	0	8.9	5329.0
0	Aphanizomenon flos-aquae	bg	44	50	30	40	3	0	0	8.3	11049.4
0	eucaryote filament	eu	48	50	30	40	4	0	0	16.1	16071.9
0	Oocystis sp.	grn	3	50	30	40	8	10	0	3.4	2394.9
0	Ankistrodesmus fractus	grn	5	50	30	40	3	20	0	0.8	1616.6
0	Staurastrum sp.	grn	1	50	30	40	10	18	20	12.2	8486.9
0	Stephanodiscus sp.	diat	165	50	10	10	30	20	0	92.6	21602.5
0	Stephanodiscus sp.	diat	103	50	10	10	40	20	0	102.7	20548.9
0	Cosmarium sp.	desm	4	50	10	10	18	30	0	1.6	473.8
2	Actinastrum Hantzschii	grn	6	50	30	40	2	10	0	0.8	1791.3
2	flagellate medium	flag	2	50	30	40	10	12	0	3.0	2350.3
2	teardrop-shaped	flag	50	50	30	40	3	5	0	2.6	3197.4
2	Aphanizomenon flos-aquae	bg	117	50	30	40	3	0	0	22.0	29381.4
2	eucaryote filament	eu	24	50	30	40	4	0	0	8.0	8035.9
2	Planktosphaeria gelatinosa	grn	6	50	30	40	5	0	0	1.3	1598.7
2	Dictyosphaerium pulchellum	grn	8	50	30	40	4	0	0	0.9	1364.2
2	Stephanodiscus sp.	diat	30	50	10	10	30	20	0	16.8	3927.7
2	Stephanodiscus sp.	diat	72	50	10	10	40	20	0	71.8	14364.3
2	Cosmarium sp.	desm	1	50	10	10	18	30	0	0.4	118.4
5	Actinastrum Hantzschii	grn	9	50	30	40	2	10	0	1.2	2686.9
5	flagellate medium	flag	9	50	30	40	10	12	0	13.6	10576.3
5	teardrop-shaped	flag	12	50	30	40	3	5	0	0.6	767.4
5	Coelosphaerium sp.	bg	91	50	30	40	2	0	0	1.3	3879.5

M	Genus and species	Fam.	Ct.	Vol. filt. ml	Flds. ctd.	Mag.	Dim1 µm	Dim2 µm	Dim3 µm	Biovol. mm³/m³	Surf.area mm²/m³
5	Synechocystis sp.	bg	8	50	30	40	10	0	0	14.2	8526.4
5	Aphanizomenon flos-aquae	bg	109	50	30	40	3	0	0	20.5	27372.4
5	eucaryote filament	eu	26	50	30	40	4	0	0	8.7	8705.6
5	Ankistrodesmus fractus	grn	6	50	30	40	3	20	0	1.0	1939.9
5	Staurastrum sp.	grn	1	50	30	40	10	18	20	12.2	8486.9
5	Oocystis sp.	grn	8	50	30	40	4	6	0	1.4	1836.7
5	Microcystis aeruginosa	bg	72	50	30	40	3	0	0	3.5	6906.4
5	flagellate large	flag	1	50	30	40	25	30	0	33.3	7565.4
5	Stephanodiscus sp.	diat	158	50	10	10	30	20	0	88.7	20686.0
5	Stephanodiscus sp.	diat	114	50	10	10	40	20	0	113.7	22743.4
5	Cosmarium sp.	desm	15	50	10	10	18	30	0	6.1	1776.7
10	Coelosphaerium sp.	bg	48	50	30	40	2	0	0	0.7	2046.3
10	Aphanizomenon flos-aquae	bg	99	50	30	40	3	0	0	18.7	24861.2
10	eucaryote filament	eu	24	50	30	40	4	0	0	8.0	8035.9
10	Ankistrodesmus fractus	grn	5	50	30	40	3	20	0	0.8	1616.6
10	Planktosphaeria gelatinosa	grn	30	50	30	40	5	0	0	6.7	7993.5
10	Microcystis aeruginosa	bg	96	50	30	40	3	0	0	4.6	9208.5
10	Schroederia setigera	grn	1	50	30	40	5	50	0	1.1	1338.9
10	Stephanodiscus sp.	diat	94	50	10	10	30	20	0	52.7	12306.9
10	Stephanodiscus sp.	diat	48	50	10	10	40	20	0	47.9	9576.2
10	Cosmarium sp.	desm	9	50	10	10	18	30	0	3.6	1066.0
15	Actinastrum Hantzschii	grn	5	50	30	40	2	10	0	0.7	1492.7
15	flagellate medium	flag	1	50	30	40	10	12	0	1.5	1175.1
15	teardrop-shaped	flag	35	50	30	40	3	5	0	1.8	2238.2
15	Aphanizomenon flos-aquae	bg	41	50	30	40	3	0	0	7.7	10296.0
15	eucaryote filament	eu	22	50	30	40	4	0	0	7.4	7366.3
15	Oocystis sp.	grn	1	50	30	40	8	10	0	1.1	798.3
15	Ankistrodesmus fractus	grn	7	50	30	40	3	20	0	1.1	2263.2
15	Staurastrum sp.	grn	1	50	30	40	10	18	20	12.2	8486.9
15	Oocystis sp.	grn	2	50	30	40	4	6	0	0.3	459.2
15	Microcystis aeruginosa	bg	48	50	30	40	3	0	0	2.3	4604.3
15	flagellate large	flag	1	50	30	40	25	30	0	33.3	7565.4
15	Schroederia setigera	grn	2	50	30	40	5	50	0	2.2	2677.8
15	Oocystis sp.	grn	4	50	30	40	10	20	0	14.2	7286.7
15	Stephanodiscus astrea	diat	2	50	30	40	4	6	0	0.5	682.1
15	Stephanodiscus sp.	diat	139	50	10	10	30	20	0	78.0	18198.5
15	Stephanodiscus sp.	diat	72	50	10	10	40	20	0	71.8	14364.3
15	Cosmarium sp.	desm	3	50	10	10	18	30	0	1.2	355.3
20	Actinastrum Hantzschii	grn	4	50	30	40	2	10	0	0.5	1194.2
20	flagellate medium	flag	1	50	30	40	10	12	0	1.5	1175.1
20	teardrop-shaped	flag	62	50	30	40	3	5	0	3.2	3964.8
20	Aphanizomenon flos-aquae	bg	53	50	30	40	3	0	0	10.0	13309.5
20	Planktosphaeria gelatinosa	grn	7	50	30	40	5	0	0	1.6	1865.2
20	Dictyosphaerium pulchellum	grn	12	50	30	40	4	0	0	1.4	2046.3
20	Stephanodiscus sp.	diat	228	50	10	10	30	20	0	127.9	29850.7
20	Stephanodiscus sp.	diat	79	50	10	10	40	20	0	78.8	15760.8
20	Cosmarium sp.	desm	3	50	10	10	18	30	0	1.2	355.3
	Jan. 28, 1981										
0	Stephanodiscus astrea	diat	366	25	30	40	5	7	0	341.3	370578.7
0	Cryptomonas erosa	cryp	4	25	30	40	18	26	0	119.7	36101.2
0	Cryptomonas erosa	cryp	13	25	30	40	15	20	0	207.8	76592.9
0	Gloebotrys sp.	chry	18	25	30	40	8	0	0	32.7	24556.0
0	Lyngbya limnetica	bg	21	25	30	40	2	0	0	3.5	7031.4
0	blue green filaments	bg	58	25	30	40	4	0	0	38.8	38840.4
0	Oocystis sp.	grn	2	25	30	40	14	24	0	33.4	12537.0
0	Chlamydomonas sp. 1	grn	75	25	30	40	5	8	0	53.3	56669.7

M	Genus and species	Fam.	Ct.	Vol. filt. ml	Flds. ctd.	Mag.	Dim1 µm	Dim2 µm	Dim3 µm	Biovol. mm³/m³	Surf.area mm²/m³
0	Stephanodiscus sp.	diat	5	25	10	10	40	20	0	10.0	1995.0
2	Stephanodiscus astrea	diat	319	20	30	40	5	7	0	371.9	403738.4
2	Cryptomonas erosa	cryp	24	20	30	40	7	9	0	47.0	37449.7
2	Gloebotrys sp.	chry	8	20	30	40	8	0	0	18.2	13642.2
2	Eudorina sp.	grn	4	20	30	40	5	6	0	2.2	2664.5
2	Cryptomonas erosa	cryp	1	20	30	40	20	40	0	71.1	18216.7
2	Schroederia setigera	grn	1	20	30	40	4	20	0	0.7	1086.9
2	Cryptomonas erosa	cryp	7	20	30	40	20	24	0	298.4	84732.1
2	Cryptomonas erosa	cryp	6	20	30	40	14	18	0	94.0	37449.7
2	Cryptomonas erosa	cryp	1	20	30	40	20	30	0	53.3	14349.0
2	Cryptomonas erosa	cryp	21	20	30	40	10	14	0	130.6	71361.8
2	Chlamydomonas sp. 1	grn	61	20	30	40	5	8	0	54.2	57614.2
2	Stephanodiscus sp.	diat	8	20	10	10	40	20	0	20.0	3990.1
4	Stephanodiscus astrea	diat	219	20	30	40	5	7	0	255.3	277174.6
4	Cryptomonas erosa	cryp	6	20	30	40	7	9	0	11.8	9362.4
4	Cryptomonas erosa	cryp	1	20	30	40	20	24	0	42.6	12104.6
4	Cryptomonas erosa	cryp	3	20	30	40	14	18	0	47.0	18724.8
4	Cryptomonas erosa	cryp	3	20	30	40	10	14	0	18.7	10194.5
4	Cryptomonas erosa	cryp	1	20	30	40	14	24	0	20.9	7835.6
4	Chlamydomonas sp. 1	grn	98	20	30	40	5	8	0	87.0	92560.6
4	Stephanodiscus sp.	diat	2	20	10	10	40	20	0	5.0	997.5
8	Stephanodiscus astrea	diat	84	20	30	40	5	7	0	97.9	106313.6
8	Cryptomonas erosa	cryp	2	20	30	40	7	9	0	3.9	3120.8
8	Gloebotrys sp.	chry	4	20	30	40	8	0	0	9.1	6821.1
8	Actinastrum gracilimum	grn	3	20	30	40	3	20	0	4.6	6564.6
8	Cryptomonas erosa	cryp	1	20	30	40	20	24	0	42.6	12104.6
8	Cryptomonas erosa	cryp	1	20	30	40	14	18	0	15.7	6241.6
8	Cryptomonas erosa	cryp	1	20	30	40	20	30	0	53.3	14349.0
8	Cryptomonas erosa	cryp	6	20	30	40	10	14	0	37.3	20389.1
8	Chlamydomonas sp. 1	grn	85	20	30	40	5	8	0	75.5	80282.1
8	Dictyosphaerium pulchellum	grn	48	20	30	40	4	0	0	13.6	20463.4
8	Stephanodiscus sp.	diat	1	20	10	10	40	20	0	2.5	498.8
12	Stephanodiscus astrea	diat	181	20	30	40	5	7	0	211.0	229080.4
12	Cryptomonas erosa	cryp	14	20	30	40	7	9	0	27.4	21845.6
12	Gloebotrys sp.	chry	14	20	30	40	8	0	0	31.8	23873.9
12	Cryptomonas erosa	cryp	1	20	30	40	20	30	0	53.3	14349.0
12	Cryptomonas erosa	cryp	12	20	30	40	10	14	0	74.6	40778.2
12	Chlamydomonas sp. 1	grn	50	20	30	40	5	8	0	44.4	47224.8
12	Dictyosphaerium pulchellum	grn	16	20	30	40	4	0	0	4.5	6821.1
12	Stephanodiscus sp.	diat	3	20	10	10	40	20	0	7.5	1496.3
23	Stephanodiscus astrea	diat	6	20	30	40	5	7	0	7.0	7593.8
23	Actinastrum gracilimum	grn	3	20	30	40	3	20	0	4.6	6564.6
23	Chlamydomonas sp. 1	grn	33	20	30	40	5	8	0	29.3	31168.4
23	Dictyosphaerium pulchellum	grn	8	20	30	40	4	0	0	2.3	3410.6
	Feb. 16, 1981										
0	Stephanodiscus astrea	diat	274	20	50	40	5	7	0	191.6	208070.8
0	Cryptomonas erosa	cryp	6	20	50	40	12	16	0	36.8	16968.3
0	Cryptomonas erosa	cryp	3	20	50	40	16	20	0	40.9	14369.1
0	Dictyosphaerium pulchellum	grn	8	20	50	40	5	0	0	2.7	3197.4
0	Chlamydomonas sp. 1	grn	27	20	50	40	5	8	0	14.4	15300.8
2	Stephanodiscus astrea	diat	490	20	50	40	5	7	0	342.7	372097.5
2	Cryptomonas erosa	cryp	12	20	50	40	12	16	0	73.7	33936.5
2	Cryptomonas erosa	cryp	3	20	50	40	16	20	0	40.9	14369.1
2	Cryptomonas erosa	cryp	10	20	50	40	10	12	0	32.0	18156.9
2	Chlamydomonas sp. 1	grn	14	20	50	40	5	8	0	7.5	7933.8
2	Chlamydomonas sp. 1	grn	24	20	50	40	4	6	0	6.1	8265.0

M	Genus and species	Fam.	Ct.	Vol. filt. ml	Flds. ctd.	Mag.	Dim1 µm	Dim2 µm	Dim3 µm	Biovol. mm^3/m^3	Surf.area mm^2/m^3
2	Gloebotrys sp.	chry	2	20	50	40	8	0	0	2.7	2046.3
2	Oocystis sp.	grn	1	20	50	40	10	18	0	4.8	2498.2
5	Stephanodiscus astrea	diat	366	20	50	40	5	7	0	256.0	277934.0
5	Cryptomonas erosa	cryp	2	20	50	40	12	16	0	12.3	5656.1
5	Cryptomonas erosa	cryp	2	20	50	40	16	20	0	27.3	9579.4
5	Cryptomonas erosa	cryp	8	20	50	40	10	12	0	25.6	14525.5
5	Cryptomonas erosa	cryp	1	20	50	40	20	35	0	37.3	9760.4
5	Chlamydomonas sp. 1	grn	14	20	50	40	5	8	0	7.5	7933.8
5	Chlamydomonas sp. 1	grn	2	20	50	40	4	6	0	0.5	688.8
5	Planktosphaeria gelatinosa	grn	12	20	50	40	5	0	0	4.0	4796.1
5	Ankistrodesmus fractus	grn	2	20	50	40	2	15	0	0.2	483.9
10	Stephanodiscus astrea	diat	475	20	50	40	5	7	0	332.2	360706.7
10	Cryptomonas erosa	cryp	5	20	50	40	12	16	0	30.7	14140.2
10	Cryptomonas erosa	cryp	3	20	50	40	16	20	0	40.9	14369.1
10	Cryptomonas erosa	cryp	14	20	50	40	10	12	0	44.8	25419.6
10	Cryptomonas erosa	cryp	2	20	50	40	20	35	0	74.6	19520.7
10	Dictyosphaerium pulchellum	grn	24	20	50	40	5	0	0	8.0	9592.2
10	Chlamydomonas sp. 1	grn	4	20	50	40	5	8	0	2.1	2266.8
10	Chlamydomonas sp. 1	grn	11	20	50	40	4	6	0	2.8	3788.1
10	Glenodinium sp.	dino	4	20	50	40	14	18	0	37.6	14979.9
10	Cryptomonas erosa	cryp	1	20	50	40	20	45	0	48.0	12113.7
15	Stephanodiscus astrea	diat	35	20	50	40	5	7	0	24.5	26578.4
15	Chlamydomonas sp. 1	grn	2	20	50	40	5	8	0	1.1	1133.4
15	Chlamydomonas sp. 1	grn	10	20	50	40	4	6	0	2.6	3443.8
15	Glenodinium sp.	dino	1	20	50	40	14	18	0	9.4	3745.0
	Feb. 26, 1981										
0	Chlamydomonas sp. 1	grn	20	20	50	40	9	11	0	47.5	29811.5
0	Chlamydomonas sp. 1	grn	37	20	50	40	8	10	0	63.1	44304.8
0	Chlamydomonas sp. 1	grn	109	20	50	40	5	6	0	43.6	49477.5
0	Chlamydomonas sp. 1	grn	42	20	50	40	3	5	0	5.0	8858.8
0	Stephanodiscus astrea	diat	43	20	50	40	5	7	0	30.1	32653.5
0	Cryptomonas erosa	cryp	2	20	50	40	30	40	0	191.8	35350.5
0	Cryptomonas erosa	cryp	1	20	50	40	10	20	0	5.3	2732.5
0	Cryptomonas erosa	cryp	9	20	50	40	12	18	0	62.2	27894.4
2	Chlamydomonas sp. 1	grn	2	20	30	40	9	11	0	7.9	4968.6
2	Chlamydomonas sp. 1	grn	14	20	30	40	8	10	0	39.8	27940.0
2	Chlamydomonas sp. 1	grn	50	20	30	40	5	6	0	33.3	37826.8
2	Chlamydomonas sp. 1	grn	47	20	30	40	3	5	0	9.4	16522.3
2	Stephanodiscus astrea	diat	66	20	30	40	5	7	0	76.9	83532.1
2	Cryptomonas erosa	cryp	1	20	30	40	30	40	0	159.9	29458.8
2	Cryptomonas erosa	cryp	9	20	30	40	10	20	0	79.9	40987.5
2	Cryptomonas erosa	cryp	18	20	30	40	12	18	0	207.2	92981.3
2	Cryptomonas erosa	cryp	6	20	30	40	10	16	0	42.6	22667.9
2	Cryptomonas erosa	cryp	4	20	30	40	20	30	0	213.2	57395.9
2	Cryptomonas erosa	cryp	15	20	30	40	8	10	0	42.6	29935.7
2	Glenodinium sp.	dino	3	20	30	40	14	18	0	47.0	18724.8
2	Dictyosphaerium pulchellum	grn	4	20	30	40	4	0	0	1.1	1705.3
5	Chlamydomonas sp. 1	grn	18	20	30	40	8	10	0	51.2	35922.8
5	Chlamydomonas sp. 1	grn	31	20	30	40	5	6	0	20.6	23452.6
5	Chlamydomonas sp. 1	grn	69	20	30	40	3	5	0	13.8	24256.1
5	Stephanodiscus astrea	diat	142	20	30	40	5	7	0	165.5	179720.5
5	Cryptomonas erosa	cryp	3	20	30	40	10	20	0	26.6	13662.5
5	Cryptomonas erosa	cryp	18	20	30	40	12	18	0	207.2	92981.3
5	Cryptomonas erosa	cryp	14	20	30	40	10	16	0	99.5	52891.8
5	Cryptomonas erosa	cryp	4	20	30	40	20	30	0	213.2	57395.9
5	Cryptomonas erosa	cryp	23	20	30	40	8	10	0	65.4	45901.4

M	Genus and species	Fam.	Ct.	Vol. filt. ml	Flds. ctd.	Mag.	Dim1 µm	Dim2 µm	Dim3 µm	Biovol. mm^3/m^3	Surf.area mm^2/m^3
5	Glenodinium sp.	dino	5	20	30	40	14	18	0	78.3	31208.0
5	Dictyosphaerium pulchellum	grn	24	20	30	40	4	0	0	6.8	10231.7
5	Actinastrum gracilimum	grn	8	20	30	40	2	10	0	2.7	5970.9
5	Cryptomonas erosa	cryp	3	20	30	40	10	14	0	18.7	10194.5
10	Chlamydomonas sp. 1	grn	12	20	30	40	8	10	0	34.1	23948.6
10	Chlamydomonas sp. 1	grn	47	20	30	40	5	6	0	31.3	35557.2
10	Chlamydomonas sp. 1	grn	24	20	30	40	3	5	0	4.8	8436.9
10	Stephanodiscus astrea	diat	128	20	30	40	5	7	0	149.2	162001.6
10	Cryptomonas erosa	cryp	13	20	30	40	12	18	0	149.6	67153.1
10	Cryptomonas erosa	cryp	2	20	30	40	10	16	0	14.2	7556.0
10	Cryptomonas erosa	cryp	1	20	30	40	20	30	0	53.3	14349.0
10	Cryptomonas erosa	cryp	22	20	30	40	8	10	0	62.5	43905.7
10	Glenodinium sp.	dino	5	20	30	40	14	18	0	78.3	31208.0
10	Cryptomonas erosa	cryp	14	20	30	40	10	14	0	87.0	47574.6
15	Chlamydomonas sp. 1	grn	4	20	50	40	8	10	0	6.8	4789.7
15	Chlamydomonas sp. 1	grn	12	20	50	40	5	6	0	4.8	5447.1
15	Chlamydomonas sp. 1	grn	12	20	50	40	3	5	0	1.4	2531.1
15	Stephanodiscus astrea	diat	46	20	50	40	5	7	0	32.2	34931.6
15	Cryptomonas erosa	cryp	4	20	50	40	12	18	0	27.6	12397.5
15	Cryptomonas erosa	cryp	1	20	50	40	20	30	0	32.0	8609.4
15	Cryptomonas erosa	cryp	10	20	50	40	8	10	0	17.1	11974.3
15	Glenodinium sp.	dino	7	20	50	40	14	18	0	65.8	26214.8
15	Cryptomonas erosa	cryp	7	20	50	40	10	14	0	26.1	14272.4
	Mar. 10, 1981										
0	Glenodinium sp.	dino	4	20	60	40	14	18	0	31.3	12483.2
0	Chlamydomonas sp. 1	grn	15	20	60	40	5	7	0	5.8	6371.6
0	Cryptomonas erosa	cryp	3	20	60	40	8	10	0	4.3	2993.6
2	Glenodinium sp.	dino	6	10	30	40	14	18	0	188.0	74899.3
2	Chlamydomonas sp. 1	grn	18	10	30	40	5	7	0	28.0	30583.6
2	Chlamydomonas sp. 1	grn	5	10	30	40	8	10	0	28.4	19957.1
2	Chlamydomonas sp. 1	grn	9	10	30	40	3	5	0	3.6	6327.7
2	Lyngbya limnetica	bg	13	10	30	40	2	0	0	5.4	10882.0
2	Cryptomonas erosa	cryp	34	10	30	40	8	10	0	193.3	135708.5
2	Cryptomonas erosa	cryp	28	10	30	40	5	7	0	43.5	47574.6
2	Cryptomonas erosa	cryp	20	10	30	40	16	20	0	909.5	319314.0
2	Cryptomonas erosa	cryp	20	10	30	40	12	14	0	358.1	170787.6
2	Cryptomonas erosa	cryp	1	10	30	40	22	30	0	129.0	32233.4
4	Glenodinium sp.	dino	11	10	30	40	14	18	0	344.7	137315.4
4	Chlamydomonas sp. 1	grn	11	10	30	40	5	7	0	17.1	18690.0
4	Chlamydomonas sp. 1	grn	5	10	30	40	3	5	0	2.0	3515.4
4	Lyngbya limnetica	bg	15	10	30	40	2	0	0	6.3	12556.2
4	Cryptomonas erosa	cryp	15	10	30	40	8	10	0	85.3	59871.4
4	Cryptomonas erosa	cryp	3	10	30	40	5	7	0	4.7	5097.3
4	Cryptomonas erosa	cryp	16	10	30	40	16	20	0	727.6	255451.2
4	Cryptomonas erosa	cryp	32	10	30	40	12	14	0	573.0	273260.1
4	Cryptomonas erosa	cryp	7	10	30	40	22	30	0	902.7	225633.9
4	Stephanodiscus sp.	diat	1	10	10	10	40	20	0	5.0	997.5
15	Glenodinium sp.	dino	6	15	30	40	14	18	0	125.3	49932.9
15	Chlamydomonas sp. 1	grn	12	15	30	40	5	7	0	12.4	13592.7
15	Chlamydomonas sp. 1	grn	5	15	30	40	8	10	0	18.9	13304.8
15	Cryptomonas erosa	cryp	1	15	30	40	25	45	0	166.5	34697.8
15	Cryptomonas erosa	cryp	17	15	30	40	8	10	0	64.4	45236.2
15	Cryptomonas erosa	cryp	2	15	30	40	5	7	0	2.1	2265.5
15	Cryptomonas erosa	cryp	17	15	30	40	16	20	0	515.4	180944.6
15	Cryptomonas erosa	cryp	14	15	30	40	12	14	0	167.1	79700.9
15	Cryptomonas erosa	cryp	2	15	30	40	22	30	0	171.9	42977.9

M	Genus and species	Fam.	Ct.	Vol. filt. ml	Flds. ctd.	Mag.	Dim1 μm	Dim2 μm	Dim3 μm	Biovol. mm³/m³	Surf.area mm²/m³
15	Actinastrum gracilimum	grn	6	15	30	40	3	18	0	11.0	15877.1
	Mar. 20, 1981										
0	Pandorina sp.	grn	59	20	30	40	6	8	0	56.6	56594.0
0	Cryptomonas erosa	cryp	31	20	30	40	16	18	0	634.4	229281.6
0	Cryptomonas erosa	cryp	55	20	30	40	10	12	0	293.1	166438.0
0	Cryptomonas erosa	cryp	13	20	30	40	18	22	0	411.5	129183.2
0	Cryptomonas erosa	cryp	8	20	30	40	8	10	0	22.7	15965.7
0	Dictyosphaerium pulchellum	grn	20	20	30	40	4	0	0	5.7	8526.4
0	Chlorococcum sp.	grn	12	20	30	40	5	0	0	6.7	7993.5
0	Chlamydomonas sp. 1	grn	29	20	30	40	5	7	0	22.5	24636.8
0	Chlamydomonas sp. 1	grn	8	20	30	40	8	10	0	22.7	15965.7
0	Chlamydomonas sp. 1	grn	12	20	30	40	3	5	0	2.4	4218.5
0	Glenodinium sp.	dino	14	20	30	40	14	18	0	219.3	87382.5
0	Stephanodiscus astrea	diat	24	20	30	40	5	7	0	28.0	30375.3
0	Actinastrum gracilimum	grn	8	20	30	40	2	10	0	2.7	5970.9
0	Lyngbya limnetica	bg	33	20	30	40	2	0	0	6.9	13811.8
2	Pandorina sp.	grn	32	15	30	40	6	8	0	40.9	40926.7
2	Cryptomonas erosa	cryp	21	15	30	40	16	18	0	573.0	207093.0
2	Cryptomonas erosa	cryp	45	15	30	40	10	12	0	319.7	181568.7
2	Cryptomonas erosa	cryp	6	15	30	40	18	22	0	253.2	79497.3
2	Cryptomonas erosa	cryp	4	15	30	40	8	10	0	15.2	10643.8
2	Chlamydomonas sp. 1	grn	77	15	30	40	5	7	0	79.8	87220.0
2	Chlamydomonas sp. 1	grn	30	15	30	40	8	10	0	113.7	79828.5
2	Chlamydomonas sp. 1	grn	60	15	30	40	3	5	0	16.0	28123.0
2	Glenodinium sp.	dino	6	15	30	40	14	18	0	125.3	49932.9
2	Stephanodiscus astrea	diat	16	15	30	40	5	7	0	24.9	27000.3
	Mar. 30, 1981										
0	Cryptomonas erosa	cryp	47	10	30	40	10	12	0	500.9	284457.6
0	Cryptomonas erosa	cryp	34	10	30	40	14	16	0	947.0	389392.7
0	Cryptomonas erosa	cryp	14	10	30	40	5	7	0	21.8	23787.3
0	Cryptomonas erosa	cryp	8	10	30	40	20	24	0	682.1	193673.3
0	Cryptomonas erosa	cryp	6	10	30	40	22	30	0	773.8	193400.5
0	Cryptomonas erosa	cryp	12	10	30	40	14	20	0	417.8	162382.8
0	Stephanodiscus astrea	diat	107	10	30	40	5	7	0	249.5	270846.4
0	Glenodinium sp.	dino	7	10	30	40	14	18	0	219.3	87382.5
0	Chlamydomonas sp. 1	grn	17	10	30	40	8	10	0	96.6	67854.2
0	Chlamydomonas sp. 1	grn	35	10	30	40	5	7	0	54.4	59468.2
0	Planktosphaeria gelatinosa	grn	23	10	30	40	5	0	0	25.5	30641.8
0	Gloebotrys sp.	chry	1	10	30	40	8	0	0	4.5	3410.6
0	Lyngbya limnetica	bg	20	10	30	40	2	0	0	8.4	16741.5
0	Oocystis sp.	grn	2	10	30	40	12	20	0	51.2	22498.4
0	Stephanodiscus sp.	diat	5	10	10	10	50	25	0	48.7	7793.1
2	Cryptomonas erosa	cryp	51	10	30	40	10	12	0	543.6	308666.8
2	Cryptomonas erosa	cryp	39	10	30	40	14	16	0	1086.3	446656.4
2	Cryptomonas erosa	cryp	14	10	30	40	5	7	0	21.8	23787.3
2	Cryptomonas erosa	cryp	1	10	30	40	20	24	0	85.3	24209.2
2	Cryptomonas erosa	cryp	4	10	30	40	22	30	0	515.8	128933.7
2	Cryptomonas erosa	cryp	26	10	30	40	14	20	0	905.2	351829.3
2	Stephanodiscus astrea	diat	193	10	30	40	5	7	0	450.0	488536.1
2	Glenodinium sp.	dino	6	10	30	40	14	18	0	188.0	74899.3
2	Chlamydomonas sp. 1	grn	18	10	30	40	8	10	0	102.3	71845.7
2	Chlamydomonas sp. 1	grn	92	10	30	40	5	7	0	143.0	156316.4
2	Planktosphaeria gelatinosa	grn	13	10	30	40	5	0	0	14.4	17319.3
2	Gloebotrys sp.	chry	9	10	30	40	8	0	0	40.9	30695.0
2	Aphanizomenon flos-aquae	bg	42	10	30	40	3	0	0	39.6	52735.8
2	Stephanodiscus sp.	diat	2	10	10	10	50	25	0	19.5	3117.2
4	Cryptomonas erosa	cryp	47	10	30	40	10	12	0	500.9	284457.6
4	Cryptomonas erosa	cryp	33	10	30	40	14	16	0	919.1	377940.0
4	Cryptomonas erosa	cryp	16	10	30	40	5	7	0	24.9	27185.5
4	Cryptomonas erosa	cryp	2	10	30	40	20	24	0	170.5	48418.3
4	Cryptomonas erosa	cryp	6	10	30	40	22	30	0	773.8	193400.5
4	Cryptomonas erosa	cryp	26	10	30	40	14	20	0	905.2	351829.3
4	Stephanodiscus astrea	diat	189	10	30	40	5	7	0	440.6	478411.0
4	Glenodinium sp.	dino	10	10	30	40	14	18	0	313.3	124832.2
4	Chlamydomonas sp. 1	grn	9	10	30	40	8	10	0	51.2	35922.8
4	Chlamydomonas sp. 1	grn	104	10	30	40	5	7	0	161.6	176705.5
4	Planktosphaeria gelatinosa	grn	18	10	30	40	5	0	0	20.0	23980.5
4	Gloebotrys sp.	chry	3	10	30	40	8	0	0	13.6	10231.7
4	Microcystis aeruginosa	bg	4	10	30	40	4	0	0	2.3	3410.6
4	Stephanodiscus sp.	diat	3	10	10	10	50	25	0	29.2	4675.9
4	Cosmarium sp.	desm	1	10	10	10	14	18	0	0.7	292.1
8	Cryptomonas erosa	cryp	55	10	30	40	10	12	0	586.2	332875.9
8	Cryptomonas erosa	cryp	62	10	30	40	14	16	0	1726.9	710069.1
8	Cryptomonas erosa	cryp	14	10	30	40	5	7	0	21.8	23787.3
8	Cryptomonas erosa	cryp	1	10	30	40	20	24	0	85.3	24209.2
8	Cryptomonas erosa	cryp	1	10	30	40	22	30	0	129.0	32233.4
8	Cryptomonas erosa	cryp	20	10	30	40	14	20	0	696.3	270637.9
8	Stephanodiscus astrea	diat	259	10	30	40	5	7	0	603.8	655600.3
8	Glenodinium sp.	dino	9	10	30	40	14	18	0	282.0	112349.0
8	Chlamydomonas sp. 1	grn	5	10	30	40	8	10	0	28.4	19957.1
8	Chlamydomonas sp. 1	grn	100	10	30	40	5	7	0	155.4	169909.2
8	Planktosphaeria gelatinosa	grn	21	10	30	40	5	0	0	23.3	27977.3
8	Gloebotrys sp.	chry	1	10	30	40	8	0	0	4.5	3410.6
8	Aphanizomenon flos-aquae	bg	25	10	30	40	3	0	0	23.5	31390.4
8	Stephanodiscus sp.	diat	5	10	10	10	50	25	0	48.7	7793.1
12	Cryptomonas erosa	cryp	34	10	30	40	10	12	0	362.4	205777.8
12	Cryptomonas erosa	cryp	33	10	30	40	14	16	0	919.1	377940.0
12	Cryptomonas erosa	cryp	12	10	30	40	5	7	0	18.7	20389.1
12	Cryptomonas erosa	cryp	1	10	30	40	20	24	0	85.3	24209.2
12	Cryptomonas erosa	cryp	3	10	30	40	22	30	0	386.9	96700.3
12	Cryptomonas erosa	cryp	9	10	30	40	14	20	0	313.3	121787.1
12	Stephanodiscus astrea	diat	130	10	30	40	5	7	0	303.1	329065.8
12	Glenodinium sp.	dino	8	10	30	40	14	18	0	250.7	99865.7
12	Chlamydomonas sp. 1	grn	18	10	30	40	8	10	0	102.3	71845.7
12	Chlamydomonas sp. 1	grn	98	10	30	40	5	7	0	152.3	166511.0
12	Planktosphaeria gelatinosa	grn	21	10	30	40	5	0	0	23.3	27977.3
12	Gloebotrys sp.	chry	7	10	30	40	8	0	0	31.8	23873.9
12	Microcystis aeruginosa	bg	6	10	30	40	4	0	0	3.4	5115.8
12	Dictyosphaerium pulchellum	grn	12	10	30	40	5	0	0	13.3	15987.0
12	Actinastrum gracilimum	grn	24	10	30	40	2	10	0	16.3	35825.3
12	Stephanodiscus sp.	diat	3	10	10	10	50	25	0	29.2	4675.9
16	Cryptomonas erosa	cryp	48	10	30	40	10	12	0	511.6	290509.9
16	Cryptomonas erosa	cryp	56	10	30	40	14	16	0	1559.8	641352.7
16	Cryptomonas erosa	cryp	9	10	30	40	5	7	0	14.0	15291.8
16	Cryptomonas erosa	cryp	7	10	30	40	20	24	0	596.8	169464.1
16	Cryptomonas erosa	cryp	6	10	30	40	22	30	0	773.8	193400.5
16	Cryptomonas erosa	cryp	13	10	30	40	14	20	0	452.6	175914.7
16	Stephanodiscus astrea	diat	154	10	30	40	5	7	0	359.0	389816.4
16	Glenodinium sp.	dino	11	10	30	40	14	18	0	344.7	137315.4
16	Chlamydomonas sp. 1	grn	32	10	30	40	8	10	0	181.9	127725.6
16	Chlamydomonas sp. 1	grn	150	10	30	40	5	7	0	233.1	254863.7
16	Planktosphaeria gelatinosa	grn	8	10	30	40	5	0	0	8.9	10658.0

M	Genus and species	Fam.	Ct.	Vol. filt. ml	Flds. ctd.	Mag.	Dim1 μm	Dim2 μm	Dim3 μm	Biovol. mm^3/m^3	Surf.area mm^2/m^3
16	Microcystis aeruginosa	bg	8	10	30	40	4	0	0	4.5	6821.1
16	Stephanodiscus sp.	diat	2	10	10	10	50	25	0	19.5	3117.2
20	Cryptomonas erosa	cryp	17	10	30	40	10	12	0	181.2	102888.9
20	Cryptomonas erosa	cryp	43	10	30	40	14	16	0	1197.7	492467.3
20	Cryptomonas erosa	cryp	7	10	30	40	5	7	0	10.9	11893.6
20	Cryptomonas erosa	cryp	1	10	30	40	22	30	0	129.0	32233.4
20	Cryptomonas erosa	cryp	9	10	30	40	14	20	0	313.3	121787.1
20	Stephanodiscus astrea	diat	68	10	30	40	5	7	0	158.5	172126.7
20	Glenodinium sp.	dino	5	10	30	40	14	18	0	156.7	62416.1
20	Chlamydomonas sp. 1	grn	20	10	30	40	8	10	0	113.7	79828.5
20	Chlamydomonas sp. 1	grn	112	10	30	40	5	7	0	174.1	190298.3
20	Planktosphaeria gelatinosa	grn	5	10	30	40	5	0	0	5.6	6661.3
20	Gloebotrys sp.	chry	2	10	30	40	8	0	0	9.1	6821.1
	Apr. 6, 1981										
0	Cryptomonas erosa	cryp	3	10	40	40	20	22	0	175.9	51188.7
0	Cryptomonas erosa	cryp	27	10	40	40	14	16	0	564.0	231917.7
0	Cryptomonas erosa	cryp	26	10	40	40	10	14	0	242.5	132529.1
0	Cryptomonas erosa	cryp	23	10	40	40	7	10	0	75.1	58356.3
0	Cryptomonas erosa	cryp	3	10	40	40	12	24	0	69.1	29511.0
0	Cryptomonas erosa	cryp	12	10	40	40	14	20	0	313.3	121787.1
0	Cryptomonas erosa	cryp	2	10	40	40	20	30	0	159.9	43046.9
0	Cryptomonas erosa	cryp	2	10	40	40	20	40	0	213.2	54650.0
0	Gloebotrys sp.	chry	8	10	40	40	8	0	0	27.3	20463.4
0	Chlamydomonas sp. 1	grn	5	10	40	40	3	5	0	1.5	2636.5
0	Chlamydomonas sp. 1	grn	261	10	40	40	5	8	0	347.7	369770.0
0	Fragillaria sp.	diat	35	10	40	40	4	24	0	171.0	99741.0
0	Planktosphaeria gelatinosa	grn	8	10	40	40	7	0	0	18.3	15667.3
0	Stephanodiscus astrea	diat	39	10	40	40	5	7	0	68.2	74039.8
0	Stephanodiscus astrea	diat	2	10	40	40	12	10	0	28.8	15347.5
0	Lyngbya limnetica	bg	14	10	40	40	2	0	0	4.4	8789.3
0	Oocystis sp.	grn	1	10	40	40	12	14	0	13.4	6404.5
0	Glenodinium sp.	dino	8	10	40	40	14	18	0	188.0	74899.3
0	Stephanodiscus sp.	diat	6	10	10	10	40	20	0	29.9	5985.1
2	Cryptomonas erosa	cryp	20	10	40	40	14	16	0	417.8	171790.9
2	Cryptomonas erosa	cryp	5	10	40	40	10	14	0	46.6	25486.4
2	Cryptomonas erosa	cryp	8	10	40	40	7	10	0	26.1	20297.8
2	Cryptomonas erosa	cryp	5	10	40	40	12	24	0	115.1	49185.0
2	Cryptomonas erosa	cryp	10	10	40	40	14	20	0	261.1	101489.2
2	Cryptomonas erosa	cryp	3	10	40	40	20	30	0	239.8	64570.3
2	Cryptomonas erosa	cryp	1	10	40	40	20	40	0	106.6	27325.0
2	Gloebotrys sp.	chry	8	10	40	40	8	0	0	27.3	20463.4
2	Chlamydomonas sp. 1	grn	133	10	40	40	5	8	0	177.2	188426.9
2	Planktosphaeria gelatinosa	grn	11	10	40	40	7	0	0	25.1	21542.5
2	Stephanodiscus astrea	diat	5	10	40	40	5	7	0	8.7	9492.3
2	Oocystis sp.	grn	1	10	40	40	14	20	0	26.1	10148.9
2	Merismopedia sp.	bg	6	10	40	40	6	0	0	8.6	8633.0
2	Actinastrum gracilimum	grn	4	10	40	40	2	10	0	2.0	4478.2
2	Glenodinium sp.	dino	7	10	40	40	14	18	0	164.5	65536.9
4	Cryptomonas erosa	cryp	11	10	40	40	20	22	0	644.8	187691.7
4	Cryptomonas erosa	cryp	59	10	40	40	14	16	0	1232.5	506783.2
4	Cryptomonas erosa	cryp	1	10	40	40	10	14	0	9.3	5097.3
4	Cryptomonas erosa	cryp	10	10	40	40	7	10	0	32.6	25372.3
4	Cryptomonas erosa	cryp	2	10	40	40	12	24	0	46.0	19674.0
4	Cryptomonas erosa	cryp	6	10	40	40	14	20	0	156.7	60893.5
4	Gloebotrys sp.	chry	2	10	40	40	8	0	0	6.8	5115.8
4	Chlamydomonas sp. 1	grn	166	10	40	40	5	8	0	221.2	235179.4

M	Genus and species	Fam.	Ct.	Vol. filt. ml	Flds. ctd.	Mag.	Dim1 μm	Dim2 μm	Dim3 μm	Biovol. mm^3/m^3	Surf.area mm^2/m^3
4	Stephanodiscus astrea	diat	31	10	40	40	5	7	0	54.2	58852.1
4	Dictyosphaerium pulchellum	grn	16	10	40	40	5	0	0	13.3	15987.0
4	Actinastrum gracilimum	grn	8	10	40	40	2	10	0	4.1	8956.3
4	Glenodinium sp.	dino	19	10	40	40	14	18	0	446.5	177885.9
4	Stephanodiscus sp.	diat	1	10	10	10	40	20	0	5.0	997.5
10	Cryptomonas erosa	cryp	12	10	40	40	20	22	0	703.4	204754.6
10	Cryptomonas erosa	cryp	40	10	40	40	14	16	0	835.6	343581.8
10	Cryptomonas erosa	cryp	3	10	40	40	10	14	0	28.0	15291.8
10	Cryptomonas erosa	cryp	9	10	40	40	7	10	0	29.4	22835.1
10	Cryptomonas erosa	cryp	13	10	40	40	14	20	0	339.5	131936.0
10	Cryptomonas erosa	cryp	2	10	40	40	20	30	0	159.9	43046.9
10	Cryptomonas erosa	cryp	1	10	40	40	20	40	0	106.6	27325.0
10	Gloebotrys sp.	chry	4	10	40	40	8	0	0	13.6	10231.7
10	Chlamydomonas sp. 1	grn	259	10	40	40	5	8	0	345.1	366936.5
10	Planktosphaeria gelatinosa	grn	4	10	40	40	7	0	0	9.1	7833.6
10	Stephanodiscus astrea	diat	5	10	40	40	5	7	0	8.7	9492.3
10	Microcystis aeruginosa	bg	5	10	40	40	4	0	0	2.1	3197.4
10	Dictyosphaerium pulchellum	grn	8	10	40	40	5	0	0	6.7	7993.5
10	Oocystis sp.	grn	1	10	40	40	14	20	0	26.1	10148.9
10	Glenodinium sp.	dino	23	10	40	40	14	18	0	540.5	215335.5
10	Staurastrum sp.	grn	1	10	10	10	20	40	60	9.5	2909.2
16	Cryptomonas erosa	cryp	4	10	40	40	20	22	0	234.5	68251.5
16	Cryptomonas erosa	cryp	21	10	40	40	14	16	0	438.7	180380.5
16	Cryptomonas erosa	cryp	4	10	40	40	10	14	0	37.3	20389.1
16	Cryptomonas erosa	cryp	8	10	40	40	7	10	0	26.1	20297.8
16	Cryptomonas erosa	cryp	13	10	40	40	14	20	0	339.5	131936.0
16	Gloebotrys sp.	chry	2	10	40	40	8	0	0	6.8	5115.8
16	Chlamydomonas sp. 1	grn	144	10	40	40	5	8	0	191.8	204011.1
16	Fragillaria sp.	diat	42	10	40	40	4	24	0	205.2	119689.2
16	Planktosphaeria gelatinosa	grn	3	10	40	40	7	0	0	6.9	5875.2
16	Stephanodiscus astrea	diat	28	10	40	40	5	7	0	49.0	53156.8
16	Lyngbya limnetica	bg	10	10	40	40	2	0	0	3.1	6278.1
16	Glenodinium sp.	dino	5	10	40	40	14	18	0	117.5	46812.1
16	Stephanodiscus sp.	diat	1	10	10	10	40	20	0	5.0	997.5
20	Cryptomonas erosa	cryp	6	10	40	40	20	22	0	351.7	102377.3
20	Cryptomonas erosa	cryp	43	10	40	40	14	16	0	898.3	369350.5
20	Cryptomonas erosa	cryp	5	10	40	40	10	14	0	46.6	25486.4
20	Cryptomonas erosa	cryp	5	10	40	40	7	10	0	16.3	12686.2
20	Cryptomonas erosa	cryp	16	10	40	40	14	20	0	417.8	162382.8
20	Cryptomonas erosa	cryp	3	10	40	40	20	30	0	239.8	64570.3
20	Cryptomonas erosa	cryp	1	10	40	40	20	40	0	106.6	27325.0
20	Gloebotrys sp.	chry	3	10	40	40	8	0	0	10.2	7673.8
20	Chlamydomonas sp. 1	grn	147	10	40	40	5	8	0	195.8	208261.3
20	Fragillaria sp.	diat	16	10	40	40	4	24	0	78.2	45595.9
20	Planktosphaeria gelatinosa	grn	6	10	40	40	7	0	0	13.7	11750.4
20	Stephanodiscus astrea	diat	9	10	40	40	5	7	0	15.7	17086.1
20	Dictyosphaerium pulchellum	grn	24	10	40	40	5	0	0	20.0	23980.5
20	Merismopedia sp.	bg	4	10	40	40	6	0	0	5.8	5755.3
20	Glenodinium sp.	dino	17	10	40	40	14	18	0	399.5	159161.0
20	Stephanodiscus sp.	diat	1	10	10	10	40	20	0	5.0	997.5
	Apr. 13, 1981										
0	Cryptomonas erosa	cryp	3	15	30	40	16	22	0	100.0	34316.4
0	Cryptomonas erosa	cryp	13	15	30	40	14	20	0	301.7	117276.4
0	Cryptomonas erosa	cryp	4	15	30	40	12	14	0	47.7	22771.7
0	Cryptomonas erosa	cryp	24	15	30	40	10	12	0	170.5	96836.6
0	Cryptomonas erosa	cryp	10	15	30	40	18	20	0	383.7	123721.9

M	Genus and species	Fam.	Ct.	Vol. filt. ml	Flds. ctd.	Mag.	Dim1 μm	Dim2 μm	Dim3 μm	Biovol. mm³/m³	Surf.area mm²/m³
0	Cryptomonas erosa	cryp	7	15	30	40	20	30	0	497.4	133923.7
0	Cryptomonas erosa	cryp	14	15	30	40	14	16	0	260.0	106892.1
0	Cryptomonas erosa	cryp	5	15	30	40	22	24	0	343.9	91230.2
0	Cryptomonas erosa	cryp	1	15	30	40	20	40	0	94.7	24288.9
0	Chlamydomonas sp. 1	grn	208	15	30	40	5	7	0	215.5	235607.4
0	Glenodinium sp.	dino	11	15	30	40	14	18	0	229.8	91543.6
0	Planktosphaeria gelatinosa	grn	11	15	30	40	8	0	0	33.3	25010.8
0	Stephanodiscus astrea	diat	15	15	30	40	12	10	0	191.8	102316.8
0	Stephanodiscus astrea	diat	12	15	30	40	9	7	0	60.4	44124.1
0	Lyngbya limnetica	bg	36	15	30	40	2	0	0	10.0	20089.8
0	Actinastrum gracilimum	grn	10	15	30	40	2	10	0	4.5	9951.5
0	Gloebotrys sp.	chry	3	15	30	40	8	0	0	9.1	6821.1
0	Dictyosphaerium pulchellum	grn	8	15	30	40	5	0	0	5.9	7105.3
2	Cryptomonas erosa	cryp	1	15	30	40	16	22	0	33.3	11438.8
2	Cryptomonas erosa	cryp	8	15	30	40	14	20	0	185.7	72170.1
2	Cryptomonas erosa	cryp	8	15	30	40	12	14	0	95.5	45543.3
2	Cryptomonas erosa	cryp	22	15	30	40	10	12	0	156.3	88766.9
2	Cryptomonas erosa	cryp	19	15	30	40	18	20	0	729.0	235071.6
2	Cryptomonas erosa	cryp	5	15	30	40	20	30	0	355.3	95659.8
2	Cryptomonas erosa	cryp	21	15	30	40	14	16	0	389.9	160338.2
2	Cryptomonas erosa	cryp	5	15	30	40	22	24	0	343.9	91230.2
2	Cryptomonas erosa	cryp	1	15	30	40	20	40	0	94.7	24288.9
2	Cryptomonas erosa	cryp	2	15	30	40	12	24	0	40.9	17488.0
2	Chlamydomonas sp. 1	grn	278	15	30	40	5	7	0	288.1	314898.3
2	Glenodinium sp.	dino	9	15	30	40	14	18	0	188.0	74899.3
2	Planktosphaeria gelatinosa	grn	2	15	30	40	8	0	0	6.1	4547.4
2	Stephanodiscus astrea	diat	9	15	30	40	12	10	0	115.1	61390.1
2	Stephanodiscus astrea	diat	9	15	30	40	9	7	0	45.3	33093.1
2	Lyngbya limnetica	bg	7	15	30	40	2	0	0	2.0	3906.4
2	Actinastrum gracilimum	grn	12	15	30	40	2	10	0	5.4	11941.8
2	Gloebotrys sp.	chry	1	15	30	40	8	0	0	3.0	2273.7
2	Dictyosphaerium pulchellum	grn	32	15	30	40	5	0	0	23.7	28421.3
4	Cryptomonas erosa	cryp	4	15	30	40	16	22	0	133.4	45755.2
4	Cryptomonas erosa	cryp	6	15	30	40	14	20	0	139.3	54127.6
4	Cryptomonas erosa	cryp	12	15	30	40	12	14	0	143.2	68315.0
4	Cryptomonas erosa	cryp	16	15	30	40	10	12	0	113.7	64557.8
4	Cryptomonas erosa	cryp	11	15	30	40	18	20	0	422.1	136094.1
4	Cryptomonas erosa	cryp	2	15	30	40	20	30	0	142.1	38263.9
4	Cryptomonas erosa	cryp	16	15	30	40	14	16	0	297.1	122162.4
4	Cryptomonas erosa	cryp	2	15	30	40	22	24	0	137.6	36492.1
4	Cryptomonas erosa	cryp	1	15	30	40	20	40	0	94.7	24288.9
4	Chlamydomonas sp. 1	grn	284	15	30	40	5	7	0	294.3	321694.7
4	Glenodinium sp.	dino	11	15	30	40	14	18	0	229.8	91543.6
4	Planktosphaeria gelatinosa	grn	4	15	30	40	8	0	0	12.1	9094.8
4	Stephanodiscus astrea	diat	11	15	30	40	12	10	0	140.7	75032.3
4	Stephanodiscus astrea	diat	7	15	30	40	9	7	0	35.3	25739.1
4	Oocystis sp.	grn	8	15	30	40	10	16	0	75.8	40298.5
4	Oocystis sp.	grn	2	15	30	40	12	20	0	34.1	14999.0
4	Fragillaria sp.	diat	28	15	30	40	4	20	0	101.3	60794.5
4	Actinastrum gracilimum	grn	32	15	30	40	2	10	0	14.5	31844.7
4	Gloebotrys sp.	chry	1	15	30	40	8	0	0	3.0	2273.7
4	Stephanodiscus sp.	diat	1	15	10	10	40	20	0	3.3	665.0
8	Cryptomonas erosa	cryp	1	15	30	40	16	22	0	33.3	11438.8
8	Cryptomonas erosa	cryp	6	15	30	40	14	20	0	139.3	54127.6
8	Cryptomonas erosa	cryp	9	15	30	40	12	14	0	107.4	51236.3
8	Cryptomonas erosa	cryp	22	15	30	40	10	12	0	156.3	88766.9
8	Cryptomonas erosa	cryp	7	15	30	40	18	20	0	268.6	86605.3
8	Cryptomonas erosa	cryp	2	15	30	40	20	30	0	142.1	38263.9
8	Cryptomonas erosa	cryp	25	15	30	40	14	16	0	464.2	190878.8
8	Cryptomonas erosa	cryp	1	15	30	40	22	24	0	68.8	18246.0
8	Cryptomonas erosa	cryp	1	15	30	40	20	40	0	94.7	24288.9
8	Chlamydomonas sp. 1	grn	258	15	30	40	5	7	0	267.3	292243.8
8	Glenodinium sp.	dino	11	15	30	40	14	18	0	229.8	91543.6
8	Planktosphaeria gelatinosa	grn	3	15	30	40	8	0	0	9.1	6821.1
8	Stephanodiscus astrea	diat	4	15	30	40	12	10	0	51.2	27284.5
8	Stephanodiscus astrea	diat	14	15	30	40	9	7	0	70.5	51478.1
8	Oocystis sp.	grn	4	15	30	40	12	20	0	68.2	29997.9
8	Fragillaria sp.	diat	16	15	30	40	4	20	0	57.9	34739.7
8	Actinastrum gracilimum	grn	20	15	30	40	2	10	0	9.0	19903.0
8	Dictyosphaerium pulchellum	grn	12	15	30	40	5	0	0	8.9	10658.0
8	Stephanodiscus sp.	diat	3	15	10	10	40	20	0	10.0	1995.0
12	Cryptomonas erosa	cryp	2	15	40	40	14	20	0	34.8	13531.9
12	Cryptomonas erosa	cryp	6	15	40	40	12	14	0	53.7	25618.1
12	Cryptomonas erosa	cryp	9	15	40	40	10	12	0	48.0	27235.3
12	Cryptomonas erosa	cryp	1	15	40	40	18	20	0	28.8	9279.1
12	Cryptomonas erosa	cryp	3	15	40	40	14	16	0	41.8	17179.1
12	Chlamydomonas sp. 1	grn	112	15	40	40	5	7	0	87.0	95149.1
12	Glenodinium sp.	dino	6	15	40	40	14	18	0	94.0	37449.7
12	Stephanodiscus astrea	diat	3	15	40	40	9	7	0	11.3	8273.3
12	Microcystis aeruginosa	bg	144	15	40	40	3	0	0	17.3	34531.9
12	Ankistrodesmus fractus	grn	1	15	40	40	8	130	0	18.5	13881.6
12	Actinastrum gracilimum	grn	4	15	40	40	2	10	0	1.4	2985.4
12	Dictyosphaerium pulchellum	grn	8	15	40	40	5	0	0	4.4	5329.0
16	Cryptomonas erosa	cryp	3	15	40	40	14	20	0	52.2	20297.8
16	Cryptomonas erosa	cryp	6	15	40	40	10	12	0	32.0	18156.9
16	Cryptomonas erosa	cryp	4	15	40	40	18	20	0	115.1	37116.6
16	Cryptomonas erosa	cryp	9	15	40	40	14	16	0	125.3	51537.3
16	Chlamydomonas sp. 1	grn	208	15	40	40	5	7	0	161.6	176705.5
16	Glenodinium sp.	dino	7	15	40	40	14	18	0	109.7	43691.3
16	Stephanodiscus astrea	diat	7	15	40	40	12	10	0	67.1	35810.9
16	Stephanodiscus astrea	diat	9	15	40	40	9	7	0	34.0	24819.8
16	Lyngbya limnetica	bg	7	15	40	40	2	0	0	1.5	2929.8
16	Pediastrum sp.	grn	1	15	40	40	5	30			
16	Oocystis sp.	grn	1	15	40	40	10	16	0	7.1	3778.0
16	Oocystis sp.	grn	1	15	40	40	12	20	0	12.8	5624.6
16	Actinastrum gracilimum	grn	14	15	40	40	2	10	0	4.7	10449.1
16	Dictyosphaerium puichellum	grn	12	15	40	40	5	0	0	6.7	7993.5
16	Stephanodiscus sp.	diat	2	15	10	10	40	20	0	6.7	1330.0
20	Dictyosphaerium pulchellum	grn	24	15	40	40	4	0	0	6.8	10231.7
20	Aphanocapsa sp.	bg	10	15	40	40	10	0	0	44.4	26645.0
20	Actinastrum gracilimum	grn	8	15	40	40	2	10	0	2.7	5970.9
20	Cryptomonas erosa	cryp	1	15	40	40	14	20	0	17.4	6765.9
20	Cryptomonas erosa	cryp	2	15	40	40	12	14	0	17.9	8539.4
20	Cryptomonas erosa	cryp	1	15	40	40	10	12	0	5.3	3026.1
20	Chlamydomonas sp. 1	grn	76	15	40	40	5	7	0	59.1	64565.5
20	Glenodinium sp.	dino	1	15	40	40	14	18	0	15.7	6241.6
20	Stephanodiscus astrea	diat	1	15	40	40	9	7	0	3.8	2757.8
20	Stephanodiscus sp.	diat	1	15	10	10	40	20	0	3.3	665.0
	Apr. 27, 1981										
0	Cryptomonas erosa	cryp	2	10	40	40	24	40	0	307.0	67495.3
0	Cryptomonas erosa	cryp	1	10	40	40	24	50	0	191.8	40763.4
0	Cryptomonas erosa	cryp	7	10	40	40	8	10	0	29.8	20955.0
0	Cryptomonas erosa	cryp	14	10	40	40	12	14	0	188.0	89663.5

M	Genus and species	Fam.	Ct.	Vol. filt. ml	Flds. ctd.	Mag.	Dim1 µm	Dim2 µm	Dim3 µm	Biovol. mm³/m³	Surf.area mm²/m³
0	Cryptomonas erosa	cryp	35	10	40	40	14	16	0	731.1	300634.1
0	Cryptomonas erosa	cryp	25	10	40	40	18	20	0	1079.1	347967.8
0	Cryptomonas erosa	cryp	1	10	40	40	20	30	0	79.9	21523.4
0	Dictyosphaerium pulchellum	grn	69	10	40	40	4	0	0	29.4	44124.1
0	Chlamydomonas sp. 1	grn	395	10	40	40	5	7	0	460.5	503355.9
0	Actinastrum gracilimum	grn	24	10	40	40	2	10	0	12.2	26869.0
0	Stephanodiscus astrea	diat	16	10	40	40	10	8	0	127.9	83132.4
0	Oocystis sp.	grn	1	10	40	40	12	18	0	17.3	7748.4
0	Stephanodiscus sp.	diat	13	10	10	10	40	30	0	97.3	16209.7
0	Stephanodiscus sp.	diat	13	10	10	10	30	20	0	36.5	8510.1
2	Cryptomonas erosa	cryp	1	10	40	40	8	10	0	4.3	2993.6
2	Cryptomonas erosa	cryp	35	10	40	40	12	14	0	470.0	224158.7
2	Cryptomonas erosa	cryp	60	10	40	40	14	16	0	1253.4	515372.7
2	Cryptomonas erosa	cryp	22	10	40	40	18	20	0	949.6	306211.7
2	Dictyosphaerium pulchellum	grn	40	10	40	40	4	0	0	17.1	25579.2
2	Chlamydomonas sp. 1	grn	313	10	40	40	5	7	0	364.9	398861.8
2	Actinastrum gracilimum	grn	14	10	40	40	2	10	0	7.1	15673.6
2	Stephanodiscus astrea	diat	2	10	40	40	10	8	0	16.0	10391.6
2	Oocystis sp.	grn	1	10	40	40	10	18	0	12.0	6245.6
2	Ankistrodesmus fractus	grn	2	10	40	40	2	45	0	1.2	3600.6
2	Lyngbya limnetica	bg	18	10	40	40	2	0	0	5.7	11300.5
2	Stephanodiscus sp.	diat	3	10	10	10	40	30	0	22.4	3740.7
2	Stephanodiscus sp.	diat	4	10	10	10	30	20	0	11.2	2618.5
4	Cryptomonas erosa	cryp	2	10	40	40	20	40	0	213.2	54650.0
4	Cryptomonas erosa	cryp	39	10	40	40	12	14	0	523.7	249776.8
4	Cryptomonas erosa	cryp	47	10	40	40	14	16	0	981.8	403708.6
4	Cryptomonas erosa	cryp	24	10	40	40	18	20	0	1036.0	334049.1
4	Dictyosphaerium pulchellum	grn	52	10	40	40	4	0	0	22.2	33253.0
4	Chlamydomonas sp. 1	grn	235	10	40	40	5	7	0	273.9	299464.9
4	Actinastrum gracilimum	grn	3	10	40	40	2	10	0	1.5	3358.6
4	Stephanodiscus astrea	diat	2	10	40	40	10	8	0	16.0	10391.6
4	Ankistrodesmus fractus	grn	1	10	40	40	2	45	0	0.6	1800.3
4	Planktosphaeria gelatinosa	grn	29	10	40	40	5	0	0	24.1	28976.4
4	Microcystis aeruginosa	bg	96	10	40	40	3	0	0	17.3	34531.9
4	Stephanodiscus sp.	diat	2	10	10	10	30	20	0	5.6	1309.2
8	Cryptomonas erosa	cryp	3	10	40	40	24	40	0	460.4	101242.9
8	Cryptomonas erosa	cryp	2	10	40	40	8	10	0	8.5	5987.1
8	Cryptomonas erosa	cryp	32	10	40	40	12	14	0	429.7	204945.1
8	Cryptomonas erosa	cryp	40	10	40	40	14	16	0	835.6	343581.8
8	Cryptomonas erosa	cryp	8	10	40	40	18	20	0	345.3	111349.7
8	Cryptomonas erosa	cryp	1	10	40	40	20	30	0	79.9	21523.4
8	Dictyosphaerium pulchellum	grn	80	10	40	40	4	0	0	34.1	51158.4
8	Chlamydomonas sp. 1	grn	187	10	40	40	5	7	0	218.0	238297.6
8	Actinastrum gracilimum	grn	36	10	40	40	2	10	0	18.3	40303.5
8	Stephanodiscus astrea	diat	1	10	40	40	10	8	0	8.0	5195.8
8	Oocystis sp.	grn	1	10	40	40	10	18	0	12.0	6245.6
8	Lyngbya limnetica	bg	22	10	40	40	2	0	0	6.9	13811.8
8	Oocystis sp.	grn	7	10	40	40	8	10	0	29.8	20955.0
8	Stephanodiscus sp.	diat	4	10	10	10	40	30	0	29.9	4987.6
8	Stephanodiscus sp.	diat	3	10	10	10	30	20	0	8.4	1963.9
12	Cryptomonas erosa	cryp	36	10	40	40	12	14	0	483.4	230563.2
12	Cryptomonas erosa	cryp	18	10	40	40	14	16	0	376.0	154611.8
12	Cryptomonas erosa	cryp	3	10	40	40	18	20	0	129.5	41756.1
12	Dictyosphaerium pulchellum	grn	16	10	40	40	4	0	0	6.8	10231.7
12	Chlamydomonas sp. 1	grn	102	10	40	40	5	7	0	118.9	129980.5
12	Actinastrum gracilimum	grn	19	10	40	40	2	10	0	9.7	21271.3
12	Fragillaria sp.	diat	25	10	40	40	4	20	0	101.8	61065.9
12	Ankistrodesmus fractus	grn	1	10	40	40	2	45	0	0.6	1800.3
12	Oocystis sp.	grn	8	10	40	40	8	10	0	34.1	23948.6
12	Stephanodiscus sp.	diat	3	10	10	10	40	30	0	22.4	3740.7
12	Stephanodiscus sp.	diat	4	10	10	10	30	20	0	11.2	2618.5
16	Cryptomonas erosa	cryp	7	10	40	40	12	14	0	94.0	44831.7
16	Cryptomonas erosa	cryp	6	10	40	40	14	16	0	125.3	51537.3
16	Dictyosphaerium pulchellum	grn	24	10	40	40	4	0	0	10.2	15347.5
16	Chlamydomonas sp. 1	grn	147	10	40	40	5	7	0	171.4	187324.9
16	Actinastrum gracilimum	grn	4	10	40	40	2	10	0	2.0	4478.2
	May 4, 1981										
0	Chlamydomonas sp. 1	grn	316	15	30	40	4	6	0	179.6	241827.8
0	Dictyosphaerium pulchellum	grn	64	15	30	40	5	0	0	47.4	56842.7
0	Cryptomonas erosa	cryp	22	15	30	40	14	16	0	408.5	167973.3
0	Cryptomonas erosa	cryp	22	15	30	40	10	12	0	156.3	88766.9
0	Cryptomonas erosa	cryp	4	15	30	40	20	22	0	208.4	60668.0
0	Actinastrum gracilimum	grn	12	15	30	40	2	10	0	5.4	11941.8
0	Stephanodiscus sp.	diat	4	15	10	10	30	20	0	7.5	1745.7
0	Stephanodiscus sp.	diat	2	15	10	10	40	20	0	6.7	1330.0
2	Chlamydomonas sp. 1	grn	301	15	30	40	4	6	0	171.1	230348.7
2	Dictyosphaerium pulchellum	grn	8	15	30	40	5	0	0	5.9	7105.3
2	Cryptomonas erosa	cryp	17	15	30	40	14	16	0	315.7	129797.6
2	Cryptomonas erosa	cryp	4	15	30	40	10	12	0	28.4	16139.4
2	Cryptomonas erosa	cryp	3	15	30	40	20	22	0	156.3	45501.0
2	Actinastrum gracilimum	grn	28	15	30	40	2	10	0	12.7	27864.1
2	Planktosphaeria gelatinosa	grn	10	15	30	40	5	0	0	7.4	8881.7
2	Fragillaria sp.	diat	20	15	30	40	4	20	0	72.4	43424.6
2	Ankistrodesmus fractus	grn	1	15	30	40	8	120	0	22.7	17090.7
2	Stephanodiscus sp.	diat	3	15	10	10	30	20	0	5.6	1309.2
4	Chlamydomonas sp. 1	grn	234	15	30	40	4	6	0	133.0	179075.1
4	Dictyosphaerium pulchellum	grn	20	15	30	40	5	0	0	14.8	17763.3
4	Cryptomonas erosa	cryp	28	15	30	40	14	16	0	519.9	213784.2
4	Cryptomonas erosa	cryp	11	15	30	40	10	12	0	78.2	44383.5
4	Cryptomonas erosa	cryp	5	15	30	40	20	22	0	260.5	75835.0
4	Actinastrum gracilimum	grn	45	15	30	40	2	10	0	20.4	44781.7
4	Stephanodiscus sp.	diat	4	15	10	10	30	20	0	7.5	1745.7
8	Chlamydomonas sp. 1	grn	163	15	30	40	4	6	0	92.7	124740.3
8	Dictyosphaerium pulchellum	grn	32	15	30	40	5	0	0	23.7	28421.3
8	Cryptomonas erosa	cryp	25	15	30	40	14	16	0	464.2	190878.8
8	Cryptomonas erosa	cryp	6	15	30	40	10	12	0	42.6	24209.2
8	Cryptomonas erosa	cryp	4	15	30	40	20	22	0	208.4	60668.0
8	Actinastrum gracilimum	grn	46	15	30	40	2	10	0	20.8	45776.8
8	Fragillaria sp.	diat	10	15	30	40	4	20	0	36.2	21712.3
	May 26, 1981										
0			68	50	50	40	3	0	0	7.7	10245.8
0	Oocystis sp.	grn	18	50	50	40	6	10	0	6.9	6074.6
0	Sphaerocystis sp.	grn	16	50	50	40	5	0	0	2.1	2557.9
0	Cryptomonas erosa	cryp	2	50	50	40	14	20	0	8.4	3247.7
0	Cryptomonas erosa	cryp	1	50	50	40	12	14	0	2.1	1024.7
2			30	30	50	40	3	0	0	5.7	7533.7
2	Oocystis sp.	grn	2	30	50	40	6	10	0	1.3	1124.9
2	Oocystis sp.	grn	14	30	50	40	8	10	0	15.9	11176.0
2	Cryptomonas erosa	cryp	2	30	50	40	14	20	0	13.9	5412.8
4			28	50	50	40	3	0	0	3.2	4218.9
4	Oocystis sp.	grn	12	50	50	40	6	10	0	4.6	4049.7
4	Oocystis sp.	grn	8	50	50	40	8	10	0	5.5	3831.8

M	Genus and species	Fam.	Ct.	Vol. filt. ml	Flds. ctd.	Mag.	Dim1 µm	Dim2 µm	Dim3 µm	Biovol. mm³/m³	Surf.area mm²/m³
4	Cryptomonas erosa	cryp	11	50	50	40	14	20	0	46.0	17862.1
4	Cryptomonas erosa	cryp	11	50	50	40	12	14	0	23.6	11272.0
4	Staurastrum sp.	grn	2	50	50	40	8	20	26	16.9	11742.3
4	Dictyosphaerium pulchellum	grn	32	50	50	40	4	0	0	2.2	3274.1
8	Oocystis sp.	grn	13	50	50	40	8	10	0	8.9	6226.6
8	Cryptomonas erosa	cryp	4	50	50	40	14	20	0	16.7	6495.3
8	Cryptomonas erosa	cryp	3	50	50	40	12	14	0	6.4	3074.2
12			70	50	60	40	3	0	0	6.6	8789.3
12	Oocystis sp.	grn	11	50	60	40	8	10	0	6.3	4390.6
12	Cryptomonas erosa	cryp	1	50	60	40	14	20	0	3.5	1353.2
12	Cryptomonas erosa	cryp	1	50	60	40	12	14	0	1.8	853.9
12	Gloebotrys sp.	chry	12	50	60	40	8	0	0	5.5	4092.7
12	Dictyosphaerium pulchellum	grn	20	50	60	40	4	0	0	1.1	1705.3
12	Pediastrum sp.	grn	40	50	60	40	4	6	12	19.5	33833.8
12	Merismopedia sp.	bg	56	50	60	40	3	0	0	1.3	2685.8
12	Chlamydomonas sp. 1	grn	4	50	60	40	4	6	0	0.3	459.2
12	Stephanodiscus sp.	diat	5	50	10	10	40	20	0	5.0	997.5
16	Oocystis sp.	grn	8	50	50	40	8	10	0	5.5	3831.8
16	Cryptomonas erosa	cryp	1	50	50	40	14	20	0	4.2	1623.8
16	Cryptomonas erosa	cryp	1	50	50	40	12	14	0	2.1	1024.7
16	Stephanodiscus sp.	diat	5	50	10	10	40	20	0	5.0	997.5
	June 1, 1981										
0	Oocystis sp.	grn	6	50	50	40	8	12	0	4.9	3306.0
0	Oocystis sp.	grn	1	50	50	40	16	20	0	5.5	1915.9
0	Oocystis sp.	grn	4	50	50	40	10	14	0	6.0	3262.3
0	Cryptomonas erosa	cryp	1	50	50	40	20	30	0	12.8	3443.8
0	Cryptomonas erosa	cryp	1	50	50	40	18	20	0	6.9	2227.0
0	Cryptomonas erosa	cryp	7	50	50	40	20	22	0	65.7	19110.4
0	Cryptomonas erosa	cryp	6	50	50	40	10	12	0	7.7	4357.6
0	Pediastrum sp.	grn	6	50	50	40	4	8	10	3.9	6208.3
0	Pediastrum sp.	grn	72	50	50	40	3	5	6	13.2	31348.0
0	Sphaerocystis sp.	grn	36	50	50	40	5	0	0	4.8	5755.3
0	Aphanizomenon flos-aquae	bg	123	50	50	40	3	0	0	13.9	18532.9
2	Oocystis sp.	grn	8	50	50	40	16	24	0	52.4	17632.0
2	Cryptomonas erosa	cryp	1	50	50	40	20	30	0	12.8	3443.8
2	Cryptomonas erosa	cryp	11	50	50	40	20	22	0	103.2	30030.7
2	Cryptomonas erosa	cryp	1	50	50	40	10	12	0	1.3	726.3
2	Cryptomonas erosa	cryp	9	50	50	40	14	16	0	30.1	12368.9
2	Sphaerocystis sp.	grn	86	50	50	40	5	0	0	11.5	13748.8
2	Aphanizomenon flos-aquae	bg	265	50	50	40	3	0	0	30.0	39928.6
2	Gloebotrys sp.	chry	24	50	50	40	8	0	0	13.1	9822.4
2	Cryptomonas erosa	cryp	1	50	50	40	22	44	0	22.7	5290.1
2	Ceratium hirundinella	dino	2	50	10	10	50	50	0	2.6	440.8
2	Staurastrum sp.	grn	18	50	10	10	10	30	40	8.6	4278.6
4	Oocystis sp.	grn	12	50	50	40	10	14	0	17.9	9786.8
4	Cryptomonas erosa	cryp	1	50	50	40	20	30	0	12.8	3443.8
4	Cryptomonas erosa	cryp	2	50	50	40	20	22	0	18.8	5460.1
4	Cryptomonas erosa	cryp	1	50	50	40	14	16	0	3.3	1374.3
4	Pediastrum sp.	grn	22	50	50	40	4	8	10	14.3	22763.6
4	Sphaerocystis sp.	grn	101	50	50	40	5	0	0	13.5	16146.9
4	Aphanizomenon flos-aquae	bg	94	50	50	40	3	0	0	10.6	14163.3
4	Gloebotrys sp.	chry	8	50	50	40	8	0	0	4.4	3274.1
4	Sphaerocystis sp.	grn	144	50	50	40	3	0	0	4.1	8287.7
4	Staurastrum sp.	grn	4	50	10	10	10	30	40	1.9	950.8
8	Oocystis sp.	grn	31	50	50	40	8	12	0	25.4	17081.0
8	Oocystis sp.	grn	6	50	50	40	10	14	0	9.0	4893.4

M	Genus and species	Fam.	Ct.	Vol. filt. ml	Flds. ctd.	Mag.	Dim1 µm	Dim2 µm	Dim3 µm	Biovol. mm³/m³	Surf.area mm²/m³
8	Cryptomonas erosa	cryp	3	50	50	40	20	30	0	38.4	10331.3
8	Cryptomonas erosa	cryp	7	50	50	40	18	20	0	48.3	15589.0
8	Cryptomonas erosa	cryp	4	50	50	40	14	16	0	13.4	5497.3
8	Sphaerocystis sp.	grn	82	50	50	40	5	0	0	10.9	13109.3
8	Gloebotrys sp.	chry	8	50	50	40	8	0	0	4.4	3274.1
8	Cosmarium sp.	desm	4	50	10	10	18	30	0	1.6	473.8
12	Cryptomonas erosa	cryp	2	50	50	40	18	20	0	13.8	4454.0
12	Cryptomonas erosa	cryp	1	50	50	40	10	12	0	1.3	726.3
12	Cryptomonas erosa	cryp	1	50	50	40	14	16	0	3.3	1374.3
12	Pediastrum sp.	grn	24	50	50	40	4	8	10	15.6	24833.1
12	Oocystis sp.	grn	16	50	50	40	5	10	0	4.3	4372.0
	June 8, 1981										
0	Pediastrum sp.	grn	76	25	30	40	4	6	8	99.0	183546.4
0	Pediastrum sp.	grn	80	25	30	40	2	4	6	26.1	79573.7
0	Cryptomonas erosa	cryp	1	25	30	40	24	34	0	69.6	15803.4
0	Cryptomonas erosa	cryp	3	25	30	40	20	22	0	93.8	27300.6
0	Cryptomonas erosa	cryp	5	25	30	40	10	12	0	21.3	12104.6
0	Sphaerocystis sp.	grn	56	25	30	40	4	0	0	12.7	19099.1
0	Aphanizomenon flos-aquae	bg	205	25	30	40	4	0	0	137.3	137280.6
0	Oocystis sp.	grn	1	25	30	40	18	20	0	23.0	7423.3
0	Dictyosphaerium pulchellum	grn	8	25	30	40	5	0	0	3.6	4263.2
0	Gloebotrys sp.	chry	2	25	30	40	15	0	0	24.0	9592.2
0	Staurastrum sp.	grn	2	25	10	10	16	34	20	1.7	855.1
0	Cosmarium sp.	desm	2	25	10	10	18	30	0	1.6	473.8
0	Ceratium hirundinella	dino	24	25	10	10	40	50	0	39.9	7664.7
2	Pediastrum sp.	grn	40	25	30	40	2	4	6	13.0	39786.9
2	Cryptomonas erosa	cryp	1	25	30	40	24	34	0	69.6	15803.4
2	Cryptomonas erosa	cryp	2	25	30	40	20	22	0	62.5	18200.4
2	Cryptomonas erosa	cryp	2	25	30	40	10	12	0	8.5	4841.8
2	Sphaerocystis sp.	grn	40	25	30	40	4	0	0	9.1	13642.2
2	Aphanizomenon flos-aquae	bg	141	25	30	40	4	0	0	94.4	94422.3
2	Gloebotrys sp.	chry	2	25	30	40	15	0	0	24.0	9592.2
2	Microcystis aeruginosa	bg	3200	25	30	40	3	0	0	307.0	613900.9
4	Cryptomonas erosa	cryp	3	25	30	40	20	22	0	93.8	27300.6
4	Cryptomonas erosa	cryp	3	25	30	40	10	12	0	12.8	7262.7
4	Sphaerocystis sp.	grn	11	25	30	40	4	0	0	2.5	3751.6
4	Aphanizomenon flos-aquae	bg	107	25	30	40	4	0	0	71.7	71653.8
4	Oocystis sp.	grn	1	25	30	40	10	20	0	7.1	3643.3
4	Staurastrum sp.	grn	1	25	10	10	16	34	50	2.2	799.0
4	Ceratium hirundinella	dino	2	25	10	10	40	50	0	3.3	638.7
8	Cryptomonas erosa	cryp	1	25	30	40	24	34	0	69.6	15803.4
8	Cryptomonas erosa	cryp	1	25	30	40	20	22	0	31.3	9100.2
8	Cryptomonas erosa	cryp	5	25	30	40	10	12	0	21.3	12104.6
8	Sphaerocystis sp.	grn	56	25	30	40	4	0	0	12.7	19099.1
8	Aphanizomenon flos-aquae	bg	17	25	30	40	4	0	0	11.4	11384.2
8	Oocystis sp.	grn	1	25	30	40	18	20	0	23.0	7423.3
8	Schroederia setigera	grn	1	25	30	40	3	30	0	0.5	964.0
8	Gloebotrys sp.	chry	16	25	30	40	8	0	0	29.1	21827.6
8	Oocystis sp.	grn	4	25	30	40	8	10	0	9.1	6386.3
8	Oocystis sp.	grn	10	25	30	40	5	10	0	8.9	9108.3
8	Staurastrum sp.	grn	1	25	10	10	20	30	50	2.4	847.7
12	Sphaerocystis sp.	grn	47	40	35	40	4	0	0	5.7	8587.3
12	Dictyosphaerium pulchellum	grn	32	40	35	40	5	0	0	7.6	9135.4
12	Oocystis sp.	grn	12	40	35	40	7	12	0	13.4	10074.4
16	Dictyosphaerium pulchellum	grn	28	40	50	40	5	0	0	4.7	5595.5
16	Closterium sp.	desm	1	40	50	40	10	170	0	11.3	6806.2

M	Genus and species	Fam.	Ct.	Vol. filt. ml	Flds. ctd.	Mag.	Dim1 µm	Dim2 µm	Dim3 µm	Biovol. mm³/m³	Surf.area mm²/m³
	June 15, 1981										
0	Anabaena sp.	bg	608	5	20	40	3	0	0	437.4	874808.7
0	Aphanizomenon flos-aquae	bg	708	5	20	40	3	0	0	2000.7	2666926.4
0	Microcystis aeruginosa	bg	1395	5	20	40	2	0	0	297.4	892074.7
0	Ceratium hirundinella	dino	8	5	10	10	50	50	0	103.9	17633.8
0	Cosmarium sp.	desm	1	5	10	10	22	30	0	6.0	1508.4
0	Staurastrum sp.	grn	2	5	10	10	16	34	50	21.6	7989.6
2	Anabaena sp.	bg	112	10	40	40	3	0	0	20.1	40287.2
2	Aphanizomenon flos-aquae	bg	620	10	40	40	3	0	0	438.0	583861.0
2	Microcystis aeruginosa	bg	224	10	40	40	2	0	0	11.9	35810.9
2	Aphanocapsa sp.	bg	24	10	40	40	4	0	0	10.2	15347.5
2	Sphaerocystis sp.	grn	72	10	40	40	5	0	0	60.0	71941.5
2	Pediastrum sp.	grn	12	10	40	40	2	6	4	7.3	22786.0
2	Ankistrodesmus fractus	grn	1	10	40	40	10	130	0	43.3	26055.6
2	Pandorina sp.	grn	16	10	40	40	5	6	0	13.3	15987.0
2	Schroederia setigera	grn	1	10	40	40	5	20	0	1.7	2059.9
2	Cryptomonas erosa	cryp	3	10	40	40	20	24	0	191.8	54470.6
2	Cryptomonas erosa	cryp	1	10	40	40	14	16	0	20.9	8589.5
2	Ceratium hirundinella	dino	197	10	10	10	50	50	0	1279.4	217116.0
2	Cosmarium sp.	desm	1	10	10	10	22	30	0	3.0	754.2
2	Staurastrum sp.	grn	1	10	10	10	16	34	50	5.4	1997.4
4	Anabaena sp.	bg	21	15	50	40	3	0	0	2.0	4028.7
4	Aphanizomenon flos-aquae	bg	155	15	50	40	3	0	0	58.4	77848.1
4	Sphaerocystis sp.	grn	69	15	50	40	5	0	0	30.6	36770.1
4	Pediastrum sp.	grn	4	15	50	40	2	6	4	1.3	4050.8
4	Pandorina sp.	grn	8	15	50	40	5	6	0	3.6	4263.2
4	Cryptomonas erosa	cryp	1	15	50	40	20	24	0	34.1	9683.7
4	Cryptomonas erosa	cryp	1	15	50	40	14	16	0	11.1	4581.1
4	Gloebotrys sp.	chry	8	15	50	40	8	0	0	14.6	10913.8
4	Oocystis sp.	grn	4	15	50	40	6	8	0	4.1	3770.7
4	Ceratium hirundinella	dino	24	15	10	10	50	50	0	103.9	17633.8
4	Cosmarium sp.	desm	1	15	10	10	22	30	0	2.0	502.8
8	Sphaerocystis sp.	grn	129	25	50	40	5	0	0	34.4	41246.5
8	Pediastrum sp.	grn	16	25	50	40	2	6	4	3.1	9722.0
8	Pandorina sp.	grn	8	25	50	40	5	6	0	2.1	2557.9
8	Cryptomonas erosa	cryp	1	25	50	40	20	24	0	20.5	5810.2
8	Gloebotrys sp.	chry	4	25	50	40	8	0	0	4.4	3274.1
8	Ceratium hirundinella	dino	2	25	10	10	50	50	0	5.2	881.7
8	Cosmarium sp.	desm	2	25	10	10	22	30	0	2.4	603.4
8	Staurastrum sp.	grn	9	25	10	10	16	34	50	19.4	7190.7
	June 22, 1981										
0	Schroederia setigera	grn	2	15	40	40	4	50	0	3.6	5346.0
0	Oocystis sp.	grn	2	15	40	40	18	20	0	57.6	18558.3
0	Dictyosphaerium pulchellum	grn	88	15	40	40	4	0	0	25.0	37516.2
0	Sphaerocystis sp.	grn	200	15	40	40	5	0	0	111.0	133225.0
0	Aphanizomenon flos-aquae	bg	83	15	40	40	3	0	0	39.1	52108.0
0	Gloebotrys sp.	chry	56	15	40	40	8	0	0	127.3	95495.7
0	Microcystis aeruginosa	bg	50	15	40	40	3	0	0	6.0	11990.3
0	Cryptomonas erosa	cryp	1	15	40	40	20	34	0	60.4	15880.8
0	Cryptomonas erosa	cryp	3	15	40	40	10	12	0	16.0	9078.4
0	Planktosphaeria gelatinosa	grn	45	15	40	40	8	0	0	102.3	76737.6
0	Ceratium hirundinella	dino	32	15	10	10	40	50	0	88.7	17032.6
0	Cosmarium sp.	desm	14	15	10	10	22	30	0	28.2	7039.3
0	Staurastrum sp.	grn	6	15	10	10	10	30	50	11.9	5678.9
0	Stephanodiscus sp.	diat	2	15	10	10	50	30	0	15.6	2286.0

M	Genus and species	Fam.	Ct.	Vol. filt. ml	Flds. ctd.	Mag.	Dim1 µm	Dim2 µm	Dim3 µm	Biovol. mm³/m³	Surf.area mm²/m³
2	Schroederia setigera	grn	8	20	50	40	4	50	0	8.5	12830.5
2	Oocystis sp.	grn	1	20	50	40	18	20	0	17.3	5567.5
2	Dictyosphaerium pulchellum	grn	28	20	50	40	4	0	0	4.8	7162.2
2	Sphaerocystis sp.	grn	196	20	50	40	5	0	0	65.3	78336.3
2	Aphanizomenon flos-aquae	bg	177	20	50	40	3	0	0	50.0	66673.2
2	Gloebotrys sp.	chry	117	20	50	40	8	0	0	159.6	119710.7
2	Planktosphaeria gelatinosa	grn	16	20	50	40	8	0	0	21.8	16370.7
2	Crucigenia Lauterbornii	grn	16	20	50	40	4	6	0	4.1	5510.0
2	Cryptomonas erosa	cryp	1	20	50	40	18	20	0	17.3	5567.5
2	Coelosphaerium sp.	bg	120	20	50	40	2	0	0	2.6	7673.8
2	Westella sp.	grn	26	20	50	40	5	0	0	8.7	10391.6
2	Oocystis sp.	grn	1	20	50	40	8	12	0	2.0	1377.5
2	Oocystis sp.	grn	4	20	50	40	4	6	0	1.0	1377.5
2	Oscillatoria sp.	bg	16	20	50	40	4	0	0	8.0	8035.9
2	Ceratium hirundinella	dino	54	20	10	10	40	50	0	112.2	21556.9
2	Cosmarium sp.	desm	7	20	10	10	22	30	0	10.6	2639.7
2	Staurastrum sp.	grn	2	20	10	10	10	30	50	3.0	1419.7
4	Schroederia setigera	grn	2	20	40	40	4	50	0	2.7	4009.5
4	Oocystis sp.	grn	1	20	40	40	18	20	0	21.6	6959.4
4	Dictyosphaerium pulchellum	grn	104	20	40	40	4	0	0	22.2	33253.0
4	Gloebotrys sp.	chry	46	20	40	40	5	0	0	19.2	22981.3
4	Sphaerocystis sp.	grn	156	20	40	40	5	0	0	64.9	77936.6
4	Aphanizomenon flos-aquae	bg	121	20	40	40	3	0	0	42.7	56973.5
4	Gloebotrys sp.	chry	76	20	40	40	8	0	0	129.6	97201.0
4	Cryptomonas erosa	cryp	2	20	40	40	10	12	0	8.0	4539.2
4	Planktosphaeria gelatinosa	grn	2	20	40	40	8	0	0	3.4	2557.9
4	Oocystis sp.	grn	10	20	40	40	12	18	0	86.3	38742.2
4	Oocystis sp.	grn	5	20	40	40	8	12	0	12.8	8609.4
4	Ceratium hirundinella	dino	42	20	10	10	40	50	0	87.3	16766.5
4	Cosmarium sp.	desm	25	20	10	10	22	30	0	37.7	9427.6
4	Staurastrum sp.	grn	4	20	10	10	10	30	50	6.0	2839.4
6	Oocystis sp.	grn	1	20	30	40	18	20	0	28.8	9279.1
6	Dictyosphaerium pulchellum	grn	8	20	30	40	4	0	0	2.3	3410.6
6	Gloebotrys sp.	chry	24	20	30	40	5	0	0	13.3	15987.0
6	Sphaerocystis sp.	grn	244	20	30	40	5	0	0	135.4	162534.5
6	Aphanizomenon flos-aquae	bg	29	20	30	40	3	0	0	13.7	18206.4
6	Gloebotrys sp.	chry	94	20	30	40	8	0	0	213.7	160296.3
6	Microcystis aeruginosa	bg	70	20	30	40	3	0	0	8.4	16786.4
6	Planktosphaeria gelatinosa	grn	16	20	30	40	8	0	0	36.4	27284.5
6	Crucigenia Lauterbornii	grn	16	20	30	40	4	6	0	6.8	9183.3
6	Oocystis sp.	grn	6	20	30	40	12	18	0	69.1	30993.8
6	Westella sp.	grn	6	20	30	40	5	0	0	3.3	3996.8
6	Oocystis sp.	grn	4	20	30	40	4	6	0	1.7	2295.8
6	Oscillatoria sp.	bg	12	20	30	40	4	0	0	10.0	10044.9
6	Ceratium hirundinella	dino	80	20	10	10	40	50	0	166.3	31936.1
6	Cosmarium sp.	desm	19	20	10	10	22	30	0	28.7	7165.0
6	Staurastrum sp.	grn	10	20	10	10	10	30	50	14.9	7098.6
8	Oocystis sp.	grn	1	20	40	40	18	20	0	21.6	6959.4
8	Dictyosphaerium pulchellum	grn	12	20	40	40	4	0	0	2.6	3836.9
8	Gloebotrys sp.	chry	17	20	40	40	5	0	0	7.1	8493.1
8	Sphaerocystis sp.	grn	370	20	40	40	5	0	0	154.0	184849.7
8	Aphanizomenon flos-aquae	bg	87	20	40	40	3	0	0	30.7	40964.4
8	Gloebotrys sp.	chry	57	20	40	40	8	0	0	97.2	72900.7
8	Microcystis aeruginosa	bg	184	20	40	40	3	0	0	16.5	33093.1
8	Cryptomonas erosa	cryp	3	20	40	40	10	12	0	12.0	6808.8
8	Crucigenia Lauterbornii	grn	16	20	40	40	4	6	0	5.1	6887.5
8	Oocystis sp.	grn	1	20	40	40	12	18	0	8.6	3874.2

M	Genus and species	Fam.	Ct.	Vol. filt. ml	Flds. ctd.	Mag.	Dim1 µm	Dim2 µm	Dim3 µm	Biovol. mm³/m³	Surf.area mm²/m³
8	Oocystis sp.	grn	20	20	40	40	4	6	0	6.4	8609.4
8	Oscillatoria sp.	bg	12	20	40	40	4	0	0	7.5	7533.7
8	Ceratium hirundinella	dino	62	20	10	10	40	50	0	128.8	24750.5
8	Cosmarium sp.	desm	25	20	10	10	22	30	0	37.7	9427.6
8	Staurastrum sp.	grn	5	20	10	10	10	30	50	7.4	3549.3
10	Schroederia setigera	grn	2	20	40	40	4	50	0	2.7	4009.5
10	Oocystis sp.	grn	2	20	40	40	18	20	0	43.2	13918.7
10	Dictyosphaerium pulchellum	grn	60	20	40	40	4	0	0	12.8	19184.4
10	Gloebotrys sp.	chry	5	20	40	40	5	0	0	2.1	2498.0
10	Sphaerocystis sp.	grn	226	20	40	40	5	0	0	94.1	112908.2
10	Gloebotrys sp.	chry	24	20	40	40	8	0	0	40.9	30695.0
10	Crucigenia Lauterbornii	grn	8	20	40	40	4	6	0	2.6	3443.8
10	Oocystis sp.	grn	4	20	40	40	12	18	0	34.5	15496.9
10	Oocystis sp.	grn	4	20	40	40	8	12	0	10.2	6887.5
10	Oocystis sp.	grn	8	20	40	40	4	6	0	2.6	3443.8
10	Aphanocapsa sp.	bg	18	20	40	40	4	0	0	3.8	5755.3
10	Ceratium hirundinella	dino	24	20	10	10	40	50	0	49.9	9580.8
12	Westella sp.	grn	17	20	50	40	5	0	0	5.7	6794.5
12	Aphanocapsa sp.	bg	10	20	50	40	4	0	0	1.7	2557.9
12	Sphaerocystis sp.	grn	250	20	50	40	5	0	0	83.3	99918.8
12	Aphanizomenon flos-aquae	bg	60	20	50	40	3	0	0	17.0	22601.1
12	Staurastrum sp.	grn	4	20	10	10	10	30	50	6.0	2839.4
12	Cosmarium sp.	desm	2	20	10	10	22	30	0	3.0	754.2
12	Ceratium hirundinella	dino	14	20	10	10	40	50	0	29.1	5588.8
	June 29, 1981										
0	Pediastrum sp.	grn	32	15	30	40	2	4	6	17.4	53049.1
0	Cryptomonas erosa	cryp	4	15	30	40	12	14	0	47.7	22771.7
0	Cryptomonas erosa	cryp	1	15	30	40	16	18	0	27.3	9861.6
0	Oocystis sp.	grn	6	15	30	40	8	12	0	27.3	18366.7
0	Gloebotrys sp.	chry	8	15	30	40	8	0	0	24.3	18189.7
0	Aphanocapsa sp.	bg	28	15	30	40	4	0	0	10.6	15915.9
0	Ankistrodesmus fractus	grn	1	15	30	40	3	40	0	1.1	2137.6
0	Chlamydomonas sp. 1	grn	17	15	30	40	6	8	0	29.0	26709.3
0	Sphaerocystis sp.	grn	8	15	30	40	5	0	0	5.9	7105.3
0	Stephanodiscus sp.	diat	3	15	30	40	10	6	0	16.0	11723.8
0	Schroederia setigera	grn	2	15	30	40	3	30	0	1.6	3213.3
0	Pandorina sp.	grn	6	15	30	40	5	8	0	4.4	5329.0
0	Coelosphaerium sp.	bg	84	15	30	40	2	0	0	4.0	11937.0
0	Aphanizomenon flos-aquae	bg	631	15	30	40	3	0	0	396.2	528195.4
0	Anabaena sp.	bg	110	15	30	40	4	0	0	41.7	62526.9
0	Microcystis aeruginosa	bg	812	15	30	40	2	0	0	38.5	115390.6
0	Microcystis aeruginosa	bg	170	15	30	40	3	0	0	27.2	54355.8
0	Dictyosphaerium pulchellum	grn	158	15	30	40	4	0	0	59.9	89811.4
0	Fragillaria sp.	diat	76	15	30	40	3	20	0	154.7	118603.5
0	Stephanodiscus sp.	diat	5	15	10	10	40	20	0	16.6	3325.1
0	Cosmarium sp.	desm	82	15	10	10	18	30	0	110.4	32375.2
0	Ceratium hirundinella	dino	154	15	10	10	40	50	0	426.7	81969.4
2	Cryptomonas erosa	cryp	1	15	40	40	12	14	0	9.0	4269.7
2	Cryptomonas erosa	cryp	4	15	40	40	16	18	0	81.9	29584.7
2	Fragillaria sp.	diat	40	15	40	40	4	50	0	271.4	146558.2
2	Aphanocapsa sp.	bg	20	15	40	40	4	0	0	5.7	8526.4
2	Chlamydomonas sp. 1	grn	2	15	40	40	6	8	0	2.6	2356.7
2	Sphaerocystis sp.	grn	12	15	40	40	5	0	0	6.7	7993.5
2	Crucigenia Lauterbornii	grn	80	15	40	40	5	7	0	62.2	67963.7
2	Oocystis sp.	grn	2	15	40	40	18	22	0	63.3	19874.3
2	Schroederia setigera	grn	3	15	40	40	3	30	0	1.8	3615.0
2	Ulothrix sp.	grn	20	15	40	40	6	0	0	37.7	25112.3
2	Pandorina sp.	grn	7	15	40	40	5	8	0	3.9	4662.9
2	Coelosphaerium sp.	bg	72	15	40	40	2	0	0	2.6	7673.8
2	Aphanizomenon flos-aquae	bg	165	15	40	40	3	0	0	77.7	103588.2
2	Anabaena sp.	bg	38	15	40	40	4	0	0	10.8	16200.2
2	Microcystis aeruginosa	bg	154	15	40	40	2	0	0	5.5	16413.3
2	Microcystis aeruginosa	bg	116	15	40	40	3	0	0	13.9	27817.4
2	Dictyosphaerium pulchellum	grn	370	15	40	40	4	0	0	105.2	157738.4
2	Stephanodiscus sp.	diat	10	15	10	10	40	20	0	33.3	6650.1
2	Cosmarium sp.	desm	104	15	10	10	18	30	0	140.1	41061.2
2	Ceratium hirundinella	dino	501	15	10	10	40	50	0	1388.2	266666.8
2	Staurastrum sp.	grn	4	15	10	10	8	20	24	2.0	1436.6
4	Pediastrum sp.	grn	27	15	40	40	2	4	6	11.0	33570.2
4	Cryptomonas erosa	cryp	1	15	40	40	12	14	0	9.0	4269.7
4	Cryptomonas erosa	cryp	1	15	40	40	16	18	0	20.5	7396.2
4	Oocystis sp.	grn	3	15	40	40	8	12	0	10.2	6887.5
4	Aphanocapsa sp.	bg	31	15	40	40	4	0	0	8.8	13215.9
4	Chlamydomonas sp. 1	grn	35	15	40	40	6	8	0	44.8	41242.3
4	Sphaerocystis sp.	grn	151	15	40	40	5	0	0	83.8	100584.9
4	Crucigenia Lauterbornii	grn	9	15	40	40	5	7	0	7.0	7645.9
4	Stephanodiscus sp.	diat	2	15	40	40	10	6	0	8.0	5861.9
4	Oocystis sp.	grn	5	15	40	40	18	22	0	158.3	49685.8
4	Schroederia setigera	grn	5	15	40	40	3	30	0	3.0	6025.0
4	Coelosphaerium sp.	bg	114	15	40	40	2	0	0	4.1	12150.1
4	Aphanizomenon flos-aquae	bg	513	15	40	40	3	0	0	241.6	322065.3
4	Anabaena sp.	bg	384	15	40	40	4	0	0	109.1	163706.9
4	Microcystis aeruginosa	bg	658	15	40	40	2	0	0	23.4	70129.6
4	Microcystis aeruginosa	bg	168	15	40	40	3	0	0	20.1	40287.2
4	Dictyosphaerium pulchellum	grn	162	15	40	40	4	0	0	46.0	69063.8
4	Stephanodiscus sp.	diat	16	15	10	10	40	20	0	53.2	10640.2
4	Cosmarium sp.	desm	80	15	10	10	18	30	0	107.7	31585.5
4	Ceratium hirundinella	dino	374	15	10	10	40	50	0	1036.3	199068.6
4	Staurastrum sp.	grn	6	15	10	10	8	20	24	3.0	2154.9
8	Oocystis sp.	grn	12	15	30	40	8	12	0	54.6	36733.3
8	Gloebotrys sp.	chry	23	15	30	40	8	0	0	69.7	52295.3
8	Chlamydomonas sp. 1	grn	4	15	30	40	6	8	0	6.8	6284.5
8	Sphaerocystis sp.	grn	8	15	30	40	5	0	0	5.9	7105.3
8	Oocystis sp.	grn	3	15	30	40	18	22	0	126.6	39748.7
8	Schroederia setigera	grn	1	15	30	40	3	30	0	0.8	1606.7
8	Coelosphaerium sp.	bg	184	15	30	40	2	0	0	8.7	26147.6
8	Aphanizomenon flos-aquae	bg	193	15	30	40	3	0	0	121.2	161555.8
8	Microcystis aeruginosa	bg	252	15	30	40	2	0	0	11.9	35810.9
8	Microcystis aeruginosa	bg	12	15	30	40	3	0	0	1.9	3836.9
8	Dictyosphaerium pulchellum	grn	106	15	30	40	4	0	0	40.2	60253.2
8	Merismopedia sp.	bg	20	15	30	40	5	7	0	14.8	17763.3
8	Stephanodiscus sp.	diat	5	15	10	10	40	20	0	16.6	3325.1
8	Cosmarium sp.	desm	83	15	10	10	18	30	0	111.8	32770.0
8	Ceratium hirundinella	dino	55	15	10	10	40	50	0	152.4	29274.8
8	Staurastrum sp.	grn	8	15	10	10	8	20	24	4.1	2873.2
10	Oocystis sp.	grn	11	15	40	40	8	12	0	37.5	25254.2
10	Gloebotrys sp.	chry	2	15	40	40	8	0	0	4.5	3410.6
10	Sphaerocystis sp.	grn	78	15	40	40	5	0	0	43.3	51957.8
10	Oocystis sp.	grn	1	15	40	40	18	22	0	31.7	9937.2
10	Schroederia setigera	grn	1	15	40	40	3	30	0	0.6	1205.0
10	Coelosphaerium sp.	bg	182	15	40	40	2	0	0	6.5	19397.6
10	Aphanizomenon flos-aquae	bg	122	15	40	40	3	0	0	57.5	76592.5
10	Anabaena sp.	bg	19	15	40	40	4	0	0	5.4	8100.1

M	Genus and species	Fam.	Ct.	Vol. filt. ml	Flds. ctd.	Mag.	Dim1 μm	Dim2 μm	Dim3 μm	Biovol. mm^3/m^3	Surf.area mm^2/m^3
10	Microcystis aeruginosa	bg	56	15	40	40	2	0	0	2.0	5968.5
10	Microcystis aeruginosa	bg	39	15	40	40	3	0	0	4.7	9352.4
10	Dictyosphaerium pulchellum	grn	361	15	40	40	4	0	0	102.6	153901.5
10	Oocystis sp.	grn	4	15	40	40	6	18	0	11.5	9432.8
10	Quadrigula sp.	grn	4	15	40	40	3	20	0	1.6	3233.2
10	Stephanodiscus sp.	diat	12	15	10	10	40	20	0	39.9	7980.1
10	Cosmarium sp.	desm	42	15	10	10	18	30	0	56.6	16582.4
10	Ceratium hirundinella	dino	15	15	10	10	40	50	0	41.6	7984.0
10	Staurastrum sp.	grn	2	15	10	10	8	20	24	1.0	718.3
12	Oocystis sp.	grn	19	15	40	40	8	12	0	64.8	43620.8
12	Fragillaria sp.	diat	22	15	40	40	4	50	0	149.3	80607.0
12	Gloebotrys sp.	chry	26	15	40	40	8	0	0	59.1	44337.3
12	Oocystis sp.	grn	1	15	40	40	18	22	0	31.7	9937.2
12	Schroederia setigera	grn	2	15	40	40	3	30	0	1.2	2410.0
12	Coelosphaerium sp.	bg	130	15	40	40	2	0	0	4.6	13855.4
12	Aphanizomenon flos-aquae	bg	10	15	40	40	3	0	0	4.7	6278.1
12	Dictyosphaerium pulchellum	grn	24	15	40	40	4	0	0	6.8	10231.7
12	Oocystis sp.	grn	2	15	40	40	8	20	0	11.4	7097.7
12	Cosmarium sp.	desm	19	15	10	10	18	30	0	25.6	7501.6
12	Ceratium hirundinella	dino	9	15	10	10	40	50	0	24.9	4790.4
12	Staurastrum sp.	grn	2	15	10	10	8	20	24	1.0	718.3
	July 6, 1981										
0	Anabaena heterocyst	bg	1	20	30	40	8	20	0	5.7	3548.9
0	Actinastrum gracilimum	grn	16	20	30	40	3	14	0	17.1	25240.6
0	Oocystis sp.	grn	5	20	30	40	12	14	0	44.8	21348.4
0	Schroederia setigera	grn	5	20	30	40	2	30	0	1.3	4005.6
0	Cryptomonas erosa	cryp	4	20	30	40	14	16	0	55.7	22905.5
0	Pediastrum sp.	grn	36	20	30	40	2	6	4	14.7	45571.9
0	Dictyosphaerium pulchellum	grn	612	20	30	40	4	0	0	173.9	260907.9
0	Aphanizomenon flos-aquae	bg	840	20	30	40	3	0	0	395.6	527358.3
0	Microcystis aeruginosa	bg	143	20	30	40	3	0	0	17.1	34292.1
0	Anabaena sp.	bg	235	20	30	40	4	0	0	66.8	100185.2
0	Aphanocapsa sp.	bg	24	20	30	40	4	0	0	6.8	10231.7
0	Chlamydomonas sp. 1	grn	20	20	30	40	4	6	0	8.5	11479.2
0	Coelosphaerium sp.	bg	28	20	30	40	2	0	0	1.0	2984.2
0	Lyngbya limnetica	bg	33	20	30	40	2	0	0	6.9	13811.8
0	Gloebotrys sp.	chry	8	20	30	40	8	0	0	18.2	13642.2
0	Cosmarium sp.	desm	6	20	10	10	18	30	0	6.1	1776.7
0	Ceratium hirundinella	dino	28	20	10	10	40	50	0	58.2	11177.7
2	Anabaena heterocyst	bg	2	20	30	40	8	20	0	11.4	7097.7
2	Actinastrum gracilimum	grn	14	20	30	40	3	14	0	15.0	22085.5
2	Dictyosphaerium pulchellum	grn	160	20	30	40	4	0	0	45.5	68211.2
2	Aphanizomenon flos-aquae	bg	794	20	30	40	3	0	0	374.0	498479.2
2	Microcystis aeruginosa	bg	327	20	30	40	3	0	0	39.2	78416.2
2	Anabaena sp.	bg	141	20	30	40	4	0	0	40.1	60111.1
2	Aphanocapsa sp.	bg	29	20	30	40	4	0	0	8.2	12363.3
2	Chlamydomonas sp. 1	grn	6	20	30	40	4	6	0	2.6	3443.8
2	Gloebotrys sp.	chry	3	20	30	40	8	0	0	6.8	5115.8
2	Planktosphaeria gelatinosa	grn	16	20	30	40	5	6	0	8.9	10658.0
2	Oocystis sp.	grn	6	20	30	40	4	6	0	2.6	3443.8
2	Cosmarium sp.	desm	3	20	10	10	18	30	0	3.0	888.3
2	Ceratium hirundinella	dino	75	20	10	10	40	50	0	155.9	29940.1
4	Schroederia setigera	grn	1	15	30	40	2	30	0	0.4	1068.2
4	Dictyosphaerium pulchellum	grn	880	15	30	40	4	0	0	333.5	500215.5
4	Aphanizomenon flos-aquae	bg	1098	15	30	40	3	0	0	689.5	919110.2
4	Microcystis aeruginosa	bg	624	15	30	40	3	0	0	99.8	199517.8
4	Anabaena sp.	bg	209	15	30	40	4	0	0	79.2	118801.2
4	Aphanocapsa sp.	bg	26	15	30	40	4	0	0	9.9	14779.1
4	Chlamydomonas sp. 1	grn	29	15	30	40	4	6	0	16.5	22193.1
4	Coelosphaerium sp.	bg	21	15	30	40	2	0	0	1.0	2984.2
4	Lyngbya limnetica	bg	29	15	30	40	2	0	0	8.1	16183.5
4	Gloebotrys sp.	chry	9	15	30	40	8	0	0	27.3	20463.4
4	Actinastrum gracilimum	grn	61	15	30	40	2	10	0	27.6	60704.0
4	Anabaena sp.	bg	27	15	30	40	6	0	0	34.5	34531.9
4	Sphaerocystis sp.	grn	16	15	30	40	5	0	0	11.8	14210.7
4	Planktosphaeria gelatinosa	grn	9	15	30	40	5	6	0	6.7	7993.5
4	Oocystis sp.	grn	5	15	30	40	10	18	0	53.3	27758.2
4	Oocystis sp.	grn	7	15	30	40	5	10	0	10.4	10626.4
4	Cosmarium sp.	desm	28	15	10	10	18	30	0	37.7	11054.9
4	Ceratium hirundinella	dino	239	15	10	10	40	50	0	662.2	127212.3
4	Stephanodiscus sp.	diat	4	15	10	10	40	20	0	13.3	2660.0
8	Dictyosphaerium pulchellum	grn	996	15	30	40	4	0	0	377.4	566153.0
8	Aphanizomenon flos-aquae	bg	171	15	30	40	3	0	0	107.4	143140.1
8	Microcystis aeruginosa	bg	437	15	30	40	3	0	0	69.9	139726.4
8	Microcystis aeruginosa	bg	728	15	30	40	2	0	0	34.5	103453.7
8	Chlamydomonas sp. 1	grn	4	15	30	40	4	6	0	2.3	3061.1
8	Coelosphaerium sp.	bg	26	15	30	40	2	0	0	1.2	3694.8
8	Gloebotrys sp.	chry	11	15	30	40	8	0	0	33.3	25010.8
8	Sphaerocystis sp.	grn	57	15	30	40	5	0	0	42.2	50625.5
8	Oocystis sp.	grn	8	15	30	40	4	6	0	4.5	6122.2
8	Cosmarium sp.	desm	54	15	10	10	18	30	0	72.7	21320.2
8	Ceratium hirundinella	dino	313	15	10	10	40	50	0	867.3	166600.2
8	Stephanodiscus sp.	diat	7	15	10	10	40	20	0	23.3	4655.1
10	Fragillaria sp.	diat	49	15	30	40	3	20	0	99.7	76468.1
10	Dictyosphaerium pulchellum	grn	840	15	30	40	4	0	0	318.3	477478.4
10	Aphanizomenon flos-aquae	bg	71	15	30	40	3	0	0	44.6	59432.4
10	Microcystis aeruginosa	bg	63	15	30	40	3	0	0	10.1	20143.6
10	Anabaena sp.	bg	35	15	30	40	4	0	0	13.3	19894.9
10	Coelosphaerium sp.	bg	86	15	30	40	2	0	0	4.1	12221.2
10	Gloebotrys sp.	chry	21	15	30	40	8	0	0	63.7	47747.8
10	Sphaerocystis sp.	grn	8	15	30	40	5	0	0	5.9	7105.3
10	Planktosphaeria gelatinosa	grn	40	15	30	40	5	6	0	29.6	35526.7
10	Cosmarium sp.	desm	26	15	10	10	18	30	0	35.0	10265.3
10	Ceratium hirundinella	dino	77	15	10	10	40	50	0	213.4	40984.7
10	Stephanodiscus sp.	diat	6	15	10	10	40	20	0	20.0	3990.1
12	Dictyosphaerium pulchellum	grn	268	15	30	40	4	0	0	101.6	152338.4
12	Aphanizomenon flos-aquae	bg	8	15	30	40	3	0	0	5.0	6696.6
12	Microcystis aeruginosa	bg	6	15	30	40	3	0	0	1.0	1918.4
12	Aphanocapsa sp.	bg	17	15	30	40	4	0	0	6.4	9663.3
12	Coelosphaerium sp.	bg	40	15	30	40	2	0	0	1.9	5684.3
12	Lyngbya limnetica	bg	70	15	30	40	2	0	0	19.5	39063.6
12	Gloebotrys sp.	chry	15	15	30	40	8	0	0	45.5	34105.6
12	Sphaerocystis sp.	grn	13	15	30	40	5	0	0	9.6	11546.2
12	Oocystis sp.	grn	1	15	30	40	10	18	0	10.7	5551.6
12	Cosmarium sp.	desm	40	15	10	10	18	30	0	53.9	15792.8
12	Ceratium hirundinella	dino	19	15	10	10	40	50	0	52.6	10113.1
	July 13, 1981										
0	Fragillaria sp.	diat	14	15	30	40	4	20	0	50.7	30397.2
0	Pediastrum sp.	grn	16	15	30	40	4	6	10	43.4	77290.2
0	Schroederia setigera	grn	2	15	30	40	2	30	0	0.7	2136.3
0	Dictyosphaerium pulchellum	grn	120	15	30	40	4	0	0	45.5	68211.2
0	Aphanizomenon flos-aquae	bg	61	15	30	40	3	0	0	38.3	51061.7

M	Genus and species	Fam.	Ct.	Vol. filt. ml	Flds. ctd.	Mag.	Dim1 μm	Dim2 μm	Dim3 μm	Biovol. mm³/m³	Surf.area mm²/m³
0	Microcystis aeruginosa	bg	1046	15	30	40	3	0	0	167.2	334448.1
0	Sphaerocystis sp.	grn	31	15	30	40	5	0	0	22.9	27533.2
0	Anabaena sp.	bg	74	15	30	40	4	0	0	28.0	42063.6
0	Chlamydomonas sp. 1	grn	28	15	30	40	4	6	0	15.9	21427.8
0	Coelosphaerium sp.	bg	37	15	30	40	2	0	0	1.8	5257.9
0	Lyngbya limnetica	bg	262	15	30	40	2	0	0	73.1	146209.4
0	Actinastrum gracilimum	grn	52	15	30	40	2	10	0	23.5	51747.7
0	Ceratium hirundinella	dino	4	15	10	10	40	50	0	11.1	2129.1
0	Cosmarium sp.	desm	2	15	10	10	18	30	0	2.7	789.6
2	Schroederia setigera	grn	3	15	30	40	2	30	0	1.1	3204.5
2	Dictyosphaerium pulchellum	grn	220	15	30	40	4	0	0	83.4	125053.9
2	Aphanizomenon flos-aquae	bg	65	15	30	40	3	0	0	40.8	54410.0
2	Microcystis aeruginosa	bg	620	15	30	40	3	0	0	99.1	198238.8
2	Sphaerocystis sp.	grn	13	15	30	40	5	0	0	9.6	11546.2
2	Anabaena sp.	bg	38	15	30	40	4	0	0	14.4	21600.2
2	Chlamydomonas sp. 1	grn	10	15	30	40	4	6	0	5.7	7652.8
2	Coelosphaerium sp.	bg	72	15	30	40	2	0	0	3.4	10231.7
2	Lyngbya limnetica	bg	398	15	30	40	2	0	0	111.1	222104.4
2	Actinastrum gracilimum	grn	36	15	30	40	2	10	0	16.3	35825.3
2	Oocystis sp.	grn	6	15	30	40	6	10	0	12.8	11249.2
4	Dictyosphaerium pulchellum	grn	55	15	30	40	4	0	0	20.8	31263.5
4	Aphanizomenon flos-aquae	bg	40	15	30	40	3	0	0	25.1	33483.1
4	Microcystis aeruginosa	bg	937	15	30	40	3	0	0	149.8	299596.4
4	Sphaerocystis sp.	grn	9	15	30	40	5	0	0	6.7	7993.5
4	Anabaena sp.	bg	7	15	30	40	4	0	0	2.7	3979.0
4	Chlamydomonas sp. 1	grn	24	15	30	40	4	6	0	13.6	18366.7
4	Coelosphaerium sp.	bg	14	15	30	40	2	0	0	0.7	1989.5
4	Lyngbya limnetica	bg	221	15	30	40	2	0	0	61.7	123329.3
4	Actinastrum gracilimum	grn	36	15	30	40	2	10	0	16.3	35825.3
4	Microcystis aeruginosa	bg	168	15	30	40	2	0	0	8.0	23873.9
4	Oocystis sp.	grn	1	15	30	40	20	30	0	71.1	19132.0
4	Cryptomonas erosa	cryp	2	15	30	40	18	20	0	76.7	24744.4
4	Ceratium hirundinella	dino	2	15	10	10	40	50	0	5.5	1064.5
4	Cosmarium sp.	desm	2	15	10	10	18	30	0	2.7	789.6
6	Dictyosphaerium pulchellum	grn	376	15	30	40	4	0	0	142.5	213728.4
6	Aphanizomenon flos-aquae	bg	14	15	30	40	3	0	0	8.8	11719.1
6	Microcystis aeruginosa	bg	686	15	30	40	3	0	0	109.7	219341.7
6	Sphaerocystis sp.	grn	28	15	30	40	5	0	0	20.7	24868.7
6	Anabaena sp.	bg	98	15	30	40	4	0	0	37.1	55705.8
6	Chlamydomonas sp. 1	grn	6	15	30	40	4	6	0	3.4	4591.7
6	Lyngbya limnetica	bg	324	15	30	40	2	0	0	90.4	180808.6
6	Actinastrum gracilimum	grn	8	15	30	40	2	10	0	3.6	7961.2
6	Cryptomonas erosa	cryp	4	15	30	40	18	20	0	153.5	49488.8
6	Merismopedia sp.	bg	28	15	30	40	4	0	0	10.6	15915.9
8	Fragillaria sp.	diat	27	15	30	40	4	20	0	97.7	58623.3
8	Dictyosphaerium pulchellum	grn	176	15	30	40	4	0	0	66.7	100043.1
8	Aphanizomenon flos-aquae	bg	108	15	30	40	3	0	0	67.8	90404.3
8	Microcystis aeruginosa	bg	444	15	30	40	3	0	0	71.0	141964.6
8	Sphaerocystis sp.	grn	7	15	30	40	5	0	0	5.2	6217.2
8	Chlamydomonas sp. 1	grn	2	15	30	40	4	6	0	1.1	1530.6
8	Coelosphaerium sp.	bg	67	15	30	40	2	0	0	3.2	9521.1
8	Lyngbya limnetica	bg	145	15	30	40	2	0	0	40.5	80917.4
8	Planktosphaeria gelatinosa	grn	18	15	30	40	5	7	0	13.3	15987.0
8	Oocystis sp.	grn	1	15	30	40	20	30	0	71.1	19132.0
8	Cryptomonas erosa	cryp	1	15	30	40	18	20	0	38.4	12372.2
10	Dictyosphaerium pulchellum	grn	72	20	40	40	4	0	0	15.3	23021.3
10	Aphanizomenon flos-aquae	bg	16	20	40	40	3	0	0	5.7	7533.7

M	Genus and species	Fam.	Ct.	Vol. filt. ml	Flds. ctd.	Mag.	Dim1 μm	Dim2 μm	Dim3 μm	Biovol. mm³/m³	Surf.area mm²/m³
10	Microcystis aeruginosa	bg	68	20	40	40	3	0	0	6.1	12230.1
10	Anabaena sp.	bg	6	20	40	40	4	0	0	1.3	1918.4
10	Lyngbya limnetica	bg	7	20	40	40	2	0	0	1.1	2197.3
10	Oocystis sp.	grn	4	20	40	40	6	10	0	4.8	4218.5
10	Ceratium hirundinella	dino	1	20	10	10	40	50	0	2.1	399.2
10	Cosmarium sp.	desm	7	20	10	10	18	30	0	7.1	2072.8

July 20, 1981

M	Genus and species	Fam.	Ct.	Vol. filt. ml	Flds. ctd.	Mag.	Dim1 μm	Dim2 μm	Dim3 μm	Biovol. mm³/m³	Surf.area mm²/m³
0	Dictyosphaerium pulchellum	grn	60	15	30	40	4	0	0	22.7	34105.6
0	Aphanizomenon flos-aquae	bg	52	15	30	40	3	0	0	32.7	43528.0
0	Microcystis aeruginosa	bg	1582	15	30	40	3	0	0	252.9	505828.7
0	Anabaena sp.	bg	53	15	30	40	4	0	0	20.1	30126.6
0	Aphanocapsa sp.	bg	41	15	30	40	4	0	0	15.5	23305.5
0	Coelosphaerium sp.	bg	85	15	30	40	2	0	0	4.0	12079.1
0	Lyngbya limnetica	bg	478	15	30	40	2	0	0	133.4	266748.4
0	Actinastrum gracilimum	grn	18	15	30	40	2	10	0	8.1	17912.7
0	Oocystis sp.	grn	16	15	30	40	5	8	0	18.9	20149.2
0	Oocystis sp.	grn	2	15	30	40	4	12	0	2.3	2794.9
0	Ceratium hirundinella	dino	2	15	10	10	40	50	0	5.5	1064.5
2	Dictyosphaerium pulchellum	grn	52	15	30	40	4	0	0	19.7	29558.2
2	Aphanizomenon flos-aquae	bg	30	15	30	40	3	0	0	18.8	25112.3
2	Microcystis aeruginosa	bg	2072	15	30	40	3	0	0	331.3	662501.3
2	Coelosphaerium sp.	bg	40	15	30	40	2	0	0	1.9	5684.3
2	Lyngbya limnetica	bg	385	15	30	40	2	0	0	107.5	214849.7
2	Actinastrum gracilimum	grn	28	15	30	40	2	10	0	12.7	27864.1
2	Cryptomonas erosa	cryp	3	15	30	40	14	20	0	69.6	27063.8
2	Sphaerocystis sp.	grn	8	15	30	40	5	0	0	5.9	7105.3
2	Gloebotrys sp.	chry	30	15	30	40	8	0	0	90.9	68211.2
2	Ceratium hirundinella	dino	2	15	10	10	40	50	0	5.5	1064.5
4	Dictyosphaerium pulchellum	grn	68	15	30	40	4	0	0	25.8	38653.0
4	Aphanizomenon flos-aquae	bg	76	15	30	40	3	0	0	47.7	63617.8
4	Microcystis aeruginosa	bg	3552	15	30	40	3	0	0	567.9	1135716.6
4	Aphanocapsa sp.	bg	38	15	30	40	4	0	0	14.4	21600.2
4	Coelosphaerium sp.	bg	47	15	30	40	2	0	0	2.2	6679.0
4	Lyngbya limnetica	bg	658	15	30	40	2	0	0	183.6	367197.7
4	Sphaerocystis sp.	grn	27	15	30	40	5	0	0	20.0	23980.5
4	Microcystis aeruginosa	bg	616	15	30	40	2	0	0	29.2	87537.7
4	Pediastrum sp.	grn	17	15	30	40	4	8	6	36.9	68319.4
4	Westella sp.	grn	4	15	30	40	5	0	0	3.0	3552.7
4	Oocystis sp.	grn	2	15	30	40	16	20	0	60.6	21287.6
4	Oocystis sp.	grn	4	15	30	40	10	14	0	33.2	18123.6
4	Crucigenia Lauterbornii	grn	16	15	30	40	5	7	0	16.6	18123.6
4	Ceratium hirundinella	dino	4	15	10	10	40	50	0	11.1	2129.1
6	Dictyosphaerium pulchellum	grn	68	15	30	40	4	0	0	25.8	38653.0
6	Microcystis aeruginosa	bg	1696	15	30	40	3	0	0	271.1	542279.1
6	Anabaena sp.	bg	14	15	30	40	4	0	0	5.3	7958.0
6	Lyngbya limnetica	bg	440	15	30	40	2	0	0	122.8	245542.5
6	Actinastrum gracilimum	grn	6	15	30	40	2	10	0	2.7	5970.9
6	Sphaerocystis sp.	grn	32	15	30	40	5	0	0	23.7	28421.3
6	Cryptomonas erosa	cryp	6	15	30	40	16	18	0	163.7	59169.4
6	Crucigenia Lauterbornii	grn	16	15	30	40	5	7	0	16.6	18123.6
6	Cosmarium sp.	desm	2	15	10	10	18	30	0	2.7	789.6
8	Microcystis aeruginosa	bg	16	15	30	40	3	0	0	2.6	5115.8
8	Aphanocapsa sp.	bg	11	15	30	40	4	0	0	4.2	6252.7
8	Lyngbya limnetica	bg	101	15	30	40	2	0	0	28.2	56363.2
8	Sphaerocystis sp.	grn	8	15	30	40	5	0	0	5.9	7105.3
8	Chlamydomonas sp. 1	grn	2	15	30	40	7	9	0	5.2	4161.1

M	Genus and species	Fam.	Ct.	Vol. filt. ml	Flds. ctd.	Mag.	Dim1 μm	Dim2 μm	Dim3 μm	Biovol. mm³/m³	Surf.area mm²/m³
8	Cryptomonas erosa	cryp	2	15	30	40	16	18	0	54.6	19723.1
8	Schroederia setigera	grn	1	15	30	40	4	30	0	1.4	2150.5
8	Oocystis sp.	grn	3	15	30	40	5	8	0	3.6	3778.0
	July 27, 1981										
0	Oocystis sp.	grn	2	15	30	40	12	16	0	27.3	12569.1
0	Actinastrum gracilimum	grn	4	15	30	40	3	24	0	9.8	13841.6
0	Cryptomonas erosa	cryp	5	15	30	40	14	18	0	104.4	41610.7
0	Planktosphaeria gelatinosa	grn	8	15	30	40	5	0	0	5.9	7105.3
0	Dictyosphaerium pulchellum	grn	64	15	30	40	4	0	0	24.3	36379.3
0	Aphanizomenon flos-aquae	bg	111	15	30	40	3	0	0	69.7	92915.5
0	Microcystis aeruginosa	bg	1624	15	30	40	3	0	0	259.6	519257.8
0	Sphaerocystis sp.	grn	31	15	30	40	5	0	0	22.9	27533.2
0	Anabaena sp.	bg	37	15	30	40	4	0	0	14.0	21031.8
0	Coelosphaerium sp.	bg	81	15	30	40	2	0	0	3.8	11510.6
0	Lyngbya limnetica	bg	534	15	30	40	2	0	0	149.0	297999.3
0	Actinastrum gracilimum	grn	28	15	30	40	2	10	0	12.7	27864.1
2	Cryptomonas erosa	cryp	4	15	30	40	14	18	0	83.6	33288.6
2	Dictyosphaerium pulchellum	grn	48	15	30	40	4	0	0	18.2	27284.5
2	Aphanizomenon flos-aquae	bg	13	15	30	40	3	0	0	8.2	10882.0
2	Microcystis aeruginosa	bg	488	15	30	40	3	0	0	78.0	156033.1
2	Sphaerocystis sp.	grn	10	15	30	40	5	0	0	7.4	8881.7
2	Anabaena sp.	bg	160	15	30	40	4	0	0	60.6	90948.3
2	Coelosphaerium sp.	bg	117	15	30	40	2	0	0	5.5	16626.5
2	Lyngbya limnetica	bg	404	15	30	40	2	0	0	112.8	225452.7
2	Actinastrum gracilimum	grn	30	15	30	40	2	10	0	13.6	29854.4
2	Microcystis aeruginosa	bg	770	15	30	40	2	0	0	36.5	109422.1
2	Aphanocapsa sp.	bg	12	15	30	40	4	0	0	4.5	6821.1
4	Cryptomonas erosa	cryp	1	15	30	40	14	18	0	20.9	8322.1
4	Planktosphaeria gelatinosa	grn	8	15	30	40	5	0	0	5.9	7105.3
4	Dictyosphaerium pulchellum	grn	18	15	30	40	4	0	0	6.8	10231.7
4	Aphanizomenon flos-aquae	bg	70	15	30	40	3	0	0	44.0	58595.4
4	Microcystis aeruginosa	bg	1256	15	30	40	3	0	0	200.8	401593.5
4	Sphaerocystis sp.	grn	39	15	30	40	5	0	0	28.9	34638.5
4	Anabaena sp.	bg	30	15	30	40	4	0	0	11.4	17052.8
4	Coelosphaerium sp.	bg	100	15	30	40	2	0	0	4.7	14210.7
4	Lyngbya limnetica	bg	167	15	30	40	2	0	0	46.6	93194.5
4	Actinastrum gracilimum	grn	32	15	30	40	2	10	0	14.5	31844.7
4	Pediastrum sp.	grn	14	15	30	40	2	6	4	7.6	23629.9
4	Stephanodiscus sp.	diat	1	15	10	10	30	30	0	2.8	561.1
4	Ceratium hirundinella	dino	8	15	10	10	40	50	0	22.2	4258.2
6	Cryptomonas erosa	cryp	3	15	30	40	14	18	0	62.7	24966.4
6	Planktosphaeria gelatinosa	grn	8	15	30	40	5	0	0	5.9	7105.3
6	Aphanizomenon flos-aquae	bg	41	15	30	40	3	0	0	25.7	34320.1
6	Microcystis aeruginosa	bg	1216	15	30	40	3	0	0	194.4	388803.9
6	Sphaerocystis sp.	grn	46	15	30	40	5	0	0	34.0	40855.7
6	Anabaena sp.	bg	50	15	30	40	4	0	0	18.9	28421.3
6	Lyngbya limnetica	bg	612	15	30	40	2	0	0	170.8	341527.3
6	Actinastrum gracilimum	grn	14	15	30	40	2	10	0	6.3	13932.1
6	Microcystis aeruginosa	bg	364	15	30	40	2	0	0	17.2	51726.8
6	Fragillaria sp.	diat	39	15	30	40	3	30	0	119.1	87324.2
6	Ceratium hirundinella	dino	4	15	10	10	40	50	0	11.1	2129.1
8	Cryptomonas erosa	cryp	3	20	30	40	14	18	0	47.0	18724.8
8	Planktosphaeria gelatinosa	grn	56	20	30	40	5	0	0	31.1	37303.0
8	Aphanizomenon flos-aquae	bg	102	20	30	40	3	0	0	48.0	64036.4
8	Microcystis aeruginosa	bg	672	20	30	40	3	0	0	80.6	161149.0
8	Sphaerocystis sp.	grn	19	20	30	40	5	0	0	10.5	12656.4
8	Anabaena sp.	bg	57	20	30	40	4	0	0	16.2	24300.2
8	Coelosphaerium sp.	bg	130	20	30	40	2	0	0	4.6	13855.4
8	Lyngbya limnetica	bg	813	20	30	40	2	0	0	170.2	340271.7
8	Actinastrum gracilimum	grn	34	20	30	40	2	10	0	11.5	25376.3
8	Gloebotrys sp.	chry	8	20	30	40	12	14	0	61.4	30695.0
8	Oocystis sp.	grn	3	20	30	40	6	10	0	4.8	4218.5
8	Oocystis sp.	grn	1	20	30	40	12	18	0	11.5	5165.6
8	Chlamydomonas sp. 1	grn	2	20	30	40	5	7	0	1.6	1699.1
10	Microcystis aeruginosa	bg	264	20	30	40	3	0	0	31.7	63308.5
10	Lyngbya limnetica	bg	199	20	30	40	2	0	0	41.7	83289.1
10	Actinastrum gracilimum	grn	12	20	30	40	2	10	0	4.1	8956.3
	Aug. 3, 1981										
0	Dictyosphaerium pulchellum	grn	32	15	30	40	4	0	0	12.1	18189.7
0	Aphanizomenon flos-aquae	bg	344	15	30	40	3	0	0	216.0	287954.4
0	Microcystis aeruginosa	bg	176	15	30	40	3	0	0	28.1	56274.2
0	Anabaena sp.	bg	1186	15	30	40	4	0	0	449.4	674154.1
0	Lyngbya limnetica	bg	704	15	30	40	2	0	0	196.5	392868.0
2	Dictyosphaerium pulchellum	grn	48	15	30	40	4	0	0	18.2	27284.5
2	Aphanizomenon flos-aquae	bg	178	15	30	40	3	0	0	111.8	148999.7
2	Microcystis aeruginosa	bg	176	15	30	40	3	0	0	28.1	56274.2
2	Anabaena sp.	bg	1206	15	30	40	4	0	0	457.0	685522.6
2	Lyngbya limnetica	bg	464	15	30	40	2	0	0	129.5	258935.7
2	Planktosphaeria gelatinosa	grn	8	15	30	40	6	0	0	10.2	10231.7
2	Actinastrum gracilimum	grn	72	15	30	40	2	10	0	32.6	71650.7
4	Aphanizomenon flos-aquae	bg	44	15	40	40	3	0	0	20.7	27623.5
4	Microcystis aeruginosa	bg	752	15	40	40	3	0	0	90.2	180333.4
4	Anabaena sp.	bg	86	15	40	40	4	0	0	24.4	36663.5
4	Lyngbya limnetica	bg	574	15	40	40	2	0	0	120.1	240241.0
4	Planktosphaeria gelatinosa	grn	8	15	40	40	6	0	0	7.7	7673.8
4	Actinastrum gracilimum	grn	8	15	40	40	2	10	0	2.7	5970.9
4	Oocystis sp.	grn	3	15	40	40	6	10	0	4.8	4218.5
	Aug. 17, 1981										
0	Microcystis aeruginosa	bg	312	15	20	40	2	0	0	22.2	66505.9
0	Aphanizomenon flos-aquae	bg	175	15	20	40	3	0	0	164.8	219732.6
0	Anabaena sp.	bg	519	15	20	40	4	0	0	295.0	442520.2
0	Actinastrum gracilimum	grn	8	15	20	40	2	10	0	5.4	11941.8
0	Lyngbya limnetica	bg	1544	15	20	40	2	0	0	646.4	1292446.5
0	Actinastrum gracilimum	grn	4	15	20	40	3	20	0	12.2	17505.6
0	Dictyosphaerium pulchellum	grn	20	15	20	40	4	0	0	11.4	17052.8
0	Oocystis sp.	grn	4	15	20	40	7	12	0	20.9	15671.3
0	Merismopedia sp.	bg	32	15	20	40	4	0	0	18.2	27284.5
2	Microcystis aeruginosa	bg	192	15	20	40	2	0	0	13.6	40926.7
2	Aphanizomenon flos-aquae	bg	143	15	20	40	3	0	0	134.7	179553.0
2	Anabaena sp.	bg	188	15	20	40	4	0	0	106.9	160296.3
2	Lyngbya limnetica	bg	1232	15	20	40	2	0	0	515.8	1031278.5
2	Actinastrum gracilimum	grn	10	15	20	40	3	20	0	30.5	43763.9
2	Fragillaria sp.	diat	16	15	20	40	4	24	0	104.2	60794.5
2	Ceratium hirundinella	dino	4	15	10	10	40	50	0	11.1	2129.1
4	Microcystis aeruginosa	bg	784	15	20	40	2	0	0	55.7	167117.5
4	Aphanizomenon flos-aquae	bg	142	15	20	40	3	0	0	133.8	178297.3
4	Anabaena sp.	bg	351	15	20	40	4	0	0	199.5	299276.7
4	Actinastrum gracilimum	grn	24	15	20	40	2	10	0	16.3	35825.3
4	Lyngbya limnetica	bg	1758	15	20	40	2	0	0	736.0	1471580.9
4	Gloebotrys sp.	chry	8	15	20	40	10	0	0	71.1	42632.0
4	Fragillaria sp.	diat	20	15	20	40	4	24	0	130.3	75993.1

M	Genus and species	Fam.	Ct.	Vol. filt. ml	Flds. ctd.	Mag.	Dim1 μm	Dim2 μm	Dim3 μm	Biovol. mm³/m³	Surf.area mm²/m³
4	Ceratium hirundinella	dino	22	15	10	10	40	50	0	61.0	11709.9
6	Microcystis aeruginosa	bg	24	15	20	40	2	0	0	1.7	5115.8
6	Aphanizomenon flos-aquae	bg	236	15	20	40	3	0	0	222.3	296325.2
6	Anabaena sp.	bg	176	15	20	40	4	0	0	100.0	150064.7
6	Actinastrum gracilimum	grn	20	15	20	40	2	10	0	13.6	29854.4
6	Lyngbya limnetica	bg	1512	15	20	40	2	0	0	633.0	1265660.0
6	Actinastrum gracilimum	grn	6	15	20	40	3	20	0	18.3	26258.3
6	Ceratium hirundinella	dino	2	15	10	10	40	50	0	5.5	1064.5
8	Microcystis aeruginosa	bg	416	15	20	40	2	0	0	29.6	88674.6
8	Aphanizomenon flos-aquae	bg	172	15	20	40	3	0	0	162.0	215965.8
8	Anabaena sp.	bg	372	15	20	40	4	0	0	211.5	317182.1
8	Lyngbya limnetica	bg	1616	15	20	40	2	0	0	676.5	1352716.0
8	Actinastrum gracilimum	grn	6	15	20	40	3	20	0	18.3	26258.3
8	Ceratium hirundinella	dino	14	15	10	10	40	50	0	38.8	7451.8
10	Microcystis aeruginosa	bg	624	15	20	40	2	0	0	44.3	133011.9
10	Aphanizomenon flos-aquae	bg	88	15	20	40	3	0	0	82.9	110494.1
10	Anabaena sp.	bg	108	15	20	40	4	0	0	61.4	92085.1
i0	Lyngbya limnetica	bg	1130	15	20	40	2	0	0	473.1	945896.7
10	Actinastrum gracilimum	grn	6	15	20	40	3	20	0	18.3	26258.3
10	Chlamydomonas sp. 1	grn	6	15	20	40	5	7	0	9.3	10194.5
10	Ceratium hirundinella	dino	3	15	10	10	40	50	0	8.3	1596.8
	Aug. 24, 1981										
0	Microcystis aeruginosa	bg	1808	15	30	40	2	0	0	85.6	256928.9
0	Aphanizomenon flos-aquae	bg	477	15	30	40	3	0	0	299.5	399285.6
0	Anabaena sp.	bg	759	15	30	40	4	0	0	287.6	431435.9
0	Actinastrum gracilimum	grn	58	15	30	40	2	10	0	26.2	57718.6
0	Actinastrum gracilimum	grn	28	15	30	40	3	20	0	57.0	81692.6
0	Lyngbya limnetica	bg	841	15	30	40	2	0	0	234.7	469321.0
0	Dictyosphaerium pulchellum	grn	4	15	30	40	4	0	0	1.5	2273.7
0	Peridinium sp.	dino	1	15	10	10	40	40	0	4.4	332.5
0	Ceratium hirundinella	dino	17	15	10	10	40	50	0	47.1	9048.6
2	Microcystis aeruginosa	bg	64	15	30	40	2	0	0	3.0	9094.8
2	Aphanizomenon flos-aquae	bg	424	15	30	40	3	0	0	266.3	354920.5
2	Anabaena sp.	bg	854	15	30	40	4	0	0	323.6	485436.4
2	Actinastrum gracilimum	grn	32	15	30	40	2	10	0	14.5	31844.7
2	Actinastrum gracilimum	grn	24	15	30	40	3	20	0	48.9	70022.2
2	Lyngbya limnetica	bg	220	15	30	40	2	0	0	61.4	122771.3
2	Coelosphaerium sp.	bg	80	15	30	40	4	0	0	30.3	45474.1
2	Peridinium sp.	dino	2	15	10	10	40	40	0	8.9	665.0
2	Ceratium hirundinella	dino	43	15	10	10	40	50	0	119.1	22887.6
4	Microcystis aeruginosa	bg	1664	15	30	40	2	0	0	78.8	236465.5
4	Aphanizomenon flos-aquae	bg	656	15	30	40	3	0	0	411.9	549122.3
4	Anabaena sp.	bg	738	15	30	40	4	0	0	279.7	419498.9
4	Actinastrum gracilimum	grn	74	15	30	40	2	10	0	33.5	73641.0
4	Actinastrum gracilimum	grn	44	15	30	40	3	20	0	89.6	128374.1
4	Lyngbya limnetica	bg	430	15	30	40	2	0	0	120.0	239962.0
4	Peridinium sp.	dino	9	15	10	10	40	40	0	39.9	2992.6
4	Ceratium hirundinella	dino	42	15	10	10	40	50	0	116.4	22355.3
6	Microcystis aeruginosa	bg	288	15	40	40	2	0	0	10.2	30695.0
6	Aphanizomenon flos-aquae	bg	92	15	40	40	3	0	0	43.3	57758.3
6	Actinastrum gracilimum	grn	24	15	40	40	2	10	0	8.1	17912.7
6	Actinastrum gracilimum	grn	6	15	40	40	3	20	0	9.2	13129.2
6	Dictyosphaerium pulchellum	grn	24	15	40	40	4	0	0	6.8	10231.7
6	Ceratium hirundinella	dino	10	15	10	10	40	50	0	27.7	5322.7

M	Genus and species	Fam.	Ct.	Vol. filt. ml	Flds. ctd.	Mag.	Dim1 μm	Dim2 μm	Dim3 μm	Biovol. mm³/m³	Surf.area mm²/m³
	Sep. 8, 1981										
0	Stephanodiscus sp.	diat	2	15	30	40	18	12	0	69.1	26858.2
0	Microcystis aeruginosa	bg	704	15	30	40	2	0	0	33.3	100043.1
0	Aphanizomenon flos-aquae	bg	1330	15	30	40	3	0	0	835.2	1113312.0
0	Anabaena sp.	bg	178	15	30	40	4	0	0	67.5	101180.0
0	Actinastrum gracilimum	grn	248	15	30	40	2	10	0	112.2	246796.7
0	Lyngbya limnetica	bg	1096	15	30	40	2	0	0	305.9	611624.1
0	Actinastrum gracilimum	grn	62	15	30	40	3	20	0	126.2	180890.8
0	Dictyosphaerium pulchellum	grn	8	15	30	40	4	0	0	3.0	4547.4
0	Cryptomonas erosa	cryp	1	15	30	40	12	14	0	11.9	5692.9
0	Ceratium hirundinella	dino	70	15	10	10	40	40	0	155.2	32916.4
0	Peridinium sp.	dino	14	15	10	10	32	40	0	39.7	6973.3
2	Stephanodiscus sp.	diat	6	15	30	40	18	12	0	207.2	80574.5
2	Microcystis aeruginosa	bg	528	15	30	40	2	0	0	25.0	75032.3
2	Aphanizomenon flos-aquae	bg	684	15	30	40	3	0	0	429.5	572560.5
2	Anabaena sp.	bg	66	15	30	40	4	0	0	25.0	37516.2
2	Actinastrum gracilimum	grn	106	15	30	40	2	10	0	47.9	105485.7
2	Lyngbya limnetica	bg	644	15	30	40	2	0	0	179.7	359384.9
2	Actinastrum gracilimum	grn	48	15	30	40	3	20	0	97.7	140044.5
2	Dictyosphaerium pulchellum	grn	24	15	30	40	4	0	0	9.1	13642.2
2	Cryptomonas erosa	cryp	1	15	30	40	12	14	0	11.9	5692.9
2	Ceratium hirundinella	dino	86	15	10	10	40	40	0	190.6	40440.2
2	Peridinium sp.	dino	24	15	10	10	32	40	0	68.1	11954.3
4	Microcystis aeruginosa	bg	544	15	30	40	2	0	0	25.8	77306.0
4	Aphanizomenon flos-aquae	bg	928	15	30	40	3	0	0	582.7	776807.2
4	Anabaena sp.	bg	90	15	30	40	4	0	0	34.1	51158.4
4	Actinastrum gracilimum	grn	110	15	30	40	2	10	0	49.8	109466.3
4	Lyngbya limnetica	bg	1210	15	30	40	2	0	0	337.7	675241.9
4	Actinastrum gracilimum	grn	14	15	30	40	3	20	0	28.5	40846.3
4	Ceratium hirundinella	dino	98	15	10	10	40	40	0	217.2	46083.0
4	Peridinium sp.	dino	24	15	10	10	32	40	0	68.1	11954.3
6	Microcystis aeruginosa	bg	120	15	30	40	2	0	0	5.7	17052.8
6	Aphanizomenon flos-aquae	bg	732	15	30	40	3	0	0	459.7	612740.2
6	Anabaena sp.	bg	48	15	30	40	4	0	0	18.2	27284.5
6	Actinastrum gracilimum	grn	102	15	30	40	2	10	0	46.1	101505.1
6	Lyngbya limnetica	bg	1026	15	30	40	2	0	0	286.4	572560.5
6	Actinastrum gracilimum	grn	9	15	30	40	3	20	0	18.3	26258.3
6	Dictyosphaerium pulchellum	grn	36	15	30	40	4	0	0	13.6	20463.4
6	Ceratium hirundinella	dino	83	15	10	10	40	40	0	184.0	39029.4
6	Peridinium sp.	dino	23	15	10	10	32	40	0	65.3	11456.2
8	Microcystis aeruginosa	bg	24	15	30	40	2	0	0	1.1	3410.6
8	Aphanizomenon flos-aquae	bg	1089	15	30	40	3	0	0	683.8	911576.5
8	Anabaena sp.	bg	105	15	30	40	4	0	0	39.8	59684.8
8	Actinastrum gracilimum	grn	114	15	30	40	2	10	0	51.6	113446.9
8	Lyngbya limnetica	bg	1263	15	30	40	2	0	0	352.5	704818.6
8	Dictyosphaerium pulchellum	grn	36	15	30	40	4	0	0	13.6	20463.4
8	Ceratium hirundinella	dino	105	15	10	10	40	40	0	232.8	49374.6
8	Peridinium sp.	dino	22	15	10	10	32	40	0	62.4	10958.1
10	Microcystis aeruginosa	bg	320	15	30	40	2	0	0	15.2	45474.1
10	Aphanizomenon flos-aquae	bg	878	15	30	40	3	0	0	551.3	734953.4
10	Anabaena sp.	bg	120	15	30	40	4	0	0	45.5	68211.2
10	Actinastrum gracilimum	grn	230	15	30	40	2	10	0	104.0	228884.0
10	Lyngbya limnetica	bg	1286	15	30	40	2	0	0	358.9	717653.8
10	Actinastrum gracilimum	grn	122	15	30	40	3	20	0	248.3	355946.3
10	Cryptomonas erosa	cryp	1	15	30	40	12	14	0	11.9	5692.9
10	Ceratium hirundinella	dino	54	15	10	10	40	40	0	119.7	25392.7

M	Genus and species	Fam.	Ct.	Vol. filt. ml	Flds. ctd.	Mag.	Dim1 µm	Dim2 µm	Dim3 µm	Biovol. mm³/m³	Surf.area mm²/m³
10	Peridinium sp.	dino	12	15	10	10	32	40	0	34.0	5977.1
12	Closterium sp.	desm	1	15	30	40	5	50	0	3.7	4463.0
12	Cryptomonas erosa	cryp	4	15	30	40	14	16	0	74.3	30540.6
12	Microcystis aeruginosa	bg	1136	15	30	40	2	0	0	53.8	161433.2
12	Aphanizomenon flos-aquae	bg	940	15	30	40	3	0	0	590.3	786852.1
12	Anabaena sp.	bg	130	15	30	40	4	0	0	49.3	73895.5
12	Actinastrum gracilimum	grn	76	15	30	40	2	10	0	34.4	75631.2
12	Lyngbya limnetica	bg	1176	15	30	40	2	0	0	328.2	656268.1
12	Actinastrum gracilimum	grn	72	15	30	40	3	20	0	146.6	210066.7
12	Dictyosphaerium pulchellum	grn	32	15	30	40	4	0	0	12.1	18189.7
12	Ceratium hirundinella	dino	46	15	10	10	40	40	0	102.0	21630.8
12	Peridinium sp.	dino	10	15	10	10	32	40	0	28.4	4980.9
	Sep. 14, 1981										
0	Microcystis aeruginosa	bg	888	15	30	40	2	0	0	42.1	126190.7
0	Aphanizomenon flos-aquae	bg	1467	15	30	40	3	0	0	921.2	1227991.5
0	Anabaena sp.	bg	228	15	30	40	4	0	0	86.4	129601.3
0	Actinastrum gracilimum	grn	15	15	30	40	2	10	0	6.8	14927.2
0	Lyngbya limnetica	bg	1245	15	30	40	2	0	0	347.5	694773.7
0	Actinastrum gracilimum	grn	27	15	30	40	3	20	0	55.0	78775.0
0	Dictyosphaerium pulchellum	grn	36	15	30	40	4	0	0	13.6	20463.4
0	Cryptomonas erosa	cryp	12	15	30	40	14	16	0	222.8	91621.8
0	Chlamydomonas sp. 1	grn	84	15	30	40	5	7	0	87.0	95149.1
0	Peridinium sp.	dino	41	15	10	10	32	40	0	116.3	20421.9
0	Ceratium hirundinella	dino	121	15	10	10	40	40	0	268.2	56898.4
2	Microcystis aeruginosa	bg	312	15	30	40	2	0	0	14.8	44337.3
2	Aphanizomenon flos-aquae	bg	951	15	30	40	3	0	0	597.2	796060.0
2	Anabaena sp.	bg	42	15	30	40	4	0	0	15.9	23873.9
2	Actinastrum gracilimum	grn	42	15	30	40	2	10	0	19.0	41796.2
2	Lyngbya limnetica	bg	1152	15	30	40	2	0	0	321.5	642874.9
2	Actinastrum gracilimum	grn	9	15	30	40	3	20	0	18.3	26258.3
2	Chlamydomonas sp. 1	grn	12	15	30	40	5	7	0	12.4	13592.7
2	Peridinium sp.	dino	39	15	10	10	32	40	0	110.7	19425.7
2	Ceratium hirundinella	dino	152	15	10	10	40	40	0	336.9	71475.6
4	Aphanizomenon flos-aquae	bg	1500	15	30	40	3	0	0	941.9	1255615.1
4	Anabaena sp.	bg	126	15	30	40	4	0	0	47.7	71621.8
4	Actinastrum gracilimum	grn	45	15	30	40	2	10	0	20.4	44781.7
4	Lyngbya limnetica	bg	606	15	30	40	2	0	0	169.1	338179.0
4	Chlamydomonas sp. 1	grn	6	15	30	40	5	7	0	6.2	6796.4
4	Peridinium sp.	dino	20	15	10	10	32	40	0	56.7	9961.9
4	Ceratium hirundinella	dino	64	15	10	10	40	40	0	141.9	30095.0
8	Aphanizomenon flos-aquae	bg	1128	15	30	40	3	0	0	708.3	944222.5
8	Anabaena sp.	bg	648	15	30	40	4	0	0	245.6	368340.5
8	Actinastrum gracilimum	grn	27	15	30	40	2	10	0	12.2	26869.0
8	Lyngbya limnetica	bg	201	15	30	40	2	0	0	56.1	112168.3
8	Peridinium sp.	dino	12	15	10	10	32	40	0	34.0	5977.1
8	Ceratium hirundinella	dino	67	15	10	10	40	40	0	148.5	31505.7
10	Microcystis aeruginosa	bg	288	15	30	40	2	0	0	13.6	40926.7
10	Aphanizomenon flos-aquae	bg	1068	15	30	40	3	0	0	670.7	893997.9
10	Anabaena sp.	bg	24	15	30	40	4	0	0	9.1	13642.2
10	Actinastrum gracilimum	grn	9	15	30	40	2	10	0	4.1	8956.3
10	Lyngbya limnetica	bg	144	15	30	40	2	0	0	40.2	80359.4
10	Actinastrum gracilimum	grn	6	15	30	40	3	20	0	12.2	17505.6
10	Merismopedia sp.	bg	40	15	30	40	2	0	0	1.9	5684.3
10	Peridinium sp.	dino	3	15	10	10	32	40	0	8.5	1494.3
10	Ceratium hirundinella	dino	13	15	10	10	40	40	0	28.8	6113.0
12	Aphanizomenon flos-aquae	bg	129	15	30	40	3	0	0	81.0	107982.9

M	Genus and species	Fam.	Ct.	Vol. filt. ml	Flds. ctd.	Mag.	Dim1 µm	Dim2 µm	Dim3 µm	Biovol. mm³/m³	Surf.area mm²/m³
12	Lyngbya limnetica	bg	372	15	30	40	2	0	0	103.8	207595.0
12	Peridinium sp.	dino	2	15	10	10	32	40	0	5.7	996.2
12	Ceratium hirundinella	dino	3	15	10	10	40	40	0	6.7	1410.7
	Sep. 21, 1981										
0	Aphanizomenon flos-aquae	bg	1485	15	30	40	3	0	0	932.5	1243058.9
0	Anabaena sp.	bg	18	15	30	40	4	0	0	6.8	10231.7
0	Actinastrum gracilimum	grn	36	15	30	40	2	10	0	16.3	35825.3
0	Lyngbya limnetica	bg	537	15	30	40	2	0	0	149.9	299673.5
0	Actinastrum gracilimum	grn	18	15	30	40	3	20	0	36.6	52516.7
0	Chlamydomonas sp. 1	grn	24	15	30	40	5	7	0	24.9	27185.5
0	Ankistrodesmus fractus	grn	1	15	30	40	4	60	0	2.8	4272.7
0	Ceratium hirundinella	dino	38	15	10	10	40	40	0	84.2	17868.9
0	Peridinium sp.	dino	5	15	10	10	32	40	0	14.2	2490.5
2	Aphanizomenon flos-aquae	bg	969	15	30	40	3	0	0	608.5	811127.3
2	Actinastrum gracilimum	grn	37	15	30	40	2	10	0	16.7	36820.5
2	Lyngbya limnetica	bg	849	15	30	40	2	0	0	237.0	473785.4
2	Actinastrum gracilimum	grn	30	15	30	40	3	20	0	61.1	87527.8
2	Ceratium hirundinella	dino	26	15	10	10	40	40	0	57.6	12226.1
2	Peridinium sp.	dino	3	15	10	10	32	40	0	8.5	1494.3
4	Aphanizomenon flos-aquae	bg	964	15	30	40	3	0	0	605.4	806942.0
4	Anabaena sp.	bg	72	15	30	40	4	0	0	27.3	40926.7
4	Actinastrum gracilimum	grn	18	15	30	40	2	10	0	8.1	17912.7
4	Lyngbya limnetica	bg	379	15	30	40	2	0	0	105.8	211501.4
4	Cryptomonas erosa	cryp	18	15	30	40	14	16	0	334.2	137432.7
4	Ceratium hirundinella	dino	41	15	10	10	40	40	0	90.9	19279.6
4	Peridinium sp.	dino	1	15	10	10	32	40	0	2.8	498.1
8	Aphanizomenon flos-aquae	bg	304	15	30	40	3	0	0	190.9	254471.3
8	Anabaena sp.	bg	22	15	30	40	4	0	0	8.3	12505.4
8	Lyngbya limnetica	bg	466	15	30	40	2	0	0	130.1	260051.8
8	Microcystis aeruginosa	bg	288	15	30	40	2	0	0	13.6	40926.7
8	Cryptomonas erosa	cryp	12	15	30	40	14	16	0	222.8	91621.8
8	Ceratium hirundinella	dino	21	15	10	10	40	40	0	46.6	9874.9
8	Peridinium sp.	dino	1	15	10	10	32	40	0	2.8	498.1
10	Aphanizomenon flos-aquae	bg	867	20	30	40	3	0	0	408.3	544309.1
10	Anabaena sp.	bg	84	20	30	40	4	0	0	23.9	35810.9
10	Actinastrum gracilimum	grn	36	20	30	40	2	10	0	12.2	26869.0
10	Lyngbya limnetica	bg	897	20	30	40	2	0	0	187.8	375428.9
10	Actinastrum gracilimum	grn	54	20	30	40	3	20	0	82.4	118162.5
10	Cryptomonas erosa	cryp	3	20	30	40	14	16	0	41.8	17179.1
10	Dictyosphaerium pulchellum	grn	72	20	30	40	4	0	0	20.5	30695.0
10	Oocystis sp.	grn	2	20	30	40	12	16	0	20.5	9426.8
10	Ceratium hirundinella	dino	59	20	10	10	40	40	0	98.1	20807.9
10	Gymnodinium sp.	dino	22	20	10	10	30	30	0	30.9	3086.1
12	Aphanizomenon flos-aquae	bg	1728	20	30	40	3	0	0	813.8	1084851.4
12	Anabaena sp.	bg	24	20	30	40	4	0	0	6.8	10231.7
12	Actinastrum gracilimum	grn	18	20	30	40	2	10	0	6.1	13434.5
12	Lyngbya limnetica	bg	402	20	30	40	2	0	0	84.1	168252.4
12	Actinastrum gracilimum	grn	24	20	30	40	3	20	0	36.6	52516.7
12	Microcystis aeruginosa	bg	1080	20	30	40	2	0	0	38.4	115106.4
12	Cryptomonas erosa	cryp	6	20	30	40	14	16	0	83.6	34358.2
12	Dictyosphaerium pulchellum	grn	36	20	30	40	4	0	0	10.2	15347.5
12	Oocystis sp.	grn	2	20	30	40	12	16	0	20.5	9426.8
12	Oocystis sp.	grn	4	20	30	40	4	10	0	2.8	3548.9
12	Ceratium hirundinella	dino	74	20	10	10	40	40	0	123.0	26098.0
12	Gymnodinium sp.	dino	17	20	10	10	30	30	0	23.8	2384.7

M	Genus and species	Fam.	Ct.	Vol. filt. ml	Flds. ctd.	Mag.	Dim1 μm	Dim2 μm	Dim3 μm	Biovol. mm³/m³	Surf.area mm²/m³
	Sep. 28, 1981										
0	Microcystis aeruginosa	bg	168	25	30	40	2	0	0	4.8	14324.4
0	Aphanizomenon flos-aquae	bg	588	25	30	40	3	0	0	221.5	295320.7
0	Lyngbya limnetica	bg	417	25	30	40	2	0	0	69.8	139624.4
0	Cryptomonas erosa	cryp	15	25	30	40	12	14	0	107.4	51236.3
0	Chlamydomonas sp. 1	grn	36	25	30	40	5	7	0	22.4	24466.9
0	Chlamydomonas sp. 1	grn	66	25	30	40			0		
0	Ceratium hirundinella	dino	21	25	10	10	40	40	0	27.9	5925.0
0	Gymnodinium sp.	dino	5	25	10	10	30	30	0	5.6	561.1
2	Microcystis aeruginosa	bg	144	25	30	40	2	0	0	4.1	12278.0
2	Aphanizomenon flos-aquae	bg	687	25	30	40	3	0	0	258.8	345043.0
2	Lyngbya limnetica	bg	419	25	30	40	2	0	0	70.2	140294.1
2	Cryptomonas erosa	cryp	6	25	30	40	12	14	0	43.0	20494.5
2	Chlamydomonas sp. 1	grn	9	25	30	40	5	7	0	5.6	6116.7
2	Chlamydomonas sp. 1	grn	12	25	30	40			0		
2	Ulothrix sp.	grn	16	25	30	40	6	0	0	24.1	16071.9
2	Ceratium hirundinella	dino	24	25	10	10	40	40	0	31.9	6771.4
2	Gymnodinium sp.	dino	4	25	10	10	30	30	0	4.5	448.9
4	Aphanizomenon flos-aquae	bg	579	25	30	40	3	0	0	218.2	290800.4
4	Lyngbya limnetica	bg	475	25	30	40	2	0	0	79.5	159044.6
4	Cryptomonas erosa	cryp	12	25	30	40	12	14	0	85.9	40989.0
4	Chlamydomonas sp. 1	grn	2	25	30	40	5	7	0	1.2	1359.3
4	Chlamydomonas sp. 1	grn	44	25	30	40			0		
4	Oocystis sp.	grn	4	25	30	40	6	12	0	6.1	5246.4
4	Ceratium hirundinella	dino	18	25	10	10	40	40	0	23.9	5078.5
4	Gymnodinium sp.	dino	2	25	10	10	30	30	0	2.2	224.4
8	Microcystis aeruginosa	bg	576	25	30	40	2	0	0	16.4	49112.1
8	Aphanizomenon flos-aquae	bg	91	25	30	40	3	0	0	34.3	45704.4
8	Lyngbya limnetica	bg	387	25	30	40	2	0	0	64.8	129579.5
8	Cryptomonas erosa	cryp	12	25	30	40	12	14	0	85.9	40989.0
8	Fragillaria sp.	diat	28	25	30	40	5	18	0	85.5	43696.0
8	Anabaena sp.	bg	10	25	30	40	4	0	0	2.3	3410.6
8	Ceratium hirundinella	dino	9	25	10	10	40	40	0	12.0	2539.3
12	Aphanizomenon flos-aquae	bg	537	25	30	40	3	0	0	202.3	269706.1
12	Lyngbya limnetica	bg	570	25	30	40	2	0	0	95.5	190853.5
12	Cryptomonas erosa	cryp	6	25	30	40	12	14	0	43.0	20494.5
12	Chlamydomonas sp. 1	grn	16	25	30	40			0		
12	Anabaena sp.	bg	42	25	30	40	4	0	0	9.5	14324.4
12	Ceratium hirundinella	dino	16	25	10	10	40	40	0	21.3	4514.2
12	Gymnodinium sp.	dino	4	25	10	10	30	30	0	4.5	448.9
16	Aphanizomenon flos-aquae	bg	102	25	30	40	3	0	0	38.4	51229.1
16	Lyngbya limnetica	bg	62	25	30	40	2	0	0	10.4	20759.5
16	Chlamydomonas sp. 1	grn	27	25	30	40			0		
16	Anabaena sp.	bg	34	25	30	40	4	0	0	7.7	11595.9
16	Ceratium hirundinella	dino	6	25	10	10	40	40	0	8.0	1692.8
	Oct. 5, 1981										
0	Aphanizomenon flos-aquae	bg	294	25	40	40	3	0	0	83.1	110745.3
0	Lyngbya limnetica	bg	776	25	40	40	2	0	0	97.5	194871.5
0	Cryptomonas erosa	cryp	10	25	40	40	12	14	0	53.7	25618.1
0	Cryptomonas erosa	cryp	4	25	40	40	16	20	0	54.6	19158.8
0	Ceratium hirundinella	dino	13	25	10	10	40	40	0	17.3	3667.8
0	Peridinium sp.	dino	3	25	10	10	26	30	0	2.5	558.3
2	Aphanizomenon flos-aquae	bg	262	25	40	40	3	0	0	74.0	98691.3
2	Lyngbya limnetica	bg	450	25	40	40	2	0	0	56.5	113005.4
2	Cryptomonas erosa	cryp	8	25	40	40	12	14	0	43.0	20494.5
2	Cryptomonas erosa	cryp	14	25	40	40	16	20	0	191.0	67055.9
2	Microcystis aeruginosa	bg	144	25	40	40	2	0	0	3.1	9208.5
2	Ceratium hirundinella	dino	25	25	10	10	40	40	0	33.3	7053.5
2	Peridinium sp.	dino	5	25	10	10	26	30	0	4.2	930.6
4	Aphanizomenon flos-aquae	bg	302	25	40	40	3	0	0	85.3	113758.7
4	Lyngbya limnetica	bg	672	25	40	40	2	0	0	84.4	168754.7
4	Cryptomonas erosa	cryp	4	25	40	40	12	14	0	21.5	10247.3
4	Ceratium hirundinella	dino	14	25	10	10	40	40	0	18.6	3950.0
4	Peridinium sp.	dino	1	25	10	10	26	30	0	0.8	186.1
4	Cosmarium sp.	desm	1	25	10	10	14	28	0	0.5	167.1
8	Aphanizomenon flos-aquae	bg	122	25	40	40	3	0	0	34.5	45955.5
8	Lyngbya limnetica	bg	606	25	40	40	2	0	0	76.1	152180.5
8	Cryptomonas erosa	cryp	2	25	40	40	12	14	0	10.7	5123.6
8	Cryptomonas erosa	cryp	4	25	40	40	16	20	0	54.6	19158.8
8	Stephanodiscus sp.	diat	4	25	40	40	20	12	0	76.7	28137.1
8	Ceratium hirundinella	dino	20	25	10	10	40	40	0	26.6	5642.8
8	Peridinium sp.	dino	5	25	10	10	26	30	0	4.2	930.6
12	Aphanizomenon flos-aquae	bg	606	25	40	40	3	0	0	171.2	228270.8
12	Lyngbya limnetica	bg	574	25	40	40	2	0	0	72.1	144144.6
12	Cryptomonas erosa	cryp	12	25	40	40	12	14	0	64.5	30741.8
12	Cryptomonas erosa	cryp	14	25	40	40	16	20	0	191.0	67055.9
12	Actinastrum gracilimum	grn	24	25	40	40	2	10	0	4.9	10747.6
12	Dictyosphaerium pulchellum	grn	28	25	40	40	4	0	0	4.8	7162.2
12	Oscillatoria rubescens	bg	190	25	40	40	5	0	0	149.1	119283.4
12	Ceratium hirundinella	dino	27	25	10	10	40	40	0	35.9	7617.8
12	Peridinium sp.	dino	7	25	10	10	26	30	0	5.9	1302.8
16	Aphanizomenon flos-aquae	bg	110	25	40	40	3	0	0	31.1	41435.3
16	Lyngbya limnetica	bg	224	25	40	40	2	0	0	28.1	56251.6
16	Cryptomonas erosa	cryp	2	25	40	40	12	14	0	10.7	5123.6
16	Microcystis aeruginosa	bg	21	25	40	40	2	0	0	0.4	1342.9
16	Oocystis sp.	grn	3	25	40	40	8	14	0	7.2	4685.0
16	Oscillatoria rubescens	bg	152	25	40	40	5	0	0	119.3	95426.7
16	Ceratium hirundinella	dino	11	25	10	10	40	40	0	14.6	3103.5
16	Peridinium sp.	dino	2	25	10	10	26	30	0	1.7	372.2
20	Lyngbya limnetica	bg	192	25	40	40	2	0	0	24.1	48215.6
20	Actinastrum gracilimum	grn	12	25	40	40	2	10	0	2.4	5373.8
	Oct. 12, 1981										
0	Stephanodiscus sp.	diat	4	25	40	40	30	12	0	172.7	51797.9
0	Stephanodiscus sp.	diat	4	25	40	40	20	12	0	76.7	28137.1
0	Stephanodiscus sp.	diat	6	25	40	40	12	12	0	41.4	20719.2
0	Lyngbya limnetica	bg	604	25	40	40	2	0	0	75.9	151678.3
0	Microcystis aeruginosa	bg	216	25	40	40	2	0	0	4.6	13812.8
0	Dictyosphaerium pulchellum	grn	32	25	40	40	4	0	0	5.5	8185.3
0	Ceratium hirundinella	dino	10	25	10	10	40	40	0	13.3	2821.4
0	Peridinium sp.	dino	2	25	10	10	26	30	0	1.7	372.2
2	Lyngbya limnetica	bg	643	25	40	40	2	0	0	80.8	161472.1
2	Oscillatoria sp.	bg	212	25	40	40	5	0	0	166.4	133095.2
2	Cryptomonas erosa	cryp	6	25	40	40	20	30	0	191.8	51656.3
2	Cryptomonas erosa	cryp	4	25	40	40	12	14	0	21.5	10247.3
2	Ceratium hirundinella	dino	4	25	10	10	40	40	0	5.3	1128.6
2	Peridinium sp.	dino	1	25	10	10	26	30	0	0.8	186.1
4	Lyngbya limnetica	bg	367	25	40	40	2	0	0	46.1	92162.1
4	Oscillatoria sp.	bg	381	25	40	40	5	0	0	299.1	239194.7
4	Cryptomonas erosa	cryp	2	25	40	40	20	30	0	63.9	17218.8
4	Cryptomonas erosa	cryp	6	25	40	40	12	14	0	32.2	15370.9
4	Aphanizomenon flos-aquae	bg	68	25	40	40	3	0	0	19.2	25614.5

M	Genus and species	Fam.	Ct.	Vol. filt. ml	Flds. ctd.	Mag.	Dim1 µm	Dim2 µm	Dim3 µm	Biovol. mm^3/m^3	Surf.area mm^2/m^3
4	Actinastrum gracilimum	grn	12	25	40	40	2	10	0	2.4	5373.8
4	Ceratium hirundinella	dino	6	25	10	10	40	40	0	8.0	1692.8
4	Peridinium sp.	dino	3	25	10	10	26	30	0	2.5	558.3
8	Lyngbya limnetica	bg	642	25	40	40	2	0	0	80.6	161221.0
8	Microcystis aeruginosa	bg	211	25	40	40	2	0	0	4.5	13493.0
8	Oscillatoria sp.	bg	54	25	40	40	5	0	0	42.4	33901.6
8	Cryptomonas erosa	cryp	2	25	40	40	20	30	0	63.9	17218.8
8	Cryptomonas erosa	cryp	6	25	40	40	12	14	0	32.2	15370.9
8	Oocystis sp.	grn	4	25	40	40	8	14	0	9.5	6246.6
8	Ceratium hirundinella	dino	1	25	10	10	40	40	0	1.3	282.1
12	Lyngbya limnetica	bg	305	25	40	40	2	0	0	38.3	76592.5
12	Oscillatoria sp.	bg	238	25	40	40	5	0	0	186.8	149418.2
12	Cryptomonas erosa	cryp	4	25	40	40	20	30	0	127.9	34437.5
12	Cryptomonas erosa	cryp	8	25	40	40	12	14	0	43.0	20494.5
12	Ceratium hirundinella	dino	3	25	10	10	40	40	0	4.0	846.4
16	Lyngbya limnetica	bg	272	25	40	40	2	0	0	34.2	68305.5
16	Dictyosphaerium pulchellum	grn	88	25	40	40	4	0	0	15.0	22509.7
16	Oscillatoria sp.	bg	182	25	40	40	5	0	0	142.9	114261.0
16	Cryptomonas erosa	cryp	2	25	40	40	20	30	0	63.9	17218.8
16	Cryptomonas erosa	cryp	8	25	40	40	12	14	0	43.0	20494.5
16	Ceratium hirundinella	dino	4	25	10	10	40	40	0	5.3	1128.6
20	Aphanizomenon flos-aquae	bg	52	25	40	40	3	0	0	14.7	19587.6
20	Lyngbya limnetica	bg	200	25	40	40	2	0	0	25.1	50224.6
20	Oscillatoria sp.	bg	40	25	40	40	5	0	0	31.4	25112.3
20	Ceratium hirundinella	dino	4	25	10	10	40	40	0	5.3	1128.6

Oct. 19, 1981

M	Genus and species	Fam.	Ct.	Vol. filt. ml	Flds. ctd.	Mag.	Dim1 µm	Dim2 µm	Dim3 µm	Biovol. mm^3/m^3	Surf.area mm^2/m^3
0	Stephanodiscus sp.	diat	18	25	40	40	20	12	0	345.3	126617.1
0	Lyngbya limnetica	bg	346	25	40	40	2	0	0	43.5	86888.6
0	Oscillatoria sp.	bg	424	25	40	40	5	0	0	332.8	266190.4
0	Dictyosphaerium pulchellum	grn	40	25	40	40	4	0	0	6.8	10231.7
0	Cryptomonas erosa	cryp	8	25	40	40	20	24	0	204.6	58102.0
0	Cryptomonas erosa	cryp	8	25	40	40	12	14	0	43.0	20494.5
0	Cryptomonas erosa	cryp	2	25	40	40	16	20	0	27.3	9579.4
0	Ceratium hirundinella	dino	4	25	10	10	40	40	0	5.3	1128.6
0	Stephanodiscus sp.	diat	16	25	10	10	40	30	0	47.9	7980.1
0	Cosmarium sp.	desm	2	25	10	10	18	32	0	1.7	499.8
0	Peridinium sp.	dino	1	25	10	10	30	40	0	1.5	275.7
2	Stephanodiscus sp.	diat	22	25	40	40	20	12	0	422.1	154754.2
2	Lyngbya limnetica	bg	288	25	40	40	2	0	0	36.2	72323.4
2	Oscillatoria sp.	bg	248	25	40	40	5	0	0	194.7	155696.3
2	Dictyosphaerium pulchellum	grn	16	25	40	40	4	0	0	2.7	4092.7
2	Ceratium hirundinella	dino	12	25	10	10	40	40	0	16.0	3385.7
2	Stephanodiscus sp.	diat	38	25	10	10	40	30	0	113.7	18952.8
2	Cosmarium sp.	desm	2	25	10	10	18	32	0	1.7	499.8
4	Stephanodiscus sp.	diat	28	25	40	40	20	12	0	537.2	196959.9
4	Lyngbya limnetica	bg	390	25	40	40	2	0	0	49.0	97938.0
4	Oscillatoria sp.	bg	564	25	40	40	5	0	0	442.7	354083.5
4	Dictyosphaerium pulchellum	grn	112	25	40	40	4	0	0	19.1	28648.7
4	Cryptomonas erosa	cryp	2	25	40	40	20	24	0	51.2	14525.5
4	Cryptomonas erosa	cryp	2	25	40	40	12	14	0	10.7	5123.6
4	Cryptomonas erosa	cryp	8	25	40	40	16	20	0	109.1	38317.7
4	Anabaena sp.	bg	8	25	40	40	4	0	0	1.4	2046.3
4	Aphanizomenon flos-aquae	bg	26	25	40	40	3	0	0	7.3	9793.8
4	Actinastrum gracilimum	grn	12	25	40	40	2	10	0	2.4	5373.8
4	Ceratium hirundinella	dino	6	10	40	10	40	40	0	5.0	1058.0
4	Stephanodiscus sp.	diat	17	10	40	10	40	30	0	31.8	5299.3
4	Cosmarium sp.	desm	3	10	40	10	18	32	0	1.6	468.6
4	Peridinium sp.	dino	3	10	40	10	30	40	0	2.8	517.0
8	Lyngbya limnetica	bg	300	25	40	40	2	0	0	37.7	75336.9
8	Oscillatoria sp.	bg	318	25	40	40	5	0	0	249.6	199642.8
8	Cryptomonas erosa	cryp	4	25	40	40	12	14	0	21.5	10247.3
8	Cryptomonas erosa	cryp	2	25	40	40	16	20	0	27.3	9579.4
8	Oocystis sp.	grn	2	25	40	40	26	30	0	108.1	23862.6
8	Fragillaria sp.	diat	106	25	40	40	3	20	0	97.1	74439.3
8	Ceratium hirundinella	dino	3	25	10	10	40	40	0	4.0	846.4
8	Stephanodiscus sp.	diat	13	25	10	10	40	30	0	38.9	6483.9
8	Cosmarium sp.	desm	2	25	10	10	18	32	0	1.7	499.8
8	Peridinium sp.	dino	3	25	10	10	30	40	0	4.5	827.1
12	Stephanodiscus sp.	diat	22	25	40	40	20	12	0	422.1	154754.2
12	Lyngbya limnetica	bg	190	25	40	40	2	0	0	23.9	47713.4
12	Oscillatoria sp.	bg	1260	25	40	40	5	0	0	989.0	791037.5
12	Dictyosphaerium pulchellum	grn	96	25	40	40	4	0	0	16.4	24556.0
12	Cryptomonas erosa	cryp	6	25	40	40	12	14	0	32.2	15370.9
12	Cryptomonas erosa	cryp	8	25	40	40	16	20	0	109.1	38317.7
12	Fragillaria sp.	diat	150	25	40	40	3	20	0	137.4	105338.7
12	Cryptomonas erosa	cryp	5	25	40	40	7	10	0	6.5	5074.5
12	Oocystis sp.	grn	4	25	40	40	5	12	0	3.2	3207.3
12	Ceratium hirundinella	dino	22	25	10	10	40	40	0	29.3	6207.1
12	Stephanodiscus sp.	diat	129	25	10	10	40	30	0	386.0	64339.9
12	Cosmarium sp.	desm	9	25	10	10	18	32	0	7.8	2249.2
12	Peridinium sp.	dino	4	25	10	10	30	40	0	6.0	1102.9
16	Lyngbya limnetica	bg	304	25	40	40	2	0	0	38.2	76341.4
16	Oscillatoria sp.	bg	724	25	40	40	5	0	0	568.3	454532.7
16	Cryptomonas erosa	cryp	2	25	40	40	12	14	0	10.7	5123.6
16	Fragillaria sp.	diat	40	25	40	40	3	20	0	36.6	28090.3
16	Ceratium hirundinella	dino	4	25	10	10	40	40	0	5.3	1128.6
16	Stephanodiscus sp.	diat	54	25	10	10	40	30	0	161.6	26933.0
16	Cosmarium sp.	desm	4	25	10	10	18	32	0	3.4	999.6
20	Lyngbya limnetica	bg	366	25	40	40	2	0	0	46.0	91911.0
20	Oscillatoria sp.	bg	650	25	40	40	5	0	0	510.2	408074.9
20	Dictyosphaerium pulchellum	grn	144	25	40	40	4	0	0	24.6	36834.1
20	Cryptomonas erosa	cryp	6	25	40	40	12	14	0	32.2	15370.9
20	Cryptomonas erosa	cryp	12	25	40	40	16	20	0	163.7	57476.5
20	Stephanodiscus sp.	diat	16	25	40	40	20	12	0	307.0	112548.5
20	Stephanodiscus sp.	diat	10	25	10	10	40	30	0	29.9	4987.6

Nov. 10, 1981

M	Genus and species	Fam.	Ct.	Vol. filt. ml	Flds. ctd.	Mag.	Dim1 µm	Dim2 µm	Dim3 µm	Biovol. mm^3/m^3	Surf.area mm^2/m^3
0	Lyngbya limnetica	bg	136	20	40	40	2	0	0	21.4	42690.9
0	Dictyosphaerium pulchellum	grn	64	20	40	40	4	0	0	13.6	20463.4
0	Fragillaria sp.	diat	176	20	40	40	3	20	0	201.5	154496.7
0	Stephanodiscus sp.	diat	217	20	10	10	40	30	0	811.7	135288.3
0	Staurastrum sp.	grn	3	20	10	10	10	40	32	3.8	1891.6
0	Cosmarium sp.	desm	2	20	10	10	18	30	0	2.0	592.2
2	Lyngbya limnetica	bg	68	20	40	40	2	0	0	10.7	21345.5
2	Chlamydomonas sp. 1	grn	98	20	40	40	5	7	0	57.1	62441.6
2	Oocystis sp.	grn	2	20	40	40	20	30	0	79.9	21523.4
2	Planktosphaeria gelatinosa	grn	32	20	40	40	4	0	0	6.8	10231.7
2	Oscillatoria sp.	bg	40	20	40	40	5	0	0	39.2	31390.4
2	Stephanodiscus sp.	diat	171	20	10	10	40	30	0	639.7	106609.7
2	Staurastrum sp.	grn	12	20	10	10	10	40	32	15.2	7566.3
2	Cosmarium sp.	desm	10	20	10	10	18	30	0	10.1	2961.1
2	Ceratium hirundinella	dino	3	20	10	10	40	40	0	5.0	1058.0
4	Lyngbya limnetica	bg	61	20	40	40	2	0	0	9.6	19148.1

M	Genus and species	Fam.	Ct.	Vol. filt. ml	Flds. ctd.	Mag.	Dim1 µm	Dim2 µm	Dim3 µm	Biovol. mm³/m³	Surf.area mm²/m³
4	Fragillaria sp.	diat	28	20	40	40	3	20	0	32.1	24579.0
4	Stephanodiscus sp.	diat	53	20	10	10	40	30	0	198.3	33042.8
4	Cosmarium sp.	desm	1	20	10	10	18	30	0	1.0	296.1
8	Lyngbya limnetica	bg	204	20	40	40	2	0	0	32.0	64036.4
8	Oocystis sp.	grn	1	20	40	40	10	12	0	4.0	2269.6
8	Stephanodiscus sp.	diat	16	20	10	10	40	30	0	59.9	9975.2
12	Lyngbya limnetica	bg	156	40	40	40	2	0	0	12.2	24484.5
12	Dictyosphaerium pulchellum	grn	16	40	40	40	4	0	0	1.7	2557.9
12	Fragillaria sp.	diat	84	40	40	40	3	20	0	48.1	36868.5
12	Merismopedia sp.	bg	48	40	40	40	2	0	0	0.6	1918.4
12	Stephanodiscus sp.	diat	134	40	10	10	40	30	0	250.6	41771.1
12	Staurastrum sp.	grn	2	40	10	10	10	40	32	1.3	630.5
12	Cosmarium sp.	desm	5	40	10	10	18	30	0	2.5	740.3
12	Ceratium hirundinella	dino	3	40	10	10	40	40	0	2.5	529.0
16	Lyngbya limnetica	bg	78	40	40	40	2	0	0	6.1	12242.2
16	Fragillaria sp.	diat	36	40	40	40	3	20	0	20.6	15800.8
16	Chlamydomonas sp. 1	grn	52	40	40	40	5	7	0	15.2	16566.1
16	Oscillatoria sp.	bg	20	40	40	40	5	0	0	9.8	7847.6
16	Stephanodiscus sp.	diat	386	40	10	10	40	30	0	722.0	120325.6
16	Staurastrum sp.	grn	3	40	10	10	10	40	32	1.9	945.8
16	Cosmarium sp.	desm	16	40	10	10	18	30	0	8.1	2368.9
16	Ceratium hirundinella	dino	6	40	10	10	40	40	0	5.0	1058.0
20	Stephanodiscus sp.	diat	59	40	10	10	40	30	0	110.4	18391.7
20	Cosmarium sp.	desm	3	40	10	10	8	30	0	0.3	181.4
20	Ceratium hirundinella	dino	5	40	10	10	40	40	0	4.2	881.7
	Dec. 1, 1981										
0	Dictyosphaerium pulchellum	grn	104	50	40	40	4	0	0	8.9	13301.2
0	Cryptomonas erosa	cryp	6	50	40	40	10	12	0	9.6	5447.1
0	Cryptomonas erosa	cryp	2	50	40	40	20	24	0	25.6	7262.7
0	Chlamydomonas sp. 1	grn	139	50	40	40	5	7	0	32.4	35426.1
0	Cosmarium sp.	desm	15	50	10	10	14	30	0	3.7	1330.5
0	Stephanodiscus sp.	diat	405	50	10	10	40	20	0	404.0	80798.9
2	Dictyosphaerium pulchellum	grn	104	50	40	40	4	0	0	8.9	13301.2
2	Cryptomonas erosa	cryp	4	50	40	40	10	12	0	6.4	3631.4
2	Chlamydomonas sp. 1	grn	150	50	40	40	5	7	0	35.0	38229.6
2	Lyngbya limnetica	bg	70	50	40	40	2	0	0	4.4	8789.3
2	Actinastrum gracilimum	grn	10	50	40	40	2	10	0	1.0	2239.1
2	Cosmarium sp.	desm	18	50	10	10	14	30	0	4.4	1596.6
2	Stephanodiscus sp.	diat	475	50	10	10	40	20	0	473.8	94764.2
4	Dictyosphaerium pulchellum	grn	96	50	40	40	4	0	0	8.2	12278.0
4	Cryptomonas erosa	cryp	1	50	40	40	10	12	0	1.6	907.8
4	Chlamydomonas sp. 1	grn	175	50	40	40	5	7	0	40.8	44601.2
4	Lyngbya limnetica	bg	60	50	40	40	2	0	0	3.8	7533.7
4	Actinastrum gracilimum	grn	24	50	40	40	2	10	0	2.4	5373.8
4	Cosmarium sp.	desm	112	50	10	10	14	30	0	27.4	9934.6
4	Stephanodiscus sp.	diat	486	50	10	10	40	20	0	484.8	96958.7
8	Dictyosphaerium pulchellum	grn	192	50	40	40	4	0	0	16.4	24556.0
8	Cryptomonas erosa	cryp	6	50	40	40	10	12	0	9.6	5447.1
8	Cryptomonas erosa	cryp	2	50	40	40	20	24	0	25.6	7262.7
8	Chlamydomonas sp. 1	grn	186	50	40	40	5	7	0	43.4	47404.7
8	Lyngbya limnetica	bg	38	50	40	40	2	0	0	2.4	4771.3
8	Cosmarium sp.	desm	43	50	10	10	14	30	0	10.5	3814.2
8	Stephanodiscus sp.	diat	609	50	10	10	40	20	0	607.5	121497.6
12	Dictyosphaerium pulchellum	grn	112	50	40	40	4	0	0	9.5	14324.4
12	Cryptomonas erosa	cryp	2	50	40	40	20	24	0	25.6	7262.7
12	Chlamydomonas sp. 1	grn	92	50	40	40	5	7	0	21.4	23447.5
12	Lyngbya limnetica	bg	12	50	40	40	2	0	0	0.8	1506.7
12	Fragillaria sp.	diat	54	50	40	40	4	36	0	79.1	43967.4
12	Staurastrum sp.	grn	5	50	10	10	10	40	32	2.5	1261.1
12	Cosmarium sp.	desm	24	50	10	10	14	30	0	5.9	2128.8
12	Stephanodiscus sp.	diat	654	50	10	10	40	20	0	652.4	130475.3
16	Cryptomonas erosa	cryp	4	50	40	40	10	12	0	6.4	3631.4
16	Chlamydomonas sp. 1	grn	372	50	40	40	5	7	0	86.7	94809.3
16	Lyngbya limnetica	bg	68	50	40	40	2	0	0	4.3	8538.2
16	Staurastrum sp.	grn	8	50	10	10	10	40	32	4.1	2017.7
16	Cosmarium sp.	desm	6	50	10	10	14	30	0	1.5	532.2
16	Stephanodiscus sp.	diat	493	50	10	10	40	20	0	491.8	98355.2

Index